Tanase G. Dobre and
José G. Sanchez Marcano

Chemical Engineering

1807–2007 Knowledge for Generations

Each generation has its unique needs and aspirations. When Charles Wiley first opened his small printing shop in lower Manhattan in 1807, it was a generation of boundless potential searching for an identity. And we were there, helping to define a new American literary tradition. Over half a century later, in the midst of the Second Industrial Revolution, it was a generation focused on building the future. Once again, we were there, supplying the critical scientific, technical, and engineering knowledge that helped frame the world. Throughout the 20th Century, and into the new millennium, nations began to reach out beyond their own borders and a new international community was born. Wiley was there, expanding its operations around the world to enable a global exchange of ideas, opinions, and know-how.

For 200 years, Wiley has been an integral part of each generations journey, enabling the flow of information and understanding necessary to meet their needs and fulfill their aspirations. Today, bold new technologies are changing the way we live and learn. Wiley will be there, providing you the must-have knowledge you need to imagine new worlds, new possibilities, and new opportunities.

Generations come and go, but you can always count on Wiley to provide you the knowledge you need, when and where you need it!

William J. Pesce
President and Chief Executive Officer

Peter Booth Wiley
Chairman of the Board

Tanase G. Dobre and
José G. Sanchez Marcano

Chemical Engineering

Modelling, Simulation and Similitude

BICENTENNIAL
1807
WILEY
2007
BICENTENNIAL

WILEY-VCH Verlag GmbH & Co. KGaA

The Authors

Prof. Dr. Ing. Tanase G. Dobre
Politechnic University of Bucharest
Chemical Engineering Department
Polizu 1-3
78126 Bucharest, Sector 1
Romania

Dr. José G. Sanchez Marcano
Institut Européen des Membranes, I. E. M.
UMII, cc 0047
2, place Bataillon
34095 Montpellier
France

■ All books published by Wiley-VCH are carefully produced. Nevertheless, authors, editors, and publisher do not warrant the information contained in these books, including this book, to be free of errors. Readers are advised to keep in mind that statements, data, illustrations, procedural details or other items may inadvertently be inaccurate.

Library of Congress Card No.: applied for

British Library Cataloguing-in-Publication Data
A catalogue record for this book is available from the British Library.

Bibliographic information published by Die Deutsche Bibliothek
Die Deutsche Nationalbibliothek lists this publication in the Deutsche Nationalbibliografie; detailed bibliographic data are available in the Internet at <http://dnb.d-nb.de>.

© 2007 WILEY-VCH Verlag GmbH & Co. KGaA, Weinheim

Typesetting Kühn & Weyh, Freiburg
Printing Strauss GmbH, Mörlenbach
Bookbinding Litges & Dopf GmbH, Heppenheim
Cover Design Adam-Design, Weinheim
Wiley Bicentennial Logo Richard J. Pacifico

Printed in the Federal Republic of Germany.
Printed on acid-free paper.

ISBN 978-3-527-30607-7

To Christine, Laura, Benjamin and Anaïs, for their love and ongoing support
To Marie, Raluca, Diana and Fineta for their confidence and love

Contents

Chemical Engineering. Tanase G. Dobre and José G. Sanchez Marcano
Copyright © 2007 WILEY-VCH Verlag GmbH & Co. KGaA, Weinheim
ISBN: 978-3-527-30607-7

Preface

Scientific research is a systematic investigation, which establishes facts, and develops understanding in many sciences such as mathematics, physics, chemistry and biology. In addition to these fundamental goals, scientific research can also create development in engineering. During all systematic investigation, modelling is essential in order to understand and to analyze the various steps of experimentation, data analysis, process development, and engineering design. This book is devoted to the development and use of the different types of mathematical models which can be applied for processes and data analysis.

Modelling, simulation and similitude of chemical engineering processes has attracted the attention of scientists and engineers for many decades and is still today a subject of major importance for the knowledge of unitary processes of transport and kinetics as well as a fundamental key in design and scale-up. A fundamental knowledge of the mathematics of modelling as well as its theoretical basis and software practice are essential for its correct application, not only in chemical engineering but also in many other domains like materials science, bioengineering, chemistry, physics, etc. In so far as modelling simulation and similitude are essential in the development of chemical engineering processes, it will continue to progress in parallel with new processes such as micro-fluidics, nanotechnologies, environmentally-friendly chemistry processes and devices for non-conventional energy production such as fuel cells. Indeed, this subject will keep on attracting substantial worldwide research and development efforts.

This book is completely dedicated to the topic of modelling, simulation and similitude in chemical engineering. It first introduces the topic, and then aims to give the fundamentals of mathematics as well as the different approaches of modelling in order to be used as a reference manual by a wide audience of scientists and engineers.

The book is divided into six chapters, each covering a different aspect of the topic. Chapter 1 provides a short introduction to the key concepts and some pertinent basic concepts and definitions, including processes and process modelling definitions, division of processes and models into basic steps or components, as well as a general methodology for modelling and simulation including the modes of model use for all the stages of the life-cycle processes: simulation, design, parameter estimation and optimization. Chapter 2 is dedicated to the difficult task of

Chemical Engineering. Tanase G. Dobre and José G. Sanchez Marcano
Copyright © 2007 WILEY-VCH Verlag GmbH & Co. KGaA, Weinheim
ISBN: 978-3-527-30607-7

classifying the numerous types of models used in chemical engineering. This classification is made in terms of the theoretical base used for the development or the mathematical complexity of the process model. In this chapter, in addition to the traditional modelling procedures or computer-aided process engineering, other modelling and simulation fields have also been introduced. They include molecular modelling and computational chemistry, computational fluid dynamics, artificial intelligence and neural networks etc.

Chapter 3 concerns the topic of mathematical models based on transport phenomena. The particularizations of the property conservation equation for mass, energy and physical species are developed. They include the usual flow, heat and species transport equations, which give the basic mathematical relations of these models. Then, the general methodology to establish a process model is described step by step – from the division of the descriptive model into basic parts to its numerical development. In this chapter, other models are also described, including chemical engineering flow models, the distribution function and dispersion flow models as well as the application of computational fluid dynamics. The identification of parameters is approached through various methods such as the Lagrange multiplicators, the gradient and Gauss-Newton, the maximum likelihood and the Kalman Filter Equations. These methods are explained with several examples including batch adsorption, stirred and plug flow reactors, filtration of liquids and gas permeation with membranes, zone refining, heat transfer in a composite medium etc.

Chapter 4 is devoted to the description of stochastic mathematical modelling and the methods used to solve these models such as analytical, asymptotic or numerical methods. The evolution of processes is then analyzed by using different concepts, theories and methods. The concept of Markov chains or of complete connected chains, probability balance, the similarity between the Fokker–Plank–Kolmogorov equation and the property transport equation, and the stochastic differential equation systems are presented as the basic elements of stochastic process modelling. Mathematical models of the application of continuous and discrete polystochastic processes to chemical engineering processes are discussed. They include liquid and gas flow in a column with a mobile packed bed, mechanical stirring of a liquid in a tank, solid motion in a liquid fluidized bed, species movement and transfer in a porous media. Deep bed filtration and heat exchanger dynamics are also analyzed.

In Chapter 5, a survey of statistical models in chemical engineering is presented, including the characteristics of the statistical selection, the distribution of frequently used random variables as well as the intervals and limits for confidence methods such as linear, multiple linear, parabolic and transcendental regression, etc. A large part of this chapter is devoted to experimental design methods and their geometric interpretation. Starting with a discussion on the investigation of the great curvature domain of a process response surface, we introduce sequential experimental planning, the second order orthogonal or complete plan and the use of the simplex regular plan for experimental research as well as the analysis of variances and interaction of factors. In the last part of this chapter, a short review of the application in the chemical engineering field of artificial neural networks is given. Throughout this chapter, the discussion is illustrated by some numerical applications, which include the relationships between the reactant conversion and the input concentration for a continuously stirred reactor and liquid–solid extraction in a batch reactor.

Chapter 6 presents dimensional analysis in chemical engineering. The Vaschy–Buckingham Pi theorem is described here and a methodology for the identification and determination of Pi groups is discussed. After this introduction, the dimensional analysis is particularized for chemical engineering problems and illustrated by two examples: mass transfer by natural convection in a finite space and the mixing of liquids in a stirred vessel. This chapter also explains how the selection of variables is imposed in a system by its geometry, the properties of the materials and the dynamic internal and external effects. The dimensional analysis is completed with a synthetic presentation of the dimensionless groups commonly used in chemical engineering, their physical significance and their relationships. This chapter finishes with a discussion of physical models, similitude and design aspects. Throughout this chapter, some examples exemplify the analysis carried out; they include heat transfer by natural convection from a plate to an infinite medium, a catalytic membrane reactor and the heat loss in a rectification column.

We would like to acknowledge Anne Marie Llabador from the Université de Montpellier II for her help with our English. José Sanchez Marcano and Tanase Dobre gratefully acknowledge the ongoing support of the Centre National de la Recherche Scientifique, the Ecole Nationale Supérieure de Chimie de Montpellier, Université de Montpellier II and Politehnica University of Bucharest.

February 2007

Tanase G. Dobre
José G. Sanchez Marcano

1
Why Modelling?

Analysis of the cognition methods which have been used since early times reveals that the general methods created in order to investigate life phenomena could be divided into two groups: (i) the application of similitude, modelling and simulation, (ii) experimental research which also uses physical models. These methods have always been applied to all branches of human activity all around the world and consequently belong to the universal patrimony of human knowledge. The two short stories told below aim to explain the fundamental characteristics of these cognition methods.

First story. When, by chance, men were confronted by natural fire, its heat may have strongly affected them. As a result of these ancient repeated encounters on cold days, men began to feel the agreeable effect of fire and then wondered how they could proceed to carry this fire into their cold caves where they spent their nights. The precise answer to this question is not known, but it is true that fire has been taken into men's houses. Nevertheless, it is clear that men tried to elaborate a scheme to transport this natural fire from outside into their caves. We therefore realize that during the old times men began to exercise their minds in order to plan a specific action. This cognition process can be considered as one of the oldest examples of the use of modelling research on life.

So we can hold in mind that the use of modelling research on life is a method used to analyze a phenomenon based on qualitative and quantitative cognition where only mental exercises are used.

Second Story. The invention of the bow resulted in a new lifestyle because it led to an increase in men's hunting capacity. After using the bow for the first time, men began to wonder how they could make it stronger and more efficient. Such improvements were repeated continually until the effect of these changes began to be analysed. This example of human progress illustrates a cognition process based on experimentation in which a physical model (the bow) was used.

In accordance with the example described above, we can deduce that research based on a physical model results from linking the causes and effects that characterize an investigated phenomenon. With reference to the relationships existing between different investigation methods, we can conclude that, before modifying

Chemical Engineering. Tanase G. Dobre and José G. Sanchez Marcano
Copyright © 2007 WILEY-VCH Verlag GmbH & Co. KGaA, Weinheim
ISBN: 978-3-527-30607-7

the physical model used, modelling research has to be carried out. The modelling can then suggest various strategies but a single one has to be chosen. At the same time, the physical model used determines the conditions required to measure the effect of the adopted strategy. Further improvement of the physical model may also imply additional investigation.

If we investigate the scientific and technical evolution for a random selected domain, we can see that research by modelling or experimentation is fundamental. The evolution of research by modelling and/or experimentation (i.e. based on a physical model) has known an important particularization in each basic domain of science and techniques. Research by modelling, by simulation and similitude as well as experimental research, have become fundamental methods in each basic scientific domain (such as, in this book, chemical engineering). However, they tend to be considered as interdisciplinary activities. In the case of modelling simulation and similitude in chemical engineering, the interdisciplinary state is shown by coupling the phenomena studied with mathematics and computing science.

1.1
Process and Process Modelling

In chemical engineering, as well as in other scientific and technical domains, where one or more materials are physically or chemically transformed, a process is represented in its abstract form as in Fig. 1.1(a). The global process could be characterized by considering the inputs and outputs. As input variables (also called "independent process variables", "process command variables", "process factors" or "simple factors"), we have deterministic and random components. From a physical viewpoint, these variables concern materials, energy and state parameters, and of these, the most commonly used are pressure and temperature. The deterministic process input variables, contain all the process variables that strongly influence the process exits and that can be measured and controlled so as to obtain a designed process output.

The random process input variables represent those variables that influence the process evolution, but they can hardly be influenced by any external action. Frequently, the random input variables are associated with deterministic input variables when the latter are considered to be in fact normal randomly distributed variables with mean $\bar{x}_j$, $j = 1, N$ ("mean" expresses the deterministic behaviour of variable x_j) and variance σ_{xj}, $j = 1, N$. So the probability distribution function of the x_j variable can be expressed by the following equation:

$$f(x_j) = \frac{1}{\sqrt{2\pi}\sigma_{xj}} \exp\left(-\frac{(x_j - \bar{x}_j)^2}{2\sigma_{xj}^2}\right) \tag{1.1}$$

The values of $\bar{x}_j$, $j = 1, N$ and σ_{xj}, $j = 1, N$ can be obtained by the observation of each x_j when the investigated process presents a steady state evolution.

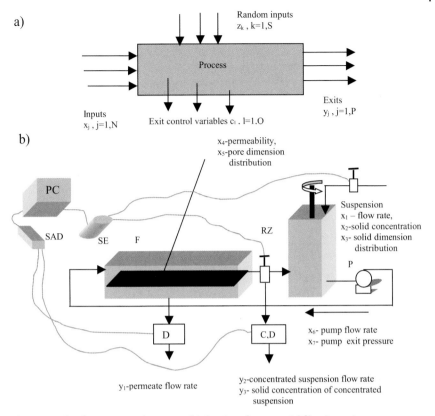

Figure 1.1 The abstract (a) and concrete (b) drawing of a tangential filtration unit.

The exit variables that present an indirect relation with the particularities of the process evolution, denoted here by $c_l, l = 1, Q$, are recognized as intermediary variables or as exit control variables. The exit process variables that depend strongly on the values of the independent process variables are recognized as dependent process variables or as process responses. These are denoted by $y_i, i = 1, P$. When we have random inputs in a process, each y_i exit presents a distribution around a characteristic mean value, which is primordially determined by the state of all independent process variables $\bar{x}_j, j = 1, N$. Figure 1.1 (b), shows an abstract scheme of a tangential filtration unit as well as an actual or concrete picture.

Here F filters a suspension and produces a clear filtrate as well as a concentrated suspension which is pumped into and out of reservoir RZ. During the process a fraction of the concentrated suspension is eliminated. In order to have a continuous process it is advisable to have working state values close to steady state values. The exit or output control variables (D and CD registered) are connected to a data acquisition system (DAS), which gives the computer (PC) the values of the filtrate flow rate and of the solid concentration for the suspension transported.

The decisions made by the computer concerning the pressure of the pump-flow rate dependence and of the flow rate of the fresh suspension, are controlled by the micro-device of the execution system (ES). It is important to observe that the majority of the input process variables are not easily and directly observable. As a consequence, a good technological knowledge is needed for this purpose. If we look attentively at the $x_1 - x_5$ input process variables, we can see that their values present a random deviation from the mean values. Other variables such as pump exit pressure and flow rate (x_6, x_7) can be changed with time in accordance with technological considerations.

Now we are going to introduce an automatic operation controlled by a computer, which means that we already know the entire process. Indeed, the values of y_1 and y_3 have been measured and the computer must be programmed with a mathematical model of the process or an experimental table of data showing the links between dependent and independent process variables. Considering each of the unit devices, we can see that each device is individually characterised by inputs, outputs and by major phenomena, such as the flow and filtration in the filter unit, the mixing in the suspension reservoir and the transport and flow through the pump. Consequently, as part of the unit, each device has its own mathematical model. The global model of the plant is then the result of an assembly of models of different devices in accordance with the technological description.

In spite of the description above, in this example we have given no data related to the dimensions or to the performance of the equipment. The physical properties of all the materials used have not been given either. These data are recognized by the theory of process modelling or of experimental process investigation as process parameters. A parameter is defined by the fact that it cannot determine the phenomena that characterize the evolution in a considered entity, but it can influence the intensity of the phenomena [1.1, 1.2].

As regards the parameters defined above, we have two possibilities of treatment: first the parameters are associated with the list of independent process variables: we will then consequently use a global mathematical model for the unit by means of the formal expression (1.2). Secondly, the parameters can be considered as particular variables that influence the process and then they must, consequently, be included individually in the mathematical model of each device of the unit. The formal expression (1.3) introduces this second mathematical model case:

$$y_i = F(x_1, x_2 \ldots, x_N, z_1, z_2 \ldots, z_S) \qquad\qquad i = 1, \ldots . P \qquad\qquad (1.2)$$

$$y_i = F(x_1, x_2 \ldots, x_N, z_1, z_2 \ldots, z_S, p_1, p_2, \ldots, p_r) \qquad i = 1, \ldots . P \qquad\qquad (1.3)$$

We can observe that the equipment is characterized by the process parameters of first order whereas process parameters of second order characterize the processed materials. The first order and second order parameters are respectively called "process parameters" and "non-process parameters".

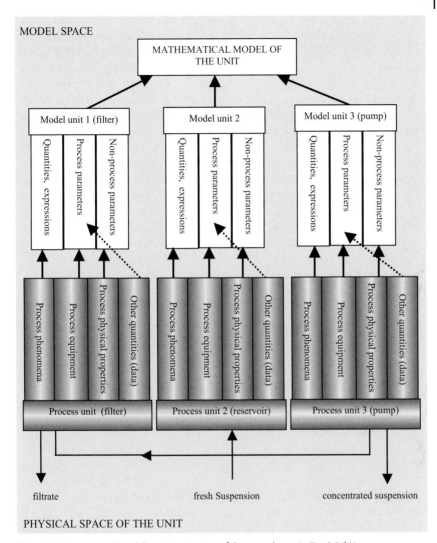

MODEL SPACE

PHYSICAL SPACE OF THE UNIT

Figure 1.2 Process and model parts (extension of the case shown in Fig. 1.1(b)).

Figure 1.2 shows a scheme of the physical space of the filtration unit and of its associated model space. The model space presents a basic level which includes the model of each device (filter, reservoir and pump) and the global model which results from the assembly of the different models of the devices.

If we establish a relation between Fig. 1.2 and the computer software that assists the operation of the filtration plant, then we can say that this software can be the result of an assembly of mathematical models of different components or/ and an assembly of experimentally characterized components.

It is important to note that the process control could be described by a simple or very complex assembly of relations of type (1.2) or (1.3). When a model of one component is experimentally characterized in an assembly, it is important to correct the experimental relationships for scaling-up because these are generally obtained by using small laboratory research devices. This problem can be solved by dimensional analysis and similitude theory. From Fig. 1.2 we can deduce that the first step for process modelling is represented by the splitting up of the process into different elementary units (such as component devices, see Fig. (1.1b)). As far as one global process is concerned, each phenomenon is characterized by its own model and each unit (part) by the model of an assembly of phenomena.

A *model* is a representation or a description of the physical phenomenon to be modelled. The physical model (empirical by laboratory experiments) or conceptual model (assembly of theoretical mathematical equations) can be used to describe the physical phenomenon. Here the word "model" refers to a mathematical model. A (mathematical) model as a representation or as a description of a phenomenon (in the physical space) is a systematic collection of empirical and theoretical equations. In a model (at least in a good model) both approaches explain and predict the phenomenon. The phenomena can be predicted either mechanistically (theoretically) or statistically (empirically).

A *process model* is a mathematical representation of an existing or proposed industrial (physical or/and chemical) process. Process models normally include descriptions of mass, energy and fluid flow, governed by known physical laws and principles. In process engineering, the focus is on processes and on the phenomena of the processes and thus we can affirm that a process model is a representation of a process. The relation of a process model and its structure to the physical process and its structure can be given as is shown in Fig. 1.2 [1.1–1.3].

A *plant model* is a complex mathematical relationship between the dependent and independent variables of the process in a real unit. These are obtained by the assembly of one or more process models.

1.2
Observations on Some General Aspects of Modelling Methodology

The first objective of modelling is to develop a software that can be used for the investigation of the problem. In this context, it is important to have more data about the modelling methodology. Such a methodology includes: (i) the splitting up of the models and the definition of the elementary modelling steps (which will then be combined to form a consistent expression of the chemical process); (ii) the existence of a generic modelling procedure which can derive the models from scratch or/and re-use existing models by modifying them in order to meet the requirements of a new context.

If we consider a model as a creation that shows the modelled technical device itself, the modelling process, can be considered as a kind of design activity [1.4, 1.5]. Consequently the concepts that characterize the design theory or those related to solving the problems of general systems [1.6, 1.7] represent a useful starting base for the evolution of the modelling methodology. Modelling can be used to create a unit in which one or more operations are carried out, or to analyse an existing plant. In some cases we, a priori, accept a split into different components or parts. Considering one component, we begin the modelling methodology with its descriptive model (this will also be described in Chapter 3). This descriptive model is in fact a splitting up procedure, which thoroughly studies the basic phenomena. Figure 1.3 gives an example of this procedure in the case of a liquid–solid extraction of oil from vegetable seeds by a percolation process. In the descriptive model of the extraction unit, we introduce entities which are endowed with their own attributes. Considering the seeds which are placed in the packed bed through which the extraction solvent is flushed, we introduce the "packed bed and mono-phase flow" entity. It is characterized by different attributes such as: (i) dynamic and static liquid hold-up, (ii) flow permeability and (iii) flow dispersion. The descriptive model can be completed by assuming that the oil from the seeds is transported and transferred to the flowing solvent. This assumption introduces two more entities: (i) the oil seed transport, which can be characterized by one of the following attributes: core model transport or porous integral diffusion model transport, and (ii) the liquid flow over a body, that can be characterized by other various attributes. It is important to observe that the attributes associated to an entity are the basis for formulation of the equations, which express the evolution or model of the entity.

The splitting up of the process to be modelled and its associated mathematical parts are not unique and the limitation is only given by the researcher's knowledge. For example, in Fig. 1.3, we can thoroughly analyse the splitting up of the porous seeds by introducing the model of a porous network and/or a simpler porous model. Otherwise we have the possibility to simplify the seed model (core diffusion model or pure diffusion model) into a model of the transport controlled by the external diffusion of the species (oil). It is important to remember that sometimes we can have a case when the researcher does not give any limit to the number of splits. This happens when we cannot extend the splitting because we do not have any coherent mathematical expressions for the associated attributes. The molecular scale movement is a good example of this assertion. In fact we cannot model this type of complex process by using the classical transport phenomena equations. Related to this aspect, we can say that the development of complex models for this type of process is one of the major objectives of chemical engineering research (see Section 1.4).

Concerning the general aspects of the modelling procedure, the definition of the modelling objectives seems largely to be determined by the researcher's pragmatism and experience. However, it seems to be useful in the development and the resulting practical use of the model in accordance with the general principles of scientific ontology [1.8, 1.9] and the general system theory [1.10]. If a model is developed by using the system theory principles, then we can observe its structure and behaviour as well as its capacity to describe an experiment such as a real experimental model.

Concerning the entities defined above (each entity together with its attributes) we introduce here the notion of basic devices with various types of connections.

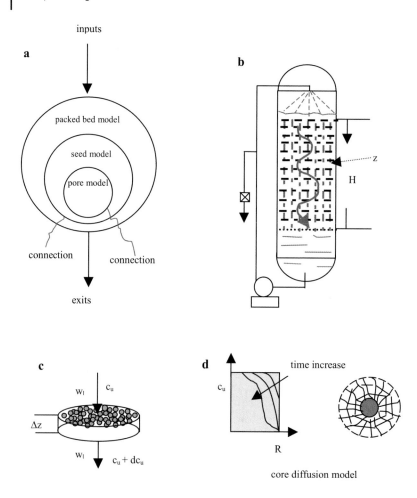

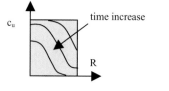

Figure 1.3 Entities and attributes for percolation extraction of oil from seeds.
(a) Hierarchy and connections of the model,
(b) percolation plant,
(c) section of elementary length in packed seeds bed,
(d) physical description of two models for oil seed transport

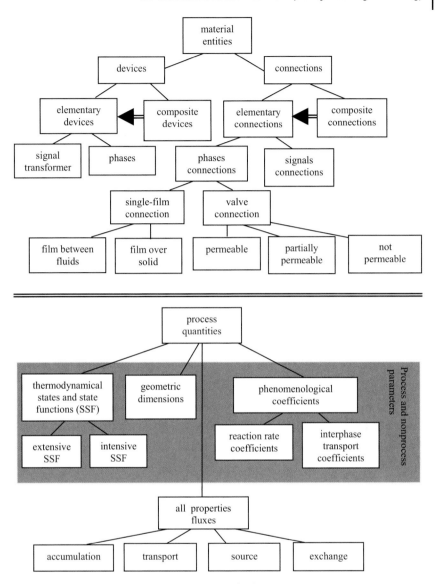

Figure 1.4 Material entities and process quantities for the development of a structured mathematical modelling.

One of the characteristics of basic devices is that they cannot be split up into parts. A basic device can also be a signal transformer (a function which transforms the input into output, such as the thermocouple that transforms the input temperature into an electrical tension). The process phases are connected and characterized quantitatively, from the viewpoint of characteristic relations (equations), as in Fig. 1.4 [1.11–1.13]. This structured mathematical modelling development corre-

sponds to the case of a modelling based on transport phenomena, which are analysed in Chapter 3. Now, we shall give some explanation concerning some aspects
of the basic chemical engineering introduced in Fig. 1.4. With respect to the generalized fluxes and their refinements, it is known that they directly correspond to
the physicochemical phenomena occurring in a phase or at its boundary according to the phase properties. Otherwise, any process quantity assigned to a particular phase may depend on one or several coordinates such as time and spatial dimensions.

Various laws restrict the values of the process quantities. These laws may represent either fundamental, empirical physicochemical relationships or experimentally identified equations (from statistical modelling or from dimensional analysis
particularizations). In contrast to statistical or dimensional analysis based models
[1.14], which are used to fix the behaviour of signal transformers, the models of
transport phenomena are used to represent generalized phases and elementary
phase connections. Here, the model equations reveal all the characterizing attributes given in the description of a structure. They include *balance equations, constitutive equations* and *constraints*.

The last introduced notions show that the modelling methodology tends to a
scientific synthetic working procedure, where the use of an abstract language is
needed to unify the very high diversity of cases that require an analysis made by
mathematical modelling. At the same time the problem discussed here has shown
that the creation of models could be considered as a special problem of design
and modelling, i.e.could be considered as an art rather than a science [1.15],
emphasizing a modeller's creativity and intuition rather than a scientific methodology.

1.3
The Life-cycle of a Process and Modelling

The life-cycle of a chemical compound production or of a chemical process development starts when a new and original idea is advanced taking into account its
practical implementation. The former concept with respect to the process life-
cycle, which imposed a rigid development from research and development to process operation, has been renewed [1.16–1.18]. It is well known that the most
important stages of the life-cycle of a process are research and development, conceptual design, detailed engineering, piloting and operation. These different steps
partially overlap and there is some feedback between them as shown in Fig. 1.5.
For example, plant operation models can be the origin of valuable tips and potential research topics, obviously these topics directly concern the research and development steps (R&D). The same models, with some changes, are preferably utilized in all the steps. The good transfer of information, knowledge and ideas is
important for successfully completion of the development of all the process
phases. For this purpose, it is important to have a proper documentation of underlying theories and assumptions about the model (models). This acts as a check list

when a problem occurs and ensures that knowledge is transferred to the people concerned. The models are an explicit way of describing the knowledge of the process and related phenomena. They provide a systematic approach to the problems in all the stages of the process life-cycle. In addition, the process of writing the theory as mathematical expressions and codes, reveals the deficiencies with respect to the form and content.

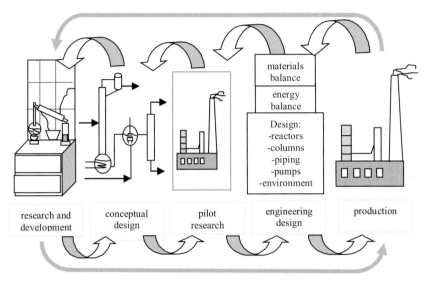

Figure 1.5 The stages of the process life-cycle and their main relationships.

Among the factors that influence the amount of work required to develop a model, we can retain the complexity, the novelty and the particular knowledge related to the process in modelling. Otherwise, commercial modelling software packages are frequently used as an excellent platform. In the following sections we detail some particularities of the models used in the process life-cycle.

1.3.1
Modelling and Research and Development Stage

The models in the R&D stage can first be simple, and then become more detailed as work proceeds. At this stage, attention has to be focused on the phenomena of phase equilibrium, on the physical properties of the materials, on chemical kinetics as well as on the kinetics of mass and heat transfer. As previously shown (see Figs 1.2 and 1.3), the decomposition of the process into different elementary units is one of the first activities. This action requires careful attention especially because, at this life-cycle stage, the process could be nothing but an idea. The work starts with the physical properties, as they act as an input to all other components. The guidelines to choose physical properties, phase equilibrium data, characteristic state equations etc. can be found in the usual literature. For each studied case, we can choose the level of detail such as the complexity of the equations and the number of parameters. If the literature information on the physical properties is restricted an additional experimental step could be necessary. As far as industrial applications are concerned, the estimation of the reaction kinetics is usually semi-empirical. Therefore, a full and detailed form of kinetics equations is not expected for the majority of the investigated cases. Some physical phenomena along with their effects can require special attention. Conventional engineering correlations may not apply, and, consequently, the research must be directed to study of these problems.

The ideal modelling and experimental work have to be realized simultaneously and are strongly related. Models provide a basis to choose, both qualitatively and quantitatively, appropriate experimental conditions. The data obtained from experimental work are used to confirm or reject the theories or the form of equations if an empirical model is being applied. Otherwise, these data are used to estimate the model parameters. This work is sequential in the sense that starting from an initial guess, the knowledge of the system grows and models get more and more accurate and detailed as the work proceeds. With a proper experimental design, the models can be used to evaluate and to rank competitive theories. Since, at the research and development (R&D) stage, models are still in the building phase, they can mainly be used for experimental design.

When the R&D steps are almost completed, the models related to the phenomena are combined into process unit models. Bench scale tests are used to check separate process ideas. The estimation of the equipment parameters can be seen as R&D work, especially if the equipment is in some way new, such as a new innovation or a new application. Based on a good knowledge of the phenomena, valuable tips concerning optimal operating parameters (such as temperature and pressure range as well as restricting phenomena) can be given to the next stages. In this stage we can meet several important sub-problems to select the appropriate models. There are often competitive theories to describe the relevant phenomena. A choice has also to be made between mechanistic and empirical approaches. The former should be favoured but an integrated solution is usually more beneficial. Then, the degree of detail has to be chosen in order to serve the model usefully. Practically, the best solution is to describe the most relevant phenomena in a detailed way, whereas the less important ones will be left approximate or in an empirical state.

1.3.2
Modelling and Conceptual Design Stage

The establishing of the optimal process structure and the best operating conditions characterizes the process development at this stage. Firstly, attention must be focused on the synthesis of the proceess. The extent to which models can be used in this phase varies. If we have a new process, information from similar cases may not be available at this stage. In the opposite situation, when the chemical components are well known, which usually means that their properties and all related parameters can be found in databanks, the models can be used to quickly check new process ideas. For example, at this stage, for a multiple-component distillation problem, models are used to identify key and non-key components, optimum distillation sequence, the number of ideal stages, the position of feed, etc. At this stage also, we always focus on the full-scale plant. Another question is how the concept will be carried out in the pilot phase. It is known that for this (piloting) stage, the equipment does not have to be a miniature of the full scale. Practice has shown that the choices made here affect both investment and operating costs later on. An image of the full-scale plant should also be obtained. The researchers who work at this level will propose some design computations which are needed by the piloting stage of process life-cycle. Their flow-sheet is the basis of the pilot design or development.

1.3.3
Modelling and Pilot Stage

The whole process concept is generally improved in the pilot plant. We can transform this stage into a process analysis made of models if enough experimental data and knowledge about the process exist (for example when we reuse some old processes). For reference, we should mention that other situations are important, such as, for example, knowing that a pilot plant provides relatively easy access to the actual conditions of the process. Some by-pass or small streams could be taken off from the pilot unit and be used in the operation of apparatuses specially designed for the experimental work. Now the models should be ready, except for the correct values of the parameters related to the equipment. A special pilot stage feature consists in adding the equations describing the non-ideal process hardware to the model in order to compute efficiency (tray efficiency, heat exchanger efficiency, non-ideality numbers, etc). This stage is strongly limited in time, so, to be efficient, researchers must prepare a careful experimental program. It may be impossible to foresee all the details, since the experimentation related to the estimation of parameters is often carried out in sequences, but still, a systematic preparation and organization of the work to be done remains useful. Since a pilot plant is rigid as far as its manœuvrability is concerned, full advantage should be taken of all the data acquired with its help. Data loggers are recommended to collect and store process data and also to provide customized data reports for modelling. If we link the process data logger with the laboratory information system, then there is a good possibility of getting a full image of the state of the process at a precise time. It is important to remember, that the goal of the pilot stage in terms of modelling is to get a valid mass and energy balance model and to validate the home-made models.

1.3.4
Modelling and Detailed Engineering Stage

In this stage, models are used for the purpose for which they have been created: the design and development of a full scale plant which is described in the detailed engineering stage. On the basis of what has been learned before, the equipment can be scaled-up, taking into consideration pilot phase and related data, as well as the concepts of similitude. Special attention should be paid to the detailed engineering of the possible technical solutions. Depending on their nature, the models can either provide a description of how the system behaves in certain conditions or be used to calculate the detailed geometric measures of the equipment. For example, we show that all the dimensions of a distillation column can be calculated when we definitively establish the separation requirements. Special consideration should be given to the process of scaling-up, because here we must appreciate whether the same phenomena occur identically occur on both scales (see Chapter 6 for similitude laws). Similarly, when equipment is being scaled up, attention should be paid to its parameters, because they can be a function of the size. When dealing with empirical models where the origin of the effects is unknown, the uncertainty of the model validity must be considered.

It is useful to have detailed documentation concerning all the assumptions and theories used in the model. The yield and energy consumption of a process are easily optimised using fine-tuned models to design a new unit or process. Depending on the process integration, pinch analysis and other similar analysis procedures can be used to find a solution of heat integration. Various data on streams and energy consumption, which are easily developed from simulation results, can be used to sustain the adopted technical solutions.

1.3.5
Modelling and Operating Stage

At this stage of the process life-cycle, the models must include all relevant physical, chemical and mechanical aspects that characterize the process. The model predictions are compared to actual plant measurements and are further tuned to improve the accuracy of the predictions. This consideration is valuable, especially for the finally adjusted models that create the conditions of use to meet the demand of this operating stage so as to guarantee optimal production. Models can also be used in many ways in order to reduce the operating costs. In the mode of parameter estimation, the model is provided with the process measurement data reflecting the current state of the process, which makes it possible, for example, to monitor the fouling of a plant heat exchanger. Once the heat transfer coefficient falls under a preset limit, it is time for maintenance work. In this way a virtual process can be kept up to date.

In simulation mode, the performance of the process can be followed. Discrepancies between the model and the process may reveal instrumentation malfunction, problems of maintenance etc. Verified flow-sheet models can be used to further analyse the process operation. In the optimising mode, the models are especially used when different grades of the product are manufactured with the process. At this point we criticize the old practices that rely on the tacit knowledge of an experimented operator and consider the models as an artificial creation, which cannot attain the operator's performance.

The importance of storing process data has been emphasized here. After all, the data are an important link in the creation cycle of the process knowledge. Future applications concerning the gathering of new data will provide a powerful tool in the use of the stored data or process memory. It is important to keep in mind that, at this stage, the process could be further improved as new ideas, capacity increasing spin-off projects, R&D projects, etc. are developed. These developments frequently require a partial implementation of the methodology described above. Therefore, the models of the existing process could act as a tool in further developments.

In practice, models are often tailor-made and their use requires expertise. Building interfaces, which take into account the special demands arising from man–computer interaction, can greatly expand the use of the models.

Table 1.1 summarizes the discussions on the modes under which the models are used, which are explained in Sections 1.1–1.3.

Table 1.1 The modes of model use for all the stages in the life-cycle process.

Mode	Input models data	Computed (exit) models data
Simulation	values of input process variables values of process parameters values of non-process parameters	values for exit process variables
Design	values of input process variables values of exit process variables values of non-process parameters	values of process parameter (those that show the equipment size)
Parameter estimation	values of input process variables values of exit process variables values of process parameters	values of non-process parameters
Optimization	all fixed input process variables – all fixed exit process variable – all process non-parameters – some fixed process parameters – optimization expressions	optimal non-fixed inputs optimal non-fixed exits – optimal non fixed parameters – values of optimized functions

Concerning the question *Why modelling?*, which is also the title of this chapter, we can assert that the use of models is important because these have the capacity to assist the solution of many important and fundamental problems in chemical engineering.

We can especially mention that modelling can be successfully used to:
- reduce manufacturing costs
- reduce time and costs in all stages of the process life-cycle
- increase process efficiency
- allow a better and deeper understanding of the process and its operation
- be used as support for the solutions adopted during the process development and exploitation
- ensure an easy technological transfer of the process
- increase the quality of process management
- reveal abilities to handle complex problems
- contribute to reducing pollution
- improve the safety of the plants
- market new products faster
- reduce waste emission while the process is being developed
- improve the quality of the products
- ensure a high quality of training of the operators.

1.4
Actual Objectives for Chemical Engineering Research

In the past, the scope of chemical engineering research ranged from process engineering to product engineering. It was firstly defined as the capacity to produce one chemical product with complex designed properties. It was occasioned by the necessity to produce one profound transformation of the existing chemical production systems. The objective then was to produce the state displacement in the vicinity of its thermodynamic efficiency [1.19, 1.20]. Many theoreticians and practitioners accept that if the researcher wants to obey all the statements described above, changes in many of the classical research procedures are required [1.21, 1.22].

Trying to discover the basic concepts that will be the keys to successful applications in the future, more and more scientists consider that the chemical engineering design and research must meet five major objectives [1.23–1.26]:

1. *The first objective* is represented by the need to increase the productivity and selectivity of both existing and new processes through intelligent operations and multiscale control of processes. This objective is sustained by the important results obtained thanks to the synthesis of a new class of engineered porous supports and catalysts. So the catalytic reactions and separation processes that use these materials can be efficiently controlled.

   Microtechnology makes it possible to produce these materials in series. Other materials with a controlled structure begin to be developed for chiral technologies.

Such approaches imply that chemical engineers should go down to the nanoscale to control events at the molecular level. At this level, manipulating supramolecular building blocks can create new functions in interacting species such as self-organization, regulation, replication and communication. Consequently a new mathematical characterisation must be produced and used to describe these discrete functions.

At the microscale level, detailed local temperature and composition control through the staged feed and supply of reactants or the removal of products would result in a higher selectivity and productivity than would the conventional approach. Indeed, this conventional approach imposes boundary conditions and lets a system operate under spontaneous reaction and transfer. To produce a local energy supply, microwave and ultrasound can be used instead of heat. To operate the relevant models on these energies, local sensors and actuators as well as close computer control will absolutely be needed.

On the other hand, on-line information on the process state and on the quality of the products should not be limited to such usual parameters as pressure, temperature, pH and composition, but should extend to more sophisticated characteristics such as colour, smoothness, odour, etc. To produce and to introduce these parameters into the current production in progress, modelling and experimental research must be combined.

2. *The second objective* is represented by the need to design novel equipment based on scientific principles corresponding to novel modes of production.

We cannot begin a short discussion about this objective without observing that despite new technological and material developments, most of the equipment used in chemical plants is based on 100-year old principles. On the other hand, past research in chemical engineering has led to a better understanding of the elementary phenomena and now we can conceive novel equipment based on these scientific principles.

Apparently, it is not difficult to imagine coupling a chemical reaction with separation or heat transfer to obtain a concept of multifunctional reactors which frequently result in higher productivity.

The scientific design of the novel equipment and of the new modes of production is also sustained by new operating modes used on the laboratory scale, such as reversed flow, cyclic processes, unsteady state operation, extreme operating conditions (very high pressure and temperature) as well as supercritical media processing. These new modes of production have proved their efficiency and capacity to be modelled and controlled.

Current production modes may also be challenged by miniaturization, modularisation, and decentralization. Recently developed microtechnologies using microreactors, microseparators and very small microanalysers show new possible ways to accurately control reaction conditions with respect to mixing, quenching and temperature profiles. These microtechnologies show that the scientific design of novel equipment begins to be a reality. Such innovative systems can be applied if these novel technologies prove to be robust, reliable, safe, cheap, easy to control, and if they provide significant gains over existing processes.

3. *The third objective* is the need to manufacture chemical products with imposed end-use properties. The consideration of this objective is given by the present and prospective market demand.

There is indeed a growing market demand for sophisticated products combining several functions and properties. As examples, we can mention coatings, cosmetics, detergents, inks, lubricants, surfactants, plastics, food, agrochemicals, and many more products the basic function of which has been excluded while two or more characterizing functions have been identified. In the past, most formulation recipes have resulted from experiments and empirical tests. A good knowledge of the characteristics of such complex media as non-Newtonian liquids, gels, foams, hydrosoluble polymers, dispersions and suspensions can be the key to revolutionizing the design of such products. At the same time, rheology and interfacial phenomena can play a major role in this design.

The prospective market shows signs indicating a great demand for special solids which can act as vehicles conveying condensed matter: this particular property is one of the most frequently demanded. These products can open the way to solvent-less processes. These so-called intelligent solids, presenting controlled reactivity or programmed release of active components, may be obtained through multiple coating on a base solid. All the operations that are related to the manufacture of these products must be reanalysed and reconsidered with respect to the micro- and nanoscale evolution. Particle-size distribution and morphology control are the central concerns in such operations as precipitation, crystallization, prilling, generation of aerosols, and nanoparticles. Agglomeration, granulation, calcination, and compaction as final shaping operations need better understanding and control.

Several questions are raised by the overall problem of manufacturing chemicals with multifunctional properties: how can the operations be scaled-up from the laboratory model to an actual plant? Will the same product be obtained and its properties preserved? What is the role of equipment design in determining the properties of the products? These questions are strongly sustained by the fact that the existing scaling-up procedures cannot show how such end-use properties such as colour, flowability, sinterability, biocompatibility and many others can be controlled.

4. *The fourth objective* includes the need to use multiscale computational chemical engineering in real-life situations.

   The computer applications of molecular modelling using the principles of statistics and quantum mechanics have been developed successfully. They are a new domain for chemical engineering research. Some basic characteristics of the materials' interaction can be calculated by molecular modelling based on information from data banks.

   Dynamic process modelling is being developed to be used on the macroscopic scale. Full complex plant models may involve up to 5.0×10^4 variables, 2.0×10^5 equations and over 1.0×10^5 optimisation variables.

   It is important to avoid confusion between modelling and numerical simulation. Modelling is an intellectual activity requiring experience, skills, judgment and the knowledge of scientific facts. For example, the main obstacle to developing good models of multiphase and complex systems consists more in understanding the physics and chemistry of all interactions than refining the numerical codes of calculations.

   Actually, a model could be divided into smaller units. For example, a global production unit could be divided into catalyst particles, droplets, bubbles, etc, and this could even be extended up to discrete molecular processes.

5. *The fifth objective* concerns the need to preserve the environment. This objective requires the use of non-polluting technologies, the reduction of harmful emissions from existing chemical sites and the development of more efficient and specialized pollutant treatment plants.

We can see that the above-mentioned objectives clearly show that, when one research problem has been fixed, the solution has to be reached taking into consideration the strong relation between the modelling and the experimental research. First both modelling and simulation must indicate the type of experiment needed for a thorough knowledge of the phenomenon. Then, the modelling must identify the best conditions for the evolution of the process phenomena. Complex models with a high hierarchy and complex part connections followed by more and more simulations can contribute to the success of this modern type of chemical engineering research.

1.5
Considerations About the Process Simulation

From the sections above, the reader can observe that the notion of a chemical process can be quite complex. The chemical reactions that take place over a broad range of temperatures and pressures are extraordinarily diverse. From the modelling viewpoint, this complexity results in a considerable number of process and non-process parameters with an appreciable quantity of internal links, as well as in very complex equations describing the process state (the relationships between input and output process variables).

When we build a model, some phenomena are simplified and consequently some parameters are disregarded or distorted in comparison with their reality. In addition, some of the relationships between the parameters could be neglected. Two ways of controlling the output or input of information are available in model building: (i) the *convergence way* which accepts the input or output information only if it preserves or accentuates the direction of the evolution with respect to the real modelled case; (ii) the *divergence way* in which we refuse the input or output of information because it results in a bad model response. To identify the direction of the model response to an input or output of information, we need to realize partial model simulations adding or eliminating mathematical relations from the original model architecture. In Table 1.1 we can see a final process model which is used for the exploitation of the process in the simulation or optimisation mode for an actual case.

One of the answers to the question *Why modelling?* could therefore be the establishment of a set of simulation process analyses. In addition to the mathematical simulation of processes described above, we have the simulation of a physical process, which, in fact, is a small-scale experimental process investigation. In other words, to simulate a process at laboratory-scale, we use the analysis of a more affordable process which is similar to experimental investigation.

1.5.1
The Simulation of a Physical Process and Analogous Computers

The simulation of a physical process consists in analysing the phenomena of the whole process or of a part of it. This is based on the use of a reduced-scale plant, which allows a selected variability of all input variables. We have to focus this analysis on the physical particularities and on the increase in the dimensions of the plant. We then treat the obtained experimental data in accordance with dimensional analysis and similitude theory (for instance, see Chapter 6). The dimensionless data arrangement, imposed by this theory, creates the necessary conditions to particularize the general similitude relationships to the analysed – physically simulated – case. As expected, these physical simulations are able to reproduce the constant values of dimensionless similitude criteria in order to scale-up an experimental plant into a larger one. Then, it makes it possible to scale-up the plant by simply modifying the characteristic dimensions of each device of the experimental plant.

At the same time, when we impose the dimensions of the plant, we can focus on obtaining one or more of the optimal solutions (maximum degree of species transformation, minimum chemical consumption, maximum degree of species transformation with minimum chemical consumption etc.) For this purpose, it is recommended to use both mathematical and physical simulations.

For physical process simulation, as well as for mathematical model development, we can use the *isomorphism* principle. This is based on the formal analogy of the mathematical and physical descriptions of different phenomena. We can detail this principle by considering the conductive flux transport of various properties, which can be written as follows:

for momentum transport $$\tau_{yx} = -\eta \frac{dw_y}{dx} \tag{1.4}$$

for heat transport: $$q_x = -\lambda \frac{dt}{dx} \tag{1.5}$$

for species A transport: $$N_{Ax} = -D_A \frac{dc_A}{dx} \tag{1.6}$$

for electric current transport: $$i_x = -\frac{1}{\rho} \frac{dU}{dx} \tag{1.7}$$

It is not difficult to observe that in all of these expressions we have a multiplication between the property gradient and a constant that characterizes the medium in which the transport occurs. As a consequence, with the introduction of a transformation coefficient we can simulate, for example, the momentum flow, the heat flow or species flow by measuring only the electric current flow. So, when we have the solution of one precise transport property, we can extend it to all the cases that present an analogous physical and mathematical description. Analogous computers [1.27] have been developed on this principle. The analogous computers, able to simulate mechanical, hydraulic and electric micro-laboratory plants, have been experimented with and used successfully to simulate heat [1.28] and mass [1.29] transport.

References

1.1 M. M. Denn, *Process Modelling*, Longman, New York, 1985.

1.2 B. P. Zeigler, *The Theory of Modelling and Simulation*, Wiley, New York, 1976.

1.3 J. M. Douglas, *Conceptual Design of Chemical Processes*, McGraw-Hill, New York, 1988.

1.4 A. W. Westerberg, H. P. Hutchinson, L. R. Motard, P. Winter, *Process Flowsheeting*, Cambridge University Press, Cambridge, 1979.

1.5 *Aspen Plus User's Guide*, Aspen Technology Inc, Cambridge, 1998.

1.6 H. A. Simon, *The Sciences of Artificial*, MIT Press, Cambridge, 1981.

1.7 J. P. Van Gligch, *System Design Modelling and Metamodelling* , Plenum Press, New York, 1991.

1.8 K. Benjamin, On Representing Commonsense Knowledge, in *The Representation Use of Knowledge by Computers*, V. N. Findler, (Ed), Academic Press, New York, 1979.

1.9 F. N. Natalya, D. L. McGuiness, *A Guide to Creating Your First Ontology*, Stanford University Press, 1995.

1.10 G. J. Klir, *Architecture of Systems Problems Solving*, Plenum Press, New York, 1985.

1.11 W. Marquardt, Dynamic Process Simulation; Recent Trends and Future Challenges, in *Chemical Process Control CPC-IV*, Y. Arkun, H. W. Ray, (Eds.), pp. 131–188, CACHE, Austin, 1991.

1.12 W. Marquardt, Towards a Process Modelling Methodology, in *Model-based Process Control*, R. Berber, (Ed.), pp. 3–40, Kluwer, Amsterdam, 1995.

1.13 W. Marquardt, *Comput. Chem. Eng.* **1996**, *20*, 67, 591.

1.14 T. Dobre, O. Floarea, *Chemical Engineering-Momentum Transfer*, Matrix Press, Bucharest, 1997.

1.15 R. Aris, *Mathematical Modelling Techniques*, Pitman, London, 1978.

1.16 C. Cohen, *Br. J. Hist. Sci.* **1996**, *29*, 101, 171.

1.17 E. A. Bratu, *Unit Operations for Chemical Engineering-Vol. 1*, Technical Press, Bucharest, 1983.

1.18 *Hysis Process User's Guide*, Hyprotech Ltd, 1998.

1.19 N. R. Amundson, *Chemical Engineering Frontiers: Research Needs and Opportunity (The Amundson Report)*, National Research Council, Academic Press, Washington, 1988.

1.20 A. R. Mashelkar, *Chem. Eng. Sci.* **1995**, *50* (1), 1.

1.21 A. Colin, J. Howell, *AIChE annual Meeting, Exchanging Ideas for Innovation*, November 12–17, 2000.

1.22 M. C. Fleming, W. R. Cahn, *Acta. Mater.* **2000**, *48*, 371.

1.23 J. Villermaux, *Chem.Eng. Sci.* **1993**, *48* (14), 2525.

1.24 J. Villermaux, *Trans. Inst. Chem. Eng.* **1993**, *71*, 45.

1.25 H. J. Krieger, *Chem. Eng. News*, **1996**, *1*, 10.

1.26 J. C. Charpentier, P. Trambouze, *Chem. Eng. Process.* **1998**, *37*, 559.

1.27 L. Levine, *Methods for Solving Engineering Problems Using Analogous Computers*, McGraw-Hill, New York, 1964.

1.28 H. S. Craslow, J. C. Jaegar, *Conduction of Heat in Solids*, Clarendon Press, Oxford, 1959.

1.29 J. Crank, *The Mathematics of Diffusion*, Clarendon Press, Oxford, 1956.

2
On the Classification of Models

The advances in basic knowledge and model-based process engineering methodologies will certainly result in an increasing demand for models. In addition, computer assistance to support the development and implementation of adequate and clear models will be increasingly used, especially in order to minimize the financial support for industrial production by optimizing global production processes. The classification of models depending on their methodology, mathematical development, objectives etc. will be a useful tool for beginners in modelling in order to help them in their search for the particular model able to solve the different and variable products synthesis.

Highly-diversified models are used in chemical engineering, consequently, it is not simple to propose a class grouping for models. The different grouping attempts given here are strongly related to the modeled phenomena. In the case of a device model or plant model, the assembly of the model parts creates an important number of cases that do not present any interest for class grouping purposes. In accordance with the qualitative process theory to produce the class grouping of one phenomenon or event, it is important to select a clear characterization criterion which can assist the grouping procedure. When this criterion is represented by the theoretical base used for the development of models, the following classification is obtained:
- mathematical models based on the laws of transport phenomena
- mathematical models based on the stochastic evolution laws
- mathematical models based on statistical regression theory
- mathematical models resulting from the particularization of similitude and dimensional analysis.

When the grouping criterion is given by the mathematical complexity of the process model (models), we can distinguish:
- mathematical models expressed by systems of equations with complex derivatives
- mathematical models containing one equation with complex derivatives and one (or more) ordinary system(s) of differential equations

Chemical Engineering. Tanase G. Dobre and José G. Sanchez Marcano
Copyright © 2007 WILEY-VCH Verlag GmbH & Co. KGaA, Weinheim
ISBN: 978-3-527-30607-7

- mathematical models promoted by a group of ordinary systems of differential equations
- mathematical models with one set of ordinary differential equations complete with algebraic parameters and relationships between variables
- mathematical models given by algebraic equations relating the variables of the process.

For the mathematical models based on transport phenomena as well as for the stochastic mathematical models, we can introduce new grouping criteria. When the basic process variables (species conversion, species concentration, temperature, pressure and some non-process parameters) modify their values, with the time and spatial position inside their evolution space, the models that describe the process are recognized as *models with distributed parameters*. From a mathematical viewpoint, these models are represented by an assembly of relations which contain partial differential equations The models, in which the basic process variables evolve either with time or in one particular spatial direction, are called *models with concentrated parameters*.

When one or more input process variable and some process and non-process parameters are characterized by means of a random distribution (frequently normal distributions), the class of *non-deterministic models* or of *models with random parameters* is introduced. Many models with distributed parameters present the state of models with random parameters at the same time.

The models associated to a process with no randomly distributed input variables or parameters are called *rigid models*. If we consider only the mean values of the parameters and variables of one model with randomly distributed parameters or input variables, then we transform a non-deterministic model into a rigid model.

The stochastic process models can be transformed by the use of specific theorems as well as various stochastic deformed models, more commonly called *diffusion models* (for more details see Chapter 4). In the case of statistical models, we can introduce other grouping criteria. We have a detailed discussion of this problem in Chapter 5.

In our opinion, one important grouping criterion is the chemical engineering domain that promotes the model. In the next section, modeling and simulation have been coupled and a summary of this classification is given.

2.1
Fields of Modelling and Simulation in Chemical Engineering

Some important chemical engineering modelling and simulation fields as well as related activities are briefly presented here. First, we can see that the traditional modelling procedures or *computer-aided process engineering* cover a much narrower range of modelling tools than those mentioned here. A broader spectrum of chemical engineering modelling and simulation fields is developed and illustrated elsewhere in this book.

2.1.1
Steady-state Flowsheet Modelling and Simulation

Process design for continuous processes is carried out mostly using steady-state simulators. In steady-state process simulation, individual process units or entire flowsheets are calculated, such that there are no time deviations of variables and parameters. Most of the steady-state flowsheet simulators use a sequential modular approach in which the flowsheet is broken into small units. Since each unit is solved separately, the flowsheet is worked through sequentially and iteration is continued until the entire flowsheet is converged. Another way to solve the flowsheet is to use the equation oriented approach, where the flowsheet is handled as a large set of equations, which are solved simultaneously.

Flowsheet simulators consist of unit operation models, physical and thermodynamic calculation models and databanks. Consequently, the simulation results are only as good as the underlying physical properties and engineering models. Many steady-state commercial simulators [2.1, 2.2] have some dynamic (batch) models included, which can be used in steady-state simulations with intermediate storage buffer tanks.

2.1.2
Unsteady-state Process Modelling and Simulation

Unsteady-state or dynamic simulation accounts for process transients, from an initial state to a final state. Dynamic models for complex chemical processes typically consist of large systems of ordinary differential equations and algebraic equations. Therefore, dynamic process simulation is computationally intensive. Dynamic simulators typically contain three units: (i) thermodynamic and physical properties packages, (ii) unit operation models, (iii) numerical solvers. Dynamic simulation is used for: batch process design and development, control strategy development, control system check-out, the optimization of plant operations, process reliability/availability/safety studies, process improvement, process start-up and shutdown. There are countless dynamic process simulators available on the market. One of them has the commercial name Hysis [2.3].

2.1.3
Molecular Modelling and Computational Chemistry

Molecular modelling is mainly devoted to the study of molecular structure. Computational chemistry is the application of all kinds of calculations, mainly numerical, to the study of molecular structure. It can be considered as a subset of the more general field of molecular modelling because its computations occur as a result of the application of the models.

In contrast to computational chemistry, molecular modelling in the sense of spatial molecular arrangement may not involve any computations [2.4]. Today molecular modelling is being used in an increasingly broad range of chemical systems and by an increasing number of scientists. This is due to the progress made in computer hardware and software, which now allows fundamental and complex calculations on a desktop computer. Computational chemistry is rapidly becoming an essential tool in all branches of chemistry as well as related fields such as biochemistry, biology, pharmacology, chemical engineering and materials science. In some cases, computational chemistry can be used to calculate such compound properties as: shapes – structure and geometry; binding energies – strengths of bonds; charge distributions – dipole, quadrapole, octapole moments; spectra – UV, IR, NMR; thermodynamic properties – energy, entropy, radial distribution functions, structural and dynamic properties – viscosity, surface tension, potential energy surfaces; reaction pathways and energy barriers; product energy distributions and reaction probabilities.

2.1.4
Computational Fluid Dynamics

Computational fluid dynamics (CFD) is the science of determining a numerical solution to governing equations of fluid flow while the solution through space or time is under progress. This solution allows one to obtain a numerical description of the complete flow field of interest. Computational fluid dynamics obtains solutions for the governing Navier–Stokes equations and, depending upon the particular application under study, it solves additional equations involving multiphase, turbulence, heat transfer and other relevant processes [2.5, 2.6]. The partial differential Navier–Stokes and associated equations are converted into algebraic form (numerically solvable by computing) on a mesh that defines the geometry and flow domain of interest. Appropriate boundary and initial conditions are applied to the mesh, and the distributions of quantities such as velocity, pressure, turbulence, temperature and concentration are determined iteratively at every point in space and time within the domain. CFD analysis typically requires the use of computers with a high capacity to perform the mathematical calculations. CFD has shown capability in predicting the detailed flow behaviour for a wide-range of engineering applications, typically leading to improved equipment or process design. CFD is used for the early conceptual studies of new designs, detailed equipment design, scaling-up, troubleshooting and retrofitting systems. Examples in chemical and process engineering include separators, mixers, reactors, pumps, pipes, fans, seals, valves, fluidised beds, bubble columns, furnaces, filters and heat exchangers [2.7, 2.8].

2.1.5
Optimisation and Some Associated Algorithms and Methods

In an optimisation problem, the researcher tries to minimise or maximise a global characteristic of a decision process such as elapsed time or cost, by exploiting certain available degrees of freedom under a set of constraints. Optimisation problems arise in almost all branches of industrial activity: product and process design, production, logistics, short planning and strategic planning. Other areas in the process industry suitable for optimisation are process integration, process synthesis and multi-component blended-flow problems.

Optimisation modelling is a branch of mathematical modelling, which is concerned with finding the best solution to a problem. First, the problem must be represented as a series of mathematical relationships. The best solution to a mathematical model is then found using appropriate optimisation software (solver). If the model has been built correctly, the solution can be applied back to the actual problem. A mathematical model in optimisation usually consists of four key objects [2.9]: data (costs or demands, fixed operation conditions of a reactor or of a fundamental unit, capacities etc.); variables (continuous, semi-continuous, and non-frequently binary and integer); constraints (equalities, inequalities); objective function. The process of building mathematical models for optimisation usually leads to structured problems such as: linear programming, mixed integer linear programming, nonlinear programming and mixed integer nonlinear programming [2.10]. In addition, a solver, i.e. a software including a set of algorithms capable of solving problems, is needed to build a model as well as to categorize the problem. To this end, a specific software can be created but some commercial ones also exist.

Heuristic methods are able to find feasible points of optimisation problems. However, the optimisation of these points can only be proved when used in combination with exact mathematical optimisation methods. For this reason, these methods could not be considered as optimisation methods in the strict meaning of the term. Such heuristic methods include simulated annealing, evolution strategy, constraint programming, neural networks and genetic algorithms. The hybrid approaches combine elements from mathematical optimisation and heuristic methods. They should have great impact on supply chain and scheduling problems in the future.

2.1.6
Artificial Intelligence and Neural Networks

Artificial intelligence is a field of study concerned with the development and use of computer systems that bear some resemblance to human intelligence, including such operations as natural-language recognition and use, problem solving, selection from alternatives, pattern recognition, generalisation based on experience and analysis of novel situations, whereas human intelligence also involves knowledge, deductive reasoning and learning from experience. Engineering and industrial applications of artificial intelligence include [2.11]: the development of more effective control strategies, better design, the explanation of past decisions, the identification of future risks as well as the manufacturing response to changes in demands and supplies. Neural networks are a rather new and advanced artificial intelligence technology that mimic the brain's learning and decision-making process. A neural network consists of a number of connected nodes which include neurons. When a training process is being conducted, the neural network learns from the input data and gradually adjusts its neurons to reflect the desired outputs.

Fuzzy logic is used to deal with concepts that are vague. Many real-world problems are better handled by fuzzy logic than by systems requiring definite true/false distinctions. In the chemical and process industry, the main application of fuzzy logic is the automatic control of complex systems. Neural networks, fuzzy logic and genetic algorithms are also called soft computing methods when used in artificial intelligence.

2.1.7
Environment, Health, Safety and Quality Models

Special models and programs are developed for such purposes as health and safety management and assessment, risk analysis and assessment, emission control and detection and quality control. Such a program may, for example, help the user to keep records regarding training, chemical inventories, emergency response plans, material safety data, sheet expiry dates and so on.

2.1.8
Detailed Design Models and Programs

Certain models and programs are available for the detailed design of processes and process equipment. For example, the process equipment manufacturers often have detailed design and performance models for their products. Engineering design involves a lot of detailed design models.

2.1.9
Process Control

Process control is a general term used to describe many methods of regulating industrial processes. The process being controlled is monitored for changes by means of sensor devices. These sensor devices provide information about the state of the system. The information provided by the sensor devices is used to calculate some type of feedback to manipulate control valves or other control devices. This provides the process with computerized automatic regulation. The essential operations are measurement, evaluation and adjustment, which form the process control loop. Process control systems operate in real-time since they must quickly respond to the changes occurring in the process they are monitoring.

2.1.10
Estimation of Parameters

Parameter estimation for a given model deals with optimising some parameters or their evaluation from experimental data. It is based on setting the best values for the parameters using experimental data. Parameter estimation is the calculation of the non-process parameters, i.e. the parameters that are not specific to the process. Physical and chemical properties are examples of such non-process parameters. Typical stages of the parameter estimation procedure are: (i) the choice of the experimental points, (ii) the experimental work, i.e. the measurement of the values, (iii) the estimation of the parameters and analysis of the accuracy of the results, (iv) if the results are not accurate enough, additional experiments are carried out and the procedure is restarted from stage (i).

In parameter estimation, the parameters are optimised, and the variables are given fixed values. Optimality in parameter estimation consists in establishing the best match between the experimental data and the values calculated by the model. All the procedures for the identification of parameters comply with the optimality requirements [2.12].

2.1.11
Experimental Design

Experimental design (also called "optimal design of experiments" or "experimental planning") consists in finding the optimal set of experiments and measured parameters. A poorly planned experiment cannot be rescued by a more sophisticated analysis of the data. Experimental design is used to maximize the likelihood of finding the effects that are wanted. Experimental design is used to identify or scan the important factors affecting a process and to develop empirical models of processes. These techniques enable one to obtain a maximum amount of information by running a series of experiments in a minimum number of runs. In experimental design, the variables (measurement points) are optimised with fixed parameters.

2.1.12
Process Integration

Process integration is the common term used for the application of system-oriented methodologies and integrated approaches to industrial process plant design for both new and retrofit applications. Such methodologies can be mathematical, thermodynamic and economic models, methods and techniques. Examples of these methods include artificial intelligence, hierarchical analysis, pinch analysis and mathematical programming. Process integration refers to optimal design; examples of these aspects are capital investment, energy efficiency, emissions levels, operability, flexibility, controllability, safety, sustainable development and

yields. Process integration also refers to some aspects of operation and maintenance.

Process integration combines processes or units in order to minimise, for example, total energy consumption (pinch analysis). Pinch analysis has been successfully used worldwide for the integrated design of chemical production processes
for over ten years. More recent techniques address efficient use of raw materials,
waste minimisation, design of advanced separation processes, automated design
techniques, effluent minimisation, power plant design and refinery processing
[2.13, 2.14]. Responding to their basic principles, the classification of the process
integration methods can be given as follows: artificial intelligence / knowledge-
based systems; hierarchical analysis / heuristic rules; thermodynamic methods
(pinch analysis and energy analysis); optimisation (mathematical programming,
simulated annealing, genetic algorithms).

2.1.13
Process Synthesis

Process synthesis tries to find the flowsheet and equipment for specified feed and
product streams. We define process synthesis as the activity allowing one to
assume which process units should be used, how those units will be interconnected and what temperatures, pressures and flow rates will be required [2.15,
2.16].

Process flowsheet generation is an important part of process synthesis. The following tasks have been established for process flowsheet generation [2.17]: (i) the
generation of alternative processing routes, {ii} the identification of the necessary
unit operations, (iii) the sequencing of unit operations into an optimal flowsheet.

2.1.14
Data Reconciliation

The main assumption in data reconciliation is that measurement values correspond to the steady state. However, process plants are rarely at steady state. Data
reconciliation is used to "manipulate" the measured plant data to satisfy the
steady-state assumption. Data reconciliation is used to detect instrument errors
and leaks and to get "smoother" data for design calculations.

2.1.15
Mathematical Computing Software

They are the mathematical computing programs that offer tools for symbolic and/
or numeric computation, advanced graphics and visualisation with easy-to-use
programming language. These programs can be used, for example, in data analysis and visualisation, numeric and symbolic computation, engineering and scientific graphics, modelling and simulation. Examples are Matlab™ and Mathema-
tica™.

2.1.16
Chemometrics

Chemometrics is the discipline concerned with the application of statistical and mathematical methods to chemical data [2.18]. Multiple linear regression, partial least squares regression and the analysis of the main components are the methods that can be used to design or select optimal measurement procedures and experiments, or to provide maximum relevant chemical information from chemical data analysis. Common areas addressed by chemometrics include multivariate calibration, visualisation of data and pattern recognition. Biometrics is concerned with the application of statistical and mathematical methods to biological or biochemical data.

2.2
Some Observations on the Practical Use of Modelling and Simulation

The observations given here are in fact commentaries and considerations about some aspects from the following topics:
- reliability of models and simulations
- role of the industry as final user of modelling and simulation research
- role of modelling and simulation in innovations
- role of modelling in technology transfer and knowledge management
- role of the universities in modelling and simulation development

2.2.1
Reliability of Models and Simulations

Correctness, reliability and applicability of models are very important. For most engineering purposes, the models must have a broad range of applicability and they must be validated. If the models are not based on these principles, their range of applicability is usually very narrow, and they cannot be extrapolated. In many modelling and simulation applications in the process industry, kinetic data and thermodynamic property methods are the most likely sources of error. Errors often occur when and because the models are used outside the scope of their applicability. With the advent and availability of cheap computer power, process modelling has increased in sophistication, and has, at the same time, come within the reach of people who previously were deterred by complex mathematics and computer programming. Simulators are usually made of a huge number of models, and the user has to choose the right ones for the desired purpose. Making correct calculations is not usually trivial and requires a certain amount of expertise, training, process engineering background and knowledge of sometimes very complex phenomena.

The problem with commercial simulators is that, since the simulations can be carried out fairly easily, choosing the wrong models can also be quite easy. Choosing a bad model can result in totally incorrect results. Moreover, with commercial simulators, there is no access to the source code and the user cannot be sure that the calculations are made correctly. The existing commercial flowsheeting packages are very comprehensive and efficient, but the possibility of misuse and misinterpretation of simulation results is high. In CFD and molecular modelling, the results are often only qualitative. The methods can still be useful, since the results are applied to pre-screen the possible experiments, the synthesis routes and to visualise a particular phenomenon.

2.2.2
The Role of Industry as Final User of Modelling and Simulation

This role is not clear, except in the cases of big companies which have their own research and development divisions. In this case, the R&D company division has specialized teams for modelling and simulation implementation. The properly developed models and simulators are then frequently used, as we have already shown, during the life-cycle of all the particular processes or fabrications that give the company its profile. At the same time, each big company's R&D division can be an important vendor of professional software. The small companies that are highly specialized in modelling and simulation, operate as independent software creators and vendors for one or more company's R&D division. The use of modelling and simulation in small and medium size manufacturing companies is quite limited. Since small manufacturing companies and university researchers do not cooperate much, awareness and knowledge about modern Computer Aided Process Engineering tools are also limited. There are of course exceptions among manufacturing companies. Some small and medium size engineering and consulting companies are active users of modelling and simulation tools, which allows them to better justify the solutions they propose to their clients.

2.2.3
Modelling and Simulation in Innovations

Modelling and simulation are usually regarded as support tools in innovative work. They allow fast and easy testing of innovations. The use of simulators also builds a good basis for understanding complex phenomena and their interactions. In addition, it also builds a good basis for innovative thinking. It is indeed quite important to understand what the simulators really do and what the limitations of the models are. As a consequence, access to source codes is the key to the innovative use of models and simulators.

Many commercial programs are usually stuck in old thinking and well-established models, and then, the in-house-made simulators are quite often better innovative tools. Molecular modelling can be used, for example, in screening potential drug molecules or synthesis methods in order to reduce their number.

The existing molecular modelling technology is already so good that there are real benefits in using it. Molecular modelling can be a very efficient and invaluable innovative tool for the industry. The terms "artificial intelligence" and "expert systems" are based on existing knowledge. The *computers* are *not* creative, which means that these tools cannot be innovative. However, they can be used as tools in innovative development work. While most of the modelling and simulation methods are just tools, in innovative work, process synthesis can be regarded as an innovation generator, i.e. it can find novel solutions by itself.

2.2.4
Role of Modelling in Technology Transfer and Knowledge Management

Models are not only made for specific problem solving. They are also important as databases and knowledge management or technology transfer tools. For example, an in-house-made flowsheet simulator is typically a huge set of models containing the most important unit operation models, reactor models, physical property models, thermodynamics models and solver models from the literature as well as the models developed in the company over the years or even decades. Ideally, a in-house-made simulator is a well-organized and well-documented historical database of models and data. A model is also a technology transfer tool through process development and process life cycle (see for instance Fig. 1.5, in Chapter 1). The problem is that the models developed in earlier stages are no longer used in manufacturing. The people in charge of control write simple models for control purposes and the useful models from earlier stages are simply forgotten. Ideally, the models developed in earlier stages should be used and evaluated in manufactoring, and they should provide information to the research stage conceptual design stage and detailed design stage. One reason for "forgetting" the model during the process life cycle is that the simulators are not integrated. Different tools are used in each process life cycle stage. However, simulators with integrated steady-state simulation, dynamic simulation and control and operator-training tools are already being developed. The problem is that the manufactoring people are not always willing to use the models, even though the advantages are clear and the models are made very easy to use.

2.2.5
Role of the Universities in Modelling and Simulation Development

The importance of modelling and simulation for industrial use is generally promoted, in each factory, by the youngest engineers. The importance of computer-aided tools to the factory level is best understood when the application of modelling and simulation has a history. The importance of modelling and simulation is not understood so well in the sectors that do not using computer-aided tools.

Technical universities have a key role in the education of engineers (so that they can work on modelling and simulation) as well as in research and development. In fact, the universities' education role is absolutely fundamental for the future development of the industry. Indeed, in the future, the work of a process engineer will be more and more concerned with modelling and computation. Moreover, the work will be all the more demanding so that process engineers will need to have an enormous amount of knowledge not only of physics and chemistry, but also of numerical computation, modelling and programming.

References

2.1 *Chemical Complex Analysis System: User's Manual and Tutorial*, MPRI, Louisiana State University, Baton Rouge, 2001.

2.2 *Aspen Plus User's Guide*, Aspen Technology Inc, Cambridge, 1998.

2.3 *Hysis Process User's Guide*, Hyprotech Ltd, Cambridge, 1998.

2.4 J. Leszczynski (Ed.), *Computational Chemistry: Reviews of Current Trends*, Vol. 1–6, World Scientific Publishers, Singapore, 1996–2001.

2.5 K. Srinivas, J. A. C. Fletcher, *Computational Techniques for Fluid Dynamics*, Springer-Verlag, Berlin, 1992.

2.6 D. J. Anderson, Jr., *Computational Fluid Dynamics*, McGraw-Hill, New York, 1995.

2.7 A. J. Robertson, T.C. Crowe, *Engineering Fluid Mechanics*, Wiley, New York, 1996.

2.8 K. V. Garg, *Applied Computational Fluid Dynamics*, Marcel Dekker, New York, 1998.

2.9 G. S. Beveridge, R. S. Shechter, *Optimization Theory and Practice*, McGraw-Hill, New York, 1970.

2.10 M. Tawarmalani, V. Sahinidis, *Global Optimization in Continuous and Mixed-Integer Nonlinear Programming: Theory, Algorithms, Software, and Applications*, Kluwer, Boston, 2001.

2.11 A.B. Bulsari (Ed.), *Neural Networks for Chemical Engineering*, Elsevier, Amsterdam, 1995.

2.12 G. Bastin, D. Dochain, *On Line Estimation and Adaptive Control of Bioreactors*, Elsevier, Amsterdam, 1999.

2.13 B. Linnhoff, D. W. Townsend, D. Boland, G. F. Hewit, B. E. A. Thomas, A. R. Guy, R. H. Marsland, *A User Guide on Process Integration for the Efficient Use of Energy*, The Institute of Chemical Engineering, Rugby, 1992.

2.14 G. Jodicke, U. Fischer, K. Hungerbuhler, *Comput. Chem. Eng.* **2001**, *25*, 203.

2.15 J. M. Douglas, *Conceptual Design of Chemical Processes*, McGraw-Hill, New York, 1988.

2.16 J. Walsh, J. Perkins, *Operability and Control in Process Synthesis* in *Advances in Chemical Engineering, Process Synthesis*, Academic Press, New York, 1996.

2.17 A.W. Westerberg, O. Wahnschafft, *The Synthesis of Distillation-Based Separation Systems* in *Advances in Chemical Engineering, Process Synthesis*, Academic Press, New York, 1996.

2.18 I. D. Massart, G. M. B. Vandeginste, C. M. L. Buydens, S. De Jong, J. P. Lewi, J. Smeyers-Verbeke, *Handbook of Chemometrics and Qualimetrics (Parts A and B)*, NIR Publications, Chichester, 1997.

3

Mathematical Modelling Based on Transport Phenomena

The transport phenomena of mass, heat or momentum, are characterized by an assembly of general equations which can be easily particularized. Each particularization of these equations to an actual example defines the mathematical model of the example.

We consider that the notion introduced with the term transfer of property makes reference to the exchange between two fluids, which are separated by a thin wall (interface or membrane). We observe that the transfer includes the motion of the property in each fluid – a process frequently called transport (transport of property) as well as the transfer of the property through the wall.

In the case of momentum transfer, we have a particular situation where the property transport occurs towards the walls and its transformation is controlled by the geometry of the wall.

The problems of mathematical modelling based on transport phenomena always begin with the establishment of equations which are all based on the general equation for the conservation of properties [3.1–3.5].

The general equation of property conservation. For a phase defined by volume V and surface A, we consider a property which crosses the volume in the direction of a vector frequently named the transport flux $\vec{J}_t$. Inside the volume of control, the property is uniformly generated with a generation rate J_v. On the surface of the volume of control, a second generation of the property occurs due to the surface vector named the surface property flux $\vec{J}_{SA}$. Figure 3.1 illustrates this and shows a cylindrical microvolume (dV) that penetrates the volume and has a microsurface dA.

Inside this microvolume and through its microsurface, the property is generated and transported as in the surface A and control volume V.

In volume V and for a small time interval $d\tau$ when the property concentration ($\Gamma = P/V$, where P is the property quantity) changes by accumulation (from Γ to $\Gamma + d\Gamma$), the values of the components that explain the property conservation are defined as follows:

• quantity of the generated property (P_G):

$$P_G = \left(\iiint_V J_V dV \right) d\tau + \left(\iint_A J_{SA} \vec{n} dA \right) d\tau \tag{3.1}$$

Chemical Engineering. Tanase G. Dobre and José G. Sanchez Marcano
Copyright © 2007 WILEY-VCH Verlag GmbH & Co. KGaA, Weinheim
ISBN: 978-3-527-30607-7

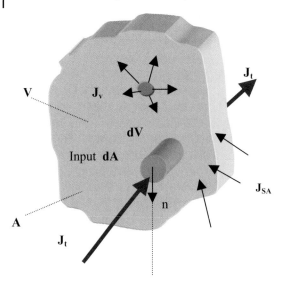

Figure 3.1 Introductory scheme for the equation of a general property balance.

- quantity of the accumulated property (P_A):

$$P_A = \iiint_V \left(\Gamma + \frac{d\Gamma}{d\tau}d\tau - \Gamma \right)dV = \left(\iiint_V \frac{d\Gamma}{d\tau}dV \right)d\tau \tag{3.2}$$

- surplus (excess) P_V of the transported property, output quantity of the property – input quantity of the property):

$$P_V = (\iint_A \vec{J_t}\vec{n}d\vec{A})d\tau \tag{3.3}$$

Based on the law of property conservation that asserts the equality between the difference of the generated and accumulated quantities and the surplus of the transported quantity, we have:

$$P_G - P_A = P_V \tag{3.4}$$

Now we can obtain the relation (3.5) that is recognized as the integral law of the conservation of a property.

$$\iint_A \vec{J_t}d\vec{A} + \iiint_V \frac{\partial \Gamma}{\partial \tau}dV = \iint_A \vec{J}_{SA}d\vec{A} + \iiint_V J_V dV \tag{3.5}$$

This relation can be transformed into its differential form if we make a random selection of the control volume V:

$$\frac{\partial \Gamma}{\partial \tau} + div\vec{J_t} = div\vec{J}_{SA} + J_V \tag{3.6}$$

This equation is similar to the relation obtained when making the property balance with respect to the microvolume dV. For a small interval of time we can write the following relations for the different classes of balance of quantity:

- quantity of the generated property (P_G):

$$P_G = (J_V dV)d\tau + (J_{SA} dA)_{srt} d\tau - (J_{SA} dA)_{ent} d\tau \tag{3.7}$$

- quantity of the accumulated property (P_A):

$$P_A = \left(\frac{\partial \Gamma}{\partial \tau} dV\right) d\tau \tag{3.8}$$

- net quantity of the transported property (P_V)

$$P_V = (J_t dA)_{srt} d\tau - (J_t dA)_{ent} d\tau \tag{3.9}$$

By coupling relations (3.7)–(3.9) with (3.4) we obtain:

$$J_V + \frac{[(J_{SA})_{srt} - (J_{SA})_{ent}]dA}{l_n dA} - \frac{\partial \Gamma}{\partial \tau} = \frac{[(J_t)_{srt} - (J_t)_{ent}]dA}{l_n dA} \tag{3.10}$$

where $l_n dA$ is the measure of the microvolume dV and l_n is the normal length of the microcylinder that defines the balance space.

Due to the random selection and random dimension of the control volume, we can assume that it is very small and so l_n approaches zero. Consequently, we can now write relation (3.11). This is identical to Eq. (3.6) that is recognized as the differential form of the property conservation law

$$\frac{\partial \Gamma}{\partial \tau} + \text{div}\vec{J}_t = \text{div}\vec{J}_{SA} + J_V \tag{3.11}$$

All the terms of relations (3.5) and (3.6) are important but special attention must be given to the transport flux vector.

Generally, this vector contains three components, which correspond to the mechanisms characterizing the behavior of the property carriers during their movement. The molecular, convective and turbulent moving mechanisms can together contribute to the vector flux formation [3.6]. In the relation below (3.12), D_Γ is the ordinary diffusion coefficient of the property. $D_{\Gamma t}$ represents the diffusion coefficient of the turbulences and $\vec{w}$ is the velocity flow vector, then the general relation of the transport flux of the property is:

$$\vec{J}_t = -D_\Gamma \overrightarrow{\text{grad}\Gamma} + \vec{w}\Gamma - D_{\Gamma t}\overrightarrow{\text{grad}\Gamma} \tag{3.12}$$

With Eqs. (3.12) and (3.6) we obtain the relation (3.13). It is recognized as the equation field of the property concentration. In fact, it represents the property conservation law for a random point from a homogeneous medium:

$$\frac{\partial \Gamma}{\partial \tau} + \text{div}(\vec{w}\Gamma) = \text{div}[(D_\Gamma + D_{\Gamma t})\overrightarrow{\text{grad}\Gamma}] + \text{div}\overrightarrow{J}_{SA} + J_V \tag{3.13}$$

Frequently the integral form of the conservation law of the property is particularized as total and partial mass balance and also as energy or thermal balance [3.7]. For each particularization, a control volume must be selected in order to have a form capable of permitting the computation of each integral from the relation (3.5). As an initial condition, we have to declare the property, the transport vector and the property generation rate. Figure 3.2 presents the way to obtain the equations of the differential balance of total mass, mass species and energy (heat). The

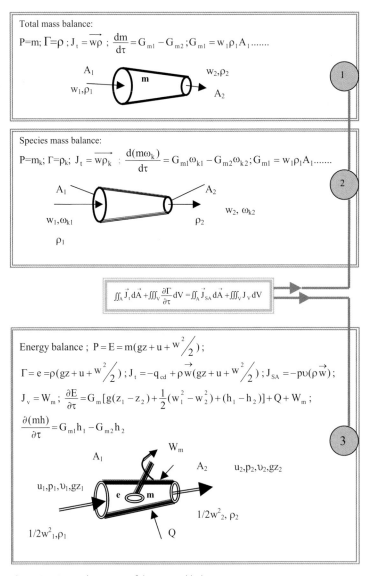

Total mass balance:

$$P=m; \; \Gamma=\rho \; ; J_t = \overrightarrow{w\rho} \; ; \; \frac{dm}{d\tau}=G_{m1}-G_{m2}; G_{m1}=w_1\rho_1 A_1 ....$$

A_1 w_1, ρ_1 m w_2, ρ_2 A_2

①

Species mass balance:

$$P=m_k; \; \Gamma=\rho_k; \; J_t = \overrightarrow{w\rho_k} \; ; \; \frac{d(m\omega_k)}{d\tau}=G_{m1}\omega_{k1}-G_{m2}\omega_{k2}; G_{m1}=w_1\rho_1 A_1 ....$$

A_1 w_1, ω_{k1} ρ_1 A_2 $w_2, \; \omega_{k2}$ ρ_2

②

$$\iint_A \vec{J}_t \, d\vec{A} + \iiint_V \frac{\partial \Gamma}{\partial \tau} dV = \iint_A \vec{J}_{SA} \, d\vec{A} + \iiint_V J_v \, dV$$

Energy balance ; $P=E=m(gz+u+ {}^w{}^2\!/_2)$;

$$\Gamma = e = \rho(gz+u+ {}^w{}^2\!/_2) \; ; J_t = -q_{cd} + \overrightarrow{\rho w}(gz+u+ {}^w{}^2\!/_2) \; ; J_{SA} = -p\upsilon(\overrightarrow{\rho w});$$

$$J_v = W_m \; ; \; \frac{\partial E}{\partial \tau}=G_m[g(z_1 - z_2)+\frac{1}{2}(w_1^2 - w_2^2)+(h_1 - h_2)]+Q+W_m \; ;$$

$$\frac{\partial (mh)}{\partial \tau}=G_{m1}h_1 -G_{m2}h_2$$

W_m A_1 A_2 $u_2, p_2, \upsilon_2, gz_2$ $u_1, p_1, \upsilon_1, gz_1$ e m $1/2w^2{}_2, \; \rho_2$ $1/2w^2{}_1, \rho_1$ Q

③

Figure 3.2 Particularization of the integral balance equation of the property (3.5) for mass and energy conservation.

symbols used in Fig. 3.2 have the following meanings: ρ_1, ρ_2 – densities; w_1, w_2 – flow velocity for the flow area A_1, A_2; G_{m1}, G_{m2} – mass flow rate; ω_{k1}, ω_{k2} – mass concentrations (fractions) for k species (component); u_1, u_2 – specific internal fluid energy; p_1, p_2 – pressure values; υ_1, υ_2 – value of the flowing fluid specific volume; z_1, z_2 – positions that characterize the local potential energy; h_1, h_2 – enthalpy for the flowing fluid; g – gravitational acceleration.

We observe from Fig. 3.2 that these transformations of integral equations of the property conservation have been obtained taking into consideration a very simple apparatus, here represented by a frustum conical pipe with flow input at the larger base and output at the smaller base.

Now we can see how the differential form of the property conservation law can generate the equations of the velocity distribution for a flowing fluid (Navier-Stokes equations), the temperature or the enthalpy distribution (Fourier second law) and the species concentration distribution inside the fluid (second Fick's law).

In all these particularization cases, we use the molecular and convective participations in the composition of the vector of transport.

The equation for the momentum transport in vectorial form, gives (by particularization) the famous Navier-Stokes equation. This equation is obtained considering the conservation law of the property of movement quantity in the differential form: $\vec{P} = m\vec{w}$. At the same time, if we consider the expression of the transport vector: $\vec{J}_t = \vec{\tau} + \vec{w}(\rho\vec{w})$ and that the molecular momentum generation rate is given with the help of one external force $\vec{F}$, which is active in the balance point, the particularization becomes: $\dfrac{\partial(\rho\vec{w})}{\partial\tau} + \text{div}[\vec{\tau} + \vec{w}(\rho\vec{w})] = \text{div}(-p\vec{n}) + \rho\vec{F}$, where p is defined as the local hydrodynamic pressure. The flux of the momentum quantity can be interpreted as a tension that characterizes the fluid deformation. Indeed, it is a tensor. When the conservable property is the mass from an infinitesimal control volume ($P = m$, $\Gamma = m/V = \rho$) where the convective flux is dominant, then the particularization of the differential form of the property conservation law becomes the flow continuity equation: $\dfrac{\partial\rho}{\partial\tau} + \text{div}(\rho\vec{w}) = 0$.

The Navier-Stokes equations and the flow continuity equation together give the general flow model; other cases associate various forms of the energy conservation equation to this model.

For the particularization of the differential conservation law to the heat transport, we consider first that the transported property is the sensible heat ($P = mc_pt$, $\Gamma = \rho c_p t$) and secondly that it is carried out by molecular and convective mechanisms ($\vec{J}_t = \vec{q}_m + \vec{w}\rho c_p t$).

When the conservable property is represented by the local quantity of the species A ($P = n_{mA}$, $\Gamma = n_{mA}/V = c_A$) transported by molecular and convective mechanisms, relation (3.6) becomes the equation of field of the species concentration. Figure 3.3 gives the three particularizations of the differential form of the property conservation law. Here we present the basic equations of momentum, heat and mass transport using their vectors and Cartesian expressions. However,

The moment transport (Flow equations): $P_i = mw_i, i=x, i=y, i=z$;

$\Gamma = H_i = \rho w_i$; $\vec{J}_{mi} = \vec{\tau}_{ij} = -\eta(\partial w_i/\partial x_j + \partial w_j/\partial x_i)$; $x_x = x..,x_y = y..,x_z = z$

$\vec{J}_{ci} = \vec{w}(\rho w_i)$; $\vec{J}_{SA} = -p\vec{n}$;

$$\frac{\partial(\rho w_i)}{\partial \tau} + \text{div}(\vec{w}(\rho w_i)) = -\frac{\partial p}{\partial x_i} + \text{div}(\frac{1}{2}v(\overrightarrow{\frac{\partial(\rho w}{\partial x_j}} + \overrightarrow{\frac{\partial(\rho w_j)}{\partial x_i}})$$

$$\frac{\partial w_x}{\partial \tau} + w_x\frac{\partial w_x}{\partial x} + w_y\frac{\partial w_x}{\partial y} + wz\frac{\partial w_x}{\partial z} = g_x - \frac{1}{\rho}\frac{\partial p}{\partial x} + \eta(\frac{\partial^2 w_x}{\partial x^2} + \frac{\partial^2 w_x}{\partial y^2} + \frac{\partial^2 w_x}{\partial z^2})$$

Flow continuity equation: $P=m$; $\vec{J}_t = \vec{J}_c = \vec{w}$; $\frac{\partial \rho}{\partial \tau} + \text{div}(\rho\vec{w}) = 0$;

①

$$\frac{\partial \Gamma}{\partial \tau} + \text{div}(\vec{w}\Gamma) = \text{div}((D_r\overrightarrow{\text{grad}\Gamma}) + J_v + \text{div}(J_{SA})$$

Heat transport: $J_{SA}=0$ $P=mc_p t$; $\Gamma = \rho c_p t$; $\vec{J}_m = \vec{q} = -a(\overrightarrow{\frac{\partial(\rho c_p t)}{\partial n}})$

$J_c = \vec{w}(\rho c_p t)$; $\frac{\partial(\rho c_p t)}{\partial \tau} + \text{div}(\vec{w}\rho c_p t) = \text{div}(a(\overrightarrow{\frac{\partial(\rho c_p t)}{\partial n}}) + Q_v$

$$\frac{\partial t}{\partial \tau} + w_x\frac{\partial t}{\partial x} + w_y\frac{\partial t}{\partial y} + w_z\frac{\partial t}{\partial z} = \lambda(\frac{\partial^2 t}{\partial x^2} + \frac{\partial^2 t}{\partial y^2} + \frac{\partial^2 t}{\partial z^2}) + \frac{Q_v}{\rho c_p}$$

- The Flow equations give here the velocity components w_x, w_y, w_z

②

Mass (species) transport: $P=n_{mA}$; $\Gamma = c_A$ $P=n_A$; $j=x,y,z$ $J_{SA}=0$

$\vec{J}_m = J_{mA} = -D_A(\overrightarrow{\frac{\partial(c_A)}{\partial x_j}})$;

$J_c = \vec{wc_A}$; $\frac{\partial c_A}{\partial \tau} + \text{div}(\vec{wc_A}) = \text{div}(D_A(\overrightarrow{\frac{\partial(c_A)}{\partial x_j}}) + v_{rA}$

$$\frac{\partial c_A}{\partial \tau} + w_x\frac{\partial c_A}{\partial x} + w_y\frac{\partial c_A}{\partial y} + wz\frac{\partial c_A}{\partial z} = D_A(\frac{\partial^2 c_A}{\partial x^2} + \frac{\partial^2 c_A}{\partial y^2} + \frac{\partial^2 c_A}{\partial z^2}) + v_{rA}$$

③

Figure 3.3 Particularization of the differential balance equation of a property (3.6) for momentum, heat and mass transport.

these equations, which characterize the fundamental properties of transport, cannot be used when we have conjugated actions. For example, if in a homogeneous system we simultaneously have gradients of species A concentration, temperature and pressure, then the molecular flux for the species A transport contains all participations and is written as follows:

$\vec{J}_{mA} = -D[\vec{\nabla}c_A + k_t\vec{\nabla}\ln(T) + k_p\vec{\nabla}\ln(p)]$, where D_Ak_t and D_Ak_p are, respectively, the thermal-diffusion coefficient and pressure-diffusion coefficient of species A. When we use the updated transport flux of species in the particularization of the balance of the differential property (Eq. (3.6)) a new expression of the field of the species concentration is obtained. The turbulence has to be considered when we have an important convective transport, which in many cases is the dominant transport mechanism. The contribution of this mechanism to the transport capacity of the medium is introduced in relation (3.3) by the addition of the coefficient of turbulent diffusion of the property.

When the transport is considered without turbulence we have, in general, D_Γ; υ is the cinematic viscosity for the momentum transport; $a = \lambda/(\rho c_p)$ is the thermal diffusivity and D_A is the diffusion coefficient of species A. Whereas with turbulence we have, in general, $D_{\Gamma t}$; υ_t is the cinematic turbulence viscosity for the momentum transport; $a_t = \lambda_t/(\rho c_p)$ is the thermal turbulence diffusivity and D_{At} is the coefficient of turbulent diffusion of species A; frequently $\upsilon_t = a_t = D_{At}$ due to the hydrodynamic origin of the turbulence.

Indeed, this is a very simplistic treatment for the general flow mechanism, so, it is important to note here that the turbulence is in fact a vast scientific domain where interdisciplinary characterization methods are frequently needed.

Chemical engineers have developed very powerful methods for the hydrodynamic characterization of flows in different regimes by using specific apparatus; this methodology allows one to model the turbulent flow in industrial or laboratory devices.

We cannot finish this short introduction on the property transport problems without some observations and commentaries about the content of Figs. 3.2 and 3.3. First, we have to note that, for the generalization of the equations, only vectorial expressions can be accepted. Indeed, considering the equations given in the figures above, some particular situations have been omitted. For example, we show the case of the vector of molecular transport of the momentum that in Fig. 3.3 has been used in a simplified form by eliminating the viscous dissipation. So, in order to generalize this vector, we must complete the τ_{ij} expression with consideration of the difference between the molecular and volume viscosities $\eta - \eta_v$:

$$\tau_{ij} = -\frac{1}{2}\upsilon\left(\frac{\partial(\rho w_i)}{\partial x_j} + \frac{\partial(\rho w_j)}{\partial x_i}\right) + \frac{2}{3}(\eta - \eta_v)\mathrm{div}(\vec{w})\delta_{ij}.$$

However, we remark here that the simplifications of the expressions in the Cartesian coordinates system have been accepted as in the case of an isotropic and non-property dependent diffusion coefficient of a property. Indeed, the independence of the general diffusion coefficient with respect to the all-internal or external solicitations of the transport medium appears unrealistic in some situations.

For actual cases, the equations from these tables have to be particularized to the geometry of the device used and to all conditions of the process including the conditions that show the process state at the wall and interphases.

Interphase transfer kinetics. At this point, we need to characterize the process that leads to the transfer of the property through the interphase. The transport of the momentum from one phase to another is spectacular when the contacting phases are deformable. Sometimes in these situations we can neglect the friction and the momentum transfer generates the formation of bubbles, drops, jets, etc. The characterization of these flow cases requires some additions to the momentum equations and energy transfer equations.

Boundary layers appear in flow situations near the walls or other non-deformable structures that exist in the flow field [3.8]. Their formation and development, stability and local thickness are of great interest to engineers and researchers because all the gradients of property concentration are concentrated here. Consequently, we can write a very simple expression for the flux of the property.

In a general case, when a property crosses the interphase, we must consider that the property flux is identical between both contacted phases. Indeed, we consider ideal behaviour of the interphase or, in other words, we must accept the interphase to be not resistive to the transfer. We can criticize this fact but frequently it is accepted as a datum.

So when we accept the ideality of the interphase and when it is positioned to the coordinate x_i we can write:

$$J_{t1} = -D_{\Gamma 1}\left[\frac{d\Gamma}{dx}\right]_{x=x_i} = J_{t2} = -D_{\Gamma 2}\left[\frac{d\Gamma_2}{dx}\right]_{x=x_i} = J_t \tag{3.14}$$

Now we have to take into account the boundary layers at the left and right sides of the interphase where we have already shown the gradient of concentration of the property of the phase. With this last consideration, we can write a set of relations (3.15) that introduce the notion of the partial coefficient of the transfer of property (3.16):

$$-D_{\Gamma 1}\left[\frac{d\Gamma_1}{dx}\right]_{x=x_i} = -D_{\Gamma 2}\left[\frac{d\Gamma_2}{dx}\right]_{x=x_i} = J_t =$$

$$\frac{-D_{\Gamma 1}\left[\frac{d\Gamma_1}{dx}\right]_{x=x_i}}{(\Gamma_{1\infty} - \Gamma_{x=x_i})}(\Gamma_{1\infty} - \Gamma_{x=x_i}) = k_{\Gamma 1}(\Gamma_{1\infty} - \Gamma_{x=x_i}) = k_{\Gamma 2}(\Gamma_{x=x_i} - \Gamma_{2\infty}) \tag{3.15}$$

$$k_{\Gamma 1} = \frac{-D_{\Gamma 1}\left[\frac{d\Gamma_1}{dx}\right]_{x=x_i}}{(\Gamma_{1\infty} - \Gamma_{x=x_i})} \quad ; \quad k_{\Gamma 2} = \frac{-D_{\Gamma 2}\left[\frac{d\Gamma_2}{dx}\right]_{x=x_i}}{(\Gamma_{x=x_i} - \Gamma_{2\infty})} \tag{3.16}$$

It is of interest that, for the computation of the transfer coefficients, various procedures have been advanced and in the past an immense quantity of data and reference materials have been collected on this subject.

3.1
Algorithm for the Development of a Mathematical Model of a Process

The relationships among the variables for a concrete process can be known through the particularization of the processes of transport phenomena. The mathematical model has to describe the state and evolution of the process while knowledge of the previous description of the operating conditions of the studied case is necessary. Indeed, terms such as flow, heat, diffusion and reaction clearly show that the transport phenomena are not absent from the investigated process. The verbal or written description must be clear and decisive with regard to identifying the effect of the independent variables on the exits of the process (dependent variables). At the same time, this description must correctly show how the dominant transport phenomena between all the unitary steps occur while the concrete process takes place. The observation spirit, a good engineering background, a good knowledge of the case and a fluent engineering language must be associated with the researcher's acute sense of responsibility in describing the process.

Indeed, the description of the process is recognized as the *first step* in the building of the mathematical modelling of a process. The result obtained here is recognized as a *descriptive model* or *model by words*. During this step, dependent and independent process variables resulting from the identification of the actions and interactions of the elementary phenomena that compose the state and evolution of the investigated process will be listed. At the same time, the effect of each independent variable on each dependent variable must be described.

The *second step* begins with a verbal or written analysis showing the coupling of the flow phenomena, heat and mass transfer, chemical reaction thermodynamics and kinetics. Here, a fraction of the factors of the process (independent variables of the process) selected by the first step will be eliminated, whereas a new limited number of factors will be added to the list. This step concerns one of the most delicate problems in mathematical modelling: the identification or creation of the mathematical clothes of the process by summation from the elementary models of the phenomena involved in the process. To finish this step, a mathematical form that characterizes the operating process is definitively established. This mathematical form is recognized as the *general mathematical model* of the process. Indeed, if the general descriptive models have been correctly decomposed into parts, then, each one of the parts will be characterized by its own general mathematical model.

The coupling of the general mathematical model with the evolution of the material and spatial conditions is given by its association with the investigated conditions of univocity of the process. This is the basis of the *third step* in the building of the mathematical model of a process. At the end of this step, we will have a *particularized mathematical model*. This step will be specified for each one of the decomposed models of the parts; i.e. for each of the particular devices in a unit. For this particularization, we use the following conditions of univocity:
- *the geometric conditions* establish the dimensions of the apparatus
  where the process is carried out from the geometric viewpoint.

Indeed, we made the choice of the coordinates system (cartesian, cylindrical or spherical) which will be used for our actual case and the model equation will be transformed for the selected coordinates system.

- *the material conditions* describe the physicochemical properties of the medium where the process takes place as well as the variation of these properties with temperature, pressure and composition using numerical values or analytic relations. Here we select the values or relations for the density, viscosity, thermal coefficient capacity, thermal conductivity, and diffusion coefficient of each component.

- *the dynamic conditions* give the initial spatial distribution and its evolution with time for each transported property. They also give the flux for each geometrical frontier as well as for each line or surface of symmetry of the system. Three major types of frontier have been established for the dynamic conditions:

  - *the boundary conditions of type I:* give the numerical values of the transported property or the function describing the variation of these values with time for each frontier of the system.

  - *the boundary conditions of type II:* give the flux values of the transported property to each frontier and to each symmetry line and surface of the system. Each flux can be described by a constant value or is dependent on time.

  - *the boundary conditions of type III:* give the values of the property state but here these values are out of the frontiers. At the same time, these conditions give the values or calculus relations for the coefficients of transfer at the interphase. With these data and using relation (3.15), we can compute the flux of the property at the frontiers. If we denote by $\Gamma_{1\infty}$ the property concentration for phase 1 and we assume a non-resistive interphase (the phases are in equilibrium with k_d, as distribution coefficient of the property) then, relation (3.15) becomes:

  $$k_{\Gamma 1}\left(\Gamma_{1\infty} - \frac{\Gamma_{2\ x=x_{int}}}{k_d}\right) = -D_{\Gamma 2}\left(\frac{d\Gamma_2}{dx}\right)_{x=x_{int}}$$

For all the situations, the *dynamic conditions for the symmetry lines and surfaces* of *the system* contain the specification that the property flux is zero. From the viewpoint of the property concentration, this fact shows that here it has a maximal or a minimal value.

  - *the tendency conditions* show the state of a dynamic process after a very long time. If a stationary state is possible for the process, then the tendency conditions show the transition from a dynamic process model to a stationary process model (steady state).

In the *fourth step* of the building of a mathematical model of a process the assemblage of the parts (if any) is carried out in order to obtain the *complete mathematical* model of the process. Now the model dimension can be appreciated and a frontal analysis can be made in order to know whether analytical solutions are possible.

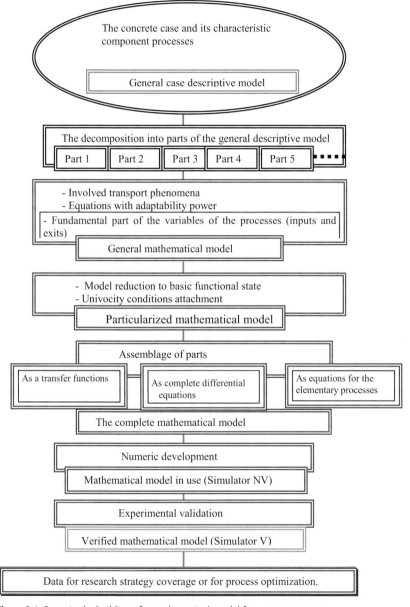

Figure 3.4 Steps in the building of a mathematical model for a concrete case.

Considering that, in this last step, we have a complete mathematical model of the process, we can now think about its valorization, by selecting the most convenient and acceptable possible solution for the model transformation into a numerical state. Indeed, the *fifth step* of this procedure results in a problem of computer software creation. Then, we have to choose the numerical solutions for the model integration as well as to select the input and output data state for the running of the computer program. We also have to select the representation of the solutions obtained with the output data processing. The final output of this step is a non-verified simulator (*Simulator NV*). The degree of sophistication of the simulator obviously depends on the model complexity. When we have some experimental data characterizing the relationship between one or more dependent variables and the independent process variables, then we can verify, after a normal calibration, whether the model produces identical or very similar data. If the model results match the experimental data, then we can affirm that we have a verified process simulator (*Simulator V*). Figure 3.4 shows this gradual development schematically, step by step, from the model establishment to simulator V. Here, we cannot logically separate the model creation part from the software creation part (numeric model transposition). It is also clear that the sense of the presented scheme is to show how we develop the model of one part of the general decomposed model. When we recompose the parts, we use the principle of maximum coupling. So, some parts will be introduced in the global model by their transfer functions, other parts with the help of their governing differential equations assemblies.

This procedure for building a mathematical model for a concrete case has also been mentioned in some scientific papers where the object is mathematical modelling by the use of transport phenomena [3.9–3.13].

3.1.1
Some Observations about the Start of the Research

Young researchers' first finished models are a source of great joy because they show their creative power. Moreover, when the models developed are successfully validated by experimentation, we can claim that the new researchers have actually stepped into real research activity.

Concerning the situation of the models that fail the test of experimental validation, we generally have two cases. The first case concerns a model that is unable to describe the whole project and which, normally, has to be rejected. The second case concerns a model that reproduces the general trends of the process but shows important differences with respect to the experimental data. This model will be again subjected to the building procedure where, with small or large modifications, it will improve its performance.

A special case occurs when some material or transport parameters are still unknown at the starting point and yet, at the same time, we have a lot of experimental data for the model validation. In this situation, we consider both data and model by formulating a *parameter identification* problem. The validation test for this type of model will be transformed into hypotheses concerning the identified values of parameters.

Another special case occurs when the model is obtained by assembling different parts, and when each part has been successfully validated. In this case, the global validation is in fact a model calibration with the experimental data available.

In an actual research programme carried out with modelling coupled with experimental work, we cannot work randomly, without a research plan. The planning research method, given in this book in Section 5.3.2, has the capacity to be used for solving the most refined requirements. For this purpose, we must accept a model simulation to be as good as an experiment. With this procedure, we can derive an indirect but complex statistical model presenting a high interest for a computer-guided process from a model of transport phenomena. In the same way, we can use the model of transport phenomena as a database for a neural network model. Therefore, the data produced by the real model will be used in the learning procedure by the neural network model. Excellent behaviour of the neural network model is expected because the learning data volume can be very rich. We point out here that in the building of a model for a concrete chemical fabrication in an industrial unit, more aspects may be considered, each requiring qualified knowledge. Indeed, the procedures and methods coming from different scientific branches have to be coupled to the basic process model.

It is evident that, in these situations, problems concerning coupling hierarchy between the different parts can appear. Generally, for a fabrication that involves a chemical reaction, the top of the hierarchy is occupied by the reactor and separator models.

Figure 3.5 shows the most important scientific branches of chemical engineering research, which have to be taken into account for the modelling. Indeed, the

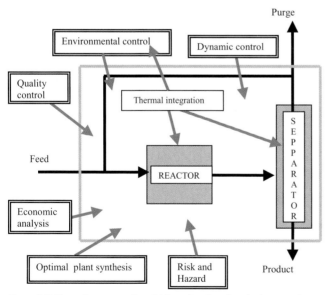

Figure 3.5 Secondary research activities related to chemical processing.

mathematical model of a process must answer the specific questions of each area division. In each case, the response comes as a result of the coupling of the computation procedures characterizing each branch. This is a complex and high-level research that gives consistency to a new scientific activity named *Advanced Processes Simulators*.

3.1.2
The Limits of Modelling Based on Transport Phenomena

Since the start of Section 3.1 we have been presenting how the transport phenomena equations are used for the mathematical modelling of a process and how we transform this model into a process simulator.

Actually, research by modelling is more and more extensively used in many applications because complex devices' models, composed of different elements, can be made by assembling models the solutions of which are frequently available. This behaviour presents an impressive growth and is sustained by the extraordinary developments in numerical calculations and by the implementation of commonly used computers with a high capacity and calculus rate. Nevertheless, modelling based on the equations of transport phenomena cannot be applied to every system, because they can present some limitations, which are summarized here.

The first limit derives from the model construction and can be called the constructive limit. It is explained by the quantity of simplifications accepted for model construction. The flow reduction by use of ideal models and the treatment of the transfer processes in equilibrium by using abstract notions – as for example, theoretical plate in distillation – represent only two of countless similar examples.

The second limit is named the cognition limit and arises from the less controlled assumptions concerning the complicated and ill acquainted phenomena involved in the process. Considering the interface as an equilibrium Gibbs interface and introducing the turbulent flow from the turbulent diffusion coefficient are two famous examples which illustrate this class of cognition limits.

The third limit is represented by the validity limits of the transfer phenomena equation. With respect to this last limitation, Fig. 3.6 shows the fixation of these limits with regard to the process scale evolution.

At this time, only a small number of nanoscale processes are characterized with transport phenomena equations. Therefore, if, for example, a chemical reaction takes place in a nanoscale process, we cannot couple the elementary chemical reaction act with the classical transport phenomena equations. However, researchers have found the keys to attaching the molecular process modelling to the chemical engineering requirements. For example in the liquid–vapor equilibrium, the solid surface adsorption and the properties of very fine porous ceramics computed earlier using molecular modelling have been successfully integrated in modelling based on transport phenomena [4.14]. In the same class of limits we can include the validity limits of the transfer phenomena equations which are based on parameters of the thermodynamic state. It is known [3.15] that the flow equations and, consequently, the heat and mass transport equations, are valid only for the

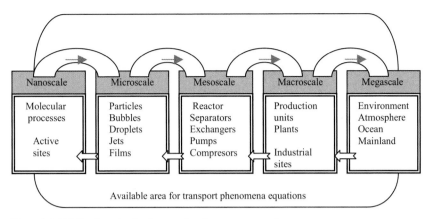

Figure 3.6 Validity domains for the transfer phenomena equations.

domain where the Truesdell criterion stays below unity. The Truesdell number given by $Tr = \dfrac{2\eta\varepsilon}{p}$, where η is the medium viscosity, ε the molecular oscillation frequency and p the medium pressure, is a combination of the Knudsen (Kd) and Mach (Ma) criteria.

Therefore, gases with very small pressure and some very viscous liquids can have a Truesdell number value over unity.

The fourth limit is the limit of contradiction. It takes place when sophisticated and complex models produced by academic and specialized research are used in industrial applications. Indeed, in industrial production, engineers can expect the current exploitation problems but they do not have any time to face new problems.

In fact, this limit depends on the standard of teaching of modelling research in technical universities. They have a key role in educating engineers capable of working with modelling and simulation as well as in research and development. The work of a process engineer in the future will be more and more concerned with modelling and using computers. Indeed, process engineers must have a considerable knowledge of physics and chemistry, as well as of processes, numerical calculation, modelling, programming and of the use of commercial programs.

The skills of graduating students are generally not very good in the fields of modelling and simulation. The goal of universities should be to produce more modelling-oriented engineers with good engineering, chemistry, physics, programming and mathematical skills.

3.2
An Example: From a Written Description to a Simulator

In this section, we will show the process of the construction of a mathematical model, step by step, in accordance with the procedure shown in Fig. 3.4. The case studied has already been introduced in Figs. 1.1 and 1.2 of Chapter 1. These figures are concerned with a device for filtration with membranes, where the gradient is given by the transmembrane pressure between the tangential flow of the suspension and the downstream flow. The interest here is to obtain data about the critical situations that impose stopping of the filtration. At the same time, it is important to, *a priori*, know the unit behaviour when some of the components of the unit, such as, for example, the type of pump or the membrane surface, are changed.

Descriptive model and its division into parts. The first steps in the model construction are related to Fig. 3.7. The pump PA assures simultaneously the suspension transport and the necessary transmembrane pressure. The excessive accumulation of the solid in the retentate is controlled by its permanent removal as a concentrated suspension from the reservoir RZ. The clear liquid (permeate) flow rate and the solid concentration in the exit suspension are permanently measured and these values are transferred to the control and command computer CE. The instantaneous values of the operation pressure and input rate of fresh suspension are established by the computer (this works with software based on the mathematical model of the process) and corrected with the command execution system CSE.

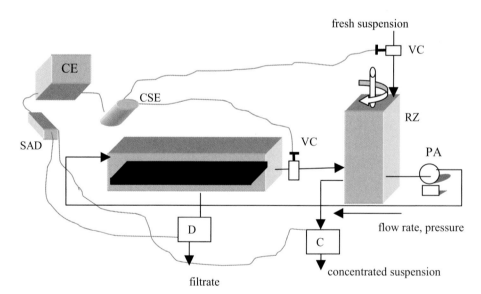

Figure 3.7 Membrane filtration plant.

If in stationary operation conditions, membrane clogging does not occur or is negligible, then the modelling case becomes banal. Nevertheless, when surface clogging cannot be eliminated by the tangential flow rate, we must introduce a continuous increase in the hydrodynamic resistance of the membrane [3.16–3.18]. In this situation, if the pressure filtration stays unchanged, the filtrate rate will decrease with time. When unacceptable values of the filtrate rate are reached, the process must be stopped and the membrane cleaned or replaced. This mode of operation is uneconomical. One solution to this problem is to increase the trans-membrane pressure in order to maintain the flow rate but, in this case, the pumping flow rate has to be reduced because pumps generally present a pre-established and characteristic flow rate–pressure relation which is, *a priori*, unchangeable. Consequently, when the pressure is continuously increased, the clogging rate will increase faster than when a high tangential velocity is used in the unit.

The clogging effect can be considered as a reduction in the value of the surface filtration constant for practical purposes. Indeed, when clogging takes place, the surface filtration constant can be given by its initial value k_0 multiplied by a decreasing time function. This assumption is frequently used when the function is obtained from experiments [3.19, 3.20]. In our example, if we do not consider the friction (and heat transfer) we can note that only a concrete mass transfer problem can be associated with the membrane separation process. The first step before starting to build the general mathematical model, concerns the division of the system into different elementary sections. Indeed, we have a model for the filtration device (i.e. the membrane and its envelope), for the pump (P) and for the reservoir of concentrated suspension (RZ) (Fig. 3.7).

General mathematical model. Considering that all we have is a mass transfer phenomenon, then, in such a system, the solid concentration changes in each plant device. With the considered coordinates system and after the notations given in Fig. 3.8, we can write the mathematical model of the filter unit as a particularization of the flow equations and the solid transport equation:

The Navier-Stokes equation in the x direction:

$$\frac{\partial w_x}{\partial \tau} + w_x \frac{\partial w_x}{\partial \tau} = -\frac{1}{\rho}\frac{\partial p}{\partial x} + \eta_{sp}\left(\frac{\partial^2 w_x}{\partial x^2} + \frac{\partial^2 w_x}{\partial y^2} + \frac{\partial^2 w_x}{\partial z^2}\right) \tag{3.17}$$

The Bernoulli equation with respect to an elementary local length dx:

$$\frac{\partial p}{\partial x} = -\frac{1}{2}\left(\rho_{susp}\partial x \frac{\partial \bar{w}_x}{\partial x}\right)^2 - \frac{\lambda}{d_e}\frac{\bar{w}_x^2}{2}\rho_{susp} \tag{3.18}$$

The formula of the definition of the suspension mean flow velocity:

$$\bar{w}_x = \frac{1}{hl}\int_0^h\int_0^l w_x(z,y)dzdy \tag{3.19}$$

The total mass balance equation with respect to the elementary local length dx:

$$\frac{dG_{pr}}{dx} = \rho_f\, lh\, \frac{\partial w_x}{\partial x} \tag{3.20}$$

The transfer equation for the permeate given by the use of its flux expression through the membrane surface:

$$\frac{dG_{pr}}{dx} = k_0 l(p_x - p_0) f(c_s \rho_s, \tau) \tag{3.21}$$

The simplified solid concentration field equation:

$$\frac{\partial c_s}{\partial \tau} + w_x \frac{\partial c_s}{\partial x} = 0 \tag{3.22}$$

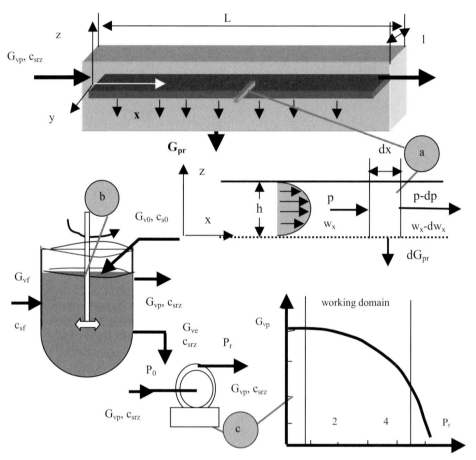

Figure 3.8 Decomposition of the filtration unit into sections and their corresponding description with relationships of the basic variables.

It is important to note that, except for the heat transfer problems, which have not been considered here, the model contains, in a particular form, all the transport phenomena relationships given at the start of this chapter. From the mathematical viewpoint, we have an assembly of differential and partly differential equations, which show the complexity of this example. However, this relative mathematical complexity can be matched with the simplicity of the descriptive model. Indeed, it will be convenient to simplify general mathematical models in order to comply with the descriptive model. Two variants can be selected to simplify the flow characterization in the membrane filtration unit.

The first variant considers that the model suspension flow corresponds to a plug flow model. In this case, the velocity w_x is a function of the coordinate x only. Its value is obtained from the ratio between the local suspension flow rate and the flow section [3.21]. With this assumption, the general mathematical model of the filter becomes:

$$w_0 = \frac{G_{vp}}{lh}, \quad w_x = \frac{G_{vx}}{lh} \tag{3.22}$$

$$p_x = p_{x-dx} - \frac{\lambda}{d_e} \frac{(w_0 + w_x)^2}{8} \rho_{susp} dx \tag{3.23}$$

$$\frac{dG_{pr}}{dx} = k_0 l(p_x - p_a) f(c_s, p_x, \tau) \tag{3.24}$$

$$G_{vx} = G_{vp} - G_{pr} \frac{\rho_f}{\rho_{susp}} \tag{3.25}$$

$$\frac{\partial c_s}{\partial \tau} + w_x \frac{\partial c_s}{\partial x} = 0 \tag{3.26}$$

In *the second variant*, the plug flow model is considered as a series of tanks with perfect mixing flow [3.22, 3.23]. In this case, the real filter will be supposedly replaced by a series of some small filters (three in this analysis) with perfect mixing flow. Figure 3.9 shows the scheme, relations and notations used. The filtrate transfer equation has been used for the mathematical characterization of each small filter for the total material balance equation and non-steady-state solid balance equation:

- *first small filter:*

$$G_{pr1} = k_0 A_1 (p_1 - p_a) f(c_{s1}, p_1, \tau) \tag{3.27}$$

$$G_{v1} = G_{vp} - G_{pr1} \frac{\rho_f}{\rho_{susp}} \tag{3.28}$$

$$\frac{dc_{s1}}{d\tau} = \frac{G_{vp}}{V_1} c_{srz} - \frac{G_{v1}}{V_1} c_{s1} \tag{3.29}$$

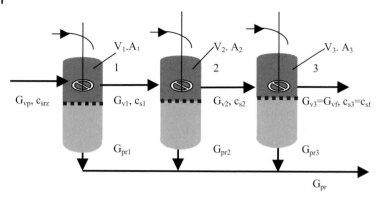

Figure 3.9 Representation of the membrane filtration unit as an ensemble of three small perfect mixing filters.

- *second small filter:*

$$G_{pr2} = k_0 A_2 (p_2 - p_a) f(c_{s2}, p_2, \tau) \tag{3.30}$$

$$G_{v2} = G_{v1} - G_{pr2} \frac{\rho_f}{\rho_{susp}} \tag{3.31}$$

$$\frac{dc_{s2}}{d\tau} = \frac{G_{v1}}{V_2} c_{s1} - \frac{G_{v2}}{V_2} c_{s2} \tag{3.32}$$

- *third small filter:*

$$G_{pr3} = k_0 A_3 (p_3 - p_a) f(c_{s3}, p_3, \tau) \tag{3.33}$$

$$G_{v3} = G_{v2} - G_{pr3} \frac{\rho_f}{\rho_{susp}} \tag{3.34}$$

$$\frac{dc_{s3}}{d\tau} = \frac{G_{v2}}{V_3} c_{s2} - \frac{G_{v3}}{V_3} c_{s3} \tag{3.35}$$

$$G_{v3} = G_{vf} \quad , \quad c_{s3} = c_{sf} \tag{3.36}$$

The general mathematical process model has to be completed with the models for the recycled suspension reservoir and for the pump. The suspension reservoir is a classical perfect mixing unit (see Fig. 3.8) so its model includes the unsteady total and solid balances. These balances are given below by relations (3.37) and (3.38). After Fig. 3.8, the mathematical model of the pump gives the relationship between the pump exit flow rate and its pressure (relation (3.39)):

$$\frac{dV_{rz}}{d\tau} = G_{v0} \frac{\rho_{s0}}{\rho_{susp}} + G_{vf} - G_{vp} - G_{ve} \tag{3.37}$$

$$\frac{dc_{srz}}{d\tau} = \frac{G_{vf}}{V_{rz}} c_{sf} + \frac{G_{v0}}{V_{rz}} - \frac{(G_{vp} + G_{ve})}{V_{rz}} c_{srz} \tag{3.38}$$

$$G_{vp} = a - bp_r^2 \tag{3.39}$$

For the whole unit we have to complete the general mathematical model with constraints that can be given by the device construction and/or operating conditions:

For the correct pump operation, the suspension level in the recycled reservoir must be within a range around a minimal value ($V_{rz\,min}$), which is lower than the geometric volume (V_0):

$$V_{rz\,min} \prec V_{rz} \prec V_0 \tag{3.40}$$

The pump cannot operate under a minimal flow rate value:

$$G_{vp} \succ G_{vp\,min} \tag{3.41}$$

The ratio of solid concentration between the recycled and fresh suspension must be limited in order to reach a good flow in the filter unit and a rational recycling; this constraint can also be applied to flow rates G_{v0} and G_{vf}:

$$1 \prec \frac{c_{srz}}{c_{s0}} \prec \beta \tag{3.42}$$

The filtrate rate or the working pressure must be limited to imposed selected constant values. For a two-dimensional model (x, τ) these constraints are given by relations (3.43) and (3.44)

$$G_{pr/x=L} = G_{primp} \tag{3.43}$$

$$p_r = p_{r0} \tag{3.44}$$

In the case of a mono-dimensional model (τ), relations (3.45) and (3.46) comply with the technological requirements

$$G_{pr1} + G_{pr2} + G_{pr3} = G_{primp} \tag{3.45}$$

$$p_r = p_{r0} \tag{3.46}$$

Table 3.1 shows all general mathematical models resulting from the analysis of the filtration plant operation. The equations include the parts assembly in the model (Fig. 3.4) and an overall formula that shows the relationships that compose each model.

Table 3.1 Mathematical models for the filtration plant analysis.

Model			
Two-dimensional x,τ		**Monodimensional τ**	
Constant filtrate flow rate operation	Constant pressure operation	Constant filtrate flow rate operation	Constant pressure operation
(3.22)–(3.26) + (3.37)–(3.43)	(3.22)–(3.26) + (3.37)–(3.42) and (3.44)	(3.27)–(3.36) + (3.37)–(3.42) and (3.45)	(3.27)–(3.36) + (3.37)–(3.42) and (3.46)

Particularized mathematical model. The univocity conditions given by the system geometry, the material conditions and the initial and frontiers state of the process variables have to be related with the models shown in Table 3.1:

- *geometrical conditions*: for the membrane and the two-dimensional model: $l = 0.15$ m; $L = 10$ m; $h = 0.075$ m; for the membrane and the monodimensional model: $A_1 = A_2 = A_3 = 0.5$ m²; $V_1 = V_2 = V_3 = 0.04$ m³ ; for the suspension reservoir : $V_0 = 1$ m³, $V_{rz\,min} = 0.15$ m³.

- *material conditions*: liquid density $\rho_f = 1000$ kg/m³, solid density $\rho_s = 1500$ kg/m³, liquid viscosity $\eta_f = 10^{-3}$ kg/(m s), initial value of the filtration constant $k_0 = 6 * 10^{-4}$ m³/(m² h Pa), solid concentration of the fresh suspension $C_{s0} = 10$ kg/m³. The remaining values of the material properties will be computed by use of suitable relations (see Fig. 3.10).

- *initial and/or boundary conditions*: for the two-dimensional model, we attach the following initial and boundary conditions to the differential and partly differential equations:

Eq. (3.24) : $x = 0$, $G_{prx} = 0$ (3.47)

Eq. (3.26) : $0 \prec x \prec L$; $\tau = 0$; $c_s = c_{s0}$, $x = 0$; $\tau \succ 0$, $c_s = c_{srz}$ (3.48)

Eq. (3.37) : $\tau = 0$, $V_{rz} = 0.5$ (3.49)

Eq. (3.38) : $\tau = 0$, $c_{sf} = c_{s0}$ (3.50)

For the monodimensional model, only initial conditions are requested. The following data express the initial model conditions and definition functions for relations (3.49) and (3.50):

Eq. (3.39): $G_{vp} = 5 * 10^{-2} - 3 * 10^{-3} p_r^2$ (3.51)

$\tau = 0$, $p_r = p_{r0} = 2$ (3.52)

$$G_{vp\ min} = 6 * 10^{-3} \tag{3.53}$$

$$F(c_s, p, \tau) = \exp\left[-0.5\left(\frac{c_s}{100}\right)\left(\frac{p}{p_{r0}}\right)\frac{\tau}{3600}\right] \tag{3.54}$$

The complete model. The parts assemblage is already given in Table 3.1. Here the result of the assembly of the models of the devices is an enumeration of the relations contained in each model.

The numerical model-Simulator NV-Simulator V. At this point, we must find the more suitable variant for passing from the differential or partly differential model equations to the numerical state. For the case of the monodimensional model, we can select the simplest numerical method – the Euler method. In order to have a stable integration, an acceptable value of the integration time increment is recommended. In a general case, a differential equations system given by relations (3.55)–(3.56) accepts a simple numerical integration expressed by the recurrent relations (3.57):

$$\begin{cases} \dfrac{dy_1}{dx} = F_1(y_1, ...y_N, x). \\ . \\ . \end{cases}$$

$$\frac{dy_N}{dx} = F_N(y_1, ...y_N, x) \tag{3.55}$$

$$y_1(x_0) = y_{10}, y_2(x_0) = y_{20}, ..., y_N(x_0) = y_{N0} \tag{3.56}$$

$$\begin{cases} y_{1k} = y_{1k-1} + F_1(y_{1k-1}, y_{2k-1}, ..., y_{Nk-1}, x_k). \\ . \\ . \end{cases}$$

$$y_{Nk} = y_{Nk-1} + F_N(y_{1k-1}, y_{2k-1}, ..., y_{Nk-1}, x_k) \tag{3.57}$$

Figure 3.10 shows the details of the numerical-solving algorithm for the monodimensional. This numerical transposition has the capacity of being related with any available software. In Fig. 3.10, we can note that only the case of constant filtrate rate has been presented. Otherwise, when we operate at constant pressure, the filtrate rate decreases with the time due to the continuous clogging phenomenon. To simulate a constant pressure filtration, some changes in the computing program of Fig. 3.10 are necessary; these modifications are shown in Fig. 3.11. It is easily observable that here the stop criterion has been completed with the decreasing of the solid concentration in the recycled suspension.

1. **Constants:** $A_1,A_2,A_3,V_1,V_2,V_3,V_0,k_0,\rho_f,\rho_{s0},\Delta\tau,G_{vp0},p_{r0},G_{vpmin},$
   $V_{rz\,min},V_{00},\eta_f,\rho_{solid},p_a,G_{v0},G_{primp},rap$

2. **Values:** $A_1=A_2=A_3=0.5;V_1=V_2=V_3=0.04;k_0=6*10^\wedge-4;c_{s0}=10;\rho_f=1000;\rho_{s0}=1005;$
   $\Delta\tau=10;G_{vp0}=38*10^\wedge-3;p_{r0}=2;V_0=1;G_{vpmin}=6*10-3;\ V_{rzmin}=0.1;V_{00}=0.5;$
   $\eta_f=10^\wedge-3\ ;\ \rho_{solid}=1500;p_a=1;\ G_{v0}=3.8*10^\wedge-3;p_{rmax}=3.7;\ G_{primp}=3.58*10^\wedge-$
   $3;\ rap=10$

3. **Functions:** $F(c_s,p,\tau)\quad exp[\ 0.5(\dfrac{c_s}{100})(\dfrac{p}{p_{r0}})\dfrac{\tau}{3600}]\ ;\ \ \rho_{susp}\,(c_s\,)=\rho_f\ +$

$$c_s\,(1-\dfrac{\rho_f}{\rho_{solid}})\ ;\ \ \eta_{susp}\,(c_s\,)=\eta_f\,(1+2.5(\dfrac{c_s}{\rho_{solid}})^{0.25}\,)$$

4. **Variables:** $n\,,k$

5. **Sequences:**

5.1	$n=0$
5.2	$c_{s10}=c_{s0};\,c_{s20}=c_{s0}\,;\,c_{s30}=c_{s0}\,;\,c_{srz0}=c_{s0}\,;\,c_{s30}=c_{sf}\,;\,V_{rz0}=V_{00}\,;\,p_r=p_{r0}\,;\,G_{vp}=G_{vp0}$
5.3	$n=1$
5.4	$\Delta p_1=0.04[(\eta_{susp}\,(c_{s1n-1})/\eta_f\,)^{0.25}\,]$
5.5	$\Delta p_2=0.04[(\eta_{susp}\,(c_{s21n-1})/\eta_f\,)^{0.25}\,]$
5.6	$\Delta p_3=0.04[(\eta_{susp}\,(c_{s3n-1})/\eta_f\,)^{0.25}\,]$
5.7	$p_1=p_r-\Delta p_1\ ;\ \tau=n\Delta\tau$
5.8	$G_{pr1}=k_0A_1(p_1-p_a)F(c_{s1n-1},p_1,\tau)$
5.9	$G_{v1}=G_{vp}-G_{pr1}\,\rho_f/\rho_{susp}(c_{s1\,n-1})$
5.10	$c_{s1\,n}=c_{s1\,n-1}+(G_{vp}c_{s\,rz\,n-1}/V_1-G_{v1}c_{s1n-1}/V_1)\Delta\tau$
5.11	$p_2=p_r-\Delta p_2\ ;$
5.12	$G_{pr2}=k_0A_2(p_2-p_a)F(c_{s2n-1},p_2,\tau)$
5.13	$G_{v2}=G_{v1}-G_{pr2}\rho_f/\rho_{susp}(c_{s2\,n-1})$
5.14	$c_{s2\,n}=c_{s2\,n-1}+(G_{v1}c_{s\,1n-1}/V_2-G_{v2}c_{s2n-1}/V_2)\Delta\tau$
5.15	$p_3=p_r-\Delta p_3\ ;$
5.16	$G_{pr3}=k_0A_3(p_3-p_a)F(c_{s3n-1},p_3,\tau)$
5.17	$G_{v3}=G_{v3}-G_{pr3}\rho_f/\rho_{susp}(c_{s3\,n-1})$
5.18	$c_{s3\,n}=c_{s3\,n-1}+(G_{v2}c_{s\,2n-1}/V_3-G_{v3}c_{s3n-1}/V_3)\Delta\tau$
5.19	$G_{vf}=G_{v3}\,;\,c_{sf}=c_{s3\,n}\,;\,k=0$
5.20	$G_{ve}=G_{v0}/rap$
5.21	$V_{rzn}=V_{rz\,n-1}+[G_{v0}\rho_{s0}/\rho_{susp}(c_{sf\,n-1})+G_{vf}-G_{vp}-G_{ve}]\Delta\tau$
5.22	$c_{srz\,n}=c_{srz\,n-1}+(G_{vf}c_{s\,f}/V_{rz}+G_{v0}c_{s0}/V_{rz}-(G_{vp}+G_{ve})c_{srz\,n-1}/V_{rz})\Delta\tau$
5.23	**Write:** $p_1,p_2,p_3,c_{s1n},\,c_{s2\,n},\,c_{s3\,n}\,,\,G_{pr1},G_{pr2},\,G_{pr3}\,,\,p_r,G_{ve},G_{vp},\tau$
5.24	**For** $V_{rz}\leq V_{rz\,min}$ **then** $k=1$ **and** $rap=rap+k$
5.25	**For** $V_{rz}\geq V_0$ **then** $k=-1$ **and** $rap=rap+k$
5.26	**For** $V_{rz\,min}\leq V_{rz}\leq V_0$ **then** $k=0$ and $rap=rap+k$
5.27	$G_{pr}=G_{pr1}+G_{pr2}+G_{pr3}$
5.28	**For** $p_r\geq p_{rmax}$ **then STOP**
5.29	**For** $G_{pr}<G_{pr\,imp}$ **then** $p_r=p_r+p_r/30$; $G_{vp}=5*10^\wedge-2-3*10^\wedge-3p_r^2;$ $n=n+1\ ;$ **Jump to 5.4**

Figure 3.10 Numerical algorithm for the monodimensional model of the membrane filtration unit. Plant operating case: Constant filtrate flow rate $G_{pr}=G_{pr\,imp}$.

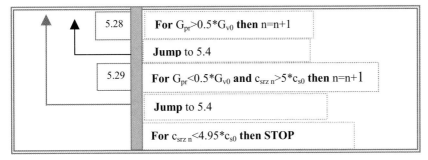

Figure 3.11 Changes to be introduced in the algorithm of Fig. 3.10 for the simulation of a constant pressure filtration (constraint $p_r = p_{r0}$ according to Eq. (3.46)).

It is obvious that the application of the two-dimensional model will introduce a supplementary mathematical diversity and complexity. Indeed, if we change the order of the relations in a given algorithm or the network integration parameters ($\Delta\tau$, Δx), the proposed integration procedure can rapidly produce integration instabilities in this concrete case. The two-dimensional integration can be maintained in the stability area, taking into consideration some observations concerning the physical meaning of the evolution of the solid concentration of the processed suspension (c_s). This model algorithm is presented in Fig. 3.12. When the intention is to use this algorithm for simulation of a constant pressure filtration, the changes given by Fig. 3.11 will be introduced. It is important to specify that the geometric plant dimensions and the flow rate of the fresh suspension are closely related. So, we cannot arbitrarily change any of these parameters independently. Once all the steps of the building of the process model have been successfully completed, the results produced with the models and their associated computer programs (software) can be presented.

Simulations and their results. It is not easy to assign the correct data to start the software running. Some of these data can be measured, others can be selected from practical design and others will be created. However, all these data must comply with the real investigated process.

First, we show that three calculation aspects seem to be interesting and must consequently be mentioned here: initially for the integration a small $\Delta\tau$ value has to be used ($\Delta\tau = 1$ s); secondly we admit that a good stability with the integration network parameters has been observed in the case of the two-dimensional model: $\Delta\tau = 1$ s and $\Delta x = 0.1$ m. Finally, we consider that the clogging rate can be selected by a careful modification of the argument of the exponential function that characterizes this process ($F(c_s, p, \tau)$).

We have selected four examples with different operating conditions: (a) a constant filtration flow rate with rapid clogging of the membrane; (b) a constant filtration flow rate with slow clogging of the membrane; (c) a constant pressure with rapid clogging of the membrane; (d) a constant pressure with slow clogging of the membrane. Each graphic representation of the simulations contains five

1. **Constants:** l , h , L , Δx , $\Delta \tau$, λ_0 ,V_0 , k_0 , ρ_f, ρ_{s0} , $\Delta \tau$, G_{vp0} , p_{r0} , G_{vpmin} , $V_{rz\,min}$, V_{00} , η_f , ρ_{solid}, p_a , G_{v0} , G_{primp} , rap

2. **Values:** l=0.15; h=0.075 ; L=10 ; Δx=0.1 ; $\Delta \tau$=1 ; λ_0=0°024 ; k_0=6*10^-4 ; c_{s0}=10; ρ_f=1000; ρ_{s0}=1005; $\Delta \tau$=10;G_{vp0}=38*10^-3;p_{r0}=2;V_0=1;G_{vpmin}=6*10-3; V_{rzmin}=0.1; V_{00}=0.5; η_f=10^-3; ρ_{solid}=1500;p_a=1;G_{v0}=3.8*10^-3; p_{rmax}=3.7;G_{primp}=3.58*10^-3; rap=10

3. **Functions:** $F(c_s,p,\tau)$=exp[-0.5(c_s/100)(p/p_{r0})τ/3600)] ; $\rho_{susp}(c_s)$=ρ_f+c_s(1-ρ_f/ρ_{solid}) ; $\eta_{susp}(c_s)$=η_f [1+2.5(c_s/ρ_{solid})^0.25]

4. **Contours:** n , k

5. **Sequences:**

5.1	M=L/Δx
5.2	n=0 : $c_{s\,0,0}$= $c_{s\,1,0}$= $c_{s\,2,0}$= $c_{s\,3,0}$= $c_{s\,4,0}$= $c_{s\,5,0}$= $c_{s\,6,0}$= $c_{s\,070}$= $c_{s\,8,0}$= $c_{s\,9,0}$= $c_{s\,10,0}$=c_{s0} ; $c_{s\,rz}$=c_{s0} ; G_{vp}=G_{vp0} ; p_r=p_{r0} ; num=25 ; de=[lh/(l+h)]^0.5
5.3	n=1 :
	$c_{s\,0,1}$=c_{s0}+0.5 ; p_0=p_r ; G_{pr0}=0 ; τ=$n\Delta\tau$
5.4	m=1
5.4.1	w_0=G_{vp}/(lh)
5.4.2	$G_{v\,m}$=G_{vp}-G_{v0}/num;
	w_m=$G_{v\,m}$/(lh) ; λ=λ_0[η_{susp}($c_{s\,m-1\,n}$)/η_f]^0.25;
	p_m=p_{m-1}-(λ/de)(((w_0+w_m)^2)/8) ρ_{susp}($c_{s\,m-1\,n}$)Δx;
	$G_{pr\,m}$=$G_{pr\,m-1}$+k_0(p_m-p_a)F($c_{s\,m-1\,n}$,p_m,τ)Δx; $G'_{v\,m}$=$G_{v\,m}$-$G_{pr\,m}$; Ere=($G'_{v\,m}$-$G_{v\,m}$)/$G_{v\,m}$
	For Ere≤0 and ABS(Ere)≥0.01 then : k=-1 ; num =num+k; **Jump to 5.42**
	For Ere≥0 and Ere≥0.01 then : k=+1 ; num =num+k; **Jump to 5.42**
	$c_{s\,m\,n}$=$c_{s\,m-1\,n}$/(1-Δx/($w_m\Delta\tau$)) ; **Write** : p_m ,$G_{pr\,m}$, $G_{v\,m}$, $c_{s\,m\,n}$
	For m≤M then : m=m+1; **Jump to 5.4.1**
5.5	$c_{sf\,n}$= $c_{s\,M\,n}$; G_{ve}=G_{v0}/rap; G_{vf}=$G_{v\,M}$
5.6	V_{rzn}=$V_{rz\,n-1}$ + [$G_{v0}\rho_{s0}$/ρ_{susp}($c_{sf\,n-1}$)+G_{vf}-G_{vp}-G_{ve}]$\Delta\tau$
5.7	$c_{srz\,n}$=$c_{srz\,n-1}$ +($G_{vf}c_{s\,f}$/V_{rz}+$G_{v0}c_{s0}$/V_{rz}-(G_{vp}+G_{ve})$c_{srz\,n-1}$/V_{rz})$\Delta\tau$
5.8	Write : $G_{pr\,M}$, p_r , $c_{s\,M\,n}$, c_{srzn} , G_{ve} , G_{vp} ,τ
5.9	**For V_{rz}≤$V_{rz\,min}$ then** : k=1 and rap=rap+k ; **For V_{rz}≥V_0 then**: k=-1 and rap=rap+k
5.10	**For $V_{rz\,min}$≤ V_{rz}≤V_0 then** : k=0 and rap=rap+k
5.11	**For p_r≥p_{r0} then STOP**
5.12	**For G_{pr}≤$G_{pr\,imp}$ then** : p_r=p_r+p_r/10 ; G_{vp}=5*10^-2 – 3*10^-3p_r^2; **Jump to 5.13**
5.13	m=0
5.14	n=2
5.15.0	$c_{s\,0\,n}$ = $c_{srz\,n}$; τ=$n\Delta\tau$; m=1
5.15.1	w_0=G_{vp}/(lh)
5.15.2	$G_{v\,m}$=G_{vp}-G_{v0}/num ; w_m=$G_{v\,m}$/(lh) ; λ=λ_0[η_{susp}($c_{s\,m-1\,n}$)/η_f]^0.25 ;
	p_m=p_{m-1}-(λ/de)(((w_0+w_m)^2)/8) ρ_{susp}($c_{s\,m-1\,n}$)Δx ;
	$G_{pr\,m}$=$G_{pr\,m-1}$+k_0(p_m-p_a)F($c_{s\,m-1\,n}$,p_m,τ)Δx; $G'_{v\,m}$=$G_{v\,m}$-$G_{pr\,m}$; Ere=($G'_{v\,m}$-$G_{v\,m}$)/$G_{v\,m}$
	For Ere≤0 and ABS(Ere)≥0.01 then : k=-1 ; num =num+k ; **Jump to 5.15.2**
	For Ere≥0 and Ere≥0.01 then : k=+1 ; num =num+k ; **Jump to 5.15.2**
	$c_{s\,m\,n}$=$c_{s\,m-1\,n}$+Δx/($w_m\Delta\tau$)($c_{s\,m\,n-1}$-$c_{s\,m\,n-2}$) ; **Write** : p_m ,$G_{pr\,m}$, $G_{v\,m}$, $c_{s\,m\,n}$
	For m≤M then : m=m+1 ; **Jump to 5.15.1**
5.16	$c_{sf\,n}$= $c_{s\,M\,n}$;G_{ve}=G_{v0}/rap;G_{vf}=$G_{v\,M}$;V_{rzn}=$V_{rz\,n-1}$+[$G_{v0}\rho_{s0}$/ρ_{susp}($c_{sf\,n-1}$)+G_{vf}-G_{vp}-G_{ve}]$\Delta\tau$
5.17	$c_{srz\,n}$=$c_{srz\,n-1}$ +($G_{vf}c_{s\,f}$/V_{rz}+$G_{v0}c_{s0}$/V_{rz}-(G_{vp}+G_{ve})$c_{srz\,n-1}$/V_{rz})$\Delta\tau$
5.18	**Write** : $G_{pr\,M}$, p_r ,$c_{s\,M\,n}$, c_{srzn} ,G_{ve} ,G_{vp} ,τ ; **For V_{rz}≤$V_{rz\,min}$ then** k=1 and rap=rap+k
5.19	**For V_{rz}≥V_0 then** k=-1 and rap=rap+k ; **For V_{rz} V_0 then** k=-1 and rap=rap+k
5.20	**For p_r≥p_{r0} then STOP**
5.21	**For G_{pr}≤$G_{pr\,imp}$ then** p_r=p_r+p_r/10; G_{vp}=5*10^-2-3*10^-3p_r^2; n=n+1
5.22	**Jump** to 5.15.0

Figure 3.12 Numerical algorithm for the two-dimensional model of the membrane filtration plant. Plant operating case: Constant filtrate rate G_{pr} = G_{primp}.

operation cases: F1: filtration type **a** where the concentrated suspension evacuation is controlled by the suspension level of the reservoir RZ; F1S: the same filtration as F1 but here the evacuation of the concentrated suspension is controlled by the instantaneous mass balance; F2/2, F2/1.8, F2/1.6: filtration type **c** with the corresponding trans-membrane pressures of 2, 1.8 and 1.6 bar. The curves that show an oscillatory state correspond to the simulations where the process control requires some intervention on the pressure pump and/or on the control of the suspension level in the recycling reservoir. Each intervention that increases and decreases the pressure to maintain the filtrate flow at a fixed value is an oscillatory process. This process is rapidly detected and processed by the model. Table 3.2 gives the oscillations that characterize the filtration with the control of the pressure. These data give the limitations of the simulation cases. At the same time, they do not reproduce reality because it is not possible to change the pressure of the pump each second. This fact imposes a condition which has to be introduced in the computation program: a change in the pressure can be produced after a minimum 30 s time interval. This constraint has been used for the simulations named F1, F1S, LF1 and LF1S.

The simulations shown in Figs. 3.10 and 3.12 were made for the following operating conditions: 1, for the monodimensional model, the filter was considered to be composed of three identical membranes with a 0.5 m² surface, the minimum permeate flow was imposed at 3.8×10^{-4} m³/s, the initial value of the filtration constant $k_0 = 33 \times 10^{-4}$ m³/m² bar; 2, in the second case, a 10 m long, 0.075 m high and 0.15 m wide filter was analyzed with a constant permeate flow rate while keeping the initial value of the filtration constant. A concentration of 10 kg/m³ was used for the fresh suspension.

It is important to specify here that complete clogging is reached between 3800 and 4200 s only in cases F1 and F1S. For the other cases – F2/2, F2/1.8 and F2/1.6 – the total clogging occurs later, between 6800 and 7300 s. However, after 2500–3000 s the filtrate flow rate becomes too low and unacceptable, as shown in Fig. 3.17 below.

As mentioned above, three factors are considered in the function which characterizes clogging: first, the time factor, which is a consequence of the Poisson distribution of the pore surface that blocks evolution; then the pressure factor, which accelerates the process of pore blocking; and finally the solid concentration factor.

The main difference between the operation at constant filtrate flow rate and at constant pressure can be observed in Fig. 3.13. In the case of a constant filtrate flow rate, the solid concentration inside the unit increases permanently, whereas, at constant pressure, the solid concentration increases very quickly initially (up to 1200 s) and then decreases for all the remaining time. If we look at both Figs. 3.13 and 3.15 we can see that it is not possible to start with the considered conditions with a 2 bar constant pressure because, in these conditions, a negative value of the exit flow rate appears for the concentrated suspension (Fig. 3.15) and the solid concentration increases tremendously from 10 to 120 kg/m³.

Tab. 3.2 Data for the exit of some variables of filtration and their evolution with time.

τ (s)		$p_1(\tau)$	$cs_1(\tau)$	$cs_2(\tau)$	$cs_3(\tau)$	$Gpr_1(\tau)$	$Gve(\tau)$	$Gvp(\tau)$
2500		1.806	108.4	111.6	115.3	0.0012857	3.2289e-05	0.039674
2501		1.713	108.0	111.5	115.0	0.0011395	0.00038912	0.039674
2502		1.801	108.4	111.6	115.3	0.0012783	1.1126e-05	0.039725
2503		1.708	108.1	111.6	115.0	0.0011325	0.00040927	0.039725
2504	Simulation case: LF1S	1.796	108.4	111.6	115.3	0.001271	9.9759e-006	0.039777
2505		1.704	108.1	111.6	115.0	0.0011256	0.00042937	0.039777
2506		1.792	108.5	111.6	115.3	0.0012637	3.1018e-005	0.039828
2507		1.700	108.1	111.6	114.9	0.0011186	0.00044941	0.039828
2508		1.787	108.5	111.6	115.3	0.0012564	5.2e-005	0.039879
2509		1.695	108.2	111.6	114.9	0.0011117	0.00046939	0.039879
2510		1.783	108.6	111.6	115.2	0.0012492	7.2924e-005	0.039929
2511		1.691	108.2	111.6	114.9	0.0011048	0.00048933	0.039929
2500		1.950	107.3	110.4	113.7	0.0010906	0.00056126	0.038
2501		1.950	107.2	110.4	113.7	0.0010906	0.00056133	0.038
2502		1.950	107.2	110.3	113.6	0.0010905	0.00056141	0.038
2503		1.950	107.2	110.3	113.6	0.0010905	0.00056148	0.038
2504	Simulation case: F2/2	1.950	107.1	110.3	113.6	0.0010905	0.00056155	0.038
2505		1.950	107.1	110.2	113.5	0.0010905	0.00056163	0.038
2506		1.950	107.0	110.2	113.5	0.0010904	0.0005617	0.038
2507		1.950	107.0	110.2	113.4	0.0010904	0.00056177	0.038
2508		1.950	107.0	110.1	113.4	0.0010904	0.00056185	0.038
2509		1.950	106.9	110.1	113.4	0.0010904	0.00056192	0.038
2510		1.950	106.9	110.0	113.3	0.0010904	0.00056199	0.038
2511		1.950	106.9	110.0	113.3	0.0010903	0.00056207	0.038
2500		2.217	95.2	98.5	102.3	0.0012667	0.00038	0.03459
2501		2.104	95.0	98.5	102.1	0.0011692	0.00038	0.03459
2502		2.211	95.2	98.5	102.2	0.0012615	0.00038	0.034667
2503		2.098	95.0	98.5	102.0	0.0011641	0.00038	0.034667
2504	Simulation case: F1	2.206	95.3	98.5	102.2	0.0012563	0.00038	0.034744
2505		2.0933	95.0	98.5	102.0	0.001159	0.00038	0.034744
2506		2.200	95.3	98.5	102.2	0.0012512	0.00038	0.03482
2507		2.088	95.0	98.5	102.0	0.001154	0.00038	0.03482
2508		2.194	95.3	98.5	102.2	0.001246	0.00038	0.034896
2509		2.082	95.1	98.5	102.0	0.001149	0.00038	0.034896
2510		2.189	95.3	98.5	102.1	0.0012409	0.00038	0.034971
2511		2.077	95.1	98.5	101.9	0.0011439	0.00038	0.034971

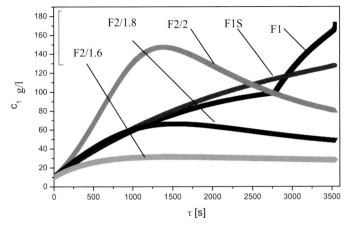

Figure 3.13 Evolution of the solid concentration in the filter unit when the membrane surface is rapidly clogged.

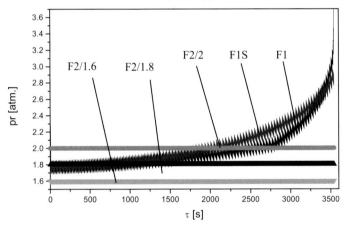

Figure 3.14 Evolution of the pressure of the pump, when a rapid clogging takes place.

When a positive exit rate of the concentrated suspension is obtained in the starting conditions, an important reduction in the filtrate flow rate will be expected, as shown in Fig. 3.17.

From Fig. 3.15 we can note that, by analogy to the 2 bar constant pressure case, example F1S shows a new special case where we have positive and small negative values in the concentrated suspension flow rate at the plant exit. This result can be explained by the background noise in the measurement of the flow of suspension. Nevertheless, the mean value of the flow rate is small but positive. If the efficiency of the filtration at constant pressure is given by the solid concentration ratio between the exit and fresh suspensions, then, as shown in Fig. 3.13, the ratio is always lower than 2 for operation case F2/1.6. For cases F1 and F1S, this ratio increases permanently, non-uniformly and attains values over 12 g/l in the proximity of the complete clogging state.

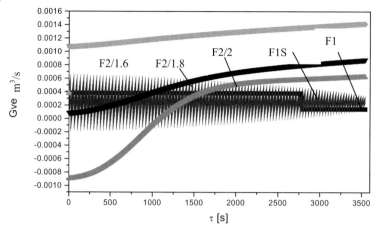

Figure 3.15 Evolution of the flow rate of the concentrated suspension when rapid clogging occurs.

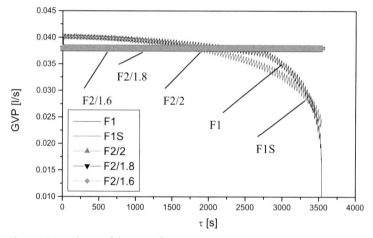

Figure 3.16 Evolution of the pump flow rate when rapid clogging occurs.

Figures 3.14 and 3.16 describe the function of the pump in the unit. When the pressure is constant, we have a constant pump exit flow rate, but, when the pressure increases to maintain the filtrate flow rate, the exit pump flow rate decreases too (see for instance relation (3.52)). In these figures, we can also observe that, for F1 and F1S, more than 110 oscillations are produced by the simulator every 30 s; these large oscillations require a pressure correction.

Figure 3.17 shows the evolution of the permeate flow rate when we work at constant pressure. We can observe (curves F1 and F1S) that controlling the pressure pump with a precision of ±0.1 bar (for instance see Table 3.2) produces a mean fluctuation of the flow rate that begins with ±20% and progressively decreases to as little as ±5% when we approach the total clogged state. In this case of slow surface clogging, it must be mentioned that the operating time before the total permeate flow rate decay is very large (40 000 s). Here, the initial filtration coefficient used was the same as that used when fast clogging occurred. Indeed, we can conclude that some properties of the suspension or interaction forces between the suspension and filter have changed and the function that describes the clogging process is not similar in both sets of operating conditions. Also, it may be noticed from Fig. 3.18 that the evolution of the concentration for different operating conditions is spectacular: (a) the solid concentration when the filtration pressure is 2 bar is unacceptable. This increase is correlated with the negative flows of the evacuated suspension (Fig. 3.21) and defines an impossible operating case; (b) the evolution of the solid concentration for the operation at constant permeate flow together with the controlled flow of evacuated concentrated suspension (through the level of the solution at the storage tank (LF1)).

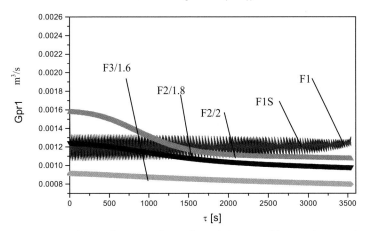

Figure 3.17 Evolution of permeate flow with rapid clogging of the membrane surface.

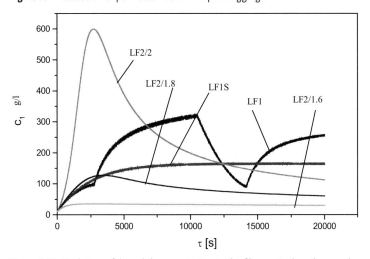

Figure 3.18 Evolution of the solid concentration in the filter unit when the membrane surface is slowly clogged.

It may be noticed that operating at 1.6 bar is not attractive from a technical or from an economic point of view. It is obvious that this state is determined by the increase in the evacuated suspension flow and the slow decrease in the permeate flow (Fig. 3.20).

Concerning Figs. 3.19 and 3.20, if we neglect the changing rate of the pump pressure and exit pump flow rate then we can appreciate that these figures are similar to Figs. 3.14 and 3.16, respectively.

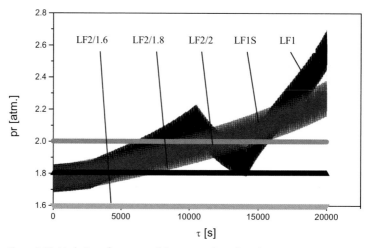

Figure 3.19 Evolution of pressure of the pump when slow clogging occurs.

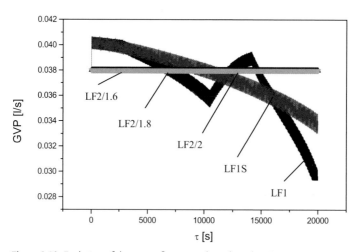

Figure 3.20 Evolution of the pump flow rate when slow clogging occurs.

Referring to the dynamics of the exits of concentrated suspension and filtrate it is interesting to observe (Figs. 3.21 and 3.22) that the cases with slow membrane clogging reproduce almost identically the corresponding cases where a rapid membrane clogging occur. Otherwise, from these representations we observe that the tendency of the operation case at 1.6 bar is near to the stationary state where all filtration dependent and independent variables stay unchanged with time. However, as explained above with respect to the solid concentration in the exit concentrated suspension (see the above definition of the filtration efficiency) this operation appears to be inefficient.

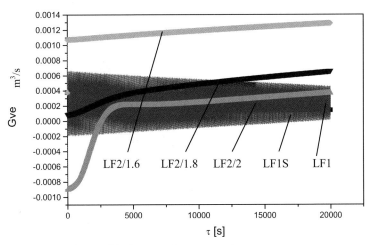

Figure 3.21 Evolution of the flow rate of concentrated suspension when slow clogging occurs during the filtration.

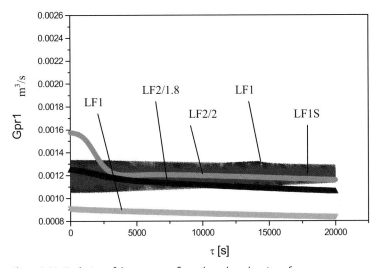

Figure 3.22 Evolution of the permeate flow when slow clogging of the membrane occurs.

To close this analysis, we note the validity of the operation with an increasing oscillatory pressure for a constant filtrate flow rate. The operation on oscillating pressure is very interesting for the enhancement of the performance of the filtration process [3.24].

To conclude this section it is important to give some general conclusions about modelling and how this example helps in the comprehension of the process of construction of a model: (1) to make a mathematical model of a process a good specialized technical knowledge of chemical engineering, software and of the actual case is necessary; (2) the model building has to be realized on a step by step basis with accurate rules for each action; (3) to pass from the complete model to the simulator it is necessary to take into account multiple factors. Among these we can mention the capacity to write a complex program according to the scientific disposable soft; to correctly assign data for the start of the simulator; to integrate the simulator with other simulators when necessary and (4) the choice of the simulation examples is a problem that can be solved only by a specialist who can also interpret the results.

From this example, we can establish some generalities about the modelling of a process:

1. The model of a process is a relation between the "outputs" and "inputs" (feed conditions, design parameters and adjustable parameters of the process) with a view to scaling-up the process from laboratory to industrial scale, predicting the process dynamics (case of this concrete example) and optimizing the operating conditions.

2. In the modelling of an actual case, the chemical engineers apply a methodology which involves establishing:
   - the conservation equations (mass, energy, momentum and electric charge);
   - the equilibrium laws at the interface(s);
   - the constitutive laws (e.g., ideal gas law);
   - the kinetic laws of transport and reaction;
   - the initial and boundary conditions;
   - the optimization criteria.

3. With this methodology, the problem is analyzed from the smallest to the largest scales, as appearing in the process description. As an example, in the case of a catalytic reactor, we consider the process on the following scales:
   - pore scale (catalyst and adsorbent): 1–1000 nm;
   - particle scale: 10 μm–1 cm;
   - reactor/separator scale: 1–10 m.

3.3
Chemical Engineering Flow Models

The modelling example of the previous section shows that to simplify the general mathematical model of the studied process, the real flow in the filter unit has been considered in terms of its own simplified model. Indeed, it is difficult to understand *why we have used a flow model, when in fact, for the flow characterization, we already have the Navier-Stokes equations and their expression for the computational fluid dynamics*. To answer this question some precisions about the general aspects of the computational fluids dynamics have to be given.

Computational fluid dynamics (CFD), is the science of determining a numerical solution to the equations governing the fluid flow. In order to obtain a numerical description of the complete flow field of interest, the solutions are obtained in a dynamic regime (i.e. continuously changing in space or time). CFD obtains solutions for the governing Navier-Stokes fluid flow equations and, depending upon the case under study, it solves additional equations involving multiphase, turbulence, heat transfer and other relevant processes. In CFD, partial differential Navier-Stokes and associated equations are converted into algebraic form (numerically solvable by computing) on a mesh that defines the geometry and flow domain of interest. Appropriate boundary and initial conditions are applied to the mesh, and the distributions of quantities such as velocity, pressure, turbulence, temperature and concentration are determined iteratively at every point in space and time within the domain. CFD analysis typically requires the use of a computer to perform the mathematical calculations. Graphical output shows the results of the analysis. Most of the CFD software available today requires more computing capability than can be obtained from a typical personal computer. CFD has proven its capability in predicting the detailed flow behaviour for a wide-range of engineering applications, typically leading to improved equipment or process design. CFD is used for early conceptual studies of new designs, detailed equipment design; scale-up, troubleshooting and system retrofitting. Examples in chemical and process engineering include separators, mixers, reactors, pumps, pipes, fans, seals, valves, fluidized beds, bubble columns, furnaces, filters and heat exchangers.

Concerning our problem of the modelling of the flow process, even if the CFD seems to be the most complete approach, the use of flow models for its characterization is sustained by the following statements:

1. For the majority of the specific apparatus, the flows present a turbulent comportment and, for such flow, a numerical solution is covered by high uncertainty because some hypotheses have to be accepted *a priori* [3.25]; in all the studied cases, the real apparatus has a complicated geometry that imposes very complex and frequently uncertain univocity conditions in the real CFD-based flow computation.

2. For many cases a flow model is, in fact, the real solution to an equivalent complicated CFD model. Other arguments can be given to recommend the use of the simplest type of flow model: (a) the simple and rapid possibility of these flow models to be qualitatively and quantitatively identified as a result of experimental measurements; (b) the accuracy and suppleness of the data produced by modelling when the flow models are adequately selected and identified; (c) in general, researchers with a large experience of these models, are able to rapidly assign a model after a verbal description of the real flow, in spite of non-identified parameters. The theoretical basis of these flow models is expressed by the possibility to characterize the flow with the residence time distribution of the fluid particles that compose the flow passing through the considered device.

3.3.1
The Distribution Function and the Fundamental Flow Models

The residence time of a signal that passes through a device, is in fact a random variable which is completely characterized by its probability distribution. This probability distribution, known as the residence time distribution, can be found for an actual apparatus after its exit response in the form of an input signal. Generally it is utilized as a signal, a substance (indicator) which is introduced in the input flow as a Dirac's impulse, unitary impulse or harmonic impulse. Figure 3.23 shows the scheme of an experiment dealing with the passing of a signal through a real or a scaled down (laboratory model) device. When a signal impulsion is given to the input flow of the device, the quantity of the substance that is contained in the signal coming out in the exit flow will be:

$$C_m = \int_0^\infty c(\tau)E(\tau)d\tau \tag{3.58}$$

where $E(\tau)$ represents the differential function of the residence time distribution and $c(\tau)$ is the instantaneous concentration of the substance (signal) in the exit flow. With respect to the flow, function $E(\tau)$ is in fact the fraction of the signal that comes out from the device after a residence time which ranges between τ and $\tau + d\tau$. The residence time can also be considered from the statistical viewpoint where it is a random variable, then $E(\tau)$ represents its density of probability, which is frequently called distribution function. Indeed, $E(\tau)$ must verify the norma condition:

$$\int_a^\infty E(\tau)d\tau = 1 \tag{3.59}$$

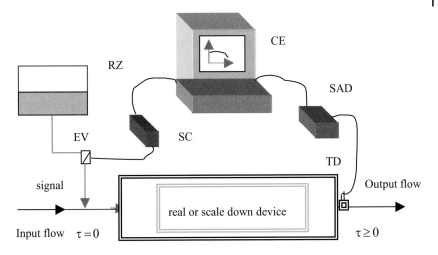

Figure 3.23 Schematic arrangement for a signal introduction in a device.
CE: computer, RZ: reservoir with signal solution, SAD: data acquisition system,
SC: command system, EV: electric valve, TD: signal concentration transducer.

We can use the experimental data that result from the measuring of the signal concentration in the exit flow, then $E(\tau)$ will be computed with relation (3.60) where N gives the number of experiments necessary for the signal concentration to disappear:

$$E(\tau) = \frac{c(\tau)}{\int_0^\infty c(\tau)d\tau} = \frac{c(\tau)}{\sum\limits_{i=1}^{N} c(\tau_i)\Delta\tau} \tag{3.60}$$

Function $F(\tau)$ is directly connected to the residence time distribution. It is recognized as the repartition function of the residence time random variable. So, $F(\tau)$ shows the fraction of the fluid elements that stayed in the device for a time less than or equal to τ. Between $F(\tau)$ and $E(\tau)$ the following integral and differential link exists:

$$F(\tau) = \int_0^\tau E(\tau^\circ)d\tau \quad \text{or} \quad E(\tau) = \frac{dF(\tau)}{d\tau} \tag{3.61}$$

Function $F(\tau)$ represents the apparatus response to a unitary impulsion signal where C_0 is the concentration in the input flow. By measuring the signal concentration in the exit flow we can write $F(\tau)$ with the relation (3.62). When the condition of "pure unitary signal" is respected, we can easily observe that $F(0) = 0$ and $F(\infty) = 1$. In this case, function $F(\tau)$ can be written as:

$$F(\tau) = \frac{c(\tau)}{c_0} \tag{3.62}$$

For ideal flow models such as perfect mixing flow, plug flow and all other ideal models, a combination of functions $E(\tau)$ and $F(\tau)$ can be obtained directly or indirectly using the model transfer function $T(p)$. Before obtaining an expression for $E(\tau)$ for the perfect mixing flow, we notice that the transfer function of a flow model is in fact the Laplace's transformation of the associated $E(\tau)$ function:

$$T(p) = L(E(\tau)) = \frac{c_{sort}(p)}{c_{ent}(p)} \tag{3.63}$$

For the computation of $E(\tau)$ and $T(p)$, in the case of a perfect mixing model, we use the representation and notation given in Fig. 3.24. Including the mass balance of the species in the signal, we derive the following differential equation:

$$\frac{dc}{d\tau} = -\frac{G_v}{V} c \tag{3.64}$$

the characteristic conditions for an impulse δ in the input flow are:

$$c_{inp} = c_0 \text{ for } \tau = 0 \text{ and } c_{inp} = 0 \text{ for } \tau \geq 0 \tag{3.65}$$

and then Eq. (3.64) becomes:

$$\frac{c}{c_0} = \exp\left(-\frac{G_v}{V}\tau\right) \tag{3.66}$$

Now, combining relations (3.66) and (3.61), we obtain the expression for the time distribution function of the perfect mixing flow model:

$$E(\tau) = \frac{c(\tau)}{\int_0^\infty c(\tau)d\tau} = \frac{c_0\exp\left(-\dfrac{G_v}{V}\tau\right)}{\int_0^\infty c_0\exp\left(-\dfrac{G_v}{V}\tau\right)d\tau} = \frac{G_v}{V}\exp\left(-\frac{G_v}{V}\tau\right) = \frac{1}{\tau_m}\exp(-\tau/\tau_m) \tag{3.67}$$

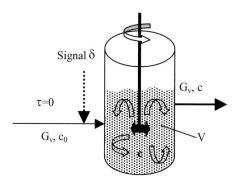

Figure 3.24 Physical model for perfect mixing (PM) flow.

We can obtain the repartition function of the residence time for the model of perfect mixing flow from relations (3.67) and (3.62). This function is:

$$F(\tau) = \int_0^\tau E(\tau)d\tau = \int_0^\tau \tau_m^{-1}\exp\left(-\frac{\tau}{\tau_m}\right)d\tau = 1 - \exp\left(-\frac{\tau}{\tau_m}\right) \tag{3.68}$$

by definition, the transfer function is:

$$T(p) = L(E(\tau)) = \frac{c_{sort}(p)}{c_{ent}(p)} = \int_0^\infty \exp(-p\tau)\tau_m^{-1}\exp\left(-\frac{\tau}{\tau_m}\right)d\tau = \frac{1}{\tau_m p + 1} \tag{3.69}$$

In a more general case where the input signal is given by a function $c_{ent}(\tau)$ the balance of the species characterizing the signal can be written as follows:

$$\frac{dc}{d\tau} = -\frac{G_v}{V}(c - c_{ent}) \tag{3.70}$$

The Laplace's transformation of the differential equation (3.70) gives relation (3.71) where p is the Laplace's argument:

$$pc_{sort}(p) - pc(0_-) = \frac{G_v}{V}(c_{sort}(p) - c_{ent}(p)) \tag{3.71}$$

In general, we have $c(0) = 0$ for all the signal types; then, we can transform the previous relation to show the transfer model function:

$$\frac{c_{sort}(p)}{c_{ent}(p)} = T(p) = \frac{1}{\tau_m p + 1} \tag{3.72}$$

From this last relation we remark that the transfer model function can be obtained from the differential model equation that, in fact, is a particularization of the balance of the concerned species in the actual model.

The species balance in a plug flow (Fig. 3.25) is carried out in an elementary dx length of the control volume; the result is the partial differential equation (3.73) where w is the velocity of the fluid moving with a plug flow pattern. Then, the relation between the flow rate and the section crossed by flow becomes:

$$\frac{\partial c}{\partial \tau} = -w\frac{\partial c}{\partial x} \tag{3.73}$$

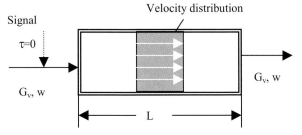

Figure 3.25 Physical model of the plug flow (PF).

The boundary conditions for a δ signal in the input associated to the plug flow model (Eq. (3.73)) are written as follows:

$$\tau = 0_- \quad , \quad 0 \prec x \prec L \quad , \quad c = 0 \tag{3.74}$$

$$\tau = 0 \quad , \quad x = 0 \quad , \quad c = c_0 \tag{3.75}$$

$$\tau = 0_+ \quad , \quad x = 0 \quad , \quad c = 0 \tag{3.76}$$

Now, we can write the plug flow model transfer function. With the Laplace's transformation of relation (3.73) and with Eqs. (3.74)–(3.76) we have:

$$pc(p) - pc(0_-) = w \frac{dc(p)}{dx}$$

$$-\frac{dc(p)}{c(p)} = p \frac{dx}{w}$$

$$\frac{c_{sort}(p)}{c_{ent}(p)} = T(p) = \exp\left(-p \frac{L}{w}\right) = \exp(-p\tau_m)$$

The residence time distribution $E(\tau)$ and the residence time repartition will be obtained starting with the inverse transformation of the transfer function $T(p)$:

$$E(\tau) = \delta(\tau - \tau_m) \tag{3.77}$$

where the δ impulse function is given by relation (3.78):

$$\delta(\tau - \tau_m) = \begin{cases} 0 & \text{for } x = L, \ \tau < \tau_m \\ 1 & \text{for } x = L, \ \tau = \tau_m \\ 0 & \text{for } x = L, \ \tau > \tau_m \end{cases} \tag{3.78}$$

Here, it is important to notice that, in the case of a combined model composed of PM and PF models, the transfer function is obtained from multiplication of the individual transfer functions:

$$T(p) = T_1(p)T_2(p)T_3(p).....T_N(p) \tag{3.79}$$

Table 3.3 shows the transfer functions that characterize the simplest and the combined models which are most commonly obtained by a combination of PM and PF models.

When we have a combination of recycled flow, by-pass connections, the presence of dead regions and a complex series and/or parallel coupling of the basic PM and PF models in a system, we have an important class of flow models recognized as combined flow models (CFM).

Table 3.3 The transfer function and model equation for some flow models

Model name	Model equation	Transfer Function	Symbol
1 Perfect mixing Flow	$\dfrac{dc}{d\tau} = -\dfrac{G_v}{V}(c - c_{ent})$	$T(p) = \dfrac{1}{\tau_m p + 1}$	$\tau_m = \dfrac{V}{G_v}$
2 Plug flow Model	$\dfrac{\partial c}{\partial \tau} = -w\dfrac{\partial c}{\partial x}$	$T(p) = \exp(-p\tau_m)$	$\tau_m = \dfrac{L}{w}$
3 Cellular perfect mixing Equal N cellules	$\dfrac{dc_i}{d\tau} = -\dfrac{G_v}{V}(c_i - c_{i-1})$	$T(p) = \dfrac{1}{(\tau_m p + 1)^N}$	V – cellule volume
4 Cellular perfect mixing Non-equal N cellules	$\dfrac{dc_i}{d\tau} = -\dfrac{G_v}{V_i}(c_i - c_{i-1})$	$T(p) = \displaystyle\prod_{i=1}^{N} \dfrac{1}{\tau_{mi} p + 1}$	V_i – cellule i volume
5 Series perfect mixing- Plug flow		$T(p) = \dfrac{\exp(-p\tau_{md})}{\tau_m p + 1}$	τ_{md} – mean residence time at PFM

3.3.2
Combined Flow Models

The construction of a combined model starts with one image (created, supposed or seeded) where it is accepted that the flow into the device is composed of distinct zones which are coupled in series or parallel and where we have various patterns of flow: flow zones with perfect mixing, flow zones with plug flow, zones with stagnant fluid (dead flow). We can complete this flow image by showing that we can have some by-pass connections, some recycled flow and some slip flow situations in the device.

The occurrence of these different types of flow can be established using the curve that shows the evolution of the species concentration (introduced as a signal at the input) at the device exit. It is important to notice that we can describe a flow process with an arbitrary number of regions and links. This procedure can result in a very complex system which makes it more difficult to identify the parameters of the CFM. In addition, we seriously increase the dimension of the problem, which results in quite a complex process of model building. Table 3.4 presents some simple combined models showing the model response by an analytic expression and by a qualitative graphic representation when we have a signal δ as input. Concerning the relations of Table 3.4, θ represents the dimensionless time τ/τ_m and $C(\theta)$ is the ratio $c(\theta)/c_0$. In fact, $C(\theta)$ is equivalent to $E(\tau)$ and, consequently, the dimensionless repartition function for the residence time $F(\theta)$ will be obtained by the integration of the function $C(\theta)$ from zero to θ. In Table 3.4, the models that result from the simplifications of three general types of combined models presented here below, are shown.

Table 3.4 The response curves to a δ signal for some simple CFM.

i	Model name	Model schedule	Response C(θ) vs. θ

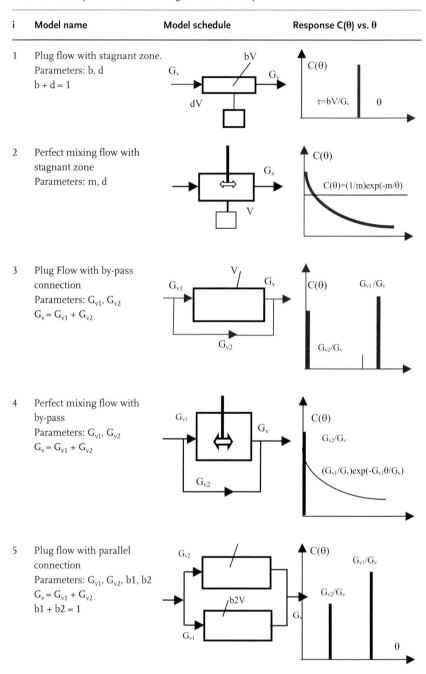

1   Plug flow with stagnant zone.
Parameters: b, d
b + d = 1

2   Perfect mixing flow with
stagnant zone
Parameters: m, d

3   Plug Flow with by-pass
connection
Parameters: G_{v1}, G_{v2}
$G_v = G_{v1} + G_{v2}$

4   Perfect mixing flow with
by-pass
Parameters: G_{v1}, G_{v2}
$G_v = G_{v1} + G_{v2}$

5   Plug flow with parallel
connection
Parameters: G_{v1}, G_{v2}, b1, b2
$G_v = G_{v1} + G_{v2}$
b1 + b2 = 1

Table 3.4 Continued

i	Model name	Model schedule	Response C(θ) vs. θ
6	Plug flow with recycling Parameters: k $a = k/(k + 1)$		
7	Plug flow with parallel perfect mixing Parameters: m, b, G_{v1}, G_{v2} $m + b = 1$		
8	Plug flow with series perfect mixing Parameters: b, m $b + m = 1$		

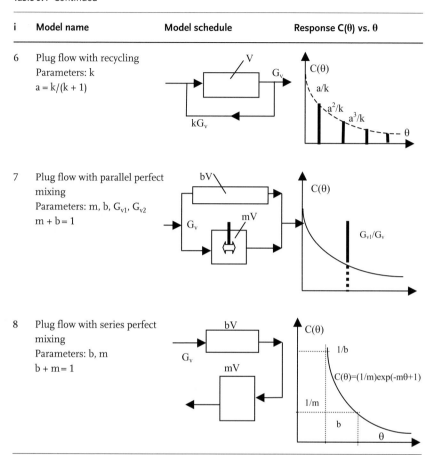

The first kind of CFM is characterized by Eqs. (3.80)–(3.82) and is shown in Fig. 3.26. For this CFM configuration, we can notice a lack of recycled flow or by-pass connections. The second type of CFM is introduced by Fig. 3.27 and is quantitatively characterized by relations (3.83)–(3.85) which show the dimensionless evolution of C(θ) and F(θ). Here we observe that we do not have any PF participants and recycled flow. The third CFM class is given in Fig. 3.28 and described by relations (3.86)–(3.88). Here the by-pass connections and PM participants are missing.

$$b1+b2+b3+d1+d2+m=1$$

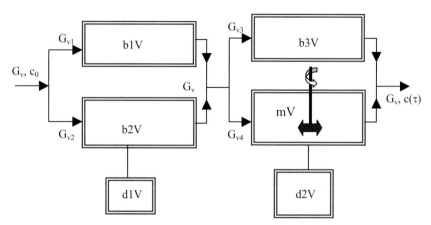

Figure 3.26 Block-scheme for the general mixing CFM.

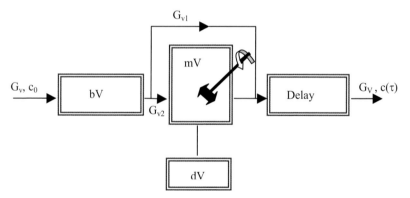

Figure 3.27 Block-scheme for the general by-passing CFM.

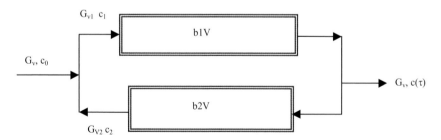

Figure 3.28 Block-scheme for the general recycling CFM.

$$T(p) = \sum_{i=1}^{2} \left\{ \frac{\dfrac{G_{v1}G_{v3}}{G_v^2} \exp\left[-\dfrac{G_v}{G_{v1}} bi(p+k)\tau_m\right]}{1 + \dfrac{mG_v}{G_{v4}}(p+k)} + \right.$$

$$\left. \frac{G_{v1}G_{v3}}{G_v^2} \exp\left[-\frac{G_v}{G_{v1}} b1(p+k)\tau_m - \frac{G_v}{G_{v3}} b3(p+k)\tau_m\right] \right\} \tag{3.80}$$

$$C(\theta) = \sum_{i=1}^{2} \left\{ \frac{G_{vi}G_{v3}}{G_v^2} \exp\left(-\frac{G_v}{G_{vi}} b1k\tau_m - \frac{G_v}{G_{v3}} b3k\tau_m\right) \right.$$

$$* \delta\left(\theta - \frac{G_v}{G_{v1}} b1 - \frac{G_v}{G_{v3}} b3\right) + \frac{G_{v1}G_{v3}}{mG_v^2} \exp\left[-\frac{G_v}{G_{vi}} b1k\tau_m - \right.$$

$$\left. \left(k\tau_m + \frac{G_{v4}}{mG_v}\right)\left(\theta - \frac{G_v}{G_{vi}} bi\right) * v\left(\theta - \frac{G_v}{G_{v1}} b1\right)\right] \right\} \tag{3.81}$$

$$F(\theta) = \sum_{i=1}^{2} \left\{ \frac{G_{vi}G_{v3}}{mG_v^2} \frac{\exp\left(-\dfrac{G_v}{G_{vi}} b1\tau_m\right)}{\left(k\tau_m + \dfrac{G_{v4}}{mG_v}\right)} * \left[1 - \exp\left(-k\tau_m + \frac{G_{v4}}{mG_v}\right)\left(\theta - \frac{G_v}{G_{vi}} bi\right)\right] \right.$$

$$* v\left(\theta - \frac{G_v}{G_{vi}} bi\right) + \frac{G_{vi}G_{v3}}{G_v^2} \exp\left(-\frac{G_v}{G_{vi}} bik\tau_m - \frac{G_v}{G_{v3}} b3k\tau_m\right)$$

$$\left. * \vartheta\left(\theta - \frac{G_v}{G_{vi}} bi - \frac{G_v}{G_{v3}} b3\right) \right\} \tag{3.82}$$

$$T(p) = \frac{\exp(-p\varepsilon - k(\varepsilon - \tau_{rt}))}{\dfrac{\tau_m}{\eta}(p+k) + 1} \tag{3.83}$$

$$C(\theta) = \eta \exp\left(-k(\varepsilon - \tau_{rt}) - (k\tau_m + \eta)\left(\theta - \frac{\varepsilon}{\tau_m}\right)\right) * \delta\left(\theta - \frac{\varepsilon}{\tau_m}\right) \tag{3.84}$$

$$F(\theta) = \frac{\exp(-k(\varepsilon - \tau_{rt}))}{k\tau + \eta}\left[1 - \exp\left(-(k\tau_m + \eta)\left(\theta - \frac{\varepsilon}{\tau_m}\right)\right)\right] * \vartheta\left(\theta - \frac{\varepsilon}{\tau_m}\right) \tag{3.85}$$

It is important to notice that all the relations characterizing these three CFMs have been established by considering that a first order chemical reaction takes place in volume V and according to the accepted structure of the flow. So, here, k represents the kinetic reaction constant. When the reaction is not taken into account, we consider $k = 0$. In relations (3.83)–(3.85),τ_{rt} is the time delay expressed in a natural value, ε describes the system phase difference in time units and η is the mixing coefficient. This last parameter equals one for a perfect mixing flow and zero for plug flow. In other cases, η can be estimated with $m/(m + b + d)$ as shown in Fig. 3.27.

$$T(p) = \frac{\dfrac{G_v}{G_{v1}} \exp\left(-(p+k)\tau_m \dfrac{b1G_v}{G_{v1}}\right)}{1 + \dfrac{G_{v2}}{G_{v1}} \exp\left(-(p+k)\tau_m \left(\dfrac{b1G_v}{G_{v1}} + \dfrac{b1G_v}{G_{v2}}\right)\right)} \tag{3.86}$$

$$C(\theta) = \frac{G_v}{G_{v1}} \exp\left(-k\tau_m \frac{b1G_v}{G_{v1}}\right) *$$

$$\sum_{N=1}^{N} \left[\frac{G_{v2}}{G_{v1}} \exp\left(-k\tau_m \left(\frac{b1G_v}{G_{v1}} + \frac{b2G_v}{G_{v2}}\right)\right)\right]^{N-1} \delta\left(\theta - \frac{Nb1G_v}{G_{v1}} + \frac{(N-1)b2G_v}{G_{v2}}\right) \tag{3.87}$$

$$F(\theta) = \frac{G_v}{G_{v1}} \exp\left(-k\tau_m \frac{b1G_v}{G_{v1}}\right) * \sum_{N=1}^{N}$$

$$\left[\frac{G_{v2}}{G_{v1}} \exp\left(-k\tau_m \left(\frac{b1G_v}{G_{v1}} + \frac{b2G_v}{G_{v2}}\right)\right)\right]^{N-1} \delta\left(\theta - \frac{Nb1G_v}{G_{v1}} + \frac{(N-1)b2G_v}{G_{v2}}\right) \tag{3.88}$$

The general problem of building a model for an actual process begins with a flow description where we qualitatively appreciate the number of flow regions, the zones of interconnection and the different volumes which compose the total volume of the device. We frequently obtain a relatively simple CFM, consequently, before beginning any computing, it is recommended to look for an equivalent model in Table 3.4. If the result of the identification is not satisfactory then we can try to assimilate the case with one of the examples shown in Figs. 3.26–3.28. If any of these previous steps is not satisfactory, we have three other possibilities: (i) we can compute the transfer function of the created flow model as explained above; (ii) if a new case of combination is not identified, then we seek where the slip flow can be coupled with the CFM example, (iii) we can compare the created model with the different dispersion flow models.

The CFM can be completed with a recycling model (the trajectory of which can be considered as a CFM, such as a PF with PM, series of PM, etc.), or with models with slip flows and models with multiple closed currents.

The next section will first show the importance of flow in a concrete modelling problem such as the slip flow effect on the efficiency of a permanent mechanically mixed reactor. Then the characterization of the combined flow models where the slip flow occurs will be presented.

3.3.3
The Slip Flow Effect on the Efficiency of a Mechanically Mixed Reactor in a Permanent Regime

In this section, we consider a permanent and mechanically mixed reactor, where a chemical transformation occurs and the consumption rate of one reactant is given by a formal kinetics of n order: $v_r = kc^n$. The flow conditions expressed by the geometric position and rotation speed of the stirrer and by the position on the reactor of the input and exit flow, define an internal flow structure with three regions: the surface region, named slip flow, where the reactants come rapidly to the exit without an important conversion; the middle region, where a perfectly mixing flow exists and consequently an important reactant conversion takes place; the bottom region, where we have a small flow intensity and which can be recognized as a stagnant region. Figure 3.29 gives a graphic presentation of the description of the reactor operation as well as the notation of the variables. The performances of this simplified and actual reactor (SPMR) example will be compared with those of a permanent perfect mixing reactor (PMR) having the same volume.

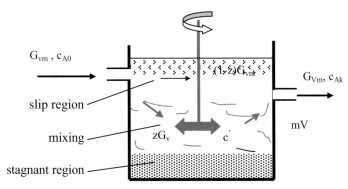

Figure 3.29 Stirred reactor with slip and stagnant flow zones.

The reactant A balance equations for these reactors (PMR and SPMR) can be written as follows:

$$V v_{r\,max} = G_{vm}(c_{A0} - c_{A\,min})$$ (3.89)

$$mVv_r = G_{vm}(c_{A0} - c_{Ak})$$ (3.90)

If we compute the ratio between Eqs. (3.89) and (3.90) we have:

$$m = \frac{c_{A0} - c_{Ak}}{c_{A0} - c_{A\,min}} \frac{v_{r\,max}}{v_r}$$ (3.91)

Here the reaction rate will be $v_{r\,max} = k(c_{A\,min})^n$ and $v_r = k(c')^n$ respectively. The value of the reactant concentration for the mixing zone of the SPMR will be obtained as a result of its comparative mass balance. If we consider that the slip flow is not present (PMR case) or when it is present but the reactant flow rate is identical, then we can write:

$$zG_{vm}c' = G_{vm}c_{Ak} - (1 - z)G_{vm}c_{A0}$$ (3.92)

So, for the reactant concentration in the mixing zone of the SPMR we establish the relation:

$$c' = \frac{c_{Ak} - (1 - z)c_{A0}}{z}$$

(3.93)

Rearranging the five previous equations (3.89) to (3.93), in order to introduce and replace the relation between v_r and $v_{r\,max}$, we have:

$$m = \left[\frac{\frac{c_{A0}}{c_{Ak}}}{\frac{c_{A0}}{c_{A\,min}}}\right]^{n-1} \left[\frac{\frac{c_{A0}}{c_{Ak}} - 1}{\frac{c_{A0}}{c_{A\,min}} - 1}\right] \cdot \left[\frac{z}{1 - (1 - z) - \frac{c_{A0}}{c_{Ak}}}\right]^{n}$$

(3.94)

If we introduce $1 - c_{Ak}/c_{A0} = X$ and $1 - c_{A\,min}/c_{A0} = X_m$ where X and X_m represent the reactant transformation degree for SPMR and PMR operation modes, then relation (3.94) becomes:

$$m = \frac{X}{X_m} \left[\frac{z(1 - X_m)}{z - X}\right]^{n}$$

(3.95)

Figure 3.30 shows clearly the effect of m and z on the reactant transformation degree for a SPMR. Only for a zero-order kinetics process, does the slip flow not affect the degree of the reactant transformation. For other X_m values, each graphic construction based on Fig. 3.30 shows the same rules of evolution (at m<0.5, z and X increase simultaneously, and, when n increases, X increases slowly; for m>0.5, X keeps a constant value determined by z). When the PM core of SPMR is exchanged with a CFM model, we obtain a special SPMR type in which the performances can be appreciated by the model developed above.

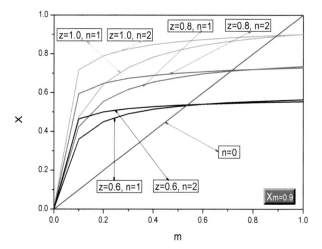

Figure 3.30 The effect of the perfect mixing region dimension and of the slip flow degree on the SPMR conversion (reaction order n = 0, 0.5, 1, 1.5, 2; X_m = 0.9).

3.3.4
Dispersion Flow Model

The models of flow dispersion are based on the plug flow model. However, in comparison with the PF model, the dispersion flow model considers various perturbation modes of the piston distribution in the flow velocity. If the forward and backward perturbations present random components with respect to the global flow direction, then we have the case of an axial dispersion flow (ADF). In addition, the axial and radial dispersion flow is introduced when the axial flow perturbations are coupled with other perturbations that induce the random fluid movement in the normal direction with respect to the global flow.

With reference to these different types of flow, there is often confusion associated with the terms: "dispersion", "diffusion" and "turbulence". When we talk about a *species in a fluid*, diffusion and turbulence produce the molecular or turbulent jumps in the existing flowing area. However, concerning dispersion, it is not conditioned by the concentration gradient (as diffusion can be) nor even by a characteristic level of the global flow velocity (as turbulence can be). The dispersion flow is a result of the effects of the basic flow interaction with various discrete fixed or mobile forms that exist or appear along the flow trajectory [3.26]. The drops moving downward or upward in a flowing or stationary fluid, the bubbles flowing within a liquid, as well as an important roughness of the pipe walls, are some of the phenomena responsible for the dispersion flow. Another case is in a fluid flowing through a packed bed. In these examples, dispersion occurs because we have a microflow situation with a completely different intensity with respect to the basic flow. It is not difficult to observe that, for all the devices where a differential contact solid–fluid or solid–fluid–fluid or fluid–fluid occurs, the dispersion flow is the characteristic flow type. As for turbulence, the dispersion characterization associates a coefficient called *dispersion coefficient* to these microflows responsible for the dispersion phenomenon. When the dispersion participation is very important, the turbulence and molecular components of the vector of total property transport can be neglected. Consequently, we can write the following expression for the vector of the property transport:

$$\overrightarrow{J_{tA}} = \overrightarrow{w}\Gamma_A - D_l\overrightarrow{grad}\Gamma_A \qquad (3.96)$$

The equation of the ADF model flow can be obtained by making a particular species mass balance, as in the case of a plug flow model. In this case, for the beginning of species balance we must consider the axial dispersion perturbations superposed over the plug flow as shown in Fig. 3.31. In the description given below, the transport vector has been divided into its convective and dispersion components.

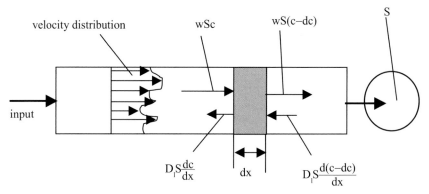

Figure 3.31 Scheme of the axial dispersion flow description.

After the particularization of the species balance in the control volume with section flow S and elementary length dx, as shown in Fig. 3.31, we obtain relation (3.97), which corresponds to the ADF model equation:

$$\frac{\partial c}{\partial \tau} = -w \frac{\partial c}{\partial \tau} + D_l \frac{\partial^2 c}{\partial x^2} \tag{3.97}$$

The axial dispersion flow model can be valid when we do not have the gradient of the property with respect to the normal flow direction. In other words, for this direction, we have a perfect mixing state. When this last condition is not met, we have to consider a flow model with two dispersion coefficients: a coefficient for the axial dispersion and another one for the radial dispersion. In this case, the flow model equation becomes:

$$\frac{\partial c}{\partial \tau} = -w \frac{\partial c}{\partial \tau} + D_l \frac{\partial^2 c}{\partial x^2} + \frac{D_r}{r} \frac{\partial}{\partial r} \left(r \frac{\partial c}{\partial r} \right) \tag{3.98}$$

The values of the dispersion coefficients will be established for most actual cases by experiments, which pursue the registration and interpretation of the exit time distribution of a signal that passes through a physical reduced model of the real device. However, in some cases, the actual device can be used. The method for identifying the dispersion coefficient [3.27, 3.28] is, in fact, the classical method of flow identification based on the introduction in the device input of a signal; (frequently as a δ impulsion or a unitary impulsion) the exit response is then recorded from its start until it disappears. It is evident that this experimental part of the method has to be completed by calculation of the dispersion model flow and by identification of the value of the dispersion coefficient. For this last objective, the sum of the square differences between the measured and computed values of the exit signal, are minimized.

For the mathematical solution of the dispersion model flow, we add the univocity conditions that include the signal input description for the initial conditions to Eq. (3.97) or (3.98). A more complete description of this mathematical model can be given with the example of the axial dispersion flow. In this case, we assign the dispersion flow conditions to the input and the exit of the apparatus Eq. (3.99) and to the model equation (3.97). The initial signal input conditions are given by relation (3.100) for the case of a δ signal and by relation (3.101) when a unitary impulse signal is used.

$$\bar{w}_1 c - D_1 \frac{\partial c}{\partial z} = 0 , \quad z = 0, \quad \tau \succ 0$$

$$D_1 \frac{\partial c}{\partial z} = 0 , \quad z = Hd, \quad \tau \succ 0 \tag{3.99}$$

$$c = 0 , \quad 0 \prec z \prec Hd, \quad \tau = 0$$

$$c = c_0 , \quad z = 0, \quad \tau = 0 \tag{3.100}$$

$$c = c_0 , \quad z = 0, \quad \tau \succ 0$$

$$c = 0 , \quad 0 \prec z \prec Hd, \quad \tau = 0 \tag{3.101}$$

For the unitary impulse signal (relation (3.100)) the axial dispersion flow model has an analytical solution:

$$\left(\frac{c}{c_0}\right)_{z=Hd} = 2 \sum_{n=1}^{\infty} \frac{\lambda_n \sin \lambda_n}{\lambda_n^2 + \left(\frac{Pe}{2}\right)^2 + \frac{Pe}{2}} \exp\left[\frac{Pe}{2} - \left(\frac{\lambda_n^2 + \left(\frac{Pe}{2}\right)^2}{Pe}\right)\theta\right] \tag{3.102}$$

where θ is the dimensionless time ($\theta = \tau/\tau_m$) and the proper values of λ_n are the solutions of Eqs. (3.103) and (3.104):

$$2\lambda_n tg\frac{\lambda_n}{2} = Pe \quad ; \quad n = 1, 3, 5, ..., 2k + 1, ... \tag{3.103}$$

$$2\lambda_n ctg\frac{\lambda_n}{2} = Pe \quad ; \quad n = 2, 4, 6, ..., 2k, ... \tag{3.104}$$

This solution can be used to set up the value of the Peclet criterion ($Pe = wH_d/D_1 = H_d^2/(D_1\tau_m)$) if we only consider the first term of Eq. (3.102). In this situation we obtain relations (3.105) and (3.106). It is not difficult to observe that, from the slope of relation (3.105), we can easily obtain the Pe value:

$$\ln\left(\frac{c}{c_0}\right)_{z=Hd} = \left[\frac{Pe}{2} - \frac{\lambda_1 + \left(\frac{Pe}{2}\right)^2}{Pe}\right]\theta + \left(\ln\frac{2\lambda_1 \sin \lambda_1}{\left(\frac{Pe}{2}\right)^2 + \frac{Pe}{2} + \lambda_1^2}\right) \tag{3.105}$$

$$2\lambda_1 tg\frac{\lambda_1}{2} = Pe \tag{3.106}$$

When the value of the mean flow velocity cannot be correctly estimated, as in the case of two or three phases contacting, the Pe number will be estimated considering the mean residence time (τ_m), the transport trajectory length (H_d) and the dispersion coefficient (D_l). For the case of a unitary signal impulse Eq. (3.100), the mean residence time will be estimated using relation (3.107):

$$\tau_m = \int_0^\infty \left(1 - \frac{c(\tau)}{c_0}\right) d\tau = \frac{\sum\limits_{i=1}^{N}(c_0 - c(\tau_i))\tau_i}{\sum\limits_{i=1}^{N}(c_0 - c(\tau_i))} \tag{3.107}$$

where N represents the number of the last appearance of $c_0 - c(\tau_i) \succ 0$ in the discrete data obtained.

Relation (3.108) gives the analytical solution of the axial dispersion model which contains relations (3.97), (3.99) and (3.100). Here the proper values of λ_k are the solutions of the transcendent equation (3.109):

$$\left(\frac{c}{c_0}\right)_{z=Hd} = \sum_{k=1}^\infty \frac{2\lambda_k \exp\left(\dfrac{Pe}{2} - \dfrac{Pe}{4}\theta - \dfrac{4}{Pe}\lambda_k^2\theta\right)}{\left(1 + \dfrac{Pe}{2}\right) + \dfrac{Pe}{2}\lambda_k \sin 2\lambda_k - \left(\dfrac{Pe}{4} - \left(\dfrac{Pe}{4}\right)^2\theta - \lambda_k^2\right)\cos 2\lambda_k} \tag{3.108}$$

$$tg2\lambda_k = \frac{\dfrac{Pe}{2}\lambda_k}{\lambda_k^2 - \left(\dfrac{Pe}{4}\right)^2} \tag{3.109}$$

If we consider the random variable theory, this solution represents the residence time distribution for a fluid particle flowing in a trajectory, which characterizes the investigated device. When we have the probability distribution of the random variable, then we can complete more characteristics of the random variable such as the non-centred and centred moments. Relations (3.110)–(3.114) give the expressions of the moments obtained using relation (3.108) as a residence time distribution. Relation (3.114) gives the two order centred moment, which is called random variable variance:

$$v_1 = 1 \tag{3.110}$$

$$v_2 = 1 + \frac{2}{Pe} + \frac{2}{Pe^2}e^{-Pe} - \frac{2}{Pe^2} \tag{3.111}$$

$$v_3 = 1 + \frac{6}{Pe} + \frac{6}{Pe^2} - \frac{24}{Pe^3} + \frac{18e^{-Pe}}{Pe^2} + \frac{24e^{-Pe}}{Pe^3} \tag{3.112}$$

$$v_4 = 1 + \frac{12}{Pe} + \frac{48}{Pe^2} - \frac{336}{Pe^4} - \frac{108e^{-Pe}}{Pe^2} + \frac{360e^{-Pe}}{Pe^3} + \frac{312e^{-Pe}}{Pe^4} + \frac{24e^{-Pe}}{Pe^4} \tag{3.113}$$

$$\sigma^2 = v_2 - v_1 = \frac{2}{Pe} - \frac{2}{Pe^2} + \frac{2}{Pe^2} e^{-Pe} \qquad (3.114)$$

Because the device response to the signal input is given by the discrete coupled data $c_i - \tau_i$, the mentioned moment, can be numerically computed as follows:

$$\tau_m = \frac{\sum\limits_i \tau_i c_i}{\sum\limits_i c_i} \qquad (3.115)$$

$$v_1 = \frac{\tau_m}{\tau_m} = 1 \qquad (3.116)$$

$$v_2 = \frac{\sum\limits_i \tau_i^2 c_i}{\tau_m^2 \sum\limits_i c_i} \qquad (3.117)$$

$$v_3 = \frac{\sum\limits_i \tau_i^3 c_i}{\tau_m^3 \sum\limits_i c_i} \qquad (3.118)$$

$$v_4 = \frac{\sum\limits_i \tau_i^4 c_i}{\tau_m^4 \sum\limits_i c_i} \qquad (3.119)$$

$$\sigma^2 = v_2 - v_1 = \frac{\sum\limits_i \tau_i^2 c_i}{\tau_m^2 \sum\limits_i c_i} - 1 \qquad (3.120)$$

The coupling of relations (3.110)–(3.114) with (3.115)–(3.120) shows four possibilities for the identification of the Pe number. With the same data, each possibility must produce an identical result for the Pe criterion.

3.3.5
Examples

Actually, it is acknowledged that all the main chemical engineering devices are well described by known equations and procedures to compute axial and radial mixing coefficients. As an example, we can remember the famous Levenspiel's equation to compute the axial mixing of a mono-phase flow in a packed bed ($Pe = wd_p/D_l = 2$), an equation verified by experiments. Undeniably, the problem of identifying a flow model can be developed using a laboratory model of the real device if the experiments can be carried out easily. For the construction of such a physical model, we must meet all the requirements imposed by the similitude laws. It is important to note that we assume that the laboratory model undergoes one or more changes in order to produce a flow model in accordance with the mathematical process simulation. Indeed, it contains a selected flow model, which produces the best results for the investigated process. Consequently, the most acceptable flow model has been indirectly established by this procedure.

In order to show how a chemical device can be scaled-up and how a solution to the problem of identifying the best flow model can be given, we suggest the following protocol:

(i) for an actual case (reaction, separation, heating, coupled transport, etc.), we can establish the best flow model (that guarantees the best exits for the fixed inputs in the modeled process) by using the mathematical modelling and simulation of a process; (ii) we design and build a laboratory device, that can easily be modified according to the results obtained from the experiments carried out to identify the flow model; (iii) with the final physical model of the device, we examine the performance of the process and, if necessary, we start the experimental research all over again to validate the model of the process; (iv) when the final physical model of the device is made, the scaled-up analysis is then started, the result of this step being the first image of the future industrial device. It is important to notice that the tricky points of an actual experimental research have to be discussed: (i) the input point of the signal and the exit point of the response must be carefully selected; (ii) the quality of the input signal must respect some requirements: indeed, if we use the expressions from Table 3.4 or those given by the assembly of Eqs. (3.80)–(3.88) to interpret the data, then the signal must be a δ or a unitary impulse; other signal types induce important difficulties for solving the flow model and for identifying the parameters; (iii) the response recording must be carried out with transducers and magnification systems which do not introduce unknown retardation times. This methodology will be illustrated in the next sections with some examples.

3.3.5.1 Mechanically Mixed Reactor for Reactions in Liquid Media

The physical model of the reactor is a 350 mm high cylindrical vessel, with a diameter of 200 mm and an elliptical bottom. The operation volume is: $V = 12 * 10^{-3}$ m^3. The entrance of the reactants is placed near the middle of the reactor, more exactly at 130 mm from the bottom. The reactor's exit is positioned on the top of the vessel but below the liquid level. At the vessel centre is placed a mixer with three helicoidal paddles with $d/D = 0.33$. It operates with a rotation speed of 150 min^{-1}. In order to establish the reactor flow model, this is filled with pure water which continuously flushes through the reactor at a flow rate of $6.6 * 10^{-5}$ m^3/s (similar to the reactants' flow rate). At time $\tau = 0$, a unitary impulse of an NaCl solution with a $c_0 = 3.6$ kg/m^3 is introduced into the reactor input. The time evolution of the NaCl concentration at the exit flow of the reactor is measured by the conductivity. Table 3.5 gives the data that show the evolution of this concentration at the reactor exit.

The proposal of an adequate flow model of the reactor and the identification of the parameters are the main requirements of this application. Solving this type of problem involves two distinct actions: first the selection of the flow model and second the computations involved in identifying the parameters.

Table 3.5 Evolution of the NaCl concentration at the reactor's exit.

i	1	2	3	4	5	6	7	8	9	10	11	12
τ_i (s)	0	1.8	3.6	4.8	18	36	72	108	144	180	216	252
$c_i = c(\tau_i)$	0	0	0	0.18	0.342	0.651	1.281	1.828	2.142	2.405	2.556	2.772

The selection of the flow model. In accordance with the description given above, we expect a flow model in which a small plug-flow region is associated with an important perfect mixing flow region, whereas a stagnant region is considered at the bottom of the reactor. If this proposal is correct, then, in the general CFM shown in Table 3.4 we must consider:

$G_{v1} = G_{v4} = G_v$; $G_{v2} = G_{v3} = 0$; $b3 = b1 = d1 = 0$; $b1 = b$; $d2 = d$ and $b + m + d = 1$. With $k = 0$ (because, here, the chemical reaction does not occur), the logarithmic transformation of relation (3.82), which gives the response $F(\theta) = c(\theta)/c_0$ for an unitary impulse, can be written as follows:

$$\ln\left(1 - \frac{c(\theta)}{c_0}\right) = -\frac{1}{m}(\theta - b) * \vartheta(\theta - b) \tag{3.121}$$

The computations to identify the parameters are given algorithmically, step by step, by the procedure below:

- We start with the computation of the value of the mean residence time: $\tau_m = V/G_v = 12 * 10^{-3}/6.66 * 10^{-5} = 180$ s.
- Using the data given in Table 3.5 we build the dependence relation $c_i = c(\theta_i)/c_0$ vs θ_i where $\theta_i = \tau_i/\tau_m$ and $C(\theta_i) = c(\theta_i)/c_0$. Table 3.6 presents this dependence. Here d_i values are also computed as $\ln(1 - c(\theta_i)/c_0)$ because they are needed for the flow model equation (3.121).

Table 3.6 Dimensionless NaCl concentration at the reactor's exit.

i	1	2	3	4	5	6	7	8	9	10	11	12
θ_i	0	0.01	0.02	0.03	0.1	0.2	0.4	0.6	0.8	1.0	1.2	1.4
$c(\theta_i)/c_0$	0	0	0	0.05	0.095	0.181	0.356	0.508	0.596	0.668	0.713	0.775
d_i	0	0	0	−0.05	−0.1	−0.2	−0.44	−0.71	−0.91	−1.1	−1.23	−1.45

- the graphic representation of $d_i = \ln(1 - C(\theta_i))$ vs θ_i in Fig. 3.32, shows that all the data do match a line with a slope equal to $-1/m$ and with the origin intersect at b/m. At the same time, function $\vartheta(\theta - b)$ shows its b value.

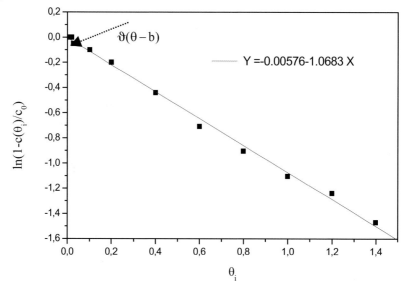

Figure 3.32 Evolution of the $\ln(1 - C(\theta_i))$ vs θ_i.

- the values of m, b and d can be calculated from Fig. 3.32 above, giving respectively m = 0.936, b = 0.00054, d = 0.06345. These results show that 6.34% of the reactor is a stagnant region and 0.054% is a plug-flow region, the remainder being a perfect mixing region.

3.3.5.2 Gas Flow in a Fluidized Bed Reactor

Catalytic butane dehydrogenation can be successfully carried out in a laboratory scale fluidized bed reactor operating at 310 °C and at atmospheric pressure. The catalytic particles have diameter 310 µm and density 2060 kg/m³. Such a reactor is 150 mm in diameter and has a fixed 500 mm long catalytic bed. When the catalyst bed is fluidized with butane blown at a velocity of 0.1 m/s, it becomes 750 mm thick.

The establishment of the flow model or a cold model of the reactor is carried out using air instead of butane and working at the same gas velocity as implemented to fluidize the catalytic bed. In these conditions, a slow motion of the solid, without any important bubbling phenomena, is observed at the bottom of the fluidized bed, while a bubbling phenomenon associated with violent solid motion occurs in the middle and upper parts of the fluidized bed. At time τ = 0, a unitary signal, which consists of replacing the air flow by an identical flow of pure nitrogen, is generated at the reactor input. Table 3.7 presents the evolution of the nitrogen concentrations at the bed exit.

Table 3.7 Evolution of the nitrogen concentration at the exit of the fluidized bed reactor.

i	1	2	3	4	5	6	7	8	9	10	11	12
τ_i sec	0	1	2	3	4	5	6	7	8	9	10	11
$c_i = c(\tau_i)$	0.79	0.79	0.79	0.79	0.874	0.924	0.947	0.973	0.981	0.984	0.994	0.996

To solve this example we use the same methodology as applied in the previous section: we begin with the flow model selection and finish by identifying the parameters.

The model selection. According to the description of the fluidization conditions given above, we can suggest a combined flow model of a plug-flow linked in series with a perfect mixing. We obtain the mathematical description of this CFM from the general CFM presented in Table 3.4 by: $G_{v1} = G_{v4} = G_v$; $G_{v2} = G_{v4} = 0$; $b_3 = b_1 = d_1 = 0$; $d_2 = d$ and $b + m + d = 1$. With this consideration, we have the simplified model characterized by relation (3.121). If the figure obtained from the graphical representation of $d_i = \ln(1 - C(\theta_i))$ vs θ_i is linear then, the proposed model can be considered as acceptable.

With *the computation* of the algorithm identifying the parameters of the model we obtain:

- The gas fraction of the bed:
  $\varepsilon = \varepsilon_0 + (H - H_0^\circ/H = 0.4 + 0.25/0.75 = 0.66$ m³ gas/m³ bed.
- The mean residence time: $\tau_m = \varepsilon H/w_f = 0.66 * 0.75/0.1 = 5$ s.
- The relation: $c(\theta_i)/c_0$ vs θ_i is computed in Table 3.8 and plotted. Here $c_{00} = 0.79$ kmol N$_2$/kmol gas and $c_0 = 1$ kmol N$_2$/ kmol gas. Table 3.8 also contains the computed line that shows the dependence of the $\ln(1 - C(\theta_i))$ vs θ_i.

Table 3.8 Evolution of the dimensionless nitrogen concentration at the reactor's output.

i	1	2	3	4	5	6	7	8	9	10	11	12
θ_i	0	0.2	0.4	0.6	0.8	1.0	1.2	1.4	1.6	1.8	2.0	2.2
$\dfrac{c(\tau_i) - c_{00}}{c_0 - c_{00}}$	0	0	0	0	0.399	0.643	0.747	0.874	0.924	0.954	0.975	0.983
$\ln(1 - C(\theta_i))$	0	0	0	0	−0.51	−1.03	−1.55	−2.07	−2.58	−3.09	−3.56	−4.03

- Figure 3.33 shows that the hypothesis of a CFM composed of a series of PF and PM is acceptable because the experimental dependence of $\ln(1 - C(\theta_i))$ vs θ_i is linear and we clearly observe the function $\vartheta(\theta - b)$.

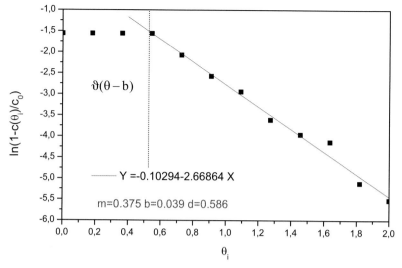

Figure 3.33 The evolution of $\ln(1 - C(\theta_i))$ vs θ_i.

- As in the case of application 3.3.5.1, we identify m = 0.375 and
  d = 0.586. The value b = 0.039, shown in Fig. 3.33, comes from
  the origin intersect of the line $Y = -2.66864X - 0.10294$, this
  value does not have a special significance. The d value can be
  increased with the b value. So it will become d = 0.625.

3.3.5.3 Flow in a Fixed Bed Catalytic Reactor

The laboratory scale physical model of the catalytic sulfur dioxide oxidation is a
0.05 m-diameter reactor containing 3 mm-diameter pellets of catalyst over a
height of 0.15 m. The bed is flushed through at 430 °C by a gas flow that contains
0.07 kmol SO_2/kmol total gas, 0.11 kmol O_2/kmol total gas and 0.82 kmol
N_2/kmol total gas. The gas spatial velocity is 0.01 m/s.

In order to obtain a reactor model flow that characterizes the gas movement
around the catalyst grains, a current of pure nitrogen is blown through the fixed
catalyst bed at the same temperature and pressure as in the reaction. At $\tau = 0$ we
apply a signal (unitary impulse) to the reactor input introducing a gas mixture
containing nitrogen and sulfur dioxide with a concentration of $c_0 = 0.1$ kmol
SO_2/kmol gas. Then, we measure the evolution of the sulfur dioxide concentration at the reactor exit. Table 3.9 gives these measured concentrations. In this
case, it is necessary to validate if the collected data verify a PF model. If they do
not, we have to identify the parameters of the axial mixing model to correct the PF
model.

Table 3.9 Evolution of the sulfur dioxide concentration at the exit of the reactor.

i	1	2	3	4	5	6	7	8	9	10	11	12	13	14
τ_i (s)	0	1.2	2.4	3.0	3.6	4.8	6	7.2	8.4	9.6	10.8	12	13.2	14.4
$c_i = c(\tau_i)$	0	0	0	0	0.01	0.03	0.06	0.08	0.085	0.09	0.095	0.097	0.099	0.1

Then, for this application, we directly start with the computations that serve to identify the parameters.

The computation of the values of the following parameters and relations are needed to solve the problem:

- the mean residence time of an elementary fluid particle in the catalyst bed $\tau_m = (\varepsilon_0 H)/w_l = 0.4 * .015/0.01 = 6$ s;
- the dimensionless dependence $C(\theta_i) = c(\theta_i)/c_0$ vs θ_i is computed and reported in Table 3.10;

Table 3.10 Evolution of the dimensionless signal at the exit of the reactor.

i	1	2	3	4	5	6	7	8	9	10	11	12	13	14
θ_i	0	0.2	0.4	0.5	0.6	0.8	1.0	1.2	1.4	1.6	1.8	2.0	2.2	2.4
$\dfrac{c(\theta_i)}{c_0}$	0	0	0	0	0.1	0.3	0.6	0.8	0.85	0.90	0.95	0.98	0.99	1.00

- the graphic representation of the dependence $C(\theta_i) = c(\theta_i)/c_0$ vs θ_i is needed to appreciate whether we have a PF or ADF flow type. Figure 3.34 clearly shows that here, an ADF flow model type can be adequate;

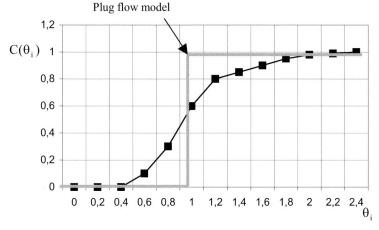

Figure 3.34 Evolution of the dimensionless concentration at the exit of the reactor.

- for the computation of the axial dispersion coefficient, we use an approximate calculation introduced by Eqs. (3.105) and (3.106). These relations are coupled with the numerical data reported in Table 3.10 and then we form the function here given by relation (3.122). It minimizes the sum of the squares of the differences between the computed and experimental values of $C(\theta_i)$. We further show that this problem of axial dispersion coefficient identification is transformed into a variant of a least squares method for parameter identification.

$$F(\lambda_1) = \sum_{i=1}^{14} \left[\sum_{i=1}^{14} \left(\frac{\lambda_1 tg^2 \frac{\lambda_1}{2} - 1}{2tg\frac{\lambda_1}{2}} \right) \theta_i + \ln \frac{2\sin\lambda_1}{\lambda_1 tg^2 \frac{\lambda_1}{2} + tg\frac{\lambda_1}{2} + \lambda_1} - \ln\left(\frac{c(\theta_i)}{c_0}\right)_{exp} \right]^2$$

$$= \min$$

$$(3.122)$$

With $a = \dfrac{\lambda_1 tg^2 \frac{\lambda_1}{2} - 1}{2tg\frac{\lambda_1}{2}}$, $b = \ln \dfrac{2\sin\lambda_1}{\lambda_1 tg^2 \frac{\lambda_1}{2} + tg\frac{\lambda_1}{2} + \lambda_1}$ and $y_i = \ln\left(\dfrac{c(\theta_i)}{c_0}\right)_{exp}$ the

minimization requested by the problem (3.122) takes the common form

$$F(a, b) = \sum_{i=1}^{14} (a\theta_i + b - y_i)^2 = \min,$$ which has the quality to accept a very simple

solution. So, in order to obtain "a" and "b", we must solve the equation system (3.123). From these values, we obtain the value of λ_1. Now, using relation (3.106) we can calculate the Peclet dispersion number.

$$\begin{cases} Nb + a \sum_{i=1}^{N=14} \theta_i = \sum_{i=1}^{N=14} y_i \\ b \sum_{i=1}^{N=14} \theta_i + a \sum_{i=1}^{N=14} \theta_i^2 = \sum_{i=1}^{N=14} y_i\theta_i \end{cases}$$

$$3.123$$

Table 3.11 shows the computation of the Pe number and of the axial dispersion coefficient by the direct minimization of relations (3.122) and (3.106). It is easily observable that this table contains in fact a MathCAD transposition of the λ_1, Pe and D_l identification.

Table 3.11 MathCAD computation of Pe, λ_1, D_l parameters.

$$\theta_i = \begin{bmatrix} 0 \\ 0.2 \\ 0.4 \\ 0.5 \\ 0.6 \\ 0.8 \\ 1.0 \\ 1.2 \\ 1.4 \\ 1.6 \\ 1.8 \\ 2.0 \\ 2.2 \\ 2.4 \end{bmatrix} \qquad \frac{c_i}{c_0} = \begin{bmatrix} 0 \\ 0 \\ 0 \\ 0 \\ 0.1 \\ 0.3 \\ 0.6 \\ 0.8 \\ 0.85 \\ 0.9 \\ 0.95 \\ 0.98 \\ 0.99 \\ 1.0 \end{bmatrix} = \frac{Pe := 1 \quad \lambda_1 := 19 \quad i := 5..14 \quad H := 0.16 \quad \tau_m = 6}{G}$$

$$\left[\sum_{i=5}^{14} \left[\frac{\lambda_1 \cdot \left(\tan\left(\frac{\lambda_1}{2}\right)\right)^2 - 1}{2 \cdot \tan\left(\frac{\lambda_1}{2}\right)} \cdot \theta_i + \ln\left[2 \frac{\sin(\lambda_1)}{\lambda_1 \left(\tan\left(\frac{\lambda_1}{2}\right)\right)^2 + \tan\left(\frac{\lambda_1}{2}\right) + \lambda_1} \right] - \ln\left(\frac{c_i}{c_0}\right) \right]^2 \right] \equiv 0$$

$$2\lambda_1 \tan\left(\frac{\lambda_1}{2}\right) \equiv Pe \qquad \binom{\lambda_1}{Pe} := Find(\lambda_1, Pe) \qquad \binom{\lambda_1}{Pe} = \binom{19.48}{12.7}$$

$$D_l = \frac{H^2}{Pe\tau_m} \qquad D_l = 2.953 * 10^{-4} \ m^2/s$$

3.3.6
Flow Modelling using Computational Fluid Dynamics

As has been shown at the beginning of this chapter, researchers have been expecting important progress on the modelling of flows in chemical reactors with the development of computational fluid dynamics (CFD). The principle of CFD is to integrate the flow equations for one particular case after dividing the flow volume into a very high number of differential elements. This volume-of-fluid technique can be used for the "*a priori*" determination of the morphology and characteristics of various kinds of flow.

Chemical engineers were not the pioneers in this field because chemical engineering flow problems can be very complex. Some of the first users of CFD were car, plane and boat designers. One of the reasons for this was that CFD could tell the designers exactly they wanted to know, that is the flow patterns obtained while their new designs moved. Indeed, the possibility to use Euler's equations for flow description has been one of the major contributions to the development of these applications. These kinds of CFD techniques have also been projected and have been successfully used to analyze heat flow from a body immersed into the flowing fluid [3.29, 3.30].

As far as chemical engineers are concerned, we must notice that there is a considerable academic and industrial interest in the use of CFD to model two-phase flows in process equipment. The problem of the single bubble rising in the fluid [3.31, 3.32] has been resolved using some simplification in the description of bubble–liquid momentum transfer. Considerable progress has been made in the CFD modelling of bubbling gas-fluidized beds and bubble columns. The CFD modelling of fluidized beds usually adopts the Eulerian framework for both dilute (bubble) and dense phases and makes use of the granular theory to calculate the rheological parameters of the dense phase [3.33–3.37].

The use of CFD models for gas–liquid bubble columns has also raised considerable interest; only Euler-Euler and Euler-Lagrange frameworks have been employed for the description of the gas and liquid phase states [3.38–3.42]. Bubble trays, considered as particular kinds of bubble columns, have lately presented enormous interest for the flow description by CFD. The flow patterns on a sieve tray have been analyzed in the liquid phase, solving the time-averaged equations of continuity and momentum [3.43].

The jump to the fully two-phase flow on a sieve tray requires the acceptance of some conditions [3.44]:
- the lift forces for the bubble must be neglected;
- the added mass forces do not have an important participation in the flow processes;
- the interphase momentum exchange must be expressed using the drag coefficient.

Then the simulation of real chemical engineering flows concerns a number of important difficulties beyond the pattern of turbulent flows. One of these complex problems concerns the description of viscosity; however, this can be resolved using rheological equations. Another difficulty is the so-called micromixing problem, which must be characterized at the level considering the integration of a very little unit.

In the case of one homogeneous reactor, where two reactants are continuously fed, mixed, reacted, and flushed out through an outlet, CFD can calculate the concentrations in each fluid element, just as it can calculate the temperature. Nevertheless, CFD cannot consider the reaction of both components as a function of the local mixing

Theoretically, CFD could quantify everything. It could predict the effect of adding reagents quickly or slowly. To achieve a specified yield, we would find out exactly how slow the addition has to be, how intense the mixing is, and what equipment would achieve that mixing. But to get a good prediction, as always, you need good input data. These data include the initial conditions, rate flows and kinetics of the reactions as well as the physical properties of the solutions. In order to get good inputs, of course, it is necessary to come back to laboratory activity.

3.4
Complex Models and Their Simulators

During a process modelling, the development of the model and the simulation of the process using a simulator, as shown in Section 2.2, represent two apparently indivisible operations. Both activities have rapidly evolved with time as a consequence of the development of basic technical sciences. Three main phases can be kept in mind with respect to this vigorous evolution:

In *the first phase,* the modelling and simulation of the apparatus was carried out considering each as independent units in the whole installation. Indeed, here, modelling was assisted by the efforts made in the high technological design of each of the specific apparatuses found in chemical plants. All types of models have been used for this purpose and the current huge computation capacities of universities and of research-design centres have sustained these scientific efforts. At present, the theoretical basis and algorithmic implementation of process modelling based on transport phenomena have been established. The general theory of computer programming has given the fundamentals of the development of easily usable means for the transposition of the models into process simulators and as guidelines for designers. Various utilitarian software languages have backed this new scientific branch and, among them, FORTRAN (FORmula TRANslation) can be considered as the most notorious. It is estimated that the full start of this phase began around 1968, when the series production of high power computers started.

The second phase began with the start of commercial activities in the modelling and simulation of processes. These commercial activities were born in the USA in 1980–1985 when the first simulators for oil distillation appeared. DistillR™, Maxstill™, and Hysim™ are some of these scientific software packages, which reach the level of the interactive simulation of a complex process model. During these years, modelling and simulation succeeded in automatically assembling the parts of a complex model according to the formal description of each part and their links. At the same time, an important data base began to be fed with the description of the different unitary operations of processes in terms of physical and chemical properties, consumption kinetics or appearance and equilibrium distribution at interphases.

The modelling tools in current commercial simulators may roughly be classified into two groups depending on their approach: block-oriented (or modular) and equation-oriented.

Block-oriented approaches mainly address the modelling at the level of flowsheets. Every process is abstracted in a block diagram consisting of standardized blocks, which model the behaviour of a process unit or a part of it. All the blocks are linked by signal-like connections representing the flow of information, material and energy, employing standardized interface and stream formats. Models of process units are pre-coded by a modelling expert and incorporated into a model library for later use. The modelling at the level of the flow-sheet is either supported by a modelling or by a graphical language. In both cases, the end user selects the models from the library, provides the model parameters and connects them to the plant model. However, the chemical engineering knowledge accumulated up to now, as well as the structure of the models, are easily accessible. Common exceptions include the models of physical properties, which can be selected in the literature independently from the process unit model.

Equation-oriented modelling tools support the implementation of the unit models and their incorporation into a model library by means of declarative modelling languages or by providing a set of subroutine templates. In this case, the tools for the modelling expert or for the end user are similar. Hence, modelling at the unit level requires a profound knowledge in such diverse areas as chemical engineering modelling and simulation, numerical mathematics, and computer science. The development of new process models is therefore often restricted to a small group of experts.

Figure 3.35 presents the page of a modern commercial simulator (Hysim™ 1995) where we can identify the different elements of the process specified in this block-oriented simulator.

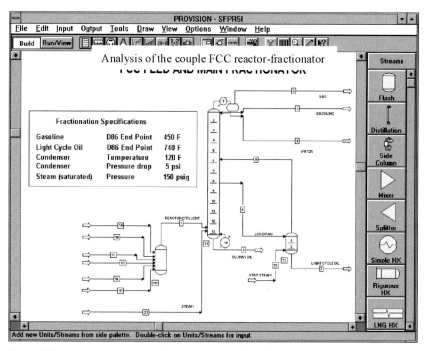

Figure 3.35 Presentation page of a block-oriented simulator for the analysis of a coupled FCC reactor-fractionator (Hysim™ 1995).

It is not difficult to observe that, in this example, we have the coupling of a specific reactor for petroleum fractionation together with a complex distillation column. If we intend to show the complexity of the process that will be simulated, then it is important to say that more than 20 components can be found in the input of this installation in the reactor, separation column, condenser, reboiler and external flash device.

Despite the considerable progress made over the last decade, when steady-state flow-sheeting with modular process simulators became routinely employed by a large community of process engineers, there is considerable incentive to extend the range of model-based applications by improving the handling of models and by increasing the level of detail for representing the processes. Many studied cases of process engineering – not only in academia but especially in the research and development laboratories of the chemical industry – have shown the potential of employing non-standard models such as dynamic models, extremely detailed models of standard equipment, or models of non-standard equipment. The modular approach to modelling and simulation, though powerful and easily accessible to many engineers for the solution of standard flow-sheet problems, does not adequately support the solution of more complicated problems. This is largely due to the lack of pre-coded models for many unit operations at an adequate level of detail. In addition, most of the coded models neglect the mass or heat transfer, assuming the equilibrium state at interphases. Examples of models that are not available in present simulators, include multiphase reactors, membrane processes, polymerization reactors, biochemical reactors, hydrodynamic separators and the majority of units involving particulates. Therefore, costly and time-consuming model development for a particular unit is often required in some projects.

Equation-oriented languages largely contribute to the implementation of models, but they do not assist the user in developing the types of models that use engineering concepts. Indeed, equation-oriented languages are not useful in providing for the documentation of the modelling process during the lifecycle of a process or for the proper design and documentation of model libraries. In consequence, the reuse of previously validated models of a unit by a new group of users is then almost impossible.

The consistency and reliability of well designed model libraries is inevitably getting lost over time. Now, even though the market for these simulators is in full evolution, spectacular progress is not expected because the basic models of the units stay at mesoscale or macroscale.

Despite this last observation, for this type of simulation and modelling research, two main means of evolution remain: the first consists in enlarging the library with new and newly coded models for unit operations or apparatuses (such as the unit processes mentioned above: multiphase reactors, membrane processes, etc.); the second is specified by the sophistication of the models developed for the apparatus that characterizes the unit operations. With respect to this second means, we can develop a hierarchy dividing into three levels. The first level corresponds to connectionist models of equilibrium (frequently used in the past). The second level involves the models of transport phenomena with heat and mass transfer kinetics given by approximate solutions. And finally, in the third level, the real transport phenomena the flow, heat and mass transport are correctly described. In this last case, oversimplifying hypotheses, such as non-resistive interfaces, are avoided

The third phase of the evolution for the modelling and simulation activities is represented by the current consumption of commercial software by scientific education. High level instruction languages such as MathCAD™, Matlab™, the CFD software, finite element softwares (for the integration of complex differential equation systems), high data volume graphic processors and softwares based on artificial intelligence represent some examples that show the important evolution of education and scientific research by modelling and simulation.

Considering the complexity and the diversity of the problems in chemical engineering research and design and taking into account the present evolution of modelling and simulation, we cannot claim that it will be possible to use universal chemical engineering simulators in the future.

The experimental researchers and the scientists that are only interested in in-depth modelling of physical phenomena are not attracted by complex simulators. The former seek models for data interpretation; the latter create models to propose solutions for a good knowledge of a concrete case. From other viewpoints, chemical engineering, because of its diversity, includes countless models. Most of them are quite interesting when they can add a lot more new situations based on particularization or modification to their starting cases. As far as the situations of this subject in chemical engineering are quite varied, it will be interesting to describe new modelling and simulation examples in the following sections. The examples shown below demonstrate firstly, how a model based on transport phenomena equations is developed and secondly, how we can extract important data for a process characterization by using a model simulation.

3.4.1
Problem of Heating in a Zone Refining Process

Among the methods of advanced purification of a crystallized of amorphous solid material, the zone refining methods occupy an important place. The principle of the method is based on the fact that an impurity from the processed material in a melting crystallization process, according to the distribution law, presents different concentrations in solid and liquid phases [3.45]. If the melting solid (liquid phase) is subjected to a movement along a stick, then the impurity will be concentrated in the position where the liquid phase is stopped. This process is also called refining. To make the solid melt and to move the melt, the solid is locally heated by means of a mobile heat inductor or a small mobile and cylindrical electrically heated oven. However, we can reach the same result by pulling the stick through a small heating source. From the heat transfer viewpoint, this example corresponds to a conductive non-steady state heating with an internal heating source (heat inductor) or with an external heating source (heating with an oven).

From the mathematical viewpoint, it is important to assume that a very rapid heat transfer occurs at the extremities of the stick, and that a rapid cooling system is activated when the heating source is stopped. In addition, as far as we only take into account the melted region, we do not consider the coupling with the liquid–solid phase change.

Figure 3.36 shows the heating principle of the zone refining purification procedure and also introduces the geometric and material conditions that characterize the process. It also shows how the stick transfers heat to the contiguous medium. For a correct introduction to this problem, we assume that the production of heat by the inductor has Gaussian behaviour, so, for the heat generation rate, we can write Eq. (3.124) where the source amplitude (watt/m³) is A, $f(\tau)$ is a dimensionless function that keeps the maximum temperature for the inductor constant and k_1 and k_2 are the constants with L^{-1} dimension:

$$Q_g = Af(\tau)\exp\left(-(k_1(z - k_2 w\tau)^2)\right) \tag{3.124}$$

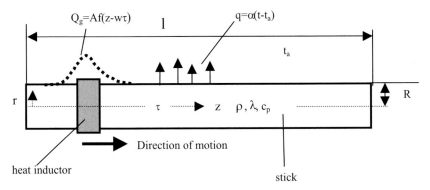

Figure 3.36 Heating scheme for a solid stick purified by a zone refining process.

From Fig. 3.36 we observe that the stick is characterized by its density ρ, thermal conductivity λ and sensible heating capacity c_p. The geometric dimensions of the stick are radius R and length l. The temperature distribution inside the stick results from relation (3.125) as a particularization of Eq. (3.6):

$$\rho c_p \frac{\partial t}{\partial \tau} = \lambda\left(\frac{\partial^2 t}{\partial r^2} + \frac{1}{r}\frac{\partial t}{\partial r} + \frac{\partial^2 t}{\partial z^2}\right) + Q_g \tag{3.125}$$

The univocity conditions that complete this general mathematical model can be written as follows:
• the initial distribution of the temperature into the stick:

$$\tau = 0 \quad -R \leq r \leq R \quad t = t_a \tag{3.126}$$

• the boundary thermal flux expression (type III conditions from the general class of boundary conditions):

$$\tau \succ 0 \quad r = -R \quad r = R \quad 0 \prec z \leq l. \quad -\lambda\frac{\partial t}{\partial r} = q = \alpha(t - t_a) \tag{3.127}$$

- the expression that gives the behaviour of the device of heat absorption placed at the stick extremity:

$$\tau \succ 0 \quad z = 1 \quad -\lambda\frac{\partial t}{\partial z} = k_{ab}(t_{ab} - t_{rab}) \tag{3.128}$$

For a correct perception of relation (3.128), we must notice that this is a heat sink that keeps its constant temperature due to a rapid heat exchange between the surface with a cooling medium maintained at constant (t_{rab}) temperature. The assembly of relations (3.124)–(3.126) represents in fact an abstract mathematical model for the above described heating case because the numerical value is given neither for the system geometry nor for the material properties. Apart from the temperature, all the other variables of the model can be transformed into a dimensionless form introducing the following dimensionless coordinates:

- the dimensionless time $T = \dfrac{\lambda\tau}{\rho c_p r^2}$ sometimes called Fourier number;
- the dimensionless radius coordinate $X = r/R$;
- the dimensionless axial coordinate $Z = z/l$;

With these transformations, the abstract model can now be described by assembling the following relations (3.129)–(3.133):

$$Q_g = Af\left(\frac{T\rho c_p R^2}{\lambda}\right)\exp\left(-k_1 Z * 1 - k_2 w \frac{T\rho c_p R^2}{\lambda}\right)^2 \tag{3.129}$$

$$T \succ 0 \quad X = -1 \quad X = 1 \quad 0 \prec Z \le 1 \quad -\frac{\lambda}{R}\frac{\partial t}{\partial X} = q = a(t - t_a) \tag{3.130}$$

$$T \succ 0 \quad Z = 1 \quad -\frac{\lambda}{1}\frac{\partial t}{\partial Z} = k_{ab}(t_{ab} - t_{rab}) \tag{3.131}$$

$$T \succ 0 \quad X = -1 \quad X = 1 \quad 0 \prec Z \le 1 \quad -\frac{\lambda}{R}\frac{\partial t}{\partial X} = q = a(t - t_a) \tag{3.132}$$

$$T \succ 0 \quad Z = 1 \quad -\frac{\lambda}{1}\frac{\partial t}{\partial Z} = k_{ab}(t_{ab} - t_{rab}) \tag{3.133}$$

Then, the heating model of the stick can simply be transposed by an adequate software for process simulation. Indeed, some conditions have to be chosen: the material properties (λ, ρ, c_p); the dimensionless stick geometry; the parameters of the heating source (A, k_1, k_2, w) and the external heat transfer parameter (a). The FlexPDE/2000™ simulator (PDE Solutions Inc. USA) based on the finite element method for integration of partial differential equations or systems has been used for the development of the simulation program. The simulator can give the results in various graphic forms. The source text of the program used to solve this model (Fig. 3.37) shows a very attractive macro language.

```
Title Heating in the zone refining
Coordinates: cylinder('Z','X')
select
cubic { Use Cubic Basis }
variables  temp(range=0,1800)
definitions  λ = 0.85 {thermal conductivity}  c_cp = 1 { heat capacity } long = 18

radius=1
   α = 0.4  {free convection boundary coupling}  Ta = 25 {ambient temperature}
   A = 4500   {amplitude}
source = A*exp(-((z-1*t)/.5)**2)*(200/(t+199))
   initial value
   temp = Ta
equations
   div( λ *grad(temp)) + source =   c_p *dt(temp)

boundaries region 1 start(0,0)
      natural(temp) = 0 line to (long,0)
      value(temp) = Ta line to (long,1)
      natural(temp) = - α *(temp - Ta) line to (0,1)
      value(temp) = Ta line to finish
   feature
      start(0.01*long,0) line to (0.01*long,1)

time -0.5 to 19 by 0.01
monitors
   for t = -0.5 by 0.5 to (long + 1)
   elevation(temp) from (0,1) to (long,1) range=(0,1800) as "Surface Temp"
   contour(temp)
plots
   for t = -0.5 by 0.5 to (long + 1)
   elevation(temp) from (0,0) to (long,0) range=(0,1800) as "Axis Temp"
histories
   history(temp) at (0,0)  (4,0)  (8,0)  (12,0)  (16,0)  (18,0)
   end
```

Figure 3.37 FlexPDE™ text for the example 3.4.1.

The first simulations present the heating dynamics along the stick, i.e. the evolution of temperature with time for two points positioned at $X = 1$ (surface of the stick) and $X = 0$ (stick centre). If we note the temperature range from Figs. 3.38–3.41 as well as the values of the material properties we see that the simulated heating case corresponds to the zone refining of a material with a very high melting point such as an inorganic material (silicium). Figure 3.38, presents the time motion of the heating front along the stick. It is easy to observe how the temperature increases in each point of the stick due to heating. After the passage of the heating inductor along the stick, the temperature rapidly decreases due to the axial and radial heat transport. This local heating dynamics (heating followed by a good cooling resulting from a high temperature difference) can also be observed at the stick extremities. Consequently, all the temperature curves present an important elongation to the right part where the heat sink at constant temperature

t_a is placed, at the end of the stick. At the same time, Fig. 3.38 shows the importance of the heat flowing in the radial direction, from the centre to the external
medium. Using Figs. 3.39 and 3.40, we can compute the radial and the axial
temperature gradient with the time values from Fig. 3.38. For example for T = 5
and 1/R = 10(middle of the stick) the radial temperature gradient is dt/dX =
330 drg/ul (ul = units of length); this value is larger than the axial temperature
gradient that, in this case, is dt/dZ = 199 drg/ul.

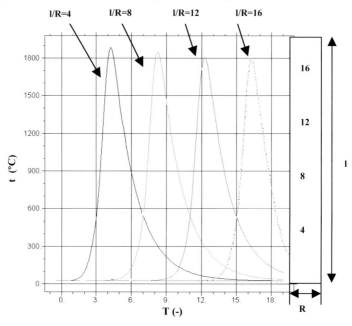

Figure 3.38 Evolution of the temperature for some points along the heated stick.

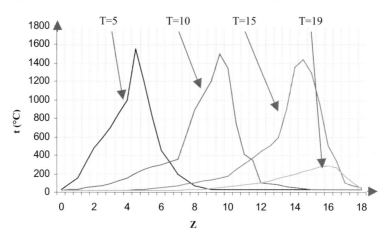

Figure 3.39 Evolution of the temperature of the surface of the stick along its
length for various dimensionless times.

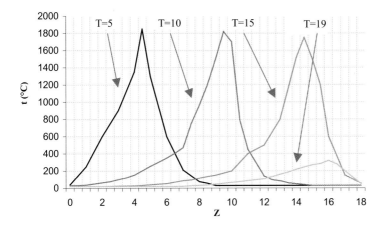

Figure 3.40 Evolution of the temperature at the centre of the stick and along its length and for various dimensionless times.

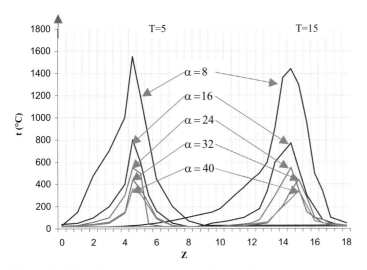

Figure 3.41 Effect of the external flow on the stick heat flow.

With respect to the data contained in Figs. 3.38–3.41, we can rapidly make a conversion to a concrete situation. For example for a stick with $\lambda = 2$ watt/(m deg), $\rho = 3000$ kg/m^3, $c_p = 2000$ j/(kg deg) and with a radius R = 0.01 m, the dimensionless time T = 5 corresponds to a real time $\tau = 1500$ s. If the stick is 100 cm long, then, with an inductor motion speed of 10 dimensionless units for 1500 s (see the distances between the maximums of the temperature in Figs. 3.38–3.40) we obtain a value of 15000 s for the time corresponding to the motion of the inductor along the stick. This shows that the zone refining purification is not an efficient method as far as time is concerned. Before closing these observations concerning heating with an inductor, we show that, in this case, the radial heat flow is permanently oriented from the stick to the adjacent medium. On the contrary, for the case of an electric oven heating, the direction of the heat flow is from the outside to the inside in the region of the stick covered by the oven.

The second simulation has been oriented to show the effect of an external flow around the stick on the thermal dynamics. It is known that the external flow around an entity, where a transport property is occurring, has a direct action on the coefficient of transfer of the property which characterizes the passing through the interface [3.3, 3.4, 3.45, 3.46]. The data obtained with the simulator when we change the values of the heat transfer coefficient from the stick to the external medium allows a quantitative estimation of the effect of this parameter. For this simulation, the temperature of the surface of the stick is considered as a dependent variable of the process. The heat source, the heat transfer coefficient α for the external fluid flow around the stick, the material properties and the stick geometry represent the independent variables of the process. Figure 3.41 shows the evolution of the stick heat flow for two values of the dimensionless time: $T = 5$ and $T = 15$. A spectacular reduction of the temperature of the stick surface occurs when the external flow becomes higher and the value of the heat transfer increases from $\alpha = 8$ w/(m²drg) to $\alpha = 40$ w/(m²drg). This phenomenon shows that an easy control of the stick-cooling rate is possible with the variation of α. Indeed, this fact can be very important for an actual process [3.45].

The third simulation example concerns the descriptive model of the cooling process of a hot stick that is maintained in a large volume of air. In the initial stages of the process, one of the stick's ends is maintained at high temperature for sufficient time for it to reach a steady state. The distribution of the temperature along the stick can then be calculated by relation (3.134). In a second step, the stick is placed in air and an unsteady cooling process starts. Concerning relation (3.134), we can notice that t_0 is the temperature of the heated end of the stick, and that all the other parameters have already been defined by the equations described above at the beginning of this section.

$$t = t_a + (t_0 - t_a)\exp\left(-\sqrt{\frac{2\alpha}{\lambda R}}z\right) \tag{3.134}$$

In this case, the simulator's text given in Fig. 3.37 has been modified, first considering $A = 0$, this statement is equivalent to the elimination of the source, and secondly by choosing the relation (3.134) as the initial value of the variable of the program named **temp**. The simulation results given in Figs. 3.42 and 3.43, show that a rapid cooling of the stick takes place, this phenomenon is mainly caused by the external conditions.

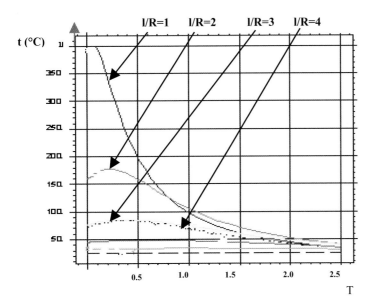

Figure 3.42 Cooling dynamics of the stick for high t_0 and small heat transfer coefficient ($\alpha = 8$ w/(m²drg).

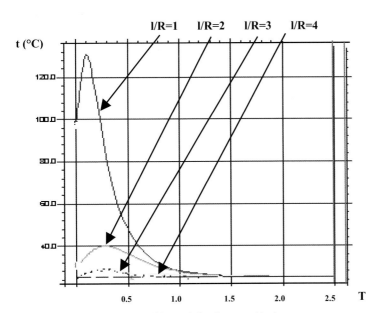

Figure 3.43 Cooling dynamics of the stick for slow t_0 and high heat transfer coefficient ($\alpha = 32$ w/(m²drg).

The zone refining process is extremely efficient for separating liquid or solid mixtures. In the old days, it was essentially applied in purifying germanium to be used in transistors. In multi-pass zone refining processes, the purification is carried out by slowly moving a series of closely spaced heaters along a relatively long solid ingot as shown in Fig. 3.36 or by restarting the heater movement when it reaches the end of the stick. The multi-pass zone refining process allows time saving because the following crystallization begins before the preceding one is completed. Many useful purifying operations can be carried out if the number and the size of the zones are properly selected. The distribution of impurities along an ingot depends on the value of the distribution coefficient, on the length of the molten zone, and on how many times the heaters move along the stick. Zone refining with a variable zone length is a topical scientific subject.

3.4.2
Heat Transfer in a Composite Medium

The description of heat transfer through a composite material can be a rather complicated task because this composite solid medium can contain various solids which are not uniformly dispersed and which have different thermal conductivity and sensible specific heat. Indeed, if we have a discrete setting of various solids in the total solid, the problems of heat transport become very complex when the number of solids and the number of agglomerations increase. These cases of totally or partly disordered composite media are not dealt with in this section. In an ordered solid composite medium, the heat can be generated or accumulated, captured or eliminated at the boundaries by a molecular-like mechanism. When the carriers pass from one zone of the solid to another, they change the frequency of discrete motion and the pathway length of each individual species because of the local properties. The heating or cooling problems of a block composed of two or more bricks (parallelepiped or other form) that exchanges energy with the adjacent medium represent the concrete case considered here. In the first modeling problem, we consider the case of a block of four bricks with different thermal conductivity, sensible specific heat and density. It is heated by a source with a Gaussian heat flow placed at the centre of the group of bricks. The group exchanges heat with the external medium at the upper and lower surface. At the surface level, the external medium is considered to be perfectly mixed and indeed, without any transfer resistance. No heat flow leaves the other block surfaces because they are completely isolated. The upper and lower contact surfaces of the bricks do not introduce any additional heat transfer resistance, so here the instantaneous heat flux equality is *a priori* accepted. The study of this model is attractive because: (i) the descriptive model given here can be explained by an interesting mathematical model; (ii) no significant problems are encountered if we carry out support modifications in order to find other important heat transfer cases; (iii) by analogy, we can obtain the data with respect to the characterization of some mass transfer cases occurring in a similar way.

For the general mathematical model construction, we consider the system of coordinates, the geometrical dimensions, the material properties and the initial temperature distribution for the block. Figure 3.44 gives a graphical introduction to the descriptive process model. We can now proceed with the particularization of the transport phenomena equations. Indeed, the concrete general mathematical process model contains:

- the partial differential equation that gives the temperature distribution in the solid block:

$$\rho c_p \frac{\partial t}{\partial \tau} = \lambda \left(\frac{\partial^2 t}{\partial x^2} + \frac{\partial^2 t}{\partial y^2} + \frac{\partial^2 t}{\partial z^2} \right) + Q_g \tag{3.135}$$

- the geometric and material conditions:

on the right region:

$$0 \prec x \prec 1 \ , \ -L \prec z \prec 0 \ , \ -\frac{h}{2} \prec y \prec \frac{h}{2} \ , \ \lambda = \lambda \ , \ c_p = c_{p1} \ , \ \rho = \rho_1 \tag{3.136}$$

$$0 \prec x \prec 1 \ , \ 0 \prec z \prec L \ , \ -\frac{h}{2} \prec y \prec \frac{h}{2} \ , \ \lambda = \lambda_2 \ , \ c_p = c_{p2} \ , \ \rho = \rho_2 \tag{3.137}$$

on the left region:

$$-1 \prec x \prec 0 \ , \ -L \prec z \prec 0 \ , \ -\frac{h}{2} \prec y \prec \frac{h}{2} \ , \ \lambda = \lambda_3 \ , \ c_p = c_{p3} \ , \ \rho = \rho_3 \tag{3.138}$$

$$0 \prec x \prec 1 \ , \ 0 \prec z \prec L \ , \ -\frac{h}{2} \prec y \prec \frac{h}{2} \ , \ \lambda = \lambda_4 \ , \ c_p = c_{p4} \ , \ \rho = \rho_4 \tag{3.139}$$

- boundary conditions:

at the top surface:

$$\tau \succ 0 \ , \ z = L \ , \ -1 \leq x \leq 1 \ , \ -\frac{h}{2} \leq y \leq \frac{h}{2} \ , \ t = t_a \tag{3.140}$$

at the bottom surface:

$$\tau \succ 0 \ , \ z = -L \ , \ -1 \leq x \leq 1 \ , \ -\frac{h}{2} \leq y \leq \frac{h}{2} \ , \ t = t_a \tag{3.141}$$

for other surfaces:

$$\tau \succ 0 \ , \ x = 1 \ , \ -\frac{h}{2} \leq y \leq \frac{h}{2} \ , \ -L \leq z \leq L \ , \ \frac{dt}{dx} = 0 \tag{3.142}$$

$$\tau \succ 0 \ , \ y = -\frac{h}{2} \ , \ -1 \leq x \leq 1 \ , \ -L \leq z \leq L \ , \ \frac{dt}{dy} = 0 \tag{3.143}$$

$$\tau \succ 0 \quad, \quad x = -1 \quad, \quad -\frac{h}{2} \leq y \leq \frac{h}{2} \quad, \quad -L \leq z \leq L \quad, \quad \frac{dt}{dx} = 0 \tag{3.144}$$

$$\tau \succ 0 \quad, \quad y = \frac{h}{2} \quad, \quad -1 \leq x \leq 1 \quad, \quad -L \leq z \leq L \quad, \quad \frac{dt}{dy} = 0 \tag{3.145}$$

- the heat flux continuity at the walls that separate the bricks:

$$\tau \succ 0 \, , \quad x = 0 \, , \quad -\frac{h}{2} \leq y \leq \frac{h}{2} \, , \quad -L \leq z \leq 0 \, , \quad \lambda_1 \left(\frac{dt}{dx}\right)_{x=0+} = \lambda_3 \left(\frac{dt}{dx}\right)_{x=0-} \tag{3.146}$$

$$\tau \succ 0 \, , \quad x = 0 \, , \quad -\frac{h}{2} \leq y \leq \frac{h}{2} \, , \quad 0 \leq z \leq L \, , \quad \lambda_2 \left(\frac{dt}{dx}\right)_{x=0+} = \lambda_4 \left(\frac{dt}{dx}\right)_{x=0-} \tag{3.147}$$

$$\tau \succ 0 \, , \quad z = 0 \, , \quad 0 \leq x \leq 1 \, , \quad -\frac{h}{2} \leq y \leq \frac{h}{2} \, , \quad \lambda_2 \left(\frac{dt}{dz}\right)_{z=0+} = \lambda_1 \left(\frac{dt}{dz}\right)_{z=0-} \tag{3.148}$$

$$\tau \succ 0 \, , \quad z = 0 \, , \quad -1 \leq x \leq 0 \, , \quad -\frac{h}{2} \leq y \leq \frac{h}{2} \, , \quad \lambda_4 \left(\frac{dt}{dz}\right)_{z=0+} = \lambda_3 \left(\frac{dt}{dz}\right)_{z=0-} \tag{3.149}$$

- temperature conditions to start the heating (initial conditions of the problem):

$$\tau = 0 \quad, \quad -1 \leq x \leq 1 \quad, \quad -\frac{h}{2} \leq y \leq \frac{h}{2} \quad, \quad -L \leq z \leq L \quad, \quad t = t_a \tag{3.150}$$

- the relation that characterizes the local value of the rate of heat production (it is the case of a small power source as, for example, a small electrical heater placed in the centre of the block):

$$Q_g = A \exp(-k(x^2 + y^2 + z^2)) \tag{3.151}$$

Now we can transform the model relations into dimensionless forms. For this purpose, we use the dimensionless temperature as a measure of a local excess with respect to the adjacent medium $T_p = (t - t_a)/t_a$; the dimensionless time recognized as the Fourier number $T = \dfrac{\lambda_1 \tau}{\rho_1 c_{p1} l^2}$; the dimensionless geometric coordinates given by $X = x/l$, $Y = y/h$, $Z = z/L$ or as $X = x/l$, $Y = y/l$, $Z = z/l$. Table 3.12 contains the dimensionless state of the mathematical model of the process.

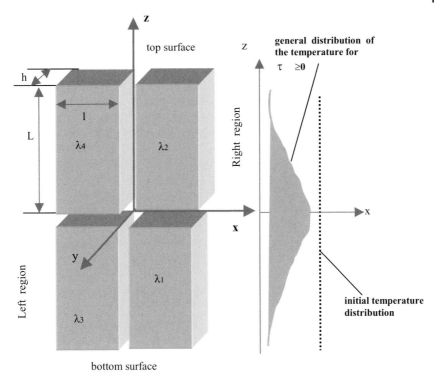

Figure 3.44 Description of the example of heating of four bricks.

Table 3.12 Dimensionless mathematical models for the case of the heating of a block of four bricks.

$$\frac{\partial T_p}{\partial T} = \left(\frac{\partial^2 T_p}{\partial X^2} + \frac{l^2}{h^2} \frac{\partial^2 T_p}{\partial Y^2} + \frac{l^2}{L^2} \frac{\partial^2 T_p}{\partial Z^2} \right) + Q_g' \; ; \; Q_g' = \frac{Q_g (4L\,lh)^{2/3}}{X_1 ta} \tag{3.135}$$

Materials conditions:

$$0 \prec X \prec 1 \;\; , \;\; -1 \prec Z \prec 0 \;\; , \;\; -\frac{1}{2} \prec Y \prec 1 \;\; , \;\; \lambda = \lambda_1 \;\; , \;\; c_p = c_{p1} \;\; , \;\; \rho = \rho_1 \tag{3.136}$$

$$0 \prec X \prec 1 \;\; , \;\; 0 \prec Z \prec 1 \;\; , \;\; -\frac{1}{2} \prec Y \prec \frac{1}{2} \;\; , \;\; \lambda = \lambda_2 \;\; , \;\; c_p = c_{p2} \;\; , \;\; \rho = \rho_2 \tag{3.137}$$

$$-1 \prec X \prec 0 \;\; , \;\; -1 \prec Z \prec 0 \;\; , \;\; -\frac{1}{2} \prec Y \prec \frac{1}{2} \;\; , \;\; \lambda = \lambda_3 \;\; , \;\; c_p = c_{p3} \;\; , \;\; \rho = \rho_3 \tag{3.138}$$

$$0 \prec X \prec 1 \;\; , \;\; 0 \prec Z \prec 1 \;\; , \;\; -\frac{1}{2} \prec Y \prec \frac{1}{2} \;\; , \;\; \lambda = \lambda_4 \;\; , \;\; c_p = c_{p4} \;\; , \;\; \rho = \rho_4 \tag{3.139}$$

Boundary conditions:

$$T \succ 0 \ , \ \ Z = 1 \ , \ \ -1 \leq X \leq 1 \ , \ \ -\frac{1}{2} \leq Y \leq \frac{1}{2} \ , \ \ Tp = 0 \qquad (3.140)$$

$$T \succ 0 \ , \ \ Z = -1 \ , \ \ -1 \leq X \leq 1 \ , \ \ -\frac{1}{2} \leq Y \leq \frac{1}{2} \ , \ \ Tp = 0 \qquad (3.141)$$

$$T \succ 0 \ , \ \ X = 1 \ , \ \ -\frac{1}{2} \leq Y \leq \frac{1}{2} \ , \ \ -1 \leq Z \leq 1 \ , \ \ \frac{dTp}{dX} = 0 \qquad (3.142)$$

$$T \succ 0 \ , \ \ Y = -\frac{1}{2} \ , \ \ -1 \leq X \leq 1 \ , \ \ -1 \leq Z \leq 1 \ , \ \ \frac{dTp}{dY} = 0 \qquad (3.143)$$

$$T \succ 0 \ , \ \ X = -1 \ , \ \ -\frac{1}{2} \leq Y \leq \frac{1}{2} \ , \ \ -1 \leq Z \leq 1 \ , \ \ \frac{dTp}{dX} = 0 \qquad (3.144)$$

$$T \succ 0 \ , \ \ Y = \frac{1}{2} \ , \ \ -1 \leq X \leq 1 \ , \ \ -1 \leq Z \leq 1 \ , \ \ \frac{dTp}{dY} = 0 \qquad (3.145)$$

$$T \succ 0 \ , \ \ X = 0 \ , \ \ -\frac{1}{2} \leq Y \leq \frac{1}{2} \ , \ \ -1 \leq Z \leq 0 \ , \ \ \lambda_1 \left(\frac{dTp}{dX}\right)_{x=0+} = \lambda_3 \left(\frac{dTp}{dX}\right)_{x=0-} \qquad (3.146)$$

$$T \succ 0 \ , \ \ X = 0 \ , \ \ -\frac{1}{2} \leq Y \leq \frac{1}{2} \ , \ \ 0 \leq Z \leq 1 \ , \ \ \lambda_2 \left(\frac{dTp}{dX}\right)_{x=0+} = \lambda_4 \left(\frac{dTp}{dX}\right)_{x=0-} \qquad (3.147)$$

$$T \succ 0 \ , \ \ Z = 0 \ , \ \ 0 \leq x \leq l1 \ , \ \ -\frac{1}{2} \leq Y \leq \frac{1}{2} \ , \ \ \lambda_2 \left(\frac{dTp}{dZ}\right)_{z=0+} = \lambda_1 \left(\frac{dTp}{dZ}\right)_{z=0-} \qquad (3.148)$$

$$T \succ 0 \ , \ \ Z = 0 \ , \ \ -1 \leq X \leq 0 \ , \ \ -\frac{1}{2} \leq Y \leq \frac{1}{2} \ , \ \ \lambda_4 \left(\frac{dTp}{dZ}\right)_{z=0+} = \lambda_3 \left(\frac{dTp}{dZ}\right)_{z=0-} \qquad (3.149)$$

$$T = 0 \ , \ \ -1 \leq X \leq 1 \ , \ \ -\frac{1}{2} \leq Y \leq \frac{1}{2} \ , \ \ -1 \leq Z \leq 1 \ , \ \ Tp = 0 \qquad (3.150)$$

$$Q_g = A \exp\left(-k(l^2 X^2 + h^2 Y^2 + L^2 Z^2)\right) \qquad (3.151)$$

In *the first simulation*, we consider a particular case of the heating dynamics of the four blocks when the heat is produced by a source at the centre of the blocks. In this example, we have different thermal conductivities for the material of each block. Figure 3.45 shows the simulation of a parallelepiped brick with its corresponding dimensionless length and width. The only difference between the dimensionless model shown in Table 3.12 and the model used in the simulator (Fig. 3.45) is the use of a partly dimensionless model in the simulator text. To show the complex dynamics of the temperature observed in Fig. 3.45, seven displaying points have been selected. These are: A – bottom right brick: $A(1/2, -h/2, -L/2)$; B – bottom right brick: $B(1/2, h/2, -L/2)$; C – bottom left brick: $C(-1/2, h/2, -L/2)$; D – bottom left brick: $D(-1/2, -h/2, -L/2)$; E – top right brick: $E(1/2, -h/2, L/2)$; F – top right brick : $F(1/2, h/2, L/2)$; G – top left brick: $G(-1/2, h/2, L/2)$; H – centre of block: $H(0, 0, 0)$.

```
title 'Simulator for the heating of a block of four bricks
select regrid=off { use fixed grid }   ngrid=5
coordinates  cartesian3
variables  Tp
definitions  long = 1  wide = 1   λ,c_P,ρ        { - values supplied later }
             Q = 3.8*10^7*exp(-x^2-y^2-z^2)          { Thermal source }
initial values  Tp = 0.
equations
   div[ λ *grad(Tp)] + Q = (ρc_P )  dt(Tp)       { the heat  transport equation }
     extrusion z = -long,0,long     { divide Z into two layers }
boundaries
 surface 1 value(Tp)=0  { bottom surface temp } surface 3 value(Tp)=0  { top surface temp }
   Region 1          { define full domain boundary in base plane }
   layer 1  λ =1   c_p =2000 ρ=2000    { bottom right brick }   layer 2   λ =0.1
 c_p=1800   ρ=1800   { top right brick }
     start(-wide,-wide)
        value(Tp) = 0         { fix all side temps }
          line to (wide,-wide)   { walk outer boundary in base plane }
             to (wide,wide)
          to (-wide,wide)        to finish
   Region 2         { overlay a second region in left half }
   layer 1  λ =0.2  c_p=1500  ρ=1200    { bottom left brick }  layer 2  λ =0.4
 c_p=.1500   ρ=1500        { top left brick }
     start(-wide,-wide)
        line to (0,-wide)          { walk left half boundary in base plane }
           to (0,wide)
           to (-wide,wide)             to finish
 time 0 to 3 by 0.01                    { establish time range and initial time steep }
 plots
    for t = endtime contour(Tp) on surface z=0  as "XY Temp" range=(0,6)
 ;contour(Tp) on surface x=0  as  "YZ Temp" range=(0,6) ;contour(Tp) on surface y=0
 as "XZ Temp" range=(0,6)
 histories
    history(Tp) at (wide/2,-wide/2,-long/2) ;(wide/2,wide/2,-long/2) ; (-wide/2,wide/2,-
 long/2) ;  (-wide/2,-wide/2,-long/2)  ;  (wide/2,-wide/2,long/2);
 (wide/2,wide/2,long/2) ;  (-wide/2,wide/2,long/2) ; (0,0,0)              range=(0,6)
 end
```

Figure 3.45 FlexPDE working text of the simulator for the heating of a block of four bricks.

Figure 3.46 presents the temperature distribution in the plane $y = 0$, which separates the left parts from the right parts of the bricks' assembly. The shape of the group of the isothermal curves shows a displacement towards the brick with the higher thermal conductivity. Using the values obtained from these isothermal curves, it is not difficult to establish that the exit heat flux for each brick from the bottom of the assembly (plane $Z = -1$) and for the top of the assembly (plane $Z = 1$) depends on its thermal conductivity and on the distribution of the isothermal curves. If we compare this figure to Fig. 3.47 we can observe that the data contained in Fig. 3.46 correspond to the situation of a steady state heat transfer.

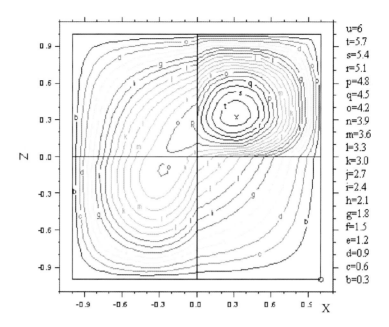

Figure 3.46 Temperature distribution at T = 3 for plane y = 0.

Figure 3.47 shows the evolution of the heating process of the composite block and how it attains a complex steady state structure with the surface zones covered by complicated isothermal curves (see also Fig. 3.46). Secondly, this figure shows how the brick with the higher thermal conductivity is at steady state and remains the hottest during the dynamic evolution. As explained above, this fact is also shown in Fig. 3.46 where all high isothermal curves are placed in the area of the brick with highest thermal conductivity. At the same time an interesting vicinity effect appears because we observe that the brick with the smallest conductivity does not present the lowest temperature in the centre (case of curve G compared with curves A and B). The comparison of curves A and B, where we have $\lambda = 0.2$, with curves C and D, where $\lambda = 0.4$, also sustains the observation of the existence of a vicinity effect. In Fig. 3.48, we can also observe the effect of the highest thermal conductivity of one block but not the vicinity effect previously revealed by Figs. 3.46 and 3.47. If we compare the curves of Fig. 3.47 with the curves of Fig. 3.48 we can appreciate that a rapid process evolution takes place between T = 0 and T = 1. Indeed, the heat transfer process starts very quickly but its evolution from a dynamic process to steady state is relatively slow.

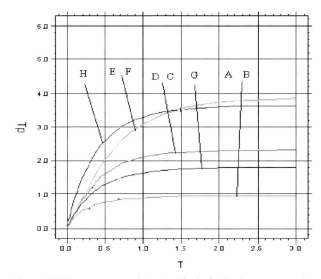

Figure 3.47 Temperature evolution inside the bricks that compose the heated block.

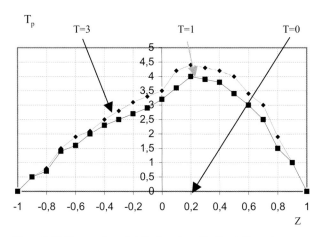

Figure 3.48 Temperature distribution along the Z axis for various dimensionless times.

The second simulation has been oriented to the analysis of the cooling process of the composite block. So, for the initial time, we have a block of four bricks heated to a constant temperature. All surfaces of the blocks except for $Z = -1$ and $Z = 1$, are isolated before placing the assembly of blocks in a cooling medium. We assume that we can use the boundary conditions of type I. To make the simulator respond to this new model with the written text shown in Fig. 3.45, we erase the generated heat ($Q = 0$) and we adequately change the initial temperature conditions. The examples given by Figs. 3.49–3.51 consider that the cooling of the composite block begins when the dimensionless temperature is $T_p = 6$. Figure 3.49 shows that each brick presents its proper thermal dynamics. We can notice that the vicinity effects are similar to those already discussed in the previous example. However, here the cooling rate of each brick does not occur in accordance with the thermal conductivity but with the thermal diffusivity.

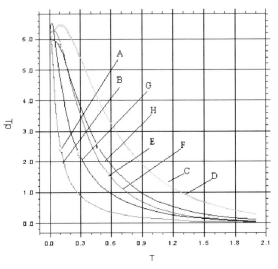

Figure 3.49 Cooling dynamics of a brick assembly for L/l = 1 and $V_a = 8.1^3$.

The curves E and F that refer to the brick with $\lambda/(\rho c_p) = 2.57 * 10^{-7}$ m²/s show a higher cooling rate than curves A and B where we have $\lambda/(\rho c_p) = 1.1 * 10^{-7}$ m²/s. At the same time, curve G, where we have $\lambda/(\rho c_p) = 0.23 * 10^{-7}$ m²/s, shows a higher cooling rate than curves C, D, E, F. Table 3.13 contains data from some simulations where the block is considered to be composed of bricks which have the same thermal diffusivity. It clearly shows that each brick presents an identical temperature field. It is obvious that, for this simulation, the temperature at the centre of each brick and at the centre of the block have the role of the dependent variables of the process when the medium temperature, the cooling temperature at the beginning, the material diffusivity and the block geometrical dimensions are the inputs or independent variables of the process. In addition, we can say that, in spite of the type I boundary conditions for the bottom and top surfaces, the data shown in Table 3.13 allow one to appreciate that the block cooling process can be characterized by the integral relation: $T_{pmean} = T_{p0}\exp\left(-\dfrac{kA_t}{mc_p}\tau\right)$ where T_{pmean} is the mean block temperature, k is the heat transfer coefficient with respect to the non-isolated surfaces, m represents the block mass and A_t is the value of the non-isolated surfaces.

Table 3.13 Evolution of the temperature at the centre of the block and at the centre of each brick.
Studied case: $\lambda_1/(\rho_1 c_{p1}) = \lambda_2/(\rho_2 c_{p2}) = \lambda_3/(\rho_3 c_{p3}) = \lambda_4/(\rho_4 c_{p4}) = 10^{-7}$ m^2 /s.

T		0	0.25	0.5	0.75	1.0	1.25	1.50	1.75	2.0
T_p – block centre		6	5	2.0	0.8	0.4	0.15	0.05	0	0
T_p – each brick centre		6	2	0.8	0.3	0.08	0.04	0	0	0

If we significantly reduce the dimension of the Z axis, then we transform the three-dimensional cooling problem into an unsteady state and monodimensional problem. Figures 3.50 and 3.51 show the results of the simulations oriented to demonstrate this fact. We can notice that all curves present the same tendency as the analytical solution or Schmidt numerical solution of the monodimensional cooling problem:

$$\frac{\partial t}{\partial \tau} = \frac{\lambda}{\rho c_p} \frac{\partial^2 t}{\partial z^2} ; \quad \frac{\lambda}{\rho c_p} = \sum_1^4 \frac{\lambda_i}{\rho_i c_{pi}} ;$$

$$\tau \geq 0, \; z = -L \text{ and } z = 1 \; , \; t = t_a; \; \tau = 0, -L \prec z \prec L, t = t_0$$

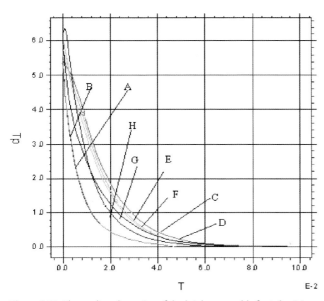

Figure 3.50 The cooling dynamics of the bricks assembly for L/l = 0.1.

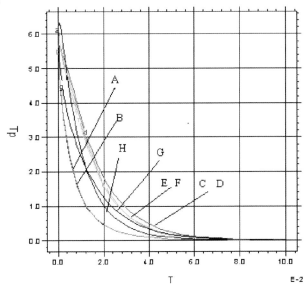

Figure 3.51 The cooling dynamics of the bricks assembly for $L/l = 0.01$.

The above-mentioned trend occurs in spite of the different conditions that characterize the calculation of each curve. As is known, the dimensionless time that characterizes cooling depends on the width of each brick. However, this dimensionless value has not changed in the simulations used for drawing Figs. 3.50 and 3.51. Consequently, these figures are characterized by the same dimensionless time axis division. In addition, the heat transfer surface used for the simulation also has the same value. Indeed, both figures are reported to use the same base of comparison.

In *the third simulation* example, we carried out an analysis of some of the aspects that characterize the case of the mass transfer of species through a membrane which is composed of two layers (the separative and the support layers) with the same thickness but with different diffusion coefficients of each entity or species. To answer this new problem the early model has been modified as follows: (i) the term corresponding to the source has been eliminated; (ii) different conditions for bottom and top surfaces have been used: for example, at the bottom surface, the dimensionless concentration of species is considered to present a unitary value while it is zero at the top surface; (iii) a new initial condition is used in accordance with this case of mass transport through a two-layer membrane; (iv) the values of the four thermal diffusion coefficients from the original model are replaced by the mass diffusion coefficients of each entity for both membrane layers; (v) the model is extended in order to respond correctly to the high value of the geometric parameter l/L.

It is clear that, for this problem, the normal trend is to use the monodimensional and unsteady state model, which is represented by the assembly of relations (3.152)–(3.156). It accepts a very simple numerical solution or an analytical solution made of one of the methods classically recommended such as the variable separation method:

$$\frac{\partial c}{\partial \tau} = D \frac{\partial^2 c}{\partial z^2} \tag{3.152}$$

$$\tau = 0 \ , \quad -L \leq z \leq L \ , \quad c = 0 \tag{3.153}$$

$$\tau \geq 0 \ , \quad z = -L \ , \quad c = c_0 \tag{3.154}$$

$$\tau \geq 0 \ , \quad z = 0 \quad D_1 \left(\frac{dc}{dz}\right)_- = D_2 \left(\frac{dc}{dz}\right)_+ \tag{3.155}$$

$$\tau \geq 0 \ , \quad z = L \ , \quad c = 0 \tag{3.156}$$

In addition, it is known that the transport of species through the membrane and its support are characterized by the coefficients of diffusion, which are experimentally determined with methods based on this model [3.47–3.51].

However, we cannot *a priori* use this model without the previous establishment of conditions which accept the transformation of the three-dimensional and unsteady state model into a one-dimensional model. These conditions can be studied using the simulations as a tool of comparison. At the same time, it is interesting to show the advantages of the dynamic (unsteady) methods for the estimation of the diffusion coefficient of the species through the porous membrane by comparison with the steady state methods.

Figures 3.52–3.54 show three cases of simulation of the process where the diffusion coefficients for the support and membrane take, respectively, the following values $D_1 = D_3 = D_{sp} = 10^{-8}$ m^2/s $D_2 = D_4 = D_{mb} = 10^{-9}$ m^2/s. All simulated cases keep the total volume of the membrane assembly constant but differ from each other due to parameter l/L which takes the values: 10, 100, and 1000.

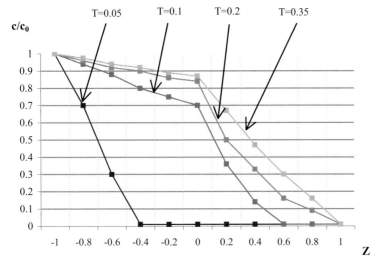

Figure 3.52 Dimensionless species concentration along the Z axis at various times (for $l/L = 10$).

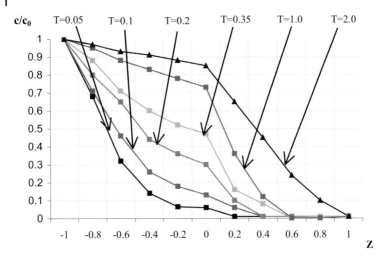

Figure 3.53 Dimensionless species concentration along the Z axis at various times (for $l/L = 1000$).

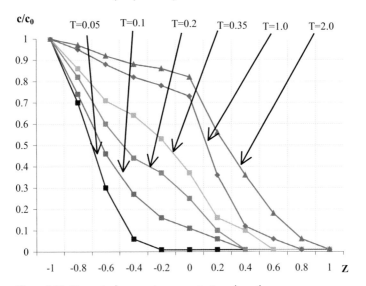

Figure 3.54 Dimensionless species concentration along the Z axis at various times (for $l/L = 100$).

Figure 3.52 significantly differs from the next two figures, especially with respect to the evolution of C/C_0 with time. For this situation, where the value of the ratio length/thickness is not very high (10) we can accept that the diffusion occurs in all directions and also that it is very rapid in the membrane support. This last fact is responsible for this apparent fast evolution with time. For the other two situations, we can observe that the diffusion process tends to attain the stationary state when the concentration profile is $C = C_1 - (1 - C_1) * Z$ for

$-1 \leq Z \leq 0$ and $C = C_1 * (1 - Z)$ for $0 \leq Z \leq 1$ respectively. Here, the dimensionless species concentration for the steady state diffusion at the plane $Z = 0$ is given by C_1. If we write the equality of the species flux for the support and for the membrane, we obtain: $\dfrac{1 - C_1}{C_1} = \dfrac{D_{mb}}{D_{sp}} \dfrac{\delta_{sp}}{\delta_{mb}}$. Unfortunately, we cannot measure the dimensionless concentration C_1 and, consequently, this relation cannot be used to determine D_{mb}, even if D_{sp} is known. At the same time, it is not simple to establish the end of the dynamic evolution and the beginning of the steady state diffusion. As an example, if we know the end of the unsteady state and the beginning of the steady state (as given in the simulations), the ratio $D_{mb}\delta_{sp}/D_{sp}\delta_{mb} = 0.1$ ($\delta_{mb} = \delta_{sp}$, $D_{mb} = 10^{-9}$ m^2/s $D_{sp} = 10^{-8}$ m^2/s) for Figs. 3.52–3.54 then, for the steady state, $(1 - C_1)/C_1$ must be 0.1.

If we observe the value of $(1-C_1)/C_1$ at $T = 2$ we have 0.135 for Fig. 3.52, 0.157 for Fig. 3.53 and 0.189 for Fig. 3.54; for all cases the persistency of the dynamic evolution is shown. Otherwise, if we insist on the development of the steady state method to determine the diffusion coefficient, then it is not difficult to observe that, from the experimental point of view, we must use the measurements of the flow rate of species (diffusion). Nevertheless, this type of experiment is not characterized by its reproducibility and simplicity.

Concerning the problem of the validity of the monodimensional and unsteady state model for the transport of an entity through the membrane, the simulations with l/L >100 show that the transport in the Z direction is dominant. At the same time, Figs. 3.55 and 3.56, which give the state of the dimensionless concentration

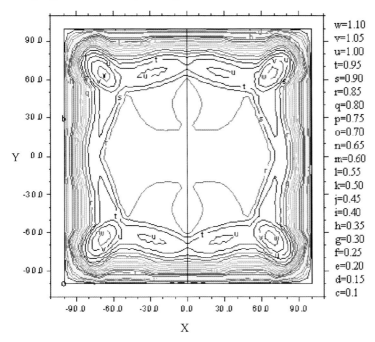

Figure 3.55 Dimensionless concentration for plane Z = 0 when l/L = 1000 and T = 2.

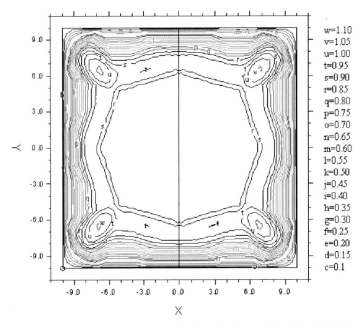

Figure 3.56 Dimensionless concentration for plane Z = 0 when l/L = 100 and T = 2.

of the species for plane Z = 0 (between the support and membrane) show that the X and Y concentration gradients are not absent in this case.

Because these gradients are not negligible, we cannot fully recommend, the monodimensional unsteady state model as a theoretical model that supports the use of dynamic methods for the characterization of the diffusion in a porous membrane. Even if over 60% of the membrane surface is covered by a constant dimensionless concentration of species, this event is not sufficient to allow the acceptance of the unsteady model. Nevertheless, looking closely at Figs. 3.55 and 3.56 we can notice that only 20% of the surface contains high concentration gradients. However, the boundary conditions chosen impose the absence of transport of the species throughout all surfaces except for Z = −1 and Z = 1. It is evident that these last observations sustain the validity of the monodimensional and unsteady state model. As a conclusion to this discussion, it is clear that the validity of the monodimensional and unsteady state model as a support for the dynamic methods to characterize the diffusion in a porous medium is not really affected. At the same time, these critical observations can be considered as a support of the various procedures that bring the necessary corrections to this model.

If both parts of the membrane (the support and the separative layers) can be characterized by values lower than 1 for the Knudsen number (Kn $= \lambda/2r_p$ where λ is the mean free path of species or molecules and r_p is the mean pore radius), then all the aspects mentioned here must be taken into consideration. To describe the motion of a species in a porous structure, other models must be used when the Knudsen number value is higher than 1 or when the support is highly porous and the separative layer is dense.

One of the most convenient ways to investigate this process is to accept the monodimensional unsteady state model of transport inside the membrane and to measure the time lag characterizing the transitional process preceding the steady-state. Traditional time lag theory has been intensively used to study gas and vapour permeation through dense films. In most cases, the derived equations describing the time lag for diffusion through composite media accept the equilibrium assumption at interfaces. This assumption is valid when the mass transfer process at the interface is much faster than the transfer within the two adjacent phases. This model has also been used in describing transport through supported liquid membranes. In some cases, the interface resistance cannot be neglected, and can be described with chemical reaction or sorption–desorption rates at the surface.

3.4.3
Fast Chemical Reaction Accompanied by Heat and Mass Transfer

The problem of the modeling of a reactor where a homogeneous reaction (in the gas or liquid phase) takes place can be relatively simple to solve after selecting the type of reactor and its corresponding flow model. It is evident that, in accordance with the accepted flow model, the reactor model will contain the particularizations of the equation of the energy conservation and of the equations of the field of the species concentration. The source term of the equation of the concentration of one species is expressed by the kinetics reaction rate. Here we consider that the homogeneous reaction is carried out in a reactor where the hydrodynamics corresponds to a plug flow (PF) model and where the reaction $A + B \rightarrow C$ occurs in the presence of an inert component D. In accordance with the descriptive model of the reactor given in Fig. 3.57, the following relations and conditions show the associated mathematical model for a steady state operation and an elementary reactor's length dz:

- the balance equations of species A consumption:

$$w \frac{dX_A}{dz} = v_{rA} \tag{3.157}$$

- the links between local concentration of B, C, D and A:

$$y_B = y_{B0}(1-X_A) \tag{3.158}$$

$$y_C = y_{C0}(1+X_A) \tag{3.159}$$

$$y_A = y_{A0}(1-X_A) \tag{3.160}$$

$$y_D = 1-y_A-y_B-y_C \tag{3.161}$$

- the heat balance equation:

$$w\frac{dt}{dz} = \frac{\Delta H_{rA}}{\rho c_p} v_{rA} - \frac{4k}{\rho c_p d}(t - t_{ex})$$ (3.162)

- the expression of the reaction kinetics:

$$v_{rA} = k_1 c_{A0}^m c_{B0}^n (1 - X_A)^{m+n} - k_2 c_{C0}^p (1 + X_A)^p$$ (3.163)

- the concentration and temperature conditions at the reactor input:

$$z = 0 \ , \ X_A = 0 \ , \ y_A = y_{A0} \ , \ y_B = y_{B0} \ , \ y_C = y_{C0} \ , \ t = t_0 \ , \ t_{ex} = t_{ex0}$$ (3.164)

- the heat balance for the fluid that flows outside of the reactor:

$$\frac{dt_{ex}}{dz} = f(G_{ex}, \rho_{ex}, c_{pex}, k, d, t)$$ (3.165)

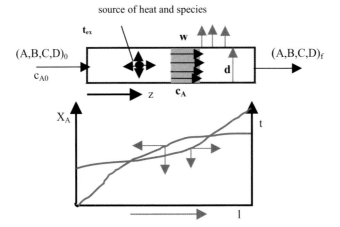

Figure 3.57 Scheme of a plug-flow reactor and homogeneous reaction.

In the PF homogeneous reactor, X_A is the conversion of species A into C, ($X_A = (c_A - c_{A0})/c_{A0}$, where c_A is the local molar concentration of A), y represents the molar fraction of species ($y_A = c_A/(c_A + c_B + c_C + c_D)$, etc.) and $k_1, k_2, m, n; p$ characterize the kinetics as reaction constants with their partial reaction orders.

The unsteady state model will be completed by adding the unsteady evolution as $\partial X_A/\partial \tau$, $\partial t/\partial \tau$ and $\partial t_{ex}/\partial \tau$ respectively on the left part of the equations (3.155), (3.160) and (3.163). At the same time, the initial conditions must be adequately changed and new univocity conditions will be attached to this new problem.

The modeling procedure explained above is not valid for all types of homogeneous reactions, for example, a very fast exothermic reaction such as the combustion of gaseous or vaporized hydrocarbon with oxygen in air or the reaction of chloride and hydrogen in inert nitrogen, etc. We will develop an example of such reactions below:

- A gaseous mixture, which contains three components – a combustible, a comburent and an inert gas, is fed into a tubular reactor which has an efficient cooling in order to maintain the walls at constant temperature;
- The gaseous mixture comes into the reactor with uniform radial velocity (plug flow) and the gas velocity increases linearly with temperature inside the reactor. Indeed, we can consider the conversion as a function of r, wτ and τ ($X_A(r, w\tau, \tau)$) and, consequently, we can build the model taking r and τ into account;
- To start the reaction in the reactor input, we have a small surface with the function of a heat inductor where the temperature of the gaseous mixture increases very rapidly to attain the inductor temperature. Inside the reactor the inductor surface operates as a stripping heat surface;
- The process occurs symmetrically with respect to a plane that contains the z axis; at the same time, the temperature and the reactant conversion will present the maximum values in the centre of the reactor due to a high speed reaction. Therefore, we permanently have the right conditions for components and heat diffusion in the reactor.

Combustion reactions are known to occur through a free radical mechanism and from this viewpoint, their kinetics is complicated. At the same time, the coupling of the reaction kinetics with the flow dynamics, as well as the species and heat diffusion is very important for most real cases [3.52–3.55]. For simplification, we consider that the formal kinetics of this reaction is first order with respect to the limiting reactant and that its expression must show a strong dependence on temperature. Now, using the descriptive model (Fig. 3.58), we can build the general mathematical model of the process. Concerning our fast highly exothermic reaction, relation (3.166) often employs the quantitative description of the reactant consumption rate. Here c is the limitative species concentration and γ is a parameter related to the reaction activation energy:

$$v_r(c, \tau) = kc * \exp[\gamma(1/t_0 - 1/t)] = kc * \exp[\gamma/t_0(1 - 1/T_p)] \tag{3.166}$$

For small γ values, we obtain small or moderate reaction rates, whereas when γ increases, it corresponds to the start of the fast reaction or high temperatures, when the reaction rate can attain a dangerous level.

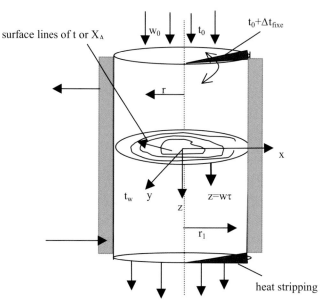

Figure 3.58 Description of a plug flow reactor with a constant wall temperature where a fast exothermic reaction occurs.

The mathematical model of the process is built only in x and y coordinates (or only in the r coordinate) because the z coordinate is self-defined using τ and w. Then, the general mathematical model contains the following equations and conditions:

- the equation of concentration field for the limitative reactant:

$$\frac{\partial c}{\partial \tau} = D_A \left(\frac{\partial^2 c}{\partial x^2} + \frac{\partial^2 c}{\partial y^2} \right) + v_r(c, t) \tag{3.167}$$

- the expression of temperature field inside the reaction mixture:

$$\frac{\partial t}{\partial \tau} = a_t \left(\frac{\partial^2 t}{\partial x^2} + \frac{\partial^2 t}{\partial y^2} \right) + v_r(c, t) \frac{\Delta H_r}{\rho c_p} \tag{3.168}$$

- the equations of the local gas velocity and plug flow reaction front position:

$$w = w_0 \frac{(271 + t)}{(271 + t_0)} \; ; z = w\tau \tag{3.169}$$

- the respective initial conditions for the temperature and concentration domains:

$$\tau = 0 \; , \quad 0 \prec x \prec r_1 \; , \quad 0 \prec y \prec r_1 \; , \quad c = 0 \; , \quad t = t_0 \tag{3.170}$$

- the univocity conditions:
  1. for the reactor input:

$$\tau \succ 0 \; , \quad 0 \prec x \vartriangleleft r_1 \; , \quad 0 \prec y \prec r_1 \; , \quad x^2 + y^2 = r^2 \; , \quad c = c_0 \; , \quad t = t_0 \tag{3.171}$$

2. for the surface of thermal process control:

$$\tau = 0 , \quad 0 \prec x \prec r_1 , \quad 0 \prec \sqrt{(x^2 + y^2)} \prec r_1 \frac{\theta}{360} , \quad t = t_0 + \Delta t , \quad \frac{dc}{dr} = 0$$

$$\tau \succ 0 , \quad 0 \prec x \prec r_1 , \quad 0 \prec \sqrt{(x^2 + y^2)} \prec r_1 \frac{\theta}{360} , \quad \frac{dt}{dr} = 0 , \quad \frac{dc}{dr} = 0 \qquad (3.172)$$

3. for the reactor walls:

$$\tau \succ 0 , \quad x = r_1 , \quad y = r_1 , \quad y_{r1} \in 360 - \theta , \quad \lambda \frac{dt}{dr} = -a(t - t_w) , \quad \frac{dc}{dr} = 0 \qquad (3.173)$$

The set of Eqs. (3.166)–(3.173) represents the general mathematical model of the described fast exothermic reactions taking place in an externally cooled plug flow reactor. If we use the dimensionless expressions of time $T = D_A \tau / r_1^2$, coordinates $X = x/r_1, Y = y/r_1, Z = z/l, R = r/r_1$, temperature $T_p = t/t_0$ and conversion X_A as dimensionless concentration of the limitative reactant, then the process model can be described by the relations contained in Table 3.14.

Table 3.14 Dimensionless mathematical model for the heat and mass transfer in a plug flow reactor for a fast exothermic reaction.

$$\frac{\partial X_A}{\partial T} = \left(\frac{\partial^2 X_A}{\partial X} + \frac{\partial^2 X_A}{\partial Y} \right) + \beta_R (1 - X_A) \exp \left[\gamma \left(1 - \frac{1}{T_p} \right) \right] \qquad (3.174)$$

$$\frac{\partial T_p}{\partial T} = \frac{Sc}{Pr} \left(\frac{\partial^2 T_p}{\partial X} + \frac{\partial^2 T_p}{\partial Y} \right) + \beta_T (1 - X_A) \exp \left[\gamma \left(1 - \frac{1}{T_p} \right) \right] \qquad (3.175)$$

$$\frac{w}{w_0} = \frac{\alpha + T_p}{\alpha + 1}; \quad Z = Pe_d \left(\frac{r_1}{l} \right) T \qquad (3.176)$$

$$T = 0 , \quad 0 \prec Z \prec 1 , \quad 0 \prec Y \prec 1 , \quad 0 \prec Y \prec 1 , \quad X_A = 0 , \quad T_p = 1 \qquad (3.177)$$

$$T \succ 0 , \quad 0 \prec X \prec 1 , \quad 0 \prec Y \prec 1 , \quad X^2 + Y^2 = R^2 , \quad X_A = 1 , \quad T_p = 1 \qquad (3.178)$$

$$T \succ 0 , \quad 0 \prec X \prec 1 , \quad 0 \prec \sqrt{X^2 + Y^2} \prec \frac{\theta}{360} , \quad T_p = 1 + \frac{\Delta t}{t_0} , \quad \frac{dX_A}{dR} = 0 \qquad (3.179)$$

$$T \succ 0 , \quad 0 \prec X \prec 1 , \quad 0 \prec \sqrt{X^2 + Y^2} \prec \frac{\theta}{360} , \quad T_p = 1 + \frac{\Delta t}{t_0} , \quad \frac{dX_A}{dR} = 0 \qquad (3.180)$$

$$T \succ 0 , \quad 0 < X \prec 1 , \quad 0 \prec Y \prec 1 , \quad Y_1 \in 360 - \theta , \quad \frac{dT_p}{dR} = Bi(T_p - T_w) , \quad \frac{dX_A}{dR} = 0 \qquad (3.181)$$

Specifications: $\beta_r = (k r_1^2)/D_A$ – Fourier number for the reaction,
$Sc = v/D_A$ – Schmidt number $Pr = (c_p v \rho)/\lambda$ – Prandtl number,
$Bi = (\alpha r_1)/\lambda$ – Biot number, $\beta_T = \beta_r (\Delta H_r c_0)/(\rho c_p t_0)$, ρ – density,
v – kinematic viscosity, c_p – specific sensible heat,
λ – thermal conductivity.

The *first simulation* program set is obtained after the model particularization using the numerical values of the parameters and the geometry and material properties as shown in Fig. 3.59. This program aims to show some characteristic aspects of this type of reactor with respect to the heat and reactant conversion dynamics using a graphic representation. In addition, the simulation allows one to know the evolution of the exit variables of the process (temperature and species concentration), when one or more of the input variables of the model are changed. For this concrete case and among these input variables, we can identify: (i) the input flow rate of reactants; (ii) the value of the limitative reactant concentration at the reactor feed; (iii) the value of the input temperature of the reactants; (iv) the temperature of the reactor walls; (v) the limitative reactant type, here introduced by the parameters that characterize the reaction kinetics.

```
Title 'PF reactor for fast exothermic reaction'
Select painted      { make color-filled contour plots }
Variables  Temp(range=0,5) X_A(range=0,1)
definitions
  Lz = 1  r1=1  heat=0      gamma = 16  beta = 0.2  betat = 0.3
  BI = 1  T0 = 1  TW = 0.92  VRS = (1-X_A)*exp(gamma-gamma/Temp)
  xev=0.96  yev=0.25 { some plot points }
  initial value  Temp=T0   X_A=0
equations
  div(grad(Temp)) + heat + betat*VR = dt(Temp)
  div(grad(X_A)) + beta*VRS = dt(X_A)
  boundaries
  region 1
  start (0,0)  natural(Temp) = 0  natural(X_A) = 0  line to (r1,0)  { a mirror plane on X-axis }
  { "Strip Heater" at fixed temperature } value(Temp)=T0 + 0.2*uramp(t,t-0.05)
   { ramp the boundary temp, because  discontinuity is costly to diffuse }
  natural(X_A)=0 { no mass flow }
  arc(center=0,0) angle 5     {.. on outer arc }
  region 2
  natural(Temp)=BI*(TW-Temp)  natural(X_A)=0   { no mass flow }
  arc(center=0,0) angle 85    { ... on outer arc }
  natural(Temp) = 0  natural(X_A) = 0  line to (0,0) finish {another mirror plane on Y- axis }
  time 0 to 1
  plots  for cycle=10              { watch the fast events by cycle }
  contour(Temp) contour(X_A)  for t= 0.2 by 0.05 to 0.3{ show some surfaces during burn }
  surface(Temp)  surface(C) as
  histories history(Temp) at (0,0) (xev/2,yev/2) (xev,yev) (yev/2,xev/2) (yev,xev)
  history(C) at (0,0)  (xev/2,yev/2) (xev,yev) (yev/2,xev/2) (yev,xev)
  end
```

Figure 3.59 FlexPDE™ software for a fast exothermic reaction in a PF reactor.

The violent runaway of the reaction is clearly shown in Figs. 3.60 and 3.61, where the evolution of the reactant conversion and of the temperature is given for some points positioned in a fourth of the reactor section because of symmetry. As we can notice, time gives supplementary information about conversion with respect to the position in the z-axis of the reactor. From these figures, we can also

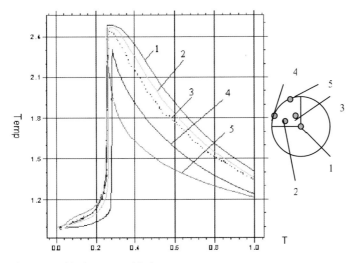

Figure 3.60 [block] caption. [block]

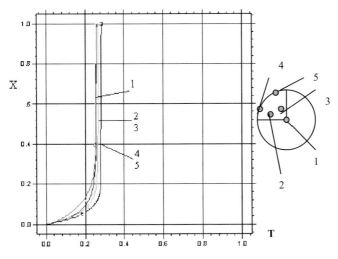

Figure 3.61 Evolution of the conversion of the reactant in the PF reactor when a fast exothermic reaction takes place.

observe that, after an induction period, when the reactant conversion reaches 0.15, the reaction becomes violent and in a very short time all the limiting reactant is consumed. The temperature evolution with time after the end of the reaction shown in Fig. 3.60 corresponds to the heat diffusion process. Indeed, the points positioned near the walls show a more rapid cooling than the points placed at the Z-axis of the reactor. With reference to the evolution through the Z-axis, the description of events is similar: after a short distance, when the conversion attains the above-mentioned value, the manifestation of the violent reaction that corresponds to a small z distance begins; in the remaining length, the gas mixture continues flowing and gas cooling takes place.

It is not difficult to observe in Fig. 3.59 that we can still change some parameters in order to increase the reaction temperature: (i) by increasing the temperature of the reactants at the reactor input (we use a T_0 value higher than 1); (ii) by increasing the input concentration of the limiting reactant (we increase the value of beta in the simulation software); (iii) by increasing the wall temperature (T_w)). In addition, we can consider an enhancement of the heat transfer through the walls by increasing the Biot number. Figures 3.62–3.64, which have been obtained with other start values for T_0, T_w, Bi, c_0, can be compared to Figs. 3.60 and 3.61. They aim to show the moments of reaction runaway more completely.

Figure 3.62 shows the temperature field of a quarter of the radial section of the reactor before the reaction firing. Combining the values of T_0, T_w and Bi results in an effective cooling of the reactor near the walls during the initial instants of the reaction (T = 0 − 0.05). In Fig. 3.63 is shown the temperature field when the dimensionless time ranges between T = 0.05 and T = 0.11. Here, the reaction runaway starts and we can observe that an important temperature enhancement occurs at the reactor centre, at the same time the reactant conversion increases (Fig. 3.64). The evolution of the reaction firing and propagation characterize this process as a very fast process. We can appreciate in real time that the reaction is completed in 10 s. It is true that the consideration of isothermal walls can be criticized but it is important to notice that the wall temperature is not a determining factor in the process evolution when the right input temperature and the right input concentrations of reactants have been selected.

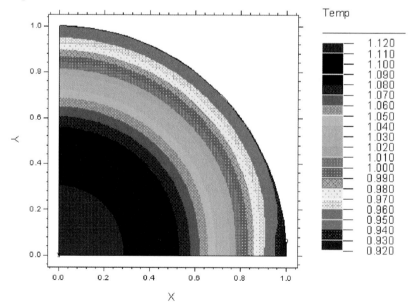

Figure 3.62 Temperature field before the reaction runaway for
T = 0.05. ($T_0 = 1.05$, $T_w = 0.9$, Bi = 10, $c_0/c_{00} = 1.4$.)

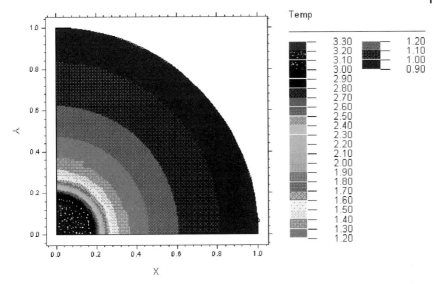

Figure 3.63 Temperature field for the time of reaction
$T = 0.11$. ($T_0 = 1.05$, $T_w = 0.9$ Bi = 10, $c_0/c_{00} = 1.4$.)

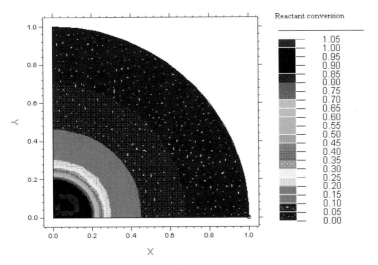

Figure 3.64 Conversion field for $T = 0.11$.
($T_0 = 1.05$, $T_w = 0.9$, Bi = 10, $c_0/c_{00} = 1.4$.)

With reference to the timing of the reaction firing, Table 3.15 presents the result of the simulations carried out with this purpose. The direction of the firing is already shown in this table. This can occur from the centre to the wall (C>W) or from the wall to the centre (W>C). Nevertheless, we will not investigate the technological problem involved with some of these input conditions now. We will mention this important fact: when the wall temperature is higher than that of the input reactants, the reaction ignition virtually starts at the input of the reactor. However, in terms of simulation, this behaviour can be modulated with the reaction kinetics.

Table 3.15 Time of reaction firing when Bi = 10 and $c_0/c_{00} = 1$.

T_0	1	1.1	1.1	0.9	0.5	0.3	0.85	0.92	0.98	0.95
Tw	0.92	0.85	0.75	2	2	2	1.1	1	0.98	0.95
T Start	0.25	0.08	0.08	0.01	0.01	0.01	0.26	0.42	0.28	0.7
Finish	0.3	0.1	0.12	0.04	0.09	0.13	0.32	0.48	0.3	0.78
Sense	C >W	C>W	C>W	W>C	W>C	W>C	W>C	W = C	C>W	C>W

As far as controlling the process is concerned, it seems to be interesting to have similar values for T_0 and T_w. In this case, after the reaction ignition in the whole radial section of the reactor input, the fast reaction propagation occurs towards the reactor centre. Before closing the discussions about these first simulations, we have to notice that modifications of the model and the associated software allow the application of this example to many other cases. For example, when the wall temperature is higher than the reactor temperature, we can simulate the cases of endothermic homogeneous reactions such as hydrocarbon cracking. Such application needs two major software modifications: a negative β_t value and a more complete kinetics. It is important to specify that the developed model can simulate the firing reaction where the limiting reactant is uniformly distributed into the reactor input. For the case when we have a jet-feed of limitative reactant where a firing reaction occurs [3.55], a new model construction is recommended.

The second set of simulations is oriented towards the analysis of the simultaneous heat and mass transfer when two fluids are separated by a porous wall (membrane). The interest here is to couple the species transport through a wall associated with the heat transfer and to consider that the wall heat conduction is higher than the heat transported by the species motion. The process takes place through a cylindrical membrane and we assume the velocity to be quite slow in the inner compartment of the membrane. The process is described schematically in Fig. 3.65. The transformation of the above general model in order to correspond to this new description gives the following set of dimensionless equations:

• the dimensionless concentration field of transferred species:

$$\frac{\partial C_A}{\partial T} = \left(\frac{\partial^2 C_A}{\partial X} + \frac{\partial^2 C_A}{\partial Y} \right)$$

(3.182)

- the dimensionless evolution of the temperature:

$$\frac{\partial T_p}{\partial T} = \frac{Sc}{Pr}\left(\frac{\partial^2 T_p}{\partial X} + \frac{\partial^2 T_p}{\partial Y}\right)$$

(3.183)

- the movement expression for the z axis:

$$\frac{w}{w_0} = \beta \frac{\alpha + T_p}{\alpha + 1}$$

(3.184)

- the initial and univocity conditions:

$$T = 0, \quad 0 \prec Z \prec 1, \quad 0 \prec Y \prec 1, \quad 0 \prec X \prec 1, \quad C_A = 0, \quad T_p = 1$$

(3.185)

$$T \succ 0, \quad 0 \prec X \prec 1, \quad 0 \prec Y \prec 1, \quad Z = 0, \quad C_A = 0, \quad T_p = 1$$

(3.186)

$$T \succ 0, \quad X = 1, \quad Y = 1, \quad \frac{dT_p}{dR} = Bi(T_p - T_w).$$

(3.187)

$$T \succ 0, \quad X = 1, \quad Y = 1, \quad \frac{dC_A}{dR} = Bi_D \frac{1}{(1 + Bi_p)}(1 - C_A)$$

(3.188)

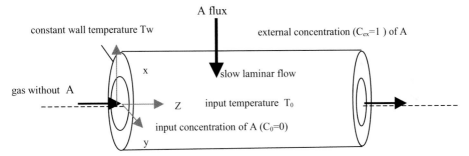

A flux

constant wall temperature Tw

external concentration (C_{ex}=1) of A

gas without A

slow laminar flow

input temperature T_0

input concentration of A (C_0=0)

Figure 3.65 Explanatory scheme for heat and mass transport through a porous wall.

In the set of relations (3.182)–(3.188), β represents the coefficient for the velocity increase due to the species transport through the wall, Bi is the heat transfer Biot number ($Bi = (\alpha r_1)/\lambda$), Bi_D is the mass transfer Biot number for the gaseous phase ($Bi_D = (kr_1)/D_A$) and Bi_p is the Biot number for the porous wall ($Bi_p = (k\delta_w)/D_{Aw}$). Two new parameters δ_w and D_{Aw}, respectively, represent the wall thickness and the wall effective diffusion coefficient of species. The model described by the set of relations (3.182)–(3.188) can easily be modified to respond to the situation of a membrane reactor when a chemical reaction occurs inside the cylindrical space and when one of the reaction products can permeate through the wall. The example particularized here concerns the heat and mass transfer of a gaseous fluid. Figures 3.66 and 3.67, respectively, present the dynamic evolution of the heat thermal field and the concentration of species A. As was explained in the previous example, the time parameter can be transformed into the z position using relation (3.184). It is interesting to observe that the steady state permeation of A is attained (the enhancement of the species concentration is linear) at dimensionless time T = 0.2 (which is relatively fast in real time). At T = 1 thermal equilibrium is reached and the heat transfer phenomenon disappears.

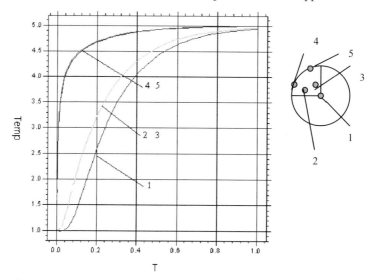

Figure 3.66 Temperature evolution inside the cylindrical membrane (Bi = 10, Bi_D = 2, Bi_p = 100 , T_w = 5).

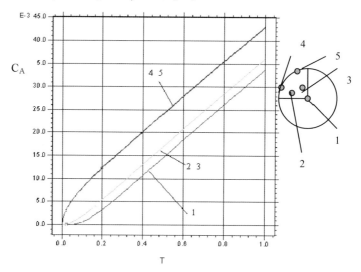

Figure 3.67 Evolution of concentration of A inside the cylindrical membrane (Bi = 10, Bi_D = 2, Bi_p = 100, T_w = 5).

In order to extract more data with respect to the gas composition in the cylindrical membrane, we carried out simulations taking a long-term gaseous permeation into account. Figure 3.68 shows this evolution for two different values of the membrane Biot number, which, in fact, is a measure of the membrane mass transfer resistance. We can observe that, over a long period, the dimensionless species concentration increases linearly, indicating that the permeate flux through the membrane wall has a constant value. This observation is in good agreement with the high value of the Biot number.

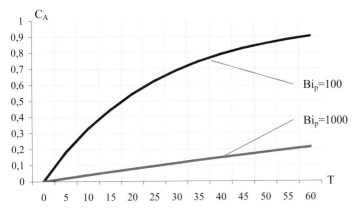

Figure 3.68 Evolution with time of the dimensionless concentration of species A.

If a light gas permeates through a 1 mm thick membrane, the Biot number corresponds to the highest value in Fig. 3.68. If we assume the pressure to be equal to 1 and normal temperature, then the gas flux will be 10^{-7} kmole/(m^2 s). With some light modifications, the software used for these simulations can be adapted to simulate the cases where the values of the Biot number of the membrane change with time. The Biot number evolution can occur in different situations when the membrane transport properties change, such as when the membrane is continuously clogged (e.g. in hydrocarbon dehydrogenation reactions in which coke is formed). The simulations presented here, as well as the observations exposed during the model presentation, show that it is not difficult to model and to simulate more complex cases such as membrane reactors. In such multifunctional chemical engineering devices, one or more reaction products or reactants can permeate through the membrane with different selectivity [3.56].

In membrane reactors, the reaction and separation processes take place simultaneously. This coupling of processes can result in the conversion enhancement of the thermodynamically-limited reactions because one or more of the product species is/are continuously removed. The performance of such reactors depends strongly on the membrane selectivity as well as on the general operating conditions which influence the membrane permeability.

3.5
Some Aspects of Parameters Identification in Mathematical Modelling

The notion of model parameters defines one or more numerical values that are contained as symbolic notations in the mathematical model of a process. These numerical values cannot be obtained without any experimental research. In reality, the most important part of experimental research is dedicated to the identification of the models' parameters. Generally, all the experimental works, laboratory methods and published papers beginning with the words "*determination of......*" are in fact particular problems of identification of parameters. The chemical and biochemical sciences use various and countless models that introduce various types of parameters into their description. We can single out the real parameters that have a physical dimension so as to accept a dimensional formula. Indeed, they are related with a process state or with material properties that characterize the process. They differ from abstract parameters, which can have a dimensional formula but are an artificial creation. The parameters characterize the investigated process and not the mathematical model in which they appear as a consequence. At the same time, they can be considered as a special class of input variables of the process. Indeed, when we start with a problem of parameter identification, then we *a priori* accept a mathematical model, which contains these parameters for the process evolution. The problem of identifying parameters is formulated schematically in Fig. 3.69.

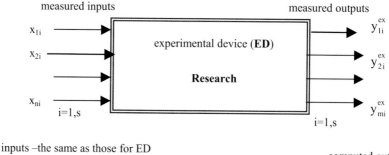

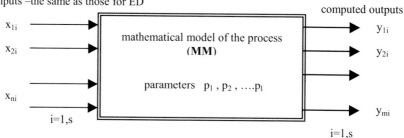

Figure 3.69 Introduction to a problem of identifying the parameters of a process.

We can notice that, when a total number of "s" experimental measures have been made for the outputs, the result of the experimental investigation of the process is given by the vector:

$$Y_i^{ex} = \begin{bmatrix} y_{1i}^{ex} \\ y_{2i}^{ex} \\ y_{mi}^{ex} \end{bmatrix} \quad , \; i = 1, \; s$$

The vector Y_i^{ex} is obtained for s coupled values of the input process variables, so we can consider the vector inputs as follows:

$$X_i = \begin{bmatrix} x_{1i} \\ x_{2i} \\ x_{ni} \end{bmatrix} \quad , \quad i = 1, s$$

At the same time, the mathematical model of the process (MM) can produce – for the established values of the vector locations of the input process – the following values of the output vector:

$$Y_i = \begin{bmatrix} y_{1i} \\ y_{2i} \\ y_{ni} \end{bmatrix} \quad , \quad i = 1, s$$

If the values of the vector of parameters $P = [p_1, p_2, \ldots\ldots p_l]$ are known and if we obtain analogous results for the measured and MM outputs for the same input variables, then we can consider that the set of parameter values of model $p_1, p_2, \ldots$ p_l are good enough.

The theory and the practice of parameter identification concern the assembly of procedures and methods showing the estimation of the values of $p_1, p_2, \ldots p_l$ with the objective of having similar values for vectors Y_i^{ex} and Y_i. Generally, the parameters of a process are linked with various types of dependences called *constraints*. Constraints show that each parameter presents a region where a minimal and a maximal value is imposed and can be classified according to equality, inequality and inclusion constraints. Inclusion constraints are frequently transformed into inequality constraints because the latter have the quality of being easily introduced into the overall identification problem.

The formulation of the mathematical problem of parameter identification for an actual case needs the use of the following general particularizations given below:

1. *One relation or an assembly of relations* that contains the condition necessary to impose the absence of important differences between the computed outputs and experimental outputs. This relation or assembly of relations frequently contains the requirement of a minimal dispersion (variance) between computed and experimental process outputs. So we need to minimize the function:

$$\Phi(p_1, p_2, ...p_l) = \sum_{k=1}^{m} \sum_{i=1}^{s} (y_{ki} - y_{ki}^{ex})^2 \qquad (3.189)$$

In the case of process exit coded by k, this equation can also be written as follows:

$$\Phi(p_1, p_2, ...p_l) = \sum_{k=1}^{m} \sum_{i=1}^{s} [(F_k(p_1, p_2, ...p_l, x_{1i}, x_{2i}, ...x_{ni}) - y_{ki}^{ex}]^2 = min \qquad (3.190)$$

where $F_k(p_1, p_2, ...p_l, x_{1i}, x_{2i}, ...x_{ni})$, $k = 1, m$ give the computed y_{ki} values of the model. If a minimal dispersion between the computed and experimental results is necessary for each process output, then we must minimize the following assembly of functions:

$$\Phi_k(p_1, p_2, ...p_l) = \sum_{i=1}^{s} [F_k(p_1, p_2, ...p_l, x_{1i}, x_{2i}, ...x_{ni}) - y_{ki}^{ex}]^2 = min \; ; \; k = 1, 2, .m \qquad (3.191)$$

In addition to the formulation that minimizes the dispersion described above, other mathematical expressions have been suggested with the purpose of obtaining the values of the parameters by requiring the model to reproduce the experimental data.

    2. *An assembly of relations* that contains the introduction of expressions for equality type constraints; this assembly links some or all of the parameters of the model. From the mathematical viewpoint we can write these relations as follows:

$$H_j(p_1, p_2, ...p_l) = 0......j = 1, ...l \qquad (3.192)$$

We observe here that we have "l" independent relations for "l" number of parameters. However, it is not strictly necessary to have the same number of parameters and relations that characterizes the equality type inter-parameter links.

    3. *An assembly of relations* that contains the inequality type constraints which are considered for all parameters of an incomplete group of parameters or for only one parameter. We write these relations as follows:

$$G_j(p_1, p_2, ...p_l) \succ 0......j = 1, ...l \qquad (3.193)$$

$$L_j(p_1, p_2, ...p_l) \prec 0......j = 1, ...l \qquad (3.194)$$

It is important to notice that the equality and inequality type constraints described above can be absent in a problem of parameter identification.

The methods for identifying the parameters of a model can be classified in terms of the complexity of the mathematical model and constraints accepted for its parameters.

All the methods used for the identification of parameters are in fact the particularizations of the general methods to determine an extreme function. This function can be simple or complicated and can be given by one or more algebraic equations or can be introduced by an assembly of relations that contains differential equations or partial differential equations. The classification for these methods is given in Table 3.16.

Table 3.16 Classification of the methods for identifying the parameters of a model.

n	Type of method	Method name	Required conditions	Examples
1	Analytical methods	Pure analytical method	1. Deterministic mathematical model given by analytical functions that are differentiable with respect to each parameter 2. Without constraints	The latest small squares method
		Lagrange coefficients method	1. Deterministic mathematical model given by analytical functions that are differentiable respect to each parameter 2. With constraints of equality type	
		Variational methods	1. Deterministic mathematical model given by analytical or numerical functions that are differentiable respect to each parameter 2. With or without constraints	
2	Mathematical programming methods	Method of geometrical programming	1. Deterministic mathematical model given as: $$\Phi(p_1...p_l) = \sum_{j=1}^{l} C_j P_j(p_1..p_l)$$ where $P_j(p_1..p_l) = \sum_{i=1}^{l} p_i^{\alpha ij}$ 2. Without constraints.	
		Methods of dynamic programming	1. Mathematical model that describe a process with sequential states 2. With or without constraints	
		Methods of linear programming	1. Deterministic mathematical model given as: $\Phi(p_1..p_l) = \alpha_1 p_1 + \alpha_2 p_2 + ..\alpha_l p_l)$ 2. With inequality type constraints	The simplex method
3	The gradient methods	Various methods	1. Deterministic process mathematical model especially given by differential equations 2. With or without inequality type constraints	The very high slope method

n	Type of method	Method name	Required conditions	Examples
4	The combined methods	Various variants	1. Process mathematical models with distributed inputs 2. Capacity to be associated with a Kalman filter 3. Without inequality type constraints	The maximum likelihood method

It is important to point out that to identify the parameters of the model, the experimental research made with physical laboratory models (apparatus) has previously established the experimental working methods that allow the identification of the actual process parameters. These experimental methods tend to be promoted as standardized methods and this reduces the dimension of the problem that is formulated for identifying the parameters of the model to the situations where $\Phi(p_1, p_2, ...p_l)$ contains one, two or a maximum of three parameters to be estimated simultaneously.

3.5.1
The Analytical Method for Identifying the Parameters of a Model

This type of method includes the classical methods, which are based on the observation that the minimal value of function $\Phi(p_1, p_2, ...p_l)$ is quite near zero. Indeed, we can derive the conditions of a minimal value of $\Phi(p_1, p_2, ...p_l)$ which can be written as a system of algebraic equations (3.195) where the unknowns are parameters $p_1, p_2, ...p_l$.

$$\begin{cases} \dfrac{\partial \Phi(p_1, p_2....p_l)}{\partial p_1} = 0 \\[2mm] \dfrac{\partial \Phi(p_1, p_2....p_l)}{\partial p_2} = 0 \\[2mm] \dfrac{\partial \Phi(p_1, p_2....p_l)}{\partial p_L} = 0 \end{cases} \tag{3.195}$$

The applicability of this method is limited by the form of function $\Phi(p_1, p_2, ...p_l)$ that must present an analytical expression with respect to each parameter $(p_1, p_2, ...p_l)$. At the same time, the dimension and the nonlinearity of system (3.195) can also be considered as the limitative factors of this method. Various concrete formulations of system (3.195) can be obtained for the mathematical model of an actual process. For example, if, in Fig. 3.69, we consider that we have only one output where y_i^{ex} and y_i $i = 1, s$ are, respectively, the measured and the computed output values of the process and if we accept that the mathematical model of the process is given by Eqs. (3.196), then the function which must be minimized is given by relation (3.197):

$$y = p_0 + p_1 f_1(x_1, x_2, \ldots x_n) + p_2 f_2(x_1, x_2, \ldots x_n) + \ldots + p_l f_l(x_1, x_2, \ldots x_n) \quad (3.196)$$

$$\Phi(p_1, p_2 \ldots p_l) = \sum_{i=1}^{s} \left([p_0 + p_1 f_1(x_{1i} \ldots x_{ni}) + \ldots p_L f_L(x_{1i}, x_{2i} \ldots x_{ni})] - y_i^{ex} \right)^2 = \min$$
$$(3.197)$$

For this case the system (3.195) leads to the following algebraic system:

$$\begin{cases} sp_0 + p_1 \sum_{i=1}^{s} f_1(x_{1i}, x_{2i} \ldots x_{ni}) + p_2 \sum_{i=1}^{s} f_2(x_{1i}, x_{2i} \ldots x_{ni}) + \ldots + p_L \sum_{i=1}^{s} f_L(x_{1i}, x_{2i} \ldots x_{ni}) = \sum_{i=1}^{s} y_i^{ex} \\[2mm] p_0 \sum_{i=1}^{s} f_1(x_{1i}, x_{2i} \ldots x_{ni}) + p_1 \sum_{i=1}^{s} [f_1(x_{1i}, x_{2i} \ldots x_{ni})]^2 + p_2 \sum_{i=1}^{s} f_2(x_{1i}, x_{2i} \ldots x_{ni}) f_1(x_{1i}, x_{2i} \ldots x_{ni}) + \ldots \\[2mm] p_L \sum_{i \gg 1}^{s} f_L(x_{1i}, x_{2i} \ldots x_{ni}) f_1(x_{1i}, x_{2i} \ldots x_{ni}) = \sum_{i=1}^{s} y_i^{ex} f_1(x_{1i}, x_{2i} \ldots x_{ni}) \\[2mm] p_0 \sum_{i=1}^{s} f_2(x_{1i}, x_{2i} \ldots x_{ni}) + p_1 \sum_{i=1}^{s} f_1(x_{1i}, x_{2i} \ldots x_{ni}) f_2(x_{1i}, x_{2i} \ldots x_{ni}) + p_2 \sum_{i=1}^{s} [f_2(x_{1i}, x_{2i} \ldots x_{ni})]^2 + \ldots \\[2mm] p_L \sum_{i \gg 1}^{s} f_L(x_{1i}, x_{2i} \ldots x_{ni}) f_1(x_{1i}, x_{2i} \ldots x_{ni}) = \sum_{i=1}^{s} y_i^{ex} f_1(x_{1i}, x_{2i} \ldots x_{ni}) \\[2mm] p_0 \sum_{i=1}^{s} f_L(x_{1i}, x_{2i} \ldots x_{ni}) + p_1 \sum_{i=1}^{s} f_1(x_{1i}, x_{2i} \ldots x_{ni}) f_L(x_{1i}, x_{2i} \ldots x_{ni}) + p_1 \sum_{i=1}^{s} f_1(x_{1i}, x_{2i} \ldots x_{ni}) f_L(..) + \ldots \\[2mm] p_L \sum_{i \gg 1}^{s} [f_L(x_{1i}, x_{2i} \ldots x_{ni})]^2 = \sum_{i=1}^{s} y_i^{ex} f_L(x_{1i}, x_{2i} \ldots x_{ni}) \end{cases}$$

$$(3.198)$$

The solution of this system allows the estimation of the numerical values of parameters $p_1, p_2, \ldots p_l$. This system is in fact the expression of the least small squares method. This method is used for the development and solution of example 3.5.1.1.

3.5.1.1 The Pore Radius and Tortuosity of a Porous Membrane for Gas Permeation

Gaseous permeation can be used for the characterization of porous membranes using an apparatus working with the technique of fixed volume–variable pressure as shown in Fig. 3.70. The technique, which was initially developed for dense polymer membranes, is based on the recording of the pressure evolution with time of a downstream compartment, which is separated from an upstream compartment filled with a pure gas by a flat membrane. Before starting the experiments, both compartments are put under very low pressure and, at the initial time of the measurements, a relatively high pressured pure gas is introduced into the upstream compartment [3.59].

The pressure evolution of the downstream compartment versus time is recorded in a curve which generally has a typical "s" shape which presents an initial small nonlinear increase and later becomes time linearly dependent. At the end of the experiment, when the gradient becomes negligible, the curve presents a nonlinear decrease [3.60].

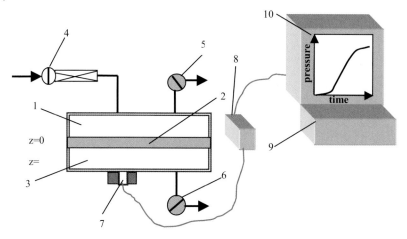

Figure 3.70 Scheme of a gas permeation apparatus. 1 – upstream compartment,
2 – membrane, 3 – downstream compartment (V_{inf}), 4 – pure gas entrance valve,
5 and 6 – vacuum valves, 7 – pressure transducer (P_{inf}), 8 – data acquisition system,
9 – computer, 10 – curve pressure vs. time.

Once the descriptive model has been realized, we need to make the mathematical model of the process, which can be used to identify the mean pore radius of the membrane pores and the associated tortuosity. Before starting with the establishment of the model, we consider that the elementary processes allowing the gas flow through the membrane are a combination of Knudsen diffusion with convective flow. If we only take into account the linear part of the curve of the pressure increase with time then we can write:

$$p_1 = \alpha_1 + \beta\tau \tag{3.199}$$

Indeed, the gas flow rate that permeates through the membrane is:

$$G_M = V_{inf}\frac{dc}{d\tau} = \frac{V_{inf}}{RT}\frac{dp_{inf}}{d\tau} = \frac{V_{inf}}{RT}\beta \tag{3.200}$$

and then we obtain for the measured gas flux:

$$N_A = \frac{G_M}{\varepsilon Sm} = \frac{V_{inf}}{RT\varepsilon Sm}\beta \tag{3.201}$$

the slope β of the pressure–time dependence is estimated with the variation of p_i with τ_i, and, consequently, we can transform the equation above as follows:

$$N_{Ai}^{ex} = \frac{V_{inf}}{RT\varepsilon Sm}\frac{p_{1i+1} - p_{1i}}{\tau_{i+1} - \tau_i} \tag{3.202}$$

As explained above, we consider that the gas transport is carried out by the Knudsen and hydrodynamic flow through the porous media, then the theoretical expression for the gas flux is given by the following relation:

$$N_A = -\left(\frac{2}{3}r\psi\sqrt{\frac{8RT}{\pi M}} + r^2\psi\frac{p}{8\eta}\right)\frac{dp}{RTdx} \tag{3.203}$$

Because N_A is constant, we can separate out the variables of the relation (3.203), then the integral relation (3.205) concerning the gas flux through the membrane is obtained after integrating the whole thickness of the membrane:

$$N_A \int_0^\delta dx = \int_{p_1}^{p_2}\left(\frac{2}{3}r\psi\sqrt{\frac{8RT}{\pi M}} + r^2\psi\frac{p}{8\eta}\right)dp \tag{3.204}$$

$$N_A = \frac{4\sqrt{2}}{3\delta}\frac{r\psi}{\sqrt{RT\pi M}}(p_2 - p_1) + \frac{r^2\psi}{16\eta\delta}(p_2^2 - p_1^2) \tag{3.205}$$

Relation (3.205) can be written in the form (3.206) where it shows the expression for the model of instantaneous gas flux through the membrane:

$$N_{Ai} = \frac{4\sqrt{2}}{3\delta}\frac{r\psi}{\sqrt{RT\pi M}}(p_{2i} - p_{1i}) + \frac{r^2\psi}{16\eta\delta}(p_{2i}^2 - p_{1i}^2) \tag{3.206}$$

If p_{1i} increases linearly with time; then p_{2i} can be calculated by $p_{2i} = p_0 - \beta_i\tau_i$ in Eq. (3.206). Here, bottom and top compartments have been considered to have the same volume. It is not difficult to observe that the parameters requiring identification are $r\psi$ and $r^2\psi$ where r is the mean pore radius and ψ the tortuosity. In this case, in accordance with relation (3.189), the function for the minimization will be written as follows:

$$\Phi(r\psi, r^2\psi) =$$

$$\sum_{i=1}^s\left[\frac{4\sqrt{2}}{3\delta}\frac{r\psi}{\sqrt{RT\pi M}}(p_{2i} - p_{1i}) + \frac{r^2\psi}{16\eta\delta}(p_{2i}^2 - p_{1i}^2) - \frac{V\inf}{RT\varepsilon Sm}\frac{p_{1i+1} - p_{1i}}{\tau_{i+1} - \tau_i}\right]^2$$

By introducing the notations $A = \dfrac{4\sqrt{2}}{3\delta}\dfrac{r\psi}{\sqrt{RT\pi M}}$, $B = \dfrac{r^2\psi}{16\eta\delta}, \alpha = \dfrac{V\inf}{RT\varepsilon Sm}$ and by computing the conditions that impose the minimal value of the function

$\Phi(r\psi, r^2\psi)$, we obtain equations system (3.207) where N represents the number of experiments considered for the parameter identification.

$$\begin{cases} A\sum_{i=1}^N(p_{2i} - p_{1i})^2 + B\sum_{i=1}^N(p_{2i}^2 - p_{1i}^2)(p_{2i} - p_{1i}) = \alpha\sum_{i=1}^N\left(\dfrac{p_{1i+1} - p_{1i}}{\tau_{i+1} - \tau_i}\right)(p_{2i} - p_{1i}) \\[4mm] A\sum_{i=1}^N(p_{2i} - p_{1i})(p_{2i}^2 - p_{1i}^2) + B\sum_{i=1}^N(p_{2i}^2 - p_{1i}^2)^2 = \alpha\sum_{i=1}^N\left(\dfrac{p_{1i+1} - p_{1i}}{\tau_{i+1} - \tau_i}\right)(p_{2i}^2 - p_{1i}^2) \end{cases} \tag{3.207}$$

The solution to equation system (3.207) will allow calculation of the values of $r\psi$ and $r^2\psi$, which are obtained after the estimation of A and B.

In our actual example, the first step for this calculation is the determination of r ψ and $r^2\psi$ by using data from Tables 3.17 and 3.18. These data have been obtained with an experimental device with $V_{inf} = V_{sup} = 7 * 10^{-5}$ m^3 $\delta = 4 * 10^{-3}$ m and for the following gases: He (M = 2 kg/kmol, η = 10^{-5} kg/ (m s)) and N$_2$ (M = 28 kg/kmol, η = $1.5*10^{-5}$ kg/ (m s)). The starting pressure at the upstream compartment is $p_0 = 2 * 10^5$ N/m^2 and $1.5*10^3$ N/m^2 for the downstream compartment.

Table 3.17 Pressure evolution in the downstream compartment for He permeation.

C .n	1	2	3	4	5	6
τ (s)	10	20	50	70	90	110
P_1 (N/m^2)	10^4	$2.5*10^4$	$4*10^4$	$5.52*10^4$	$7*10^4$	$8.41*10^4$

Table 3.18 Pressure evolution in the downstream compartment for N$_2$ permeation.

C .n	1	2	3	4	5	6
τ (s)	40	80	120	160	200	240
P_1 (N/m^2)	$2.06*10^4$	$3.41*10^4$	$4.72*10^4$	$6.06*10^4$	$7.41*10^4$	$8.7*10^4$

The algorithm for the experimental data processing follows the steps:
1. We introduce the fixed data of the problem:
   $V_{inf}, V_{sup}, \delta, S_m, \varepsilon, M, \eta, p_0, R; T; N$;
2. We give the evolution of p_{1i} versus τ_i, i = 1,N;
3. We compute the mean slope of the p_{1i} versus τ_i dependence:

$$\beta = \frac{p_{1N-1} - p_{12}}{\tau_{N-1} - \tau_2};$$

4. We establish the corresponding p_{2i} value for each τ_i:
   $p_{2i} = p_0 - \beta_i\tau_i$, i = 1,N;
5. We compute: $\alpha = V_{inf}/(RT\varepsilon S_m)$;
6. We obtain the values of the following sums:

$$S_1 = \sum_{i=1}^{N} (p_{2i} - p_{1i})^2, \quad S_2 = \sum_{i=1}^{N} (p_{2i}^2 - p_{1i}^2)(p_{2i} - p_{1i}),$$

$$S_3 = \alpha \sum_{i=1}^{N} \left(\frac{p_{1i+1} - p_{1i}}{\tau_{i+1} - \tau_I}\right)(p_{2i} - p_{1i}), \quad S_4 = \sum_{i=1}^{N} (p_{2i}^2 - p_{1i}^2)^2,$$

$$S_5 = \alpha \sum_{i=1}^{N} \left(\frac{p_{1i+1} - p_{1i}}{\tau_{i+1} - \tau_I} \right) (p_{2i}^2 - p_{1i}^2);$$

7. We solve system (3.207) for A and B which is written as follows:

$$\begin{cases} AS_1 + BS_2 = S_3 \\ AS_2 + BS_4 = S_5 \end{cases} \tag{3.208}$$

8. We compute rψ and r²ψ by using the computed A and B.

Figure 3.71 contains the MathCAD™ working text of this problem in the case of N_2 permeation. The values obtained for rψ $= 1.2 * 10^{-10}$ m and r²ψ $= 0.85 * 10^{-20}$ m² are almost the same as those calculated for He permeation.

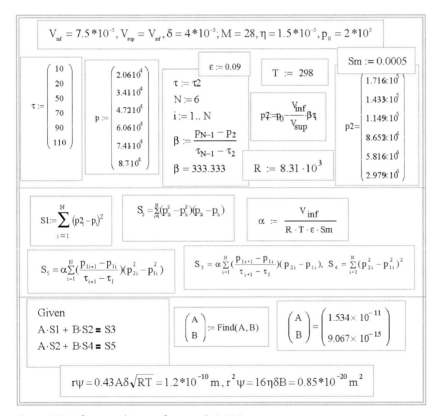

Figure 3.71 Software working text for example 3.6.1.1.

3.5.2
The Method of Lagrange Multiplicators

This high confidence method is used when one or more equality constraints are imposed on the parameters of the process [3.61]. So, if our problem is to obtain the minimal value of function $\Phi(p_1, p_2, ...p_L)$ and the parameters' requirement to verify the constraints $f_i(p_1, p_2, ..p_L)$, i = 1, g simultaneously, then the solution is obtained from the formulation of the following auxiliary function:

$$L(p_1, p_2 ...p_L, \lambda_1, ...\lambda_g) = \Phi(p_1, p_2, ...p_L) + \sum_{i=1}^{g} \lambda_i f_i(p_1, p_2 .....p_L) \qquad (3.209)$$

Function $L(p_1, p_2, ...p_L, \lambda_1, ...\lambda_g)$ supports the same minimization as $\Phi(p_1, p_2, ...p_L)$ but here, the number of parameters is increased with $\lambda_1, \lambda_2, .., \lambda_g$ which are called the Lagrange multiplicators. Then, the equation system that must be solved here is written as:

$$\begin{cases} \dfrac{\partial L(p_1, p_2, ..p_L, \lambda_1, ...\lambda_g)}{\partial p_k} = 0 & , \quad k = 1, L \\[4mm] \dfrac{\partial L(p_1, p_2, ..p_L, \lambda_1, ...\lambda_g)}{\partial \lambda_i} = 0 & , \quad i = 1, g \end{cases} \qquad (3.210)$$

From Eq. (3.210) we can obtain one or more set(s) of values for $p_1, p_2, ..., p_L$ which give for function $\Phi(p_1, p_2, ...p_L)$ one or more extreme values. If we have various extremes for this function, we choose the set of parameters which gives a physical meaning of the problem. With respect to the Lagrange method, we can observe that each equality type constraint introduces its parameter in the building of $L(p_1, p_2, ...p_L, \lambda_1, ...\lambda_g)$. Indeed, this method is strictly recommended when we have equality type constraints.

3.5.2.1   One Geometrical Problem

A chemical engineer who is designing a drug factory has to solve a problem which concerns the building of a spherical reservoir over a conical support as shown in Fig. 3.72. For this construction, the total volume must not exceed 5 m^3. At the same time, the relation between the sphere diameter and the cone height is imposed in accordance with the golden section principle ($D_s = I/5$).

The building process must be carried out minimizing the operations such as surface finishing, colouring, etc. After a few days, the engineer concludes that it is not possible to build such a structure, even though, the motivation for the response remains unknown. In order to know the reasons that motivated the engineer's decision we will use the parameters identification method of Lagrange multiplicators. To do so, we have to minimize the function that represents the surface of the building:

$$F(D, G, I) = \pi DG + \pi DI/2 + \pi(I/5)^2 \qquad (3.211)$$

with the constraint that imposes a fixed building volume :

$$C(D, G, I) = \pi G(D/2)^2 + \pi(I/3)(D/2)^2 + \pi/6(I/5)^3 - V \qquad (3.212)$$

The associated Lagrange function $L(G, I, D, \lambda) = F(G, H, D) + \lambda C(G, H, I)$ gives the following equation system:

$$\begin{cases} G + 0.5\ I + 0.5\lambda G\ D + 0.166\lambda I\ D = 0 \\ 0.5D + 0.08I + 0.0833\lambda D^2 + 0.004\lambda I^2 = 0 \\ 0.25\lambda D = -1 \\ 0.25GD^2 + 0.0833ID^2 + 0.001333\ I^3 = V/\pi \end{cases} \qquad (3.213)$$

Considering that the fixed volume is $V = 5$ m³, the solution to this system results in the following values for the heights and diameter: $G = -1.45$ m, $I = 8.8$ m and $D = 1.43$ m. We can notice here that the value of G is not realistic.

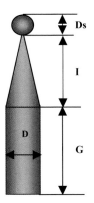

Figure 3.72 Scheme of example 3.6.2.1.

3.5.3
The Use of Gradient Methods for the Identification of Parameters

Among the methods used to identify the parameters of a process, the gradient methods play an important role because of their excellent adaptation to software making. These methods are generally quite efficient for solving problems that require the establishment of extreme positions for the assemblies of linear or non-linear functions. This statement is especially true when the functions of assemblies are given through differential or partial differential equations. This is the major reason why these methods are widely used. They are based on the establishment of the values of momentary parameters that produce the highest variation of the minimized or maximized function. From the geometrical viewpoint, this fact is equivalent to a displacement of the function along its gradient towards the extreme position. It is known that the gradient in a point of the surface of response has an orthogonal state with respect to the surface. The different gradient methods are classified depending on: (i) the procedure for the localization of the calculation point; (ii) the length of the step that characterizes the motion of the calculation point; (iii) the number of tests along the established direction; (iv) the criteria used to stop the calculation; (v) the global method simplicity.

The most important aspect of these methods, which follow the localization of an extreme for a given function, is represented by the identification of the most rapid variation of the function for each calculation point on the direction. For this problem of parameter identification, the function is given by the expression $\Phi(p_1, p_2, ...p_L)$. The graphic representation of Fig. 3.73 shows the function–gradient relation when the vector gradient expression is written as in relation (3.214).

$$\vec{grad}\Phi = \frac{\partial\Phi}{\partial p_1}\vec{i_1} + \frac{\partial\Phi}{\partial p_2}\vec{i_2} + ... + \frac{\partial\Phi}{\partial p_L}\vec{i_L} = vect\left(\frac{\partial\Phi}{\partial p_1}, \frac{\partial\Phi}{\partial p_2}, ...\frac{\partial\Phi}{\partial p_L}\right) \tag{3.214}$$

Here $\vec{i_1}, \vec{i_2}, ..., \vec{i_L}$ are the axis vectors expressed in unitary coordinates.

From the theoretical viewpoint, the scalar value of the partial derivate $\partial\Phi/\partial p_i$ is the vector gradient projection to the axis p_i. Indeed, it can be described with their module and the spatial angle between the vector of the gradient and the axis p_i as in relation (3.215):

$$\vec{grad}\Phi = \sqrt{\sum_{i=1}^{L}\left(\frac{\partial\Phi}{\partial p_i}\right)^2} \cos(\vec{grad}\Phi, \vec{i_i}) \tag{3.215}$$

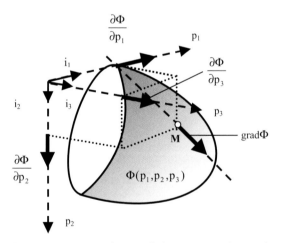

Figure 3.73 Function $\Phi(p_1, p_2, p_3)$, the M point gradient and its axis projections.

Concerning the problem of the direction of the advancement, it is important to select the length of the step of progression. It is evident that this selection first depends on the relations characterizing the response surface. When the processes are described with a complex model, this surface can only be given in a numerical form. A small length of the advancing step imposes a long and difficult computation. When the advancing step is big, it is possible to cross over the wanted extreme. These problems can be avoided using a variable length step; at the beginning, we use a big length step and, when an extreme neighbourhood is detected, the length of the step is progressively decreased. The step dimension $\Delta p_i, i = 1, ..$ L must verify first the condition that all calculation points are placed onto the gradient line beginning at the starting point and, secondly, if the constraints are active, the step dimension must respect them. The length of the step for the variable (parameter) p_i is computed using the partial derivates of the problem function with respect to the current calculation point:

$$\Delta p_i = \frac{k \frac{\partial \Phi}{\partial p_i}}{\sqrt{\sum_{i=1}^{L} \left(\frac{\partial \Phi}{\partial p_i}\right)^2}} \qquad (3.216)$$

Here k is a constant value, which is the same for the displacement of all variables (parameters). It is not difficult to appreciate that the value of k is very important for the step length.

Concerning the requirement to have an orthogonal gradient to the surface response of the process, we can notice that it first imposes the base point; if, for one of these 'L' directions, the length of the step is too big, then the vector that starts from the base point should not respect the orthogonal condition between the surface of the response and the new point where we will stop the motion. The selection of an adequate step length presumes that the derivates of the function related to the new point stay close to the derivates of the base point.

Despite the differences that exist between various gradient methods, the algorithm to determine the extreme point for a given function remains identical with respect to some general common guidelines [3.62, 3.63]:

1. we choose a base point;
2. starting from this point we establish the direction of the development;
3. we find the step length to prepare the motion along the gradient line;
4. we establish the position of the new point and consider whether it is a current point or must be transformed into a new base point;
5. we compare the value of the function for the new point and for the new base point with the value of the function of the former base point; the value of the new function is normally lower;
6. we select the new development direction for the new base point and the computation gives the area of the minimum function value; here small motion steps are recommended.

3.5.3.1 Identification of the Parameters of a Model by the Steepest Slope Method

The steepest slope method is a particular class of gradient method. This method will be illustrated with an example of the minimization of a function with two parameters with an explicit graphic interpretation. Figure 3.74 gives an excellent introduction to the steepest slope method (SSM) by showing some curves with constant Φ. They are placed around the minimum function and near the first base point. The motion progresses along the gradient line which is localized for each selected point. The development of this method starts at the first base point $M_0(p_1^0, p_2^0)$ beginning the exploration of the steepest slope by computation. We increase p_1^0 by δp_1^0 and establish the value of $\Phi(p_1^0 + \delta p_1^0, p_2^0)$. Now we repeat the computation by increasing p_2^0 by δp_2^0 and then $\Phi(p_1^0, p_2^0 + \delta p_2^0)$ is obtained. With $\Phi(p_1^0 + \delta p_1^0, p_2^0)$ and $\Phi(p_1^0, p_2^0 + \delta p_2^0)$ we can estimate the values of the two partial derivates:

$$\left(\frac{\partial \Phi}{\partial p_1}\right)_0 = \frac{\Phi(p_1^0 + \delta p_1^0, p_2^0) - \Phi(p_1^0, p_2^0)}{\delta p_1^0} \tag{3.217}$$

$$\left(\frac{\partial \Phi}{\partial p_2}\right)_0 = \frac{\Phi(p_1^0, p_2^0 + \delta p_2^0) - \Phi(p_1^0, p_2^0)}{\delta p_2^0} \tag{3.218}$$

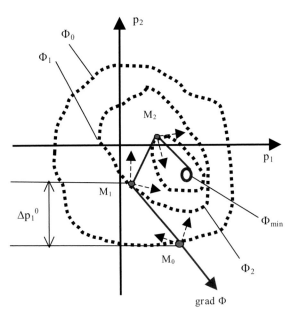

Figure 3.74 Scheme for the SSM graphic introduction.

These derivates allow the selection of the length of the step displacement or, in other words, they control the computation of each parameter modification. Δp_1^0 and Δp_2^0 are recognized when we finish the exploring computation so when we have established the vector $\overrightarrow{\text{grad}}\Phi$. The values of Δp_1^0 and Δp_2^0 are proportional to the module of the vector $\vec{\text{grad}}\Phi$ but they are in the opposite direction. If the modifications of the values of these parameters can be determined using the partial derivates shown in relation (3.219), then we can assert that the observations advanced previously were correct.

$$\Delta p_1^0 = -\alpha \left(\frac{\partial \Phi}{\partial p_1} \right)_0, \quad \Delta p_2^0 = -\alpha \left(\frac{\partial \Phi}{\partial p_2} \right)_0 \tag{3.219}$$

For a more general case, we can write the relation (3.219) in the form shown in Eq. (3.220), α is a constant in both relations:

$$\Delta p_i^0 = -\alpha \left(\frac{\partial \Phi}{\partial p_i} \right)_0 \quad i = 1, L \tag{3.220}$$

If, after each calculation step, we compute the net change ($\Delta\Phi$) of function Φ and it is negative, then we are progressing and we can continue (look at the line which join the points M_0 and M_1 in Fig. 3.74). If $\Delta\Phi \succ 0$, then the displacement has to be stopped and we begin a new exploration considering the last point position until we can establish a new good direction (such as $M_1 M_2$ in Fig. 3.74). Step by step the computation tends to approach the minimum value of the investigated function. This fact is observed by the decrease in the values of the current advancing factor ε_n.

$$\varepsilon_n = \text{abs} \left(\frac{\partial \Phi}{\partial p_1} \right)_n + \text{abs} \left(\frac{\partial \Phi}{\partial p_2} \right)_n \tag{3.221}$$

In addition to SSM, other methods like the total gradient method (TGM) present the capacity to localize the minimal value of function $\Phi(p_1, p_2, ... p_L)$. The TGM operates like an SSM but it establishes the direction of the gradient at each calculation point; at the same time, it progressively decreases the length of the step. If we have one or more constraints in the SSM formulation, then these will be represented in Fig. 3.74 by lines or curves, which cannot be affected or crossed by the gradient line. In the case of a two-parameter problem these constraints can result in a closed surface that includes the minimum function value. For this situation, it is important for the first base point to be inside the constraint surface. Otherwise we cannot move the calculation point to the minimum value of the function due to what the displacement along the constraint becomes imposed so a free gradient line is not detected. In other words, this situation determines an infinite displacement around the constraint surface.

We can extend the observations given here for a two-parameter problem to cases with more parameters. As an example, when the problem has three parameters, then, the problem of closed constraint surface becomes the problem of closed constraint volume.

Fortunately, the problems where one or more parameters are identified towards experimental research, we do not have those constraints; consequently these precautions with respect to the base point selection and with respect to the length of displacement are not important.

Some important considerations have to be taken into account in order to efficiently use the SSM and all other gradient methods with rapid displacement towards a minimum function value [3.64]: (i) the good selection of the base point; (ii) the modification of the parameters' dimension from one step to another; (iii) the complexity of the process surface response; (iv) the number of constraints imposed on the parameters. In some cases we can couple the minimization of the function with the constraint relations in a more complex function, which will be analyzed again. In this case, the problem is similar to the Lagrange problem but it is much more complex.

3.5.3.2 Identifying the Parameters of an Unsteady State Perfectly Mixed Reactor

We carried out a decomposition reaction in the experimental device shown in Fig. 3.75. The reaction is endothermic and takes place in a permanently perfectly mixed (PM) reactor. As shown in Fig. 3.75, reactant A is fed at the reactor input in a liquid flow at constant concentration value. The heat necessary for the endothermic decomposition is supplied by an oil bath, which is electrically heated in order to maintain a constant temperature (t_e). The reactor operates at constant volume because input and output flows are similar.

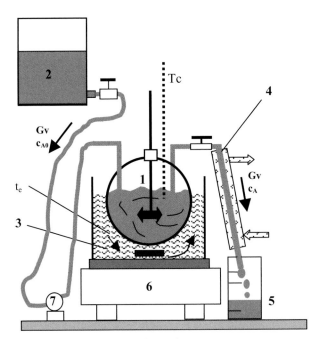

Figure 3.75 Laboratory scale plant with a continuous PM reactor.
1 – reactor, 2 – reservoir, 3 – oil bath, 4 – syphon and cooling device,
5 – collector, 6 – electrical heating device, 7 – pump.

The following input process variables are given or have been measured:
- the input and output flow rate are: $G_v = 10^{-6}$ m^3/s
- the value of the reaction enthalpy: $\Delta H_r = 60\,000$ kj/kg of A species;
- the activation energy of the reaction: $E = 1\,400\,000$ j/kmole of A species;
- the temperature of the feed flow: $t_0 = 140\,°C$;
- the initial temperature of the liquid in the reactor: $t_{00} = 200\,°C$;
- the concentration of species A in the reactor feed: $c_{A0} = 50$ kg/m^3;
- the initial concentration of species A in the reactor: $c_{A00} = 0$ kg/m^3;
- the sensible heat capacity of the mass of liquid reacting: $c_p = 3000$ j/(kg °C);
- the density of the liquid media: $\rho = 800$ kg/m^3;
- the value of the heat transfer area between the oil bath and the reactor's vessel: $A_t = 0.03$ m^2

When the oil bath reaches the set point of constant temperature, the experiment begins by starting the pump and activating the sample collecting device at the reactor's exit. During the experiments, the temperature of the liquid reacting mixture is continuously recorded. We measure the concentration of species A for each collected sample. The result of one set of experiments is given in Table 3.19. In this example, the reaction constant k_0, the reaction order n and the heat transfer coefficient from oil to liquid mass reaction k are poorly estimated and then their values are calculated from the obtained experimental data.

Table 3.19 Evolution of the temperature and concentration of A with time at the exit flow of the reactor.

τ (s)	150	300	600	1500
c_A (kg/m^3)	3.4	4.3	4.7	4.95
t (°C)	185	135	80	64

To make the calculation for the identification of parameters k_0, n, k by using the highest slope method (MHSM), we must determine the function that will be minimized and the mathematical model of the process which correlates these parameters with the computed values of c_A and t. Because we have the dependences $c_A - \tau$ and $t - \tau$ we can consider for the minimization the functions below written as:

$$\Phi_1(k_0, n, k) = \sum_{j=1}^{4} \left(c_A(k_0, n, k, \tau_j) - c_A^{exp}(\tau_j) \right)^2 \tag{3.222}$$

$$\Phi_2(k_0, n, k) = \sum_{j=1}^{4} \left(t(k_0, n, k, \tau_j) - t^{exp}(\tau_j) \right)^2 \tag{3.223}$$

The relations defining functions $\Phi_1(k_0, n, k)$ and $\Phi_2(k_0, n, k)$ contain the values of $c_A(k_0, n, k, \tau_j)$ and $t(k_0, n, k, \tau_j)$, which are obtained here from the mathematical model of the reactor. The MHSM supply the values of k_0, n, k for the mathematical model.

The process is described by the mathematical model of a nonisothermal, unsteady state, continuous and perfectly mixed reactor. It is defined by the below differential equations:

$$\frac{dc_A}{d\tau} = \frac{Gv}{V}(c_{A0} - c_A) - k_0 \exp\left(-\frac{E}{RT}\right) c_A^n \tag{3.224}$$

$$\frac{dt}{d\tau} = \frac{Gv}{V}(t_0 - t) - \frac{kA}{V\rho c_p}(t - t_r) + \frac{k_0 \exp\left(-\dfrac{E}{RT}\right)}{\rho c_p} c_A^n(-\Delta Hr) \tag{3.225}$$

$$\tau = 0 \quad c_A = c_{A00} \quad t = t_{00} \tag{3.226}$$

The solution to the problem of identifying parameters frequently needs the integration of the mathematical model of the process. The software used for this purpose is shown in Fig. 3.76. The starting point of MHSM is M_0; it has the corresponding coordinates $M_0 = M_0(k_0^0 = 0.03, n^0 = 0.5, k^0 = 250)$ in the k_0, n, k axis system. If it is not possible to operate MHSM while simultaneously minimizing functions $\Phi_1(k_0, n, k)$ and $\Phi_2(k_0, n, k)$, we have to introduce a unique and dimensionless minimizing function.

```
Title Unsteady state continuous PM reactor
Select ngrid=1  variables ca(range=0,50) tt(range=20,400)
definitions
   Gv=10**(-6)  V=10**(-3) ca0=50    k0=unknown (will be established by MHSM)
   E=1400000       deltaH=60000   k= unknown (will be established by MHSM)
   A=0.03   ro=880  cp=3000    n= unknown (will be established by MHSM)
   t00=200        tr=340       R=8310              t0=140          c00=0
initial values ca = c00   tt=t00
equations
   dt(ca) = Gv*(ca0-ca)/V-k0*exp(-E/(R*(tt+273)))*(ca**n)  { The ODE }
   dt(tt)=Gv*(t0-tt)/V-k*A*(tt-tr)/(V*ro*cp)+k0*exp(-E/(R*(tt+273)))*(ca**n)*(-deltaH*1000)/ro/cp
boundaries region 1      { define a fictitious spatial domain }
start (0,0) line to (1,0) to (1,1) to (0,1) to finish
time 0 to 2000      { define the time range }
histories { Plot the solution: } history(ca) at (0.5,0.5)  history(tt) at (0.5,0.5
end
```

Figure 3.76 Numerical FlexPDE™ state of the mathematical model of the process (CPM reactor).

The example presented in Table 3.20 shows that the numerical transposition of a concrete example with the MHSM is not a straightforward problem. We can also notice that to start the MHSM, the selection of the first base point must be the result of a primary selection process. Here, we can intuitively suggest the values of the parameters and, using the mathematical model of the process, we can find the proposal that shows a likeness and proximity between the computed and the experimentally measured values for the dependent process variables. The accepted proposal will be considered as the base point for the MHSM:

1. To identify the problems in which the process presents various exits, the use of a function for the minimization of each exit frequently attains contradictory situations when, for example, we must increase the value of a parameter in one function whereas the same parameter must be decreased in a second function. Nevertheless, if we can suggest a global function with rational participation of each partial function we can easily go on.

2. For the case when the dimensionless state of a global function is preferred for the computation, then we can operate with dimensionless process variables and with a dimensionless mathematical model of the process. However, we can also operate with dimensional variables but with partly dimensionless functions.

Table 3.20 The MHSM particularization developed to solve application 3.6.3.2.

STARTING POINT :$M_0(k_0^0 = 0.03, n^0 = 0.5, k^0 = 250)$

The computed values for dependences $c_A - \tau$ and $t - \tau$

τ (s)	150	300	600	1500
c_A (kg/m³)	3.0	4.2	5.2	6.2
t (°C)	165	120	70	35

Φ_1 and Φ_2 values:
$\Phi_1(k_0^0 = 0.03, n^0 = 0.5, k^0 = 250) =$
$(3 - 3.4)^{**}2 + (4.2 - 4.3)^{**}2 + (5.2 - 4.5)^{**}2 + (6.2 - 4.95)^{**}2 = 2.2225$;
$\Phi_2(k_0^0 = 0.03, n^0 = 0.5, k^0 = 250) =$
$(165 - 185)^{**}2 + (120 - 135)^{**}2 + (70 - 80)^{**}2 + (35 - 60)^{**}2 = 1350$

$E_1//k_0: k_0^{01} = 1.1 * k_0^0 = 0.033; M_0^{01}(k_0^0 = 0.033, n^0 = 0.5, k^0 = 250)$
The computed values for dependences $c_A - \tau$ and $t - \tau$ (**first exploration**)

τ (s)	150	300	600	1500
c_A (kg/m³)	2.8	3.8	4.6	5.6
t (°C)	160	110	62	26

Φ_1 and Φ_2 values:

$\Phi_1(k_0^{01} = 0.033, n^0 = 0.5, k^0 = 250) = 1.045$; $(\Phi_2(k_0^{01} = 0.033, n^0 = 0.5, k^0 = 250) = 2685$

Φ_1 decrease but Φ_2 increase and so increasing k_0 is not recommended

$E_2//$ n: $n^{01} = 1.1 n^0 = 0.55$; $M_0^{02}(k_0^0 = 0.03, n^{01} = 0.55, k^0 = 250)$

The computed values for dependences $c_A - \tau$ and $t - \tau$ (second exploration))

τ (s)	150	300	600	1500
c_A (kg/m³)	3.	4	4.8	5.6
t (°C)	165	115	62	30

Φ_1 and Φ_2 values:

$\Phi_1(k_0^0 = 0.03, n^{01} = 0.55, k^0 = 250) = 0.685$

$\Phi_2(k_0^0 = 0.03, n^{01} = 0.55, k^0 = 250) = 2024$

Φ_1 decrease but Φ_2 increase and so increasing n is not recommended

$E_3//$ k: $k^{01} = 1.1 k^0 = 275$; $M_0^{03}(k_0^0 = 0.03, n^{01} = 0.5, k^0 = 275)$

The computed values for dependences $c_A - \tau$ and $t - \tau$ (third exploration)

τ (s)	150	300	600	1500
c_A (kg/m³)	3.	4.2	5.2	6.0
t (°C)	172	128	77	50

Φ_1 and Φ_2 values:

$\Phi_1(k_0^0 = 0.03, n^0 = 0.5, k^{01} = 275) = 1.525$;

$\Phi_2(k_0^0 = 0.03, n^0 = 0.5, k^{01} = 275)) = 327.$

Φ_1 increase and Φ_2 decrease and so increasing k is not recommended

$E_4//k_0$: $k_0^{02} = 0.9 * k_0^0 = 0.027$; $M_0^{04}(k_0^{02} = 0.027, n^0 = 0.5, k^0 = 250)$

The computed values for dependences $c_A - \tau$ and $t - \tau$ (fourth exploration)

τ (s)	150	300	600	1500
c_A (kg/m³)	3.1	4.7	6.1	7.15
t (°C)	176	125	75	40

Φ_1 and Φ_2 values:

$\Phi_1(k_0^{02} = 0.027, n^0 = 0.5, k^0 = 250) = 7.12$

$\Phi_2(k_0^{02} = 0.027, n^0 = 0.5, k^0 = 250) = 606$

Φ_1 increase and Φ_2 decrease and so decreasing k is not recommended

$E_5//n$: $n^{02} = 0.9 * n_0 = 0.45$; $M_0^{05}(k_0^0 = 0.03, n^{02} = 0.5, k^0 = 250)$

The computed values for dependences $c_A - \tau$ and $t - \tau$ (fifth exploration)

τ (s)	150	300	600	1500
c_A (kg/m³)	3.	4.5	5.9	7.1
t (°C)	170	120	72	40

Φ_1 and Φ_2 values:

$\Phi_1(k_0^0 = 0.03, n^{02} = 0.45, k^0 = 250)) = 6.25$

$(\Phi_2(k_0^0 = 0.03, n^{02} = 0.45, k^0 = 250)) = 914$

Φ_1 increase and Φ_2 decrease and so decreasing n is not recommended

All exploring essays have produced contradictory conclusions. So we decide to unify Φ_1 and Φ_2 as a dimensionless function

$\Phi(k_0, n, k) = \Phi_1(k_0, n, k)/c_{A0}^2 + \Phi_2(k_0, n, k)/(t(\tau) - t_{00})^2$

The values of dimensionless function for exploration cases

	base	Increasing k_0	Increasing n	Increasing K	Reduction k_0	Reduction n
Φ_1	0.000889	0.00418	0.000274	0.00061	0.002800	0.002505
Φ_2	0.01167	0/01588	0.01750	0.002828	0.005240	0.007907
Φ	0.012559	0.01588	0.01774	0.003438	0.00804	0/0104
Decision :		reject	reject	Accepted	accepted	accepted

The computation of the partial derivates :

$$\left(\frac{\partial\Phi}{\partial k_0}\right)_0 = \frac{\Phi_{K_0}^0 - \Phi_0^0}{\Delta k_0} = \frac{0.00804 - 0.012559}{(0.027 - 0.03)/0.03} = 0.04519$$

$$\left(\frac{\partial\Phi}{\partial n}\right)_0 = \frac{\Phi_n^0 - \Phi_0^0}{\Delta n} = \frac{0.0104 - 0.012559}{(0.45 - 0.5)/0.5} = 0.02159$$

$$\left(\frac{\partial\Phi}{\partial k}\right)_0 = \frac{\Phi_k^0 - \Phi_0^0}{\Delta k} = \frac{0.003438 - 0.012559}{(275 - 250)/250} = -0.09079.$$

Point advancing : $i_{k0} = 0.498$; $i_n = 0.0219/0.09079 = 0.238$; $i_k = -1$.

Increase k by 3 units: $k^{(1)} = 250 + 3 = 253$ w/(m²grd).

$dn = i_k*(dk/k)/i_n*n = -1(3/250/0.238)*0.5 = -0.025$;$n^{(1)} = 0.5 - (-0.025) = 0.525$;

$dk_0 = -1(3/250/0.498)*0.03 = -0.00072; k_0^{(1)} = 0.03 - (-0.00072) = 0.0307$

The computed values for dependences $c_A - \tau$ and $t - \tau$ for $M_1(0.0307, 0.525, 253)$

τ (s)	150	300	600	1500
c_A (kg/m³)	3.0	4	4.85	5.6
t (°C)	185	115	65	30

Φ value: $\Phi_1^1 = 0.0003009$; $\Phi_2^1 = 0.011712$; $\Phi^{(1)} = 0.0120129$; $\Phi^{(1)} \prec \Phi^{(0)}$. New point M_2 coordinates: $k_0^{(2)} = 0.0307 + 0.0007 = 0.314$, $n^{(2)} = 0.525 + 0.25 = 0.55$, $k^{(2)} = 253 + 3 = 256$.

The computed values for dependences $c_A - \tau$ and $t - \tau$ for $M_2(0.0314, 0.55, 256)$

τ (s)	150	300	600	1500
c_A (kg/m³)	2.8	3.7	4.5	5.2
t (°C)	160	110	60	32

Φ value: $\Phi_1^{(2)} = 0.000329$; $\Phi_2^{(2)} = 0.021055$; $\Phi^{(2)} = 0.02189$; $\Phi^{(2)} \succ \Phi^{(1)}$.
New exploring start
Decreasing k_0: $k_0^1 = 0.9k_0^0 = 0.9 \cdot 0.03146 = 0.02826$
The computed values for dependences $c_A - \tau$ and $t - \tau$ for $M_2^{(1)}(0.0286, 0.55, 256)$

τ (s)	150	300	600	1500
c_A (kg/m³)	3.1	4.2	5.2	5.95
t (°C)	165	120	65	38

Φ value: $\Phi_{1k0}^{(2)} = 0.00054$; $\Phi_{2k0}^{(2)} = 0.011539$; $\Phi_{k0}^{(2)} = 0.012079$; Correct advancing direction.
Increasing n:: $n^1 = 0.55 + 0.05 = 0.6$
The computed values for dependences $c_A - \tau$ and $t - \tau$ for $M_2^{(2)}(0.0314, 0.6, 256)$

τ (s)	150	300	600	1500
c_A (kg/m³)	2.8	3.55	4.2	4.7
t (°C)	168	108	59	29

Φ value: $\Phi_{1n}^{(2)} = 0.000485$; $\Phi_{2n}^{(2)} = 0.02093$; $\Phi_n^{(2)} = 0.0211315$; Correct advancing direction.
Increasing k: $k^1 = 256 + 24 = 280$
The computed values for dependences $c_A - \tau$ and $t - \tau$ for $M_2^{(3)}(0.0314, 0.6, 280)$

τ (s)	150	300	600	1500
c_A (kg/m³)	2.95	3.7	4.4	4.95
t (°C)	166	120	75	50

Φ value: $\Phi_{1k}^{(2)} = 0.000261$; $\Phi_{2n}^{(2)} = 0.00615$; $\Phi_k^{(2)} = 0.00641315$; Correct advancing direction.

The computation of the partial derivates : $\left(\dfrac{\partial \Phi}{\partial k_0}\right)_2 = \dfrac{\Phi_{k0}^{(2)} - \Phi_0^{(2)}}{\Delta k_0} = 0.09612$

$\left(\dfrac{\partial \Phi}{\partial n}\right)_0 = \dfrac{\Phi_n^{(2)} - \Phi_0^{(2)}}{\Delta n} = 0.006325$; $\left(\dfrac{\partial \Phi}{\partial k}\right)_0 = \dfrac{\Phi_k^{(2)} - \Phi_0^{(2)}}{\Delta k} = -0.16510$.

Point advancing : $i_{k0} = 0.58$; $i_n = -0.38$; $i_k = -1$.
Increase k by 3 units: $k^{(3)} = 256 + 3 = 259$
$dn = i_k*(dk/k)/i_n*n = -1(3/256/(-0.38)*0.5 = 0.16$; $n^{(3)} = 0.55 + 0.16 = 0.71$
$dk_0 = -1(3/256/0.58)*0.0314 = -0.006$; $k_0^{(3)} = 0.0314 + (-0.006) = 0.0254$
The computed values for dependences $c_A - \tau$ and $t - \tau$ for $M_3(0.0254, 0.71, 259)$

τ (s)	150	300	600	1500
c_A (kg/m³)	3.05	3.84	4.5	4.95
t (°C)	170	115	61	32

Φ value: $\Phi_1^{(3)} = 0.00024$; $\Phi_2^{(3)} = 0.011712$; $\Phi^{(3)} = 0.0153$; $\Phi^{(1)} \prec \Phi^{(0)}$. New point M_4
$k_0^{(4)} = 0.0254 - 0.006 = 0.0194$, $n^{(4)} = 0.71 + 0.16 = 0.87$, $k^{(4)} = 259 + 3 = 262$
The computed values for dependences $c_A - \tau$ and $t - \tau$ for $M_4(0.0194, 0.87, 262)$

τ (s)	150	300	600	1500
c_A (kg/m³)	3.1	4.1	4.6	4.95
t (°C)	180	120	62	32

Φ value: $\Phi_1^{(3)} = 0.00056$; $\Phi_2^{(3)} = 0.0077$; $\Phi^{(3)} = 0.00776$; $\Phi^{(1)} \prec \Phi^{(0)}$. New point M_5
$k_0^{(5)} = 0.0194 - 0.006 = 0.0134$, $n^{(5)} = 0.87 + 0.16 = 1.03$, $k^{(5)} = 265 + 3 = 268$
Running the algorithm allows the estimation of the most favourable parameters which are:
$k_0 = 0.0106, n = 0.97, k = 301$ w/(m² deg)

3.5.4
The Gauss–Newton Gradient Technique

The process of parameter identification using the Gauss–Newton gradient technique is especially meant for the cases where we have a complex mathematical model of a process that imposes an attentive numerical processing.

For a process with a complex mathematical model, where the exits can be described by partial differential equations, we have the general form below:

$$Y(z, \tau, P) = F\left(Y, z, \tau, \frac{\partial Y}{\partial z}, \frac{\partial^2 Y}{\partial z^2}, \frac{\partial Y}{\partial \tau}, P\right) \qquad (3.227)$$

where $Y(z, \tau)$ and $F(z, \tau)$ are the columns of N-dimensional vectors (Y – responses vector, F – functions vector) while z and τ show the space and time where and when the process takes place. The unknown parameters are contained in the M-dimensional vector P. In addition, the model must be completed with the univocity conditions expressed by the following vectors:
- the vector of initials conditions:

$$Y(z, 0) = Y_0(z) \qquad (3.228)$$

- the vectors of the limitative conditions for z = 0 and z = z_f:

$$G\left(Y_{z=0}, 0, \tau, \frac{\partial Y}{\partial z}/_{z=0}, \frac{\partial^2 Y}{\partial z^2}/_{z=0}, P\right) = 0 \qquad (3.229)$$

$$H\left(Y_{z=zf}, 0, \tau, \frac{\partial Y}{\partial z}/_{z=zf}, \frac{\partial^2 Y}{\partial z^2}/_{z=zf}, P\right) = 0 \qquad (3.230)$$

The identification of the unknown M parameters requires supplementary conditions, which are obtained with experimental research. They given in the column vector $Y_{exp}(z, \tau)$. This column has the same dimension as $Y(z, \tau)$. As is known, the identification of the parameters requires minimization of the dispersion vector that contains the square of the differences between the observed and computed exits of the process:

$$\Phi(P) = \sum_{r=1}^{R} \sum_{s=1}^{S} [Y_{exp}(z_s, \tau_r) - Y(z_s, \tau_r, P)]^2 \tag{3.231}$$

In the relation (3.231) s represents the number of experimental points located on the z_s coordinate while r characterizes the time position when a measure is executed. The base of the development of the Newton–Gauss gradient technique resides in the Taylor expansion $Y(z, \tau, P)$ near the starting vector of parameters P_0:

$$Y(z, \tau, P) = Y(z, \tau, P_0) + (P - P_0) \left[\frac{\partial(Y(z, \tau, P_0))}{\partial P} \right]^T + \ldots \ldots \ldots \tag{3.232}$$

If we replace relation (3.232) in Eq. (3.231) we have:

$$\Phi(P) = \sum_{r=1}^{R} \sum_{s=1}^{S} \left[Y_{exp}(z_s, \tau_r) - Y(z_s, \tau_r, P_0) - (P - P_0) \left[\frac{\partial(Y(z_s, \tau_r, P_0))}{\partial P} \right]^T \right]^2 \tag{3.233}$$

The dispersions vector attains its minimal value with respect to vector P when its derivate has a zero value with respect to this vector; it is written as follows:

$$\frac{\partial \Phi(P)}{\partial P} = \sum_{r=1}^{R} \sum_{s=1}^{S} 2 \left[Y_{exp}(z_s, \tau_r) - Y(z_s, \tau_r, P_0) - (P - P_0) \left[\frac{\partial(Y(z_s, \tau_r, P_0))}{\partial P} \right]^T \right] = (0) \tag{3.234}$$

The relation (2.334) can be also written as:

$$\sum_{r=1}^{R} \sum_{s=1}^{S} Y_{exp}(z_s, \tau_r) - \sum_{r=1}^{R} \sum_{s=1}^{S} Y(z_s, \tau_r, P_0) - (P - P_0) \sum_{r=1}^{R} \sum_{s=1}^{S} \left[\frac{\partial(Y(z_s, \tau_r, P_0))}{\partial P} \right]^T = 0 \tag{3.235}$$

and, in addition, for non-repeated measures it can be particularized as:

$$Y_{exp}(z_s, \tau_r) - Y(z_s, \tau_r, P_0) - (P - P_0) \left[\frac{\partial(Y(z_s, \tau_r, P_0))}{\partial P} \right]^T = 0 \tag{3.236}$$

After the separation of vector P, the last relation can be written in a state that announces the iterative process of Gauss–Newton:

$$P = P_0 + [Y_{exp}(z_s, \tau_r) - Y(z_s, \tau_r, P_0)] \left[\left[\frac{\partial(Y(z_s, \tau_r, P_0))}{\partial P} \right]^T \left[\frac{\partial(Y(z_s, \tau_r, P_0))}{\partial P} \right] \right]^{-1} \tag{3.237}$$

The most important form of relation (3.237) is given by its transposition as an iterative Gauss–Newton procedure:

$$P_{i+1} = P_i + m_i [Y_{exp}(z_s, \tau_r) - Y(z_s, \tau_r, P_i)] \left[\left[\frac{\partial(Y(z_s, \tau_r, P_i))}{\partial P} \right]^T \left[\frac{\partial(Y(z_s, \tau_r, P_i))}{\partial P} \right] \right]^{-1} \tag{3.238}$$

where m_i represents a multiplicator which is selected in order to respect the movement of vector $\Phi(P)$ towards the minimum direction. This condition is written as $\Phi(P_{i+1}) \leq \Phi(P_i)$. In accordance with the established relations, the computation adjusts the selected starting vector P_0, by an iterative procedure. The computation can be finished when the convergence condition is attained. It requires a vector of accepted errors:

$$|P_{i+1} - P_i| \leq E_r \tag{3.239}$$

When the vector of accepted errors contains dimensionless values with respect to each parameter, we have a special case where these values can all be equalled, with a small ε. With this condition we can write the relation (3.239) as follows:

$$\varepsilon \geq \sum_{j=1}^{M} \left| (p_{ij+1} - p_{ij})/p_{ij} \right| \tag{3.240}$$

Here p_{ij} gives the value of the parameter having the number i for the iteration with the number j. The parameter m_i of the relation (3.238) can be estimated using a variation of the Gauss–Newton gradient technique. The old procedure for the estimation of m_i starts from the acceptance of the vector of parameters being limited between a minimal and maximal *a priori* accepted value: $P_{min} \prec P \prec P_{max}$. Here we can introduce a vector of dimensionless parameters $P_{nd} = (P - P_{min})/(P_{max} - P_{min})$, which is ranged between zero and one for the minimal and the maximal values, respectively. With these limit values, we can compute the values of the dimensionless function for $P_{nd} = 0, 0.5, 1$ as $\Phi(0)$, $\Phi(0.5)$ and $\Phi(1)$ and then they can be used for the estimation of m_i:

$$m_i = \begin{cases} 1 \text{ when } \Phi(1) \prec \Phi(0) \\ 0.5 \text{ when } \Phi(1) \succ \Phi(0) \text{ but } \Phi(0.5) \prec \Phi(0) \\ m_i = \left| \dfrac{3\Phi(0) - 2\Phi(0.5) + \Phi(1)}{4\Phi(0) - 8\Phi(0.5) + 4\Phi(1)} \right| \end{cases} \tag{3.241}$$

It is important to notice that the modern methods of Gauss–Newton gradient operate with variable m_i values, which are obtained for each calculation step by using a more or less complicated particular procedure.

It is easily observable that, for the case of identification of only one parameter, relation (3.238) becomes the famous Newton method for solving a transcendent equation of type $\Phi(p) = 0$. Indeed, this particularization gives the following chain of iteration:

$$p_{j+1} = p_j - \frac{y(z_s, \tau_r, p_j) - y_{exp}(z_s, \tau_r)}{\left[\dfrac{dy(z_s, \tau_r, p_j)}{dp}\right]^2} \tag{3.242}$$

In this particular case of only one-parameter identification, the identification method by research is in competition with the Gauss–Newton method. However, the choice of one particular method depends on different conditions. For instance, when the mathematical model of the process presents a simple form it is limited to an algebraic or a simple differential equation, the Newton method will then be preferred because it is the most rapid method to identify the parameters. All other methods available for solving transcendent equations can be used to identify a single parameter. These methods, along with the Newton method, become difficult to operate when we have many different experimental results because we will have many s and r in the relation (3.242). In these cases, a particularization of the relation (3.235) can be a good solution.

3.5.4.1 The Identification of Thermal Parameters for the Case of the Cooling of a Cylindrical Body

The problem analysed here considers the case of a cylinder made of an unknown material. Its dimensions are R = 0.02 m and H = 0.3 m, and it is maintained in an oven at a constant temperature of 250 °C. After a long time, it is take out of the oven and kept in air at 20 °C, where the cooling process starts. A reservoir of boiling water is placed on the top of the cylinder as shown in Fig. 3.77:

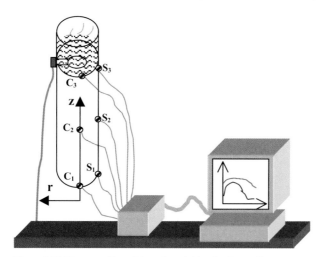

Figure 3.77 The recording of the exit variables for the cooling of a cylindrical body.

During the cooling process, the temperature at points $C_1(0.05, 0)$, $S_1(0.05, 0.02)$, $C_2(0.15, 0)$, $S_2(0.15, 0.02)$, $C_3(0.25, 0)$ et $S_3(0.25, 0.02)$ is measured and recorded. The points marked C are placed in the centre of the cylinder along its axis and the points marked S are placed at the surface. The specific sensible heat of the material of the cylinder is $c = c_p = 870$ j/ (kg deg). The evolution of the temperature at points S and C is given in Figs. 3.78 and 3.79. The heat exchange between the cylinder and the adjacent air is characterized by the evolution of the measured temperature. The heat transfer coefficient from the cylinder to the air and the thermal conductivity of the material must be determined. If we consider the Gauss–Newton gradient technique, the vector that contains the measured exits is written as shown in relation (3.243). The vector of the computed exits of the process has a similar expression but here the list of arguments of the vector of variables will be completed with the parameters λ(thermal material conductivity) and α(heat transfer coefficient from cylinder to air).

$$Y_{exp}(z_s, \tau_r) = \begin{bmatrix} t_c(z_s, \tau_r) \\ t_{sf}(z_s, \tau_r) \end{bmatrix} \tag{3.243}$$

To begin the identification of the parameters with the Gauss–Newton method, the mathematical model of the process must be available. This model allows computation of the values of the temperature at the centre and the surface of the cylinder. At the same time, to estimate the starting vector of parameters (P_0), the method needs a first evaluation of the thermal conductivity λ_0 and of the heat transfer coefficient α_0.

The mathematical model of the process is given by the assembly of relations (3.244)–(3.248) that represent the particularization of the transport phenomena to the descriptive model introduced by Fig. 3.77. It is not difficult to observe that this model is a case of a three-dimensional unsteady heat conduction (τ, r, z) cylinder.

$$\frac{\partial t}{\partial \tau} = \frac{\lambda}{\rho c_p} \left(\frac{\partial^2 t}{\partial r^2} + \frac{2}{r} \frac{\partial t}{\partial r} + \frac{\partial^2 t}{\partial z^2} \right) \tag{3.244}$$

$$\tau = 0 \ , \ 0 \leq z \leq H \ , \ 0 \leq r \leq R \ , \ t = t_0 \tag{3.245}$$

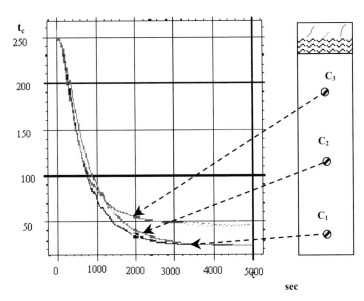

Figure 3.78 Evolution of the measured temperature at points C_1, C_2 and C_3.

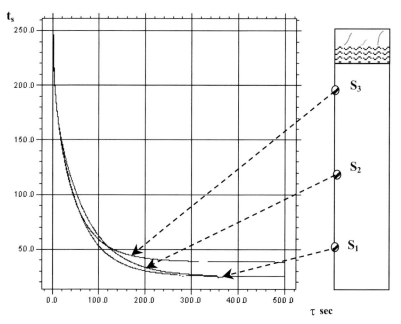

Figure 3.79 The evolution of the measured temperature at points S_1, S_2 and S_3.

$$\tau \succ 0 \;,\; 0 \prec z \leq H \;,\; r = R \;,\; \lambda \frac{dt}{dr} = \alpha(t - t_e) \tag{3.246}$$

$$\tau \succ 0 \;,\; z = H \;,\; 0 < r < R \;,\; t = t_F \tag{3.247}$$

$$\tau > 0 \;,\; z = 0 \;,\; 0 < r < R \;,\; \lambda \frac{dt}{dr} = \alpha(t - t_e) \tag{3.248}$$

The numerical values of all material properties (except thermal conductivity), geometric data and all initial and boundary conditions required by the process have been established by the mathematical model. These values are $\rho = 6100$ kg/m^3, $c_p = 870$ j/(kg deg), $R = 0.02$ m, $H = 0.3$ m, $t_0 = 250\,°C$ and $t_F = 100\,°C$.

The relation (3.249) used for the iterative calculation allowing the identification of the unknown parameters is given here below. It is a particularization of the general Gauss–Newton algorithm (3.238):

$$\left|\begin{array}{c}\lambda \\ \overline{\alpha}\end{array}\right|_{i+1,j} = \left|\begin{array}{c}\lambda \\ \overline{\alpha}\end{array}\right|_{i,j} + m_i \left[\left[\begin{array}{cc}\left(\frac{\partial t_c}{\partial\lambda}\right)_{i,j} & \left(\frac{\partial t_c}{\partial\alpha}\right)_{i,j} \\ \left(\frac{\partial t_s}{\partial\lambda}\right)_{i,j} & \left(\frac{\partial t_s}{\partial\alpha}\right)_{i,j}\end{array}\right] \cdot \left[\begin{array}{cc}\left(\frac{\partial t_c}{\partial\lambda}\right)_{i,j} & \left(\frac{\partial t_s}{\partial\lambda}\right)_{i,j} \\ \left(\frac{\partial t_c}{\partial\alpha}\right)_{i,j} & \left(\frac{\partial t_s}{\partial\alpha}\right)_{i,j}\end{array}\right] \right]^{-1} \left|\begin{array}{c}t_{cexp.j} - t_{c.calc.i,j} \\ t_{s.expj} - t_{s.calc.i,j}\end{array}\right| \tag{3.249}$$

The computation obtained with this particularization is given in Fig. 3.80. In this example we consider that the experimental data of the first group (C_1, S_1) is the starting point.

1 **Position** : 1-top (C_1 ,S_1), 2-mean (C_2 ,S_2), 3-bottom (C_3 ,S_3);r=5

2 **Starting data**: ρ =6100kg/m^3, c_p=870 j/(kg deg), R=0.02 m, H=0.3 m, t_0=250 ^{0}c,

For j=1,r and k=1,3 input $t_{c\,expk}$ (j) , $t_{s\,expk}$(j) , Vector of parameters / $\left|\dfrac{\lambda}{\alpha}\right|_0 = \left|\dfrac{30}{15}\right|$

3 **k=1**

4 **i=0**   **j=1**

5 **Model simulation** $\left|\dfrac{t_{c..calc..i,j}}{t_{s..cal..i,j}}\right| = \left|\dfrac{t_{c.i,j}}{t_{s.i,j}}\right|$ **Increase** λ_i as $\lambda_i + d\lambda_i$

Model simulation : $\left|\dfrac{t_{c\lambda}}{t_{s\lambda}}\right|$

6 **Compute**: $\left(\dfrac{\partial t_C}{\partial \lambda}\right)_{i,j} = \dfrac{t_{c\lambda} - t_{ci,j}}{\lambda - \lambda_{i,j}}$ and $\left(\dfrac{\partial t_s}{\partial \lambda}\right)_{i,j} = \dfrac{t_{s\lambda} - t_{si,j}}{\lambda - \lambda_{ij}}$

7 **Increase** α_i as $\alpha_i + d\alpha_i$ **Model simulation** : $\left|\dfrac{t_{c\alpha}}{t_{s\alpha}}\right|$

8 **Compute**: $\left(\dfrac{\partial t_C}{\partial \alpha}\right)_{i,j} = \dfrac{t_{c\alpha} - t_{ci,j}}{\alpha - \alpha_{i,j}}$ and $\left(\dfrac{\partial t_s}{\partial \alpha}\right)_{i,j} = \dfrac{t_{s\alpha} - t_{si,j}}{\alpha - \alpha_{ij}}$;

9 **Compute**: $A = \left[\begin{bmatrix} (\frac{\partial t_c}{\partial \lambda})_{i,j} & (\frac{\partial t_c}{\partial \alpha})_{i,j} \\ (\frac{\partial t_s}{\partial \lambda})_{i,j} & (\frac{\partial t_s}{\partial \alpha})_{i,j} \end{bmatrix} \cdot \begin{bmatrix} (\frac{\partial t_c}{\partial \lambda})_{i,j} & (\frac{\partial t_s}{\partial \lambda})_{i,j} \\ (\frac{\partial t_c}{\partial \alpha})_{i,j} & (\frac{\partial t_s}{\partial \alpha})_{i,j} \end{bmatrix} \right]^{-1}$

10 **Compute**: $\left|\dfrac{\lambda}{\alpha}\right|_{i+1,j} = \left|\dfrac{\lambda}{\alpha}\right|_{i,j} + A \left|\dfrac{t_{c.exp.j} - t_{c..calc..i,j}}{t_{s.exp.j} - t_{s..calc..i,j}}\right|$,

11 **Compute**: $Er_{i,j}$ $\left|\dfrac{(\lambda_{i+1,j} - \lambda_{i,j})/\lambda_{i,J}}{(\alpha_{i+1,j} - \alpha_{i,j})/\alpha_{i,j}}\right|$

12 **Convergence**: For $Er_{i+1,j} < Er_{i,j}$ retain $\alpha_i + d\alpha_i = \alpha_i$ and $\lambda_i + d\lambda_i = \lambda_i$

For j>r go to 13

For j<r then j=j+1 go to 5

For $Er_{i+1,j} > Er_{i,j}$ retain $\alpha_i = \alpha_i - d\alpha_i$ and $\lambda_i = \lambda_i - d\lambda_i$; i=i+1 go to 5

13 **Mean values**: $\alpha_k = \sum_1^r \alpha_j / r$, $\sigma_{\alpha k} = \sqrt{(\alpha_j - \alpha)^2 /(r-1)}$.

$\lambda_k = \sum_1^r \lambda_j / r$, $\sigma_{\lambda k} = \sqrt{(\lambda_j - \lambda)^2 /(r-1)}$.

14 **Control**: For k<3 then k=k+1 $\left|\dfrac{\lambda}{\alpha}\right|_0 = \left|\dfrac{\lambda_k}{\alpha_k}\right|$ go to 4

15 **Global mean** $\alpha_{mean} = \sum_1^3 \alpha_k /3$ $\lambda_{mean} = \sum_1^3 \lambda_k /3$

16 **End**

Figure 3.80 The particularization of the Gauss–Newton algorithm for the application.

We can observe that it is important to have a simulator of the model of the process ((3.244)–(3.248)) in order to estimate the value of the vector $\left|\begin{smallmatrix} t,i,j \\ t_s,i,j \end{smallmatrix}\right|$. The simulator allows the computation of the matrix of derivates $\begin{bmatrix} \left(\frac{\partial t_c}{\partial \lambda}\right)_{i,j} & \left(\frac{\partial t_c}{\partial a}\right)_{i,j} \\ \left(\frac{\partial t_s}{\partial \lambda}\right)_{i,j} & \left(\frac{\partial t_s}{\partial a}\right)_{i,j} \end{bmatrix}$, used in the iteration processes, as shown in the numerical example given here.

The model simulator of the process is based on the description given in Fig. 3.77; it considers the transformations recommended earlier as well as an adaptation to the model conditions ((3.43)–(3.46)). In the following example the FRC(λ,a) gives the values of $t_{ci,j}$ and $t_{si,j}$ respectively.

The positions of the points are: C_1, S_1; time = 1000 s/ $t_{c\,exp}$ = 69, $t_{s\,exp}$ = 52/ P_0: λ = 30, a = 15/FRC(30,15), $t_{c\,calc}$ = 65, $t_{s\,calc}$ = 39/ $\lambda + d\lambda$ = 35/ FRC(35,15) $t_{c\lambda}$ = 81, $t_{s\lambda}$ = 57, $dt_c/d\lambda$ = (81–65)/5 = 3.2, $dt_s/d\lambda$ = (57–39)/5 = 3.6/$a + da$ = 20/ FRC(30,20) t_{ca} = 49, t_{sa} = 32 ,dt_c/da = (49–65)/5 = –3.2, dt_s/da = (32–39)/5 = –1.4 / B =

$$\begin{vmatrix} 3.2 & 3.6 \\ -3.2 & -1.4 \end{vmatrix} * \begin{vmatrix} 3.2 & -3.2 \\ 3.6 & -1.4 \end{vmatrix} = \begin{vmatrix} 12.24 & 16 \\ 16 & 4.9 \end{vmatrix}$$, det B = –73, A = (1 /detB)*min(B)

$$= \begin{vmatrix} -0.2 & -0.21 \\ -0.21 & -0.16 \end{vmatrix}; \quad \begin{vmatrix} \lambda \\ a \end{vmatrix} = \begin{vmatrix} 30 \\ 15 \end{vmatrix} + \begin{vmatrix} -0.2 & -0.21 \\ -0.21 & -0.16 \end{vmatrix} * \begin{vmatrix} 69-65 \\ 52-39 \end{vmatrix} = \begin{vmatrix} 30-1.64 \\ 15-4.81 \end{vmatrix} =$$

$$\begin{vmatrix} 28.46 \\ 10.19 \end{vmatrix} / E = \begin{vmatrix} 0.055 \\ 0.32 \end{vmatrix} /$$ Go on with FRC(28.46,10.19) $t_{c\,calc}$ = 75, $t_{s\,calc}$ = 38 /$\lambda + d\lambda$ =

33.16/ FRC(33.16,10.19) $t_{c\lambda}$ = 77, $t_{s\lambda}$ = 43 , $dt_c/d\lambda$ = (77–75)/5 = 0.4 , $dt_s/d\lambda$ = (43 – 38)/5 = 1/a +da = 15.16/ FRC(28.46,15.16) t_{ca} = 65, t_{sa} = 32, dt_c/da = (65 – 75)/5

$$= -0.2, dt_s/da = (32 - 38)/5 = -1.2 / B = \begin{vmatrix} 0.4....1 \\ -2...-1.2 \end{vmatrix} * \begin{vmatrix} 0.4...-2 \\ 1....-1.2 \end{vmatrix} = \begin{vmatrix} 4.16...2.8 \\ 2.8....2.41 \end{vmatrix},$$

det B = –2.16 , A = (1/debt)*min(B) = $\begin{vmatrix} 1.9....1.28 \\ 1.28...1.09 \end{vmatrix}$ / $\begin{vmatrix} \lambda \\ a \end{vmatrix} = \begin{vmatrix} 28.46 \\ 10.19 \end{vmatrix} - 0.5 *$

$$\begin{vmatrix} 1.9.....1.28 \\ 1.28....1.09 \end{vmatrix} * \begin{vmatrix} 69-65 \\ 52-39 \end{vmatrix} = \begin{vmatrix} 28.46+8.5 \\ 10+15.4 \end{vmatrix} = \begin{vmatrix} 36.96 \\ 25.4 \end{vmatrix} / E = \begin{vmatrix} 0.32 \\ 1.5 \end{vmatrix} /.$$

We used here a too large displacement of λ and a for the construction of the matrices (relation (3.242)). However, the real computation uses a small displacement of the parameters. This explains the differences between both values of the vectors of errors as well as the evolution of the vector of the parameters along both iterations. The software of the mathematical model of the process is given by FRC(λ, a). In this specific computation, we introduced definite values of λ and a for the calculation of the corresponding temperatures ($t_{c\,calc}$, $t_{s\,calc}$, $t_{c\lambda}$, $t_{s\lambda}$ etc) for points C_1 and S_1 respectively .

The final result of identification allows the estimation of a = 9.7±0.88 w/(m² deg) and λ = 49.8±2.35 w/(m deg). Considering the value of λ, we can appreciate that the cylinder is certainly made of a type of steel, whereas the value of a shows that the occurring heat transfer is the natural convection between the cylinder and the adjacent air. This last observation is in good agreement with the descriptive model of the process given at the beginning of this section.

3.5.4.2 **Complex Models with One Unknown Parameter**

The identification of the parameters of a process can be examined from two completely different viewpoints. The former is given by laboratory researchers, who consider the identification of parameters together with a deep experimental analysis; it is then frequently difficult to criticize the experimental working methods, the quality and quantity of the experimental data. The latter is given by researchers specialized in mathematical modelling and simulation. These researchers consider that the mathematical aspects in the identification of parameters are prevailing. Nevertheless, this last consideration has some limits because, in all cases, a similar number of parameters and independent experimental data are necessary for a correct identification.

It is important to notice that, from both viewpoints, as well as in all working procedures, experimental data are required and that, at the same time, mathematical models are absolutely needed for data processing. Generally, when the mathematical model of a process is relatively complex, a good accuracy and an important volume of experimental data are simultaneously required. Therefore, in these cases the quality of the determination of parameters is the most important factor to ensure model relevance. The strategy adopted in these cases is very simple: for all the parameters of the process that accept an indirect identification, the research procedure of identification is carried out separately from the real process; whereas for the very specific process parameters that are difficult to identify indirectly, experiments are carried out with the actual process.

When we have N measures for the exit variables in a process, the technical problem of identification of the unknown parameter resides in solving the equation $\Phi(p) = 0$. From the theoretical viewpoint, all the methods recommended for the solution of the transcendent equation can be used to determine parameter p. The majority of these methods are of iterative type and require an expression or an evaluation of the $\Phi(p)$ derivate. When we evaluate the derivate numerically, as in the case of a complex process model, then important deviations can be introduced into the iteration chain. Indeed, the deviation propagation usually results in an increasing and non-realistic value of the parameter. This problem can be avoided by solving the equation $\Phi(p) = 0$ by integral methods such as the method of minimal function value (MFV). When $\Phi(p)$ values are only obtained in the area of influence of parameter p, the MFV method is reduced to a dialogue with the mathematical model of the process and then the smallest $\Phi(p)$ value gives the best value for the parameter.

The following example details how the MFV method is used to identify the diffusion coefficient of species with respect to their motion in a particle of activated carbon.

Diffusion of Species Inside a Particle of Activated Carbon

Among the inexhaustible plant resources for the production of activated carbon, we have the nutshell, which can be transformed by pyrolysis and activation with overheated water vapour. In this example, activated carbon has been used to retain some hydrocarbon traces from water using a batch reactor. The interest here is to characterize the diffusion of the chemical species that are adsorbed on the activated carbon and its dependence on the operating conditions.

Experiment. Nutshells, granulated to a maximum 6 mm diameter, were used as raw material. The reaction was carried out in a pyrolysis reactor heated with an electrical resistance and the temperature evolution inside the bed of solid was measured with a thermocouple. The integral expression of the pyrolysis dynamics was determined by the loss of weight with respect to the initial quantity of nutshells loaded into the reactor. The experimental data are presented in Fig. 3.81. The final weight loss corresponds to the removal of some non-oxygenated compounds, (which burned with a blue flame, color indicating the absence of oxygen) from the raw material. 100 g of pyrolysed material, divided into three parts, was prepared for each batch loaded into the reactor. One third, which followed a different activation treatment, was used for comparison with the other samples. The activation was performed by flowing the overheated steam through the fixed bed of pyrolysed material. Two different activation treatments, which differ in the overheated steam temperature and flow were used. The nine different activated carbons prepared are reported in Table 3.21. They were identified depending on the operating conditions as: S1, S2, S3, S1A1, S1A2, S2A1, S2A2, and S3A1 and S3A2.

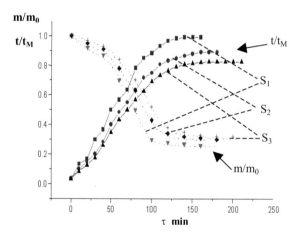

Figure 3.81 Dynamics of the weight loss and of the increase in the temperature for the nutshell pyrolysis (dimensionless).

The samples called S were used without any activation whereas the samples called SxAy were used after activation.

Characterization of the activated carbon by adsorption: nearly saturated water with benzene and activated carbon were introduced into a batch reactor with a 0.06 and 0.14 ratio of solid/liquid phases. In each experiment, the evolution of the concentration of benzene in water was determined spectrophotometrically and by potassium permanganate titration. Tables 3.22 and 3.23 as well as Figs. 3.82–3.84 show the dynamics of the concentration of the organic compound in the water for the different types of activated carbon. The mass balance of species allows one to know the mean benzene concentration adsorbed by the activated carbon particles for each experiment and in each time interval.

Table 3.21 Activation conditions for the pyrolysed material.

No.	Time (min)	A1		A2	
		t (°C)	Steam flow (kg/s)	t (°C)	Steam flow (kg/s)
1	0	300	$0.83 \cdot 10^{-4}$	300	$0.83 \cdot 10^{-4}$
2	20	700	$1.25 \cdot 10^{-4}$	600	$1.25 \cdot 10^{-4}$
3	40	700	$1.25 \cdot 10^{-4}$	600	$1.25 \cdot 10^{-4}$
4	60	700	$2.83 \cdot 10^{-4}$	600	$2.83 \cdot 10^{-4}$
5	80	700	$2.83 \cdot 10^{-4}$	600	$2.83 \cdot 10^{-4}$
6	100	700	$0.83 \cdot 10^{-4}$	600	$0.83 \cdot 10^{-4}$

Table 3.22 Evolution of the concentration of benzene in water for a solid/liquid ratio of $s/l = 0.06$.

No.	Time (min)	c/c_o								
		S1	S2	S3	S1A1	S1A2	S2A1	S2A2	S3A1	S3A2
1	0	1	1	1	1	1	1	1	1	1
2	10	0.81	0.86	0.91	0.8	0.76	0.85	0.77	0.89	0.83
3	25	0.73	0.80	0.86	0.7	0.66	0.77	0.69	0.82	0.75
4	50	0.59	0.71	0.75	0.53	0.51	0.62	0.54	0.74	0.63
5	80	0.56	0.59	0.62	0.50	0.49	0.58	0.51	0.61	0.59
6	120	0.55	0.57	0.59	0.49	0.48	0.55	0.50	0.59	0.57
7	160	0.54	0.56	0.57	0.49	0.48	0.54	0.50	0.57	0.54
8	210	0.53	0.55	0.56	0.49	0.48	0.53	0.50	0.55	0.54
9	260	0.53	0.55	0.56			0.53		0.55	0.54
10	360		0.55	0.56						

Table 3.23 Evolution of the concentration of benzene in water for an s/l = 0.14 solid/liquid ratio.

No.	Time (min)	c/c_o								
		S1	S2	S3	S1A1	S1A2	S2A1	S2A2	S3A1	S3A2
1	0	1	1	1	1	1	1	1	1	1
2	10	0.75	0.82	0.84	0.7	0.68	0.79	0.77	0.83	0.85
3	25	0.69	0.73	0.74	0.50	0.48	0.66	0.49	0.71	0.75
4	50	0.57	0.62	0.64	0.41	0.38	0.55	0.40	0.59	0.65
5	80	0.49	0.59	0.61	0.32	0.30	0.46	0.32	0.51	0.56
6	120	0.39	0.50	0.52	0.27	0.25	0.38	0.27	0.45	0.47
7	160	0.34	0.41	0.43	0.25	0.23	0.32	0.24	0.39	0.39
8	210	0.30	0.34	0.36	0.25	0.23	0.29	0.23	0.34	0.32

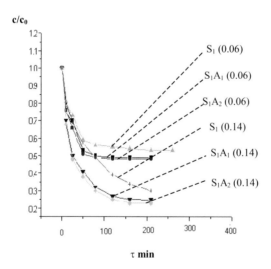

Figure 3.82 Evolution of the benzene concentration in water for the adsorption with activated carbon S_1, S_1A_1 and S_1A_2.

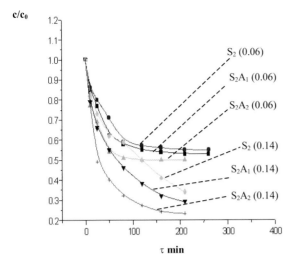

Figure 3.83 Dynamics of the benzene adsorption for the S_2, S_2A_1 and S_2A_2 activated carbon.

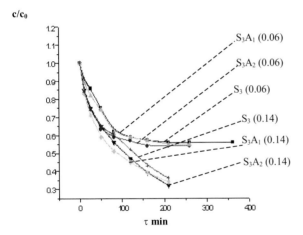

Figure 3.84 Dynamics of the benzene adsorption for the S_3, S_3A_1 and S_3A_2 activated carbon.

From these representations we can notice that: (i) *the 600 °C pyrolysis* results in an activated carbon with the best adsorption speed; (ii) *the activation* increases the speed of adsorption of the organic compound, probably as a consequence of the increase in the effective diffusion coefficient and not as a result of the opening of new pores (in which case the final equilibrium concentrations for S1, S2, S3, S1A1, S1A2, S2A1, S2A2, S3A1, S3A2 should be different); (iii) *the use* of more intense steam activation conditions (higher temperature) leads to a small increase in the speed of adsorption of the organic compound.

We can consider that the measured values of the concentration at the end of the experience at constant temperature represent the equilibrium concentrations. These data are given in Tables 3.22 and 3.23. The equilibrium constant is determined according to its definition by the following relation:

$$k_d = \frac{\dfrac{m_l c_{ech}}{m_s}}{c_0 - c_{ech}} = \frac{r \cdot c_{ech}}{c_0 - c_{ech}} = \frac{r}{\dfrac{c_0}{c_{ech}} - 1} \tag{3.250}$$

Table 3.24 shows the computed data for k_d, for both solid/liquid ratios and the mean values if we consider the hypothesis of a linear equilibrium isotherm.

Table 3.24 Adsorption equilibrium constants for all activated carbon species at 25 °C.

	S1	S2	S3	S1A1	S1A2	S2A1	S2A2	S3A1	S3A2
$k_d(0.06)$	0.0677	0.0733	0.0764	0.0576	0.0676	0.0677	0.0600	0.0733	0.070
$k_d(0.14)$	0.0600	0.0721	0.0787	0.0466	0.0418	0.0572	0.0418	0.0721	0.066
$k_d(\text{mean})$	0.0639	0.0727	0.0775	0.0521	0.0547	0.0625	0.0509	0.0727	0.068

From this table, the weak dependence between the distribution constant, the pyrolysis and activation conditions can be noticed.

Identification of the effective diffusion coefficient with the mathematical model of batch adsorption. The model assumes that the carbon particles are spherical and porous (ε_p– voids fraction). Using c (kg A/m^3 fluid inside the pores) and q (kg A/kg adsorbent) to express the concentration of the transferable species through the pores and through the particle respectively, we can write the following expression for transport flux:

$$J_{A,r} = -\left[D_{ef} \frac{\partial c}{\partial r} + \rho_p D_s \frac{\partial q}{\partial r} \right] \tag{3.251}$$

where D_{ef} represents the effective diffusion coefficient through the pores, D_s is the surface diffusion coefficient and ρ_p is the particle density. When the adsorption flow of species has been defined, it is necessary to give the net speed of adsorption using a general expression such as: $v_{ad} = G(c, q)$.

The unit for the net speed of adsorption is kmoles or kg of A by unit of solid weight and by unit of time. For example, for the net speed of adsorption, the Langmuir model gives:

$$G(c, q) = k_a(q_\infty - q) - k_{ds} \cdot q \tag{3.252}$$

Here q is the maximal concentration of the adsorbed species into the solid, k_a is the rate of adsorption; k_{ds} is the rate of desorption. The ratio $k_d = k_a/k_{ds}$ is usually called the equilibrium constant. With the considerations given above, we can now write the expression for the concentration fields c and q:

$$\varepsilon_p \frac{\partial c}{\partial \tau} = \frac{1}{r^2} \frac{\partial}{\partial r} \left(r^2 D_{ef} \frac{\partial c}{\partial r} \right) - \rho_p G(c, q) \tag{3.253}$$

$$\frac{\partial q}{\partial r} = \frac{1}{r^2} \frac{\partial}{\partial r} \left(r^2 D_s \frac{\partial q}{\partial r} \right) + G(c, q) \tag{3.254}$$

For a full definition of the model of transport through the particle, it is necessary to set up the univocity conditions for the above equations:
- the concentration fields c and q inside of the particle at the start of the process:

$$\tau = 0, \ 0 < r < R, \ c = q = 0 \tag{3.255}$$

- the absence of transport of the species into the centre of the particle:

$$\tau > 0, \ r = 0, \ \frac{\partial c}{\partial r} = \frac{\partial q}{\partial r} = 0 \tag{3.256}$$

- the equality of the convection and conduction flux at the surface of the particle:

$$\tau > 0 \ , \ r = R \ , \ k(c_l - c_R) = D_{ef} \frac{\partial c}{\partial r}\bigg|_R + \rho_p D_s \frac{\partial q}{\partial r}\bigg|_R \tag{3.257}$$

For the adsorbed species on the external surface of the particle, the next condition has to be fulfilled:

$$\tau > 0 \ , \ r = R \ , \ \frac{\partial q}{\partial r}\bigg|_{r=R} = G(c_R, q_R) \tag{3.258}$$

The next equation presents the balance of the adsorbable species for the fluid outside the particle:

$$V \frac{\partial c_l}{\partial \tau} = -\left(\frac{m_p}{\rho_p} \right) \frac{3}{R} \left[D_{ef} \frac{\partial c}{\partial \tau} + \rho_p D_s \frac{\partial q}{\partial r} \right]_R \tag{3.259}$$

Here m_p is the total mass of the particles with radius R placed in the contactor with a useful volume V. In some cases, the surface diffusion is considered the slowest process because organic components such as hydrocarbons are generally strongly adsorbed on activated carbon [3.65, 3.66]. Indeed, we can consider here that, at the surface of the particle, the adsorption equilibrium is achieved faster than the surface diffusion process. In these conditions the batch model equations are:

$$\frac{\partial q}{\partial \tau} = \frac{1}{r^2} \frac{\partial}{\partial r} \left(r^2 \overline{D_s} \frac{\partial q}{\partial r} \right) \tag{3.260}$$

$$r = R, \ k_1(c_1 - c_R) = \rho_p \overline{D}_s \frac{\partial q}{\partial r} \tag{3.261}$$

$$r = R, \ G(c, q) = 0, \ r = 0, \ \frac{\partial q}{\partial r} = 0 \tag{3.262}$$

$$V \frac{dc_1}{d\tau} = \frac{3}{R} \left(\frac{m_p}{\rho_p} \right) \rho_p \overline{D}_s \frac{\partial q}{\partial r} \bigg|_R \tag{3.263}$$

$$\tau = 0, \ c_1 = c_{1,0}, \ q = 0 \tag{3.264}$$

where $\overline{D}_s$ is the effective coefficient of the surface diffusion. The minimization of the squares of the differences between the experimental and the theoretical values of the transferable species allows the identification of $\overline{D}_s$. The calculation is made following the next steps:

1. Suggest a value for $\overline{D}_s$;
2. Propose $c_R \prec c_1$;
3. Determine q_R from Eq. (3.262);
4. Numerical integration of Eq. (3.261) and determination of $q(r, \tau), \ 0 < r < R$.;
5. Calculation of $\dfrac{\partial q}{\partial r}\bigg|_R$ and verification of the condition given by

   the relation (3.261). If it is not verified go back to 2;
6. Determine $c_1(\tau) = c_{1,th}$ from Eq. (3.263);
7. Calculate $\left(c_{1,exp} - c_{1,th} \right)_i^2$;

8. Increase τ to cover the entire period of the experiment and go to step 2;
9. Calculate $\sigma(D_s) = \frac{1}{n-1} \sqrt{\sum\limits_{l=1}^{n} \left(c_{1,exp} - c_{1,th} \right)_i^2}$;
10. Propose a new value for $\overline{D}_s$ and go to step 2;
11. Identify the minimum value of the dispersion, $\sigma(D_s)$, in order to obtain the best value of the effective coefficient of the surface diffusion.

Figure 3.85 and Table 3.25 show the identified values of the effective diffusion coefficient for all adsorption experiments. The activation technique applied can be shown to allow the enhancement of $\overline{D}_s$, so the speed of the transport process will be higher. Table 3.25 also contains the values of $\overline{D}_s$ identified by the Newton–Gauss method.

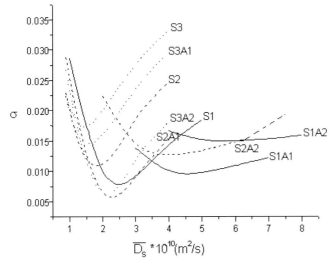

Figure 3.85 The state of the dispersion between the experimental and the theoretical values of c/c_0 versus $\overline{D}_s \cdot 10^{10}$ m^2/s.

Table 3.25 Identified $\overline{D}_s$ values for all activated carbons (first line – from Fig. 3.85, second line – by Newton–Gauss method.

	S1	S2	S3	S1A1	S1A2	S2A1	S2A2	S3A1	S3A2
σ		0.009	0.011	0.020	0.001	0.015	0.006	0.013	0.016	0.0065
$\overline{D}_s \cdot 10^{10}$ m^2/s	3	2	2	5	6	2	4	2	2
$\overline{D}_s \cdot 10^{10}$ m^2/s	3.12 ± 0.42	2.09 ± 0.31	1.92 ± 0.27	5.57 ± 0.62	6.15 ± 0.76	2.11 ± 0.29	3.88 ± 0.25	1.98 ± 0.11	2.06 ± 0.19

The increase in the activation time (A$_2$ regime) results in the best values of $\overline{D}_s$ with respect to all other cases. It can be observed that the best speed of adsorption is reached with an activated carbon produced by the first type of pyrolysis treatment (S1 samples), whereas no improvement is observed in the adsorption properties, when the activated carbon has been produced by a process in which the raw material (S3) presented the lowest loss of mass.

In conclusion, we can assert that the pyrolysis and activation process applied for the manufacture of activated carbons from nutshells resulted in good quality adsorbents. We have demonstrated the influence of both processes on the speed of the benzene adsorption from water solutions. The hypothesis that the effective surface diffusion is the slowest step of the global process was used and the estimation of the effective diffusion coefficient resulted in values ranging between 2 and 6×10^{-10} m^2/s.

3.5.5
Identification of the Parameters of a Model by the Maximum Likelihood Method

The maximum likelihood method (MLM) is used effectively to identify the unknown parameters of mathematical models when the parameters are distributed. If we consider Fig. 3.1, the actions of the normal distributed perturbations on the process cannot be neglected. Indeed, all process exits will be distributed with individual parameters that depend on the distribution functions associated to the perturbations.

In order to show the effect of the distributed perturbations on the model exits, we begin the analysis by writing the mathematical model of the process as:

$$X = f(X, U, V, \tau) \tag{3.258}$$

where X is the state vector (internal characterization of the process), U the control vector (for all or for the most important inputs of the process) and V is defined as the disturbance (perturbation) vector of the process. In all cases, all the experimental measurements have been affected by the errors which are distributed normally; the process contains one or more variables with probability distributed actions, etc. The formal measurements of vectors depend on the vectors themselves and on their state, they will be given by:

$$Y = g(X, U, V, \tau) \tag{3.259}$$

They are composed of the numerical values contained in the following sequence:

$$Y_N = \langle y_1, y_2, ... y_N \rangle \tag{3.260}$$

This sequence shows the instant values of the exit of the process conditioned by the vector parameter $P = P(p_1, p_2, ... p_L)$. Indeed, Y_N/P is the exit random vector conditioned by the vector parameter P. In this case $p(Y_N/P)$, which is the probability density of this variable, must be a maximum when the parameter vector P is quite near or superposed on the exact or theoretical vector P. Therefore, the maximum likelihood method (MLM) estimates the unknown parameter vector P as $\hat{P}$, which maximizes the likelihood function given by:

$$L = \ln[p(Y_N/P)] \tag{3.261}$$

Now we can consider that the prediction error (measurement errors) given for the k exit is shown by:

$$\varepsilon_k = (y_k - \bar{y}_k) \tag{3.262}$$

where $\bar{y}_k$ is the expected mean value. When the mean value is "white" or zero $E(\varepsilon_k) = 0$, (E operator to calculate the mean value). It can be shown that the maximization of L from relation (3.261) is equivalent to the minimizations of the L_{MDL} given by:

$$L_{MDL} = \frac{1}{2N} \sum_{k=1}^{N} \left[\varepsilon_K^T R_k^{-1} \varepsilon_K + \ln(\det(R_k)) \right] \qquad (3.263)$$

Here, R_k is the prediction error covariance matrix (it will be chosen from the start of the minimization) and N represents the number of the experiments carried out. The Kalman filter approach [3.66, 3.67] must be used in different situations: (i) in the estimation of the ε_k vector when $\bar{y}_k$ is unknown in Eq. (3.262), (ii) in the correction of the model state when it is possible to compensate some inaccuracies due to the model deficiencies and experimental plant disturbances [3.68]. In this latter case, if the model state is not corrected, considerable errors in parameter estimation can arise, even if the model structure is very close to the *correct* one. The extended Kalman Filter approach is also necessary [3.69, 3.70] when the mathematical models of the process show a nonlinear state.

The principle of the MLM is shown in Fig. 3.86. It is important to specify that the minimization of the MLM can be carried out by various techniques. The MLM algorithm works as follows:

1. It makes an initial choice of the estimated vector parameter (indeed, for i = 0 it chooses $\hat{P}_0$).
2. It uses the Kalman filter and performs a simulation with $\hat{P} = \hat{P}_i$ and then it computes $L_{MDL} = L_{MDLi}$.
3. It uses a minimization technique to update a new estimation of vector parameter: $\hat{P}_{i+1} = \hat{P}_i + \Delta\hat{P}_i$.
4. It verifies whether convergence has happened: if not, it goes back to point 2 and adds i = i + 1.

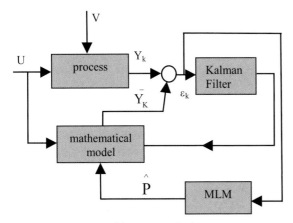

Figure 3.86 Structure of the MLM method.

When a model state is described by nonlinear equations, the extended Kalman filter has been applied using the well-known Kalman filter equations for the linearization of equations. If the state vector is enlarged with the parameter vector P_k (P_k is used because it corresponds to the discrete version of the state model) and if it is considered to be constant or varying slowly, then it is possible to transform the problem of parameters estimation into a problem of state estimation. The $P_{k+1} = P_k + n_k$ with n_k white noise correction represents the model suggested for P. It will introduce $\overline{X}_k = \begin{vmatrix} X_k \\ \overline{P_k} \end{vmatrix}$ in the augmented state vector. Then, the discrete version of the state model will be written as:

$$\overline{X}_{k+1} = \overline{X}_k + \Delta\tau * F(\overline{X}_k, U_k, V_k) \tag{3.264}$$

The structure of the augmented (extended) Kalman filter is shown in Fig. 3.87, which also presents the schematic methodology for obtaining the exit-computed vector $\overline{Y}_k$. It can be observed that coupling the process with computation procedures allows parameter identification and control of the process.

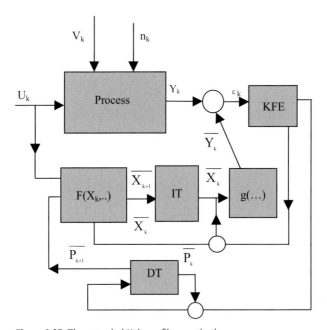

Figure 3.87 The extended Kalman filter method.

The vector $\overline{Y}_k$ can be calculated either with the normal Kalman Filter (KF) which gives $\overline{X}_k$ for the discrete equation state ($F(\overline{X}_k, U_k, V_k)$) or with the extended Kalman filter (KFE) which gives $\hat{P}_{k+1}$ in the calculation system. For this estimation, it is also necessary to obtain the state of the system $\overline{X}_k$ from the next state $\overline{X}_{k+1}$. This estimation is made by block IT (inversion translator); another IT block gives the $\hat{P}_{k+1}$ used for state system estimation with $\hat{P}_k$ from KFE. Both methods need the Kalman filter to be started. Indeed, an introduction of the Kalman filter equations is required in order to correctly appreciate how the MLM and KFE methods operate.

3.5.5.1 The Kalman Filter Equations

This method is frequently used for filtering, smoothing and identifying parameters in the case of a dynamic time process. It has been developed taking into account the following conditions: (i) acceptance of the gaussian distribution of the disturbances and exits of the variables of the process; (ii) there is a local linear dependence between the exit vector and the state vector in the mathematical model of the process.

The Kalman filter problem. Considering the relations (3.258) and (3.259)) we can write the following discrete-time system:

$$\begin{cases} X_{k+1} = F_k X_k + G_k W_k \\ Y_k = H_k X_k + V_k \end{cases} \tag{3.265}$$

where the input disturbance vector W_k is $N(0, Q_k)$, the exit disturbance vector V_k is $N(0, R_k)$ and the initial input vector X_0 is $N(m_0, S_0)$. In the expression $N(\alpha, \beta)$, α represents the mean value and β is the dispersion or covariance with respect to the mean value.

The dimensions of the state vector X_k and of the observation vector (exit vector) are N and M respectively. This short introduction is completed by assuming that R_k is positive ($R_k > 0$).

The problem considered here is the estimation of the state vector X_k (which contains the unknown parameters) from the observations of the vectors $Y_k = [y_0, y_1 \ldots y_k]$. Because the collection of variables $Y_k = (y_0, y_1, \ldots y_k)$ is jointly gaussian, we can estimate X_k by maximizing the likelihood of conditional probability distributions $p(X_k/Y_k)$, which are given by the values of conditional variables. Moreover, we can also search the estimate $\hat{X}_k$, which minimizes the mean square error $\varepsilon_k = X_k - \hat{X}_k$. In both cases (maximum likelihood or least squares), the optimal estimate for the jointly gaussian variables is the conditional mean and the error in the estimate is the conventional covariance.

In what follows, we will develop the conditional mean and covariance for the couple X_k and Y_k. This is followed by a description of the Kalman filter and a rapid and practical method for a recursive or iterative calculation of the conditional mean and covariance for the random variable vector X_k/Y_k.

In many different softwares such as SCILAB©, computational programs are available for calculating: (i) the steady-state Kalman filter which can be used when the matrices of the systems in (3.265) do not vary with time; (ii) the unsteady-state Kalman filter which can be used when the matrices of the systems in (3.265) vary with time; (iii) the square-root Kalman filter for time or non-time-varying matrices of the systems when high numerical accuracy is required.

Mean and covariance for conditional gaussian random vector. The minimum mean square estimate of a gaussian random vector when we only have observations of some of its elements is the conditional mean of the remaining elements. The error covariance of this estimate is the conditional covariance. Consequently, if Z is a random gaussian vector composed of sub-vectors x and y, then we may write:

$$Z = \begin{bmatrix} x \\ y \end{bmatrix} \text{ is } N\left(\begin{bmatrix} m_x \\ m_y \end{bmatrix}, \begin{bmatrix} C_x & C_{xy} \\ C_{yx} & C_y \end{bmatrix}\right) \tag{3.266}$$

where m_x and m_y are the mean of x and y, C_x is the covariance of x with itself, C_{xy} is the covariance of x with y, etc. It is know that the marginal and conditional distributions of a gaussian random vector are also gaussian. Indeed, the distribution of x for a given y has a probability density p(x/y) of normal type:

$$p(x/y) = N(m_{x/y}, C_{x/y}) \tag{3.267}$$

In this case the conditional mean $(m_{x/y})$ and the conditional covariance $(C_{x/y})$ may be calculated as follows:

$$m_{x/y} = m_x + C_{x/y} C_y^{-1}(y - m_y) \tag{3.268}$$

$$C_{x/y} = C_x - C_{xy} C_y^{-1} C_{yx} \tag{3.269}$$

These two relations are the basis for other important developments of the Kalman filter equations. Concerning the problem considered above, the calculation of the minimum mean square error can be carried out either:

1. By considering the individual observations on the concentrated vector Y_k. Because X_k and Y_k are both gaussian, then Eqs. (3.268) and (3.269) represent the vector used to obtain the conditional mean and covariance of X_k for a given Y_k. However, when the dimension of the vector is too large, problems with matrix multiplication and inversion can appear.

   or

2. By developing a special recursive update for the estimation of x_k from X_k based on the linear system (3.265) and a special property derived from Eqs. (3.268) and (3.269). More precisely, if the best estimate of x_k based on the observations Y_k is given (denote this estimate $\hat{x}_{k/k}$) with a new observation y_{k+1}, it is shown how to obtain the best estimate $\hat{x}_{k+1/k+1}$ and its error covariance matrix $CE_{k+1/k+1}$.

Linear systems over a gaussian random vector. If x or X is a gaussian vector with mean value m_x and covariance C_x (the minimum square error estimate for x is $\hat{x}$ and $\hat{x} = m_x$) which is considered to be in a formal linear system completed with a zero-mean gaussian vector (v is N (0, R)) then we have:

$$y = Hx + v \tag{3.270}$$

The mean and covariance of y are calculated, by their definition, as follows:

$$m_y = E|y| = E|Hx + v| = Hm_x \tag{3.271}$$

$$C_y = E \Big| y - m_y \Big| * \Big| y - m_y \Big|^T = E|H(x - m_x) + v| * |H(x - m_x) + v|^T = HC_x H^T + R \tag{3.272}$$

Consequently, the minimum mean square error estimate for y is $\hat{y} = Hm_x$ and the associated covariance of this is $C_y = HC_x H^T + R$.

Recursive estimation of gaussian random vectors. We consider here a gaussian random vector composed of three sub-vectors x,y and z:

$$\begin{bmatrix} x \\ y \\ z \end{bmatrix} \text{ is } N \left(\begin{bmatrix} m_x \\ m_y \\ m_z \end{bmatrix}, \begin{bmatrix} C_x & C_{xy} & C_{xz} \\ C_{yx} & C_y & C_{yz} \\ C_{zx} & C_{zy} & C_z \end{bmatrix} \right)$$

From Eqs. (3.268) and (3.269) the minimum mean square estimate of x for a given y is:

$$\hat{x}(y) = m_x + C_{xy} C_y^{-1}(y - m_y) \tag{3.273}$$

and the associated error covariance can be computed as follows:

$$C_x(y) = C_x - C_{xy} C_y^{-1} C_{yx} \tag{3.274}$$

It is important to note that $E[\hat{x}(y)] = m_x$

Now if z is also observed, then the minimum mean square error estimate of x for a given y and z is:

$$\hat{x}(y, z) = m_x + \begin{bmatrix} C_{xy} & C_{xz} \end{bmatrix} * \begin{bmatrix} C_y & C_{yz} \\ C_{zy} & C_z \end{bmatrix}^{-1} \begin{bmatrix} y - m_y \\ z - m_z \end{bmatrix} \tag{3.275}$$

and the error covariance:

$$C_x(y, z) = C_x - \begin{bmatrix} C_{xy} & C_{xz} \end{bmatrix} * \begin{bmatrix} C_y & C_{yz} \\ C_{zy} & C_z \end{bmatrix}^{-1} \begin{bmatrix} C_{yx} \\ C_{zx} \end{bmatrix} \tag{3.276}$$

When y and z stay independent then $C_{yz} = C_{zy} = 0$, and the relation (3.275) can be simplified as follows:

$$\hat{x}(y, x) = m_x + C_{xy} C_y^{-1}(y - m_y) + C_{xz} C_z^{-1}(z - m_z) = \hat{x}(y) + C_{xz} C_z^{-1}(z - m_z) \tag{3.277}$$

The use of Eq. (3.277) needs a recursive method to calculate $\hat{x}(y, z)$ for a given $\hat{x}(y)$ and z. The problem is that Eq. (3.275) depends on y and z, which are independent vectors. Fortunately, changing variables makes it possible to change the estimation procedure for Eq. (3.275) and then Eq. (3.277) can be modified considering the random vector v defined by:

$$v = z - \hat{z}(y) = z - [m_z + C_{xy}C_y^{-1}(y - m_y)] = (z - m_z) - C_{xy}C_y^{-1}(y - m_y) \quad (3.278)$$

Here $\hat{z}(y)$ is the minimum mean square estimate of z for a given observation of y and this is used in Rel. (3.278) by means of Rel. (3.268). This new random vector, v, has several interesting properties, which are important for the development of the Kalman filter equations:

1. Because m_v is zero, v, is a zero-mean random value:
   $$m_v = E[(z - m_z) - C_{xy}C_y^{-1}(y - m_y)] = 0.$$
2. Since
   $$C_{vy} = E\left[v(ym_y{}^T)\right] = E[(z - m_z)(y - m_y)^T - C_{zy}C_y^{-1}(y - m_y) \times (y - m_y)^T] = C_{zy} - C_{zy}C_yC_y^{-1} = 0$$
   we consider that v and y are independent .
3. Because $C_{vx(\hat{y})}$ is given by relation
   $$C_{vx(\hat{y})} = E[v(m_x + C_{xy}C_y^{-1}(y - m_y)^T] = E[v(y - m_y)^T C_y^{-1} C_{yx}] = 0,$$
   and considering the previous property (2.), we obtain that v is independent with respect to y and $\hat{x}(y)$.

Now, if we replace z by v in Eq. (3.275), we can rewrite the result as follows:

$$\hat{x}(y, z) = \hat{x}(y, v) = m_x + \begin{bmatrix} C_{xy} & C_{xv} \end{bmatrix} * \begin{bmatrix} C_y & 0 \\ 0 & C_v \end{bmatrix}^{-1} \begin{bmatrix} y - m_y \\ v \end{bmatrix}$$

$$= m_x + C_{xy}C_y^{-1}(y - m_y) + C_{xv}C_v^{-1}v = \hat{x}(y) + C_{xv}C_v^{-1}\hat{v} \quad (3.279)$$

It is easy to observe that from Eq. (3.274) we can obtain:

$$C_{xv} = E[(x - m_x)(z - m_z - C_{xy}C_y^{-1}(y - m_y) = C_{xz} - C_{xy}C_y^{-1}C_{yz} \quad (3.280)$$

and the variable correlation for v is then:

$$C_v = E[(z - m_z - C_{zy}C_y^{-1}(y - m_y))(z - m_z - C_{zy}C_y^{-1}(y - m_y))^T]$$
$$= C_z - C_{zy}C_y^{-1}C_{yz} \quad (3.281)$$

It may be noticed that the equality of $\hat{x}(y, z)$ and $\hat{x}(y, v)$ is the result of the conservation of all information while the variables in Eq. (3.278) are being replaced. Indeed, we are simply adding a constant vector to z, and this vector makes v and y independent of each other. The error covariance here noted as $CC_x(y, v)$ associated with Eq.(3.277) is:

$$CC_x(y, v) = C_x - \begin{bmatrix} C_{xy} & C_{xv} \end{bmatrix} * \begin{bmatrix} C_y & 0 \\ 0 & C_v \end{bmatrix}^{-1} * \begin{bmatrix} C_{yx} \\ C_{vx} \end{bmatrix}$$

$$= C_x - C_{xy}C_y^{-1}C_{yx} - C_{xv}C_y^{-1}C_{vx} = C_x(y) - C_{xv}C_v^{-1}C_{vx} \tag{3.281}$$

The Kalman Filter Equations are here obtained from the formulation of the Kalman filter with the purpose of finding a recursive estimation procedure for the solution of a problem (estimation of state vector). Before detailing the procedure, we have to introduce other new notations. The minimum square estimate of x_k for the given observations $Y_1 = [y_0, y_1, \dots y_l]$ is defined by $\hat{x}_{kl}$. Furthermore $CC_{k/l}$ represents the error covariance associated with $\hat{x}_{k/l}$.

With these notations, we can now explain the estimation of $\hat{x}_{k/k}$ from the estimate $\hat{x}_{k/k-1}$ and the new observation y_k. From Eqs. (3.279) and (3.281) we obtain:

$$\hat{x}_{k/k} = \hat{x}_{k/k-1} + C_{xkvk}C_{vk}^{-1}v_k \tag{3.282}$$

$$CC_{k/k} = CC_{k/k-1} - C_{xkvk}C_{vk}^{-1}C_{vkxk} \tag{3.283}$$

If v_k is extracted from Eqs. (3.265), (3.270) and (3.271) then we have:

$$v_k = y_k - H_k\hat{x}_{k/k-1} \tag{3.284}$$

The covariance matrices from Eqs. (3.282) and (3.283) may be calculated using the definition of some established relations. The following relations are then obtained:

$$C_{vk} = E\left[(y_k - H_k\hat{x}_{k/k-1})(y_k - H_k\hat{x}_{k/k-1})^T\right]$$

$$= E\left[[(H_k(x_k - \hat{x}_{k/k-1}) + v_k][(H_k(x_k - \hat{x}_{k/k-1}) + v_k]^T\right] \tag{3.285}$$

$$= H_kCC_{k/k-1}H_k^T + R_k$$

$$C_{xkvk} = E\left[(x_k - E(x_k))v_k^T\right] = E\left[(x_k - E(x_k) + E(x_k) - \hat{x}_{k/k-1})v_k^T\right]$$

$$= E\left[(x_k - \hat{x}_{k/k-1})v_k^T\right] = E\left[(x_k - \hat{x}_{k/k-1})(y_k - H_k\hat{x}_{k/k-1})^T\right] \tag{3.286}$$

$$= E\left[(x_k - \hat{x}_{k/k-1})(x_k - \hat{x}_{k/k-1})^T H_k^T\right] = CC_{k/k-1}H_k^T$$

Substituting Eqs. (3.286), (3.285) and (3.284) into Eqs.(3.283) and (3.282) we have:

$$\hat{x}_{k/k} = \hat{x}_{k/k-1} + Kg_k(y_k - H_k\hat{x}_{k/k-1}) \tag{3.287}$$

$$CC_{k/k} = CC_{k/k-1} - Kg_kH_kCC_{k/k-1} \tag{3.288}$$

where the Kalman gain of the filter is given by $Kg_k = CC_{k/k-1} H_k^T$ $[H_k CC_{k/k-1} H_k^T + R_k]^{-1}$. It is important to observe the subtraction of $R_k > 0$ considered in the definition of the disturbance vector. If $R_k > 0$, Kg_k always exists. However, if we accept that R_k is not necessarily positive, we can have problems making the inverse matrix necessary to calculate Kg_k.

Using Eqs. (3.271), (3.272) and (3.265) to complete Eqs. (3.287) and (3.288), we can establish the next two auxiliary equations:

$$\hat{x}_{k+1/k} = F_k \hat{x}_{k/k} \quad , \quad CC_{k+1/k} = F_k CC_{k/k} F_k^T + G_k Q_k G_k^T \tag{3.289}$$

Combining relations (3.287), (3.288) and (3.289) results in a set of recursive equations, which are called the *Kalman filter equations*:

$$\hat{x}_{k+1/k} = F_k \hat{x}_{k/k-1} + F_k Kg_k(y_k - H_k \hat{x}_{k/k-1}) \tag{3.290}$$

$$CC_{k+1/k} = F_k CC_{k/k-1} F_k^T - F_k Kg_k H_k CC_{k/k-1} F_k^T + G_k Q_k G_k^T \tag{3.291}$$

To be operational, the Kalman filter equations must be completed with the starting conditions $\hat{x}_{0/-1}$ and $CC_{0/-1}$, which correspond to $k = 0$ in relations (3.290) and (3.291). These conditions are obtained from the statistical starting of the initial state vector:

$$\hat{x}_{0/-1} = m_0 \quad CC_{0/-1} = s_0 \tag{3.292}$$

The optimized values for the unknown model state parameters are obtained by coupling the Kalman filter equations with the mathematical models of the process that give matrices F_k, H_k and G_k. The coupling above has to be completed step by step: we must first use an initial estimator vector for the process state (which is the starting point in the previous method) and then consider a judicious error covariance matrix and experimental data. Depending on the problem formulation stated by Eq. (3.265) there are situations where the Kalman procedure does not give satisfactory results. Indeed, in some cases, the Kalman filter can provide state estimates which diverge from the actual evolution of the vector. However, generally, this divergence is not the result of a fault of the Kalman filter, but, rather, is due to the process model provided by the user. In such cases, the user may re-examine the model formulation in order to give a better version of the model used for the estimation problem. From the mathematical viewpoint, the Kalman filter is only valid for time invariant formulations of the model in Eq. (3.265). This assertion implies that the studied system must be controllable and observable. These two conditions allow the matrix of error covariance $CC_{k/k-1}$ to converge towards a finite and positive constant matrix. Consequently, the error in the estimate $\hat{x}_{k/k-1}$ is bounded as $k \to \infty$ because of the $CC_{k/k-1}$ bounding.

Another consequence of the steady state analysis of the Kalman filter is that we can use the steady state Kalman gain instead of the time varying Kalman gain. The advantage of such an approach is that considerable computational savings are possible because we do not need to recalculate the Kalman gain for each new observation.

3.5.5.2 Example of the Use of the Kalman Filter

The example analyzed here is one of the simplest problems because it is two-dimensional with respect to the vectors state. The example illustrates the Kalman tracking for a system model, which is controllable and observable. To this aim, we use the following system model and prior statistics:

- the process state vector:

$$x_{k+1} = \begin{bmatrix} 1.1 & 0.1 \\ 0 & 0.8 \end{bmatrix} x_K + \begin{bmatrix} 1 & 0 \\ 0 & 1 \end{bmatrix} w_k$$

- the measurement vector:

$$y_k = \begin{bmatrix} 1 & 0 \\ 0 & 1 \end{bmatrix} x_k + v_k$$

- prior statistics:

$$E(w_k w_k^T) = \begin{bmatrix} 0.03 & 0.01 \\ 0.01 & 0.03 \end{bmatrix}; E(v_k v_k^T) = \begin{bmatrix} 2 & 0 \\ 0 & 2 \end{bmatrix}; E(x_0) = \begin{bmatrix} 10 \\ 10 \end{bmatrix};$$

$$E[(x - m_0)(x - m_0)^T] = \begin{bmatrix} 2 & 0 \\ 0 & 2 \end{bmatrix}$$

For actual cases, the model of the process is given *a priori* and the process state vector can be built according to the rule presented here for the case of the Gauss–Newton method. The dispersion due to the normal disturbances of the process input or the dispersion that characterizes the errors of the exit measurements must be appreciated and proposed, as shown in our case. In this example, it is not difficult to observe that the model presents two exit variables: y_1 (that decreases with the state parameters) and y_2 (that increases with the state parameters). The observations of the process (measured data of y_1 and y_2) have been generated using the system formulation and values for x_k from a random number generator that add the random multiplication of its dispersion to the mean value. Ten observations have been generated by using adequate software and have been exploited as input for the Kalman filter. The result of these observations and the estimations carried out by the Kalman procedure are illustrated in Fig. 3.88. In this figure, we can observe the actual state path and the Kalman estimation of the state path as solid and dashed lines respectively. The actual locations of the state and estimated values are indicated as white and black circles. The ellipses in the figure are centred near the positions of the actual state path and their borders represent the two standard deviations of the estimation error calculated from the matrices of error covariance. The numerical values of the standard deviations are also given in the figure.

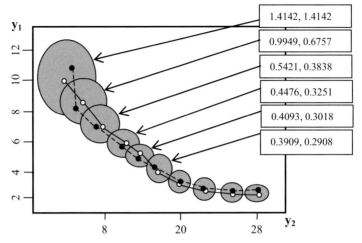

Figure 3.88 Numerical example of the Kalman filter tracking.

3.6
Some Conclusions

All theoretical aspects, engineering observations and commentaries and actual examples presented in this chapter illustrate some basic or particular aspects of the modelling and simulation of processes based on transfer phenomena in the field of chemical engineering

The essential features of the presented aspects can be described as follows:

1. Mathematical models are developed for teaching, for engineering calculations and for finding solutions to the technical problems of design using rigorous procedures where the core resides in the particularization of the transport phenomena equations to the actual case modelled. The main purpose of modelling is to provide engineers and practitioners with prediction parameters of direct practical interest, i.e. the value of concentrations and temperature, shear rates, heat and mass transfer rates, as functions of the operating conditions including equipment geometry and dimensions, the properties of the media and the process features.

2. Mathematical models have been developed by considering classical flow models. At the same time, the capacity of computational fluid dynamics to be coupled with heat and mass transfer processes and with a reaction has been considered.

3. Every mathematical model is a simplified mirror image of a real phenomenon. To sustain the modeling assumptions, all

exemplified models are characterized by experimental or simulated data. This characterization also shows the strongly cognitive capacity of the developed models.

4. Modelling usually includes several consecutive steps of calculations; therefore, to make the method practical, the software simplification of the main equations has to be accepted with respect to the practical application. In many cases, we can reduce the simulation complications without impairing the reliability of the obtained results.

5. Model-based estimation techniques applied to identify or simply estimate parameters are presented as mathematical formulations and are sustained by practical applications.

References

3.1 R. B. Bird, W. E. Steward, E. N. Lightfoot, *Transport Phenomena*, John Wiley, New-York, 1960.

3.2 V.A. Levich, *Physico-Chemical Hydrodynamics*, McGraw-Hill, New York, 1962.

3.3 A. Foust, *Principles of Unit Operations*, John Wiley, New York, 1966.

3.4 V.A. Luikov, *Heat and Mass Transfer*, Mir Publishers, Moscow, 1980.

3.5 E. A. Bratu, *Processes and Apparatus for Chemical Industry, Vol. 1*, Technical Book, Bucharest, 1969.

3.6 J. C. Slatery, *Momentum, Energy and Mass transfer in Continua*, McGraw-Hill, New York, 1972.

3.7 V. V. Veverka, F. Madron, *Material and Energy balancing in the Process Industries*, Elsevier, Amsterdam, 1997.

3.8 H. Schlichting, *Boundary Layer Theory*, 7th Edn., McGraw-Hill, New York, 1979.

3.9 R. G. E. Franks, *Modeling and Simulation in Chemical Engineering*, Wiley-Interscience, New York, 1972.

3.10 J. N. Kapur, *Mathematical Modeling*, John Wiley, New York, 1988.

3.11 S. Kotake, K. Hijikata, *Numerical Simulation of Heat Transfer and Fluid Flow on a Personal Computer*, Elsevier Science, North-Holland, Amsterdam, 1999.

3.12 B. Volesky, J. Votruba, *Modeling and Optimization of Fermentation Processes*, Elsevier Science, North-Holland, Amsterdam,1999.

3.13 W. L. Luyben, *Process Modeling, Simulation and Control for Chemical Engineering*, 4th Edn., McGraw-Hill, New York, 1990.

3.14 A. Hinchliffe, *Chemical Modeling: From Atoms to liquids*, Wiley-VCH, Weinheim,1999.

3.15 C. Truesdell, *Rational Thermodynamics*, McGraw-Hill, New York, 1969.

3.16 Th. L. Bott, E. Ehrfeld, *Chem. Eng. J.* **2000**, *80*,163, 245.

3.17 G. Barralla, M. Mattea, V. Gekas, *Sep. Pur. Technol.* 2001, 22–23, 1–3, 489.

3.18 M. Hamachi, M. Cabassud, A. Davin, M. Peucho, *Chem. Eng. Process.* **1999**, *38*, 3, 200.

3.19 W. B. Richard, N. S. Hartham, I. J. Calco, *Sep. Pur. Technol.* **2001**, *24*, 162, 297.

3.20 L. Yonghum, M. M. Clarke, *J. Membr. Sci.* **1998**, *149*, 2, 181.

3.21 V. Nassachi, *Chem. Eng. Sci.* **1998**, *53*, 6, 1253.

3.22 T. Dobre, J. Sanchez, V. S. Dinu, G. Iavorschi, Modeling and Simulation of a Tangential Filter Unit, *Proc. of the 12th International Romanian Chemistry and Chemical Engineering Conference*, S5, P29, 2001.

3.23 O. Levenspiel, *Chemical Reaction Engineering*, Wiley, New York, 1999.

3.24 C. T. Dickenson, *Filters and Filtration Handbook*, 4th Edn., Elsevier, North-Holland, Amsterdam, 1999.

3.25 A. J. Reynolds, *Turbulent Flows in Engineering*, Wiley, New York, 1974.

3.26 D. W. Deckwer, *Bubble Column Reactors*, John Wiley, New York, 1991.

3.27 G. Hebrand, D. Bastoul, M. Roustan, P. M. Comte, C. Beck, *Chem. Eng. J.* **1999**, *72*, 109.

3.28 S. Najarian, J. B. Bellhouse, *Chem. Eng. J.* **1999**, *75*, 105.

3.29 D. W. Deckwer, *Bubble Column Reactors*, John Wiley, New York, 1991.

3.30 J. H. Ferzinger, M. Peric, *Computational Method for Fluid Dynamics*, Springer-Verlag, Berlin, 1996.

3.31 V. Novozhilov, *Prog. Energ. Combust. Sci.* **2001**, *27*, 611.

3.32 R. Krisna, M. J. Van Baten, *Nature*, **1999**, *398*, 208.

3.33 R. Krisna, M. I. Urseanu, M. J. Van Baten, J. Ellenberger, *Int. Commun. Heat. Mass Transfer* **1999**, *26*, 781.

3.34 A. H. Boerner, U. R. Qi, *Int. J. Multiphase Flow* **1997**, *23*, 523.

3.35 J. K. Marschall, L. Mlezko, *Chem. Eng. Sci.* **1999**, *54*, 2085.

3.36 J. Dong, D. Gidaspow, *AIChE J.* **1990**, *36*, 523.

3.37 M. G. B. Van Wachem, C. J. Schouten, R. Krisna, M. C. Van der Bleek, *Comput. Chem. Eng.* **1998**, *22*, 299.

3.38 M. G. B. Van Wachem, C. J. Schouten, R. Krisna, M. C. Van der Bleek, *Chem. Eng. Sci.* **1999**, *54*, 2141.

3.39 A. Lapin, A. Lubert, *Chem. Eng. Sci.* **1994**, *49*, 3661.

3.40 J. T. Lin, J. Reese, T. Hong, S. L. Fan, *AIChE J.* **1996**, *42*, 301.

3.41 A. Sokolichin, G. Eingenberger, *Chem. Eng. Sci.* **1994**, *49*, 5375.

3.42 A. Sokolichin, G. Eingenberger, *Chem. Eng. Sci.* **1997**, *52*, 611.

3.43 S. S. Thakre, B. J. Joshi, *Chem. Eng. Sci.* **1999**, *54*, 5055.

3.44 B. Mehta, T. K. Chang, K. Nandakumar, *Chem. Eng. Res. Des. Trans.* **1998**, *7*, 843.

3.45 M. J. Van Baten, R. Krisna, *Chem. Eng. J.* **2000**, *77*, 143.

3.46 C. Reid, *Zone Refining*, McGraw-Hill, New York, 1956.

3.47 J. Welty, E. Ch. Wicks, E. R. Wilson, L. G. Rorrer, *Fundamentals of Momentum Heat and Mass Transfer*, Wiley, New York, 2000.

3.48 H. K. Keneth, R. T. Stein, *Math. Biosci.* **1967**, *1*, 412.

3.49 H. L. Frisch, *J. Phys.Chem.* **1959**, *63*, 1249.

3.50 R. Ash, M. R. Barrer, T. H. Chio, *J. Phys. Eng. Sci. Instrum.* **1978**, *11*, 262.

3.51 J. Sanchez, L. Gijiu, V. Hynek, O. Muntean, A. Julbe, *Sep. Pur. Technol.* **2001**, *7*, 116.

3.52 H. P. Nelson, M. S. Auerbach, *Chem. Eng. J.* **1999**, *74*, 43.

3.53 K. Kuo, *Principle of Combustion*, John Wiley, New York, 1986.

3.54 S. Candel, J. P. Martin, Coherent Flame Modeling of Chemical Reactions in Turbulent Mixing Layer, in *Complex Chemical Reaction Systems*, J. Warnatz, W. Joger (Eds.), Springer-Verlag, Berlin, 1987.

3.55 P. H. Thomas, The Growth of Fire-Ignition to Full Involvements, in *Combustion Fundamentals of Fire*, G. Cox (Ed.), Academic Press, London, 1995.

3.56 A. Cavaliere, R. Ragucci, *Prog. Energ. Combust. Sci.* **2001**, *27*, 547.

3.57 J. G. Sanchez Marcano, T. T. Tsotsis, *Catalytic Membranes and Membrane Reactors*, Wiley-VCH, Weinheim, 2002.

3.58 M. Niedzwiecki, *Identification of Time-Varying Processes*, Wiley, New York, 2000.

3.59 G. Maria, E. Heinzle, *J. Loss. Prev. Process. Ind.* **1998**, *11*, 187.

3.60 J. C. Nash, M. Walker-Smith, *Nonlinear Parameter Estimation: An Integrated System in Basic*, Marcel Dekker, New York, 1987.

3.61 W. S. Rutherford, D. D. Do, *Chem. Eng. J.* **1999**, *74*, 155.

3.62 V. V. Kafarov, *Cybernetic Methods for Technologic Chemistry*, Himia, 1969.

3.63 R. Fletcher, *Practical Method of Optimization*, 2nd Edn., Wiley, New York, 2000.

3.64 E. K. Atkinson, *Introduction to Numerical Analysis*, Wiley, New York, 2000.

3.65 G. Bastin, D. Dochain, *On Line Estimation and Adaptive Control of Bioreactors*, Elsevier, North-Holland, Amsterdam,1999.

3.66 I. Riekert, *AIChE J.* **1985**, *14*, 19.

3.67 M. Suzuki, K. Kwazoe, *J. Chem. Eng. Jpn.* **1975**, *8*, 379.

3.68 M. R. Onsiovici, L. S. Cruz, *Chem. Eng. Sci.* **2000**, *55*, 20, 4667.

3.69 S. G. Mohinder, P. A. Angus, *Kalman Filtering: Theory and Practice Using MATLAB*, 2th Edn., Wiley-VCH, Weinheim, 2001.

3.70 S. Saelid, N. A. Jensen, T. Lindstad, L. Kolbeisen, Modeling, Identification and Control of a Fluidized Bed Reactor, in *Identification and System Parameters Estimation-Proceedings of 5th IFAC Symposium*, Iserman R. (Ed.), Darmstadt, 1979.

3.71 X. Wu, H. K. Bellgardt, *J. Biotechnol.* **1998**, *62*, 11.

4
Stochastic Mathematical Modelling

Stochastic mathematical modelling is, together with transfer phenomena and statistical approaches, a powerful technique, which can be used in order to have a good knowledge of a process without much tedious experimental work. The principles for establishing models, which were described in the preceding chapter, are still valuable. However, they will be particularized for each example presented below.

4.1
Introduction to Stochastic Modelling

As analyzed in the preceding chapters concerning the description of a process evolution, stochastic modelling follows the identification of principles or laws related to the process evolution as well as the establishment of the best mathematical equations to characterize it.

The first approaches to compare stochastic models and chemical engineering were made in 1950, with the Higbie [4.1] and Dankwerts transfer models [4.2]. Until today, the development of stochastic modelling in chemical engineering has been remarkable. If we made an inventory of the chemical engineering modelling studies we could see that a stochastic solution exists or complements all the cases [4.3–4.8].

In many modelling studies, the model establishment is made in relation to the transfer and balance of a property (for instance see Chapter 3, Section 3.1). Nevertheless, a property evolution from the initial to the final state can vary randomly as a result of the stochastic combination of different elementary processes. This statement is in good agreement with the unitary concept of transfer phenomena [4.9–4.11] and was reported by Bratu [4.11] in the following assertion:

> "Each transformation or phenomenon results from one or many
> elementary steps or processes. The equilibrium state results from
> similar but contrary transport fluxes."

This statement can also be obtained when a transport process evolution is analyzed by the concept of Markov chains or completely connected chains. The math-

Chemical Engineering. Tanase G. Dobre and José G. Sanchez Marcano
Copyright © 2007 WILEY-VCH Verlag GmbH & Co. KGaA, Weinheim
ISBN: 978-3-527-30607-7

ematical theory of completely connected chains [4.12–4.16] can be described with this condensed statement:

> *"The state of a system at time **n** is a random variable A_n with values in a finite space (A **A**) (measurable). The state evolution at time n+1 results from the arrival of a B_{n+1} result, which is also a random variable with values in a finite space (B **B**) (measurable). The arrival of a result signaling the state evolution can be represented considering a **u** application of A ×B in A and introducing the following statement: $A_{n+1} = \mathbf{u}(A_n, B_{n+1})$ for all n ≥ 0. The B_{n+1} probability distribution is conditioned by $B_n, A_n, B_{n-1}, A_{n-1}, \ldots\ldots B_1, A_1, A_0$, and symbolized as $(P(B_{n+1}/B_n, A_n, \ldots))$, it depends only on state A_n. The group $[(A,\underline{A}), (B,\underline{B}), \mathbf{u}:A \times B \to A, P]$ defines a random system with complete connections."*

Some of the examples shown in the following paragraphs present the characteristics of a random system with complete connections. However, other examples do not concern a completely connected system but present only some Markov unitary processes [4.6, 4.17].

The stochastic modelling of the phenomena studied here can be described by one standard physical model (descriptive model) which can be defined by the following statement:

> *"The property carriers, such as elementary particles of fluids or molecules, evolve during their displacement through one or more elementary processes (called process components), their passage from one process to another is made by one stochastic process called connection or connection process."*

It is important to notice the similitude of this descriptive model to the complete definition of connections of a system given before. For chemical engineering processes, the model needs to be particularized and then the assertions written below have to be taken into account [4.4–4.7]:

- If one elementary particle is participating in a process of transport phenomena within a medium with random characteristics (granular medium, porous solid, etc.), the medium will be responsible for the random velocity changes of the particle. In this case, the transport process concerning the local velocity is the so-called "process component", whereas the transport process changes given by the random properties of the medium are called "connection process".
- The transport phenomenon occurs when the displacement of the carriers through different media ("process components") and the passage from one medium to another are realized by a random commutation process ("connection process").

- During their displacement, elementary particles are constantly encountering obstacles, other moving elements, oscillation states etc. The particle evolution is randomly chosen among the different presented possibilities ("process components").
- A particle (molecule, group of molecules, turbulent group etc.) evolves in a medium which produces its own transformation. This means that the process exchange characterizes the particle evolution. The process occurring before and after the transformation is called the "process component", whereas the transformation itself which represents the stochastic evolution is called the "connection process".
- The elementary particles randomly pass from one compartment to another; the process of swapping compartments forms the "connection process" whereas the transformation realized in each compartment represents the "process components".
- When phenomena result in the formation of various structures, the passage from one structure to another occurs randomly, in this case the structure formation is the "process component" and the transitory steps correspond to the "connection process".

When a stochastic process takes place, the passage from one elementary process to another is caused by external effects. These effects are related to the medium by the process evolution itself. We can assert that a process can be adapted to stochastic modelling if we can identify the elementary "process components". In addition, for the "connection process" the number of states has to be same as the "process components". This very abstract introduction will be better explained in the next paragraph by including a practical example.

4.1.1
Mechanical Stirring of a Liquid

Studies of the mixing effectiveness of stirring devices are quite numerous; they generally analyze the effects produced by:
- the turn rate or frequency of the stirrer,
- the configuration and the distribution of the stirring paddles in the apparatus,
- the physicochemical properties of the medium,
- the position of the input and the output of the currents in the apparatus tank.

It has been very difficult to develop a general model able to describe the influence of all the different parameters on the mixing effectiveness [4.18–4.20]. However, many researchers have tried to develop models as complete as possible, among them a special mention has to be given to the research team who developed the first commercial program called *Visimix* 1999 [4.21, 4.22].

As an example, we analyze here, step by step, a continuous mixing apparatus provided with stirring paddles. The first step is to uniformly express the state of the currents, which characterize the different flow patterns inside the apparatus (see Section 3.3). The second step concerns the development of a stochastic model for characterizing the mixing. For this purpose, we will use the procedure of analysis initially developed by Kafarov [4.4], which has been completed and modified since by other authors [4.5, 4.6, 4.23]. The apparatus considered here is shown in Fig. 4.1. In this apparatus provided with stirring paddles, the main current of flow is radial and it separates into two different currents closed the walls. The size of these flow currents depends on the stirrer position, the number of turns of the stirrer and the medium properties. Indeed, we can consider that the stirrer divides the apparatus into two regions (the higher region and the lower region with respect to the paddles) with different and independent currents.

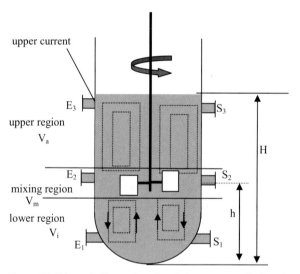

Figure 4.1 Schematic illustration of a stirring apparatus for liquid mixing.

According to this topology, the flow in the apparatus is described by two circuits (the upper and the lower region), each of which contains a variable number of ideal mixed cells but meets in the mixing region (near the paddles). This region constitutes a cell with ideal mixing.

Figure 4.1 also shows the position of the input flow (E_1, E_2, E_3) which has to be coupled with the positions of the output flow (positions S_1, S_2, S_3) and the general current of circulation inside the vessel. We can observe from Fig. 4.1 that the volumes placed in the higher and lower regions depend on H and h (thus on the position of the stirrer paddle in the apparatus) as well as on the size of the stirring region. Indeed, the corresponding volumes can be described as follows:

$$V_a = \frac{\pi D^2}{4}(H - h) - \frac{V_m}{2} \qquad (4.1)$$

$$V_a = \frac{\pi D^2}{4} h - \frac{V_m}{2} = V_u - V_s - \frac{V_m}{2} \tag{4.2}$$

We can expect the volume in the pumping region to depend on the dimensions of the stirrer paddle (diameter d, height of the pallet b) and on the dimensions of the tank:

$$V_m = \frac{\pi b}{60} \left(D^2 + \frac{Dd}{5} + \frac{d^2}{5} \right) \tag{4.3}$$

According to the topological description, we can consider a cell with ideal mixing in the stirring region; other cells can be considered in the upper region (with number: n_a) and in the lower region (with number: n_i). The respective number of cells in the regions can be calculated by the following algorithm:

1. The starting conditions are established.
2. V_a, V_m, V_i are calculated with Eqs. (4.1)–(4.3).
3. If $V_a > V_i$ then $r = V_a/V_i$; for the reverse case $r = V_i/V_a$.
4. The number of cells is chosen in the smaller region; (this consideration is frequently used) then, $n_i = n_{ch} = 1$ if the lower region is the smaller one, when the upper region is the smaller one we have $n_a = n_{ch} = 1$.
5. If $h/H = 0.5$ we can consider $n_i = n_a = n_{ch}$; if $h/H > 0.5$, which is equivalent to $V_i > V_a$ we can write $n_a = n_{ch}$ and $n_i = r * n_{ch}$; however, if $h/H < 0.5$ and $V_a > V_i$ we can consider $n_i = n_{ch}$ and $n_a = r * n_{ch}$

Once the topology has been established, it must be supplemented with the flows of the currents, which convey between the cells. Many solutions have been suggested to solve this problem. They differ by the mode of calculation of the main current produced by the stirrer. It is a function of the geometry, the number of the turns of the stirrer and the properties of the medium (density, viscosity). For the stirrer considered here, the flow rate of the main current and the flows in the higher and lower regions are calculated with the assistance of relations (4.4) and (4.5). Here f (ρ, η) expresses a function depending on the density and viscosity of the mixed medium:

$$Q = Q_1 + Q_2 = 10.5 d^2 b.n.f(\rho, \eta) \tag{4.4}$$

$$\frac{Q_1}{Q_2} = \frac{h}{H - h} \quad \text{for } h/H > 0.5 \text{ or } h/H = 0.5 \tag{4.5}$$

Now, the system contains N − 1 cells with ideal mixing, each one with a known volume; the cells are connected by the different currents. Here, N corresponds to the system exit, V_k is the volume of a k cell and Q_{kj} is the current (flow rate) from the k cell to the j cell.

If we consider a marked particle placed inside of our cellular system, then we can define this by the vector E(n) = [e_0 (n), e_1 (n);e_2 (n)... e_k (n)... e_{N-1} (n),e_N (n)] where $e_k(n)$ expresses the probability of occurrence of the marked particle in the k cell after time n. Because the incidence of the marked particle inside a system is an undoubted event, we can write:

$$\sum_{k=1}^{N} e_k(n) = 1 \qquad \forall\, n = 0, 1, 2, \ldots \tag{4.6}$$

In the elementary processes (components), we establish that, in the small interval of time $\Delta\tau$, the particle can either pass to another cell or remain within its cell. The $\Delta\tau$ interval must be chosen in such a way that the particle can pass into a close cell during this interval, but not through it. Moreover, this passage can be regarded as instantaneous.

As far as the behaviour of the particle in such a system respects the rules of the Markov process, it will be controlled by a Markov connection. This means that the probability of the particle occurrence within the k cell after n + 1 time (i.e. $\tau = n.\Delta\tau$) is given only by its probability of occurrence in the j cell after time n and by its probability of transfer from the j cell to the k cell denoted p_{jk}. Now we can write:

$$e_k(n+1) = \sum_{j=1}^{N} e_j(n) p_{jk} \tag{4.7}$$

For j = 1,N and k = 1,N the probability p_{jk} is denoted as a matrix P which is called the stochastic matrix of the process, the matrix of passing or "the stochastic one":

$$P = \begin{bmatrix} p_{11} & p_{12} & p_{13} & \cdot & p_{1N} \\ p_{21} & p_{22} & p_{23} & \cdot & p_{2N} \\ \cdot & \cdot & \cdot & \cdot & \cdot \\ p_{N-11} & p_{N-12} & \cdot & \cdot & p_{N-1N} \\ p_{N1} & p_{N2} & p_{N3} & \cdot & p_{NN} \end{bmatrix} \tag{4.8}$$

During the building of the stochastic matrix, it is necessary to make sure that

$$\sum_{j=1}^{N} p_{ij} = 1$$ (the total of the probabilities according to one line equals one) and if

$$\sum_{i=1}^{N} p_{ij} = 1$$ (the total of the probabilities according to a column equals one). How-

ever, it should be specified that $\sum_{i=1}^{N} p_{ij}$ is not always one. Before going any further,

it should be specified that:

- p_{ii} represents the probability for the marked particle to be and to remain in cell i in the interval $\Delta\tau$
- p_{ij} represents the probability for the marked particle to be in cell i and to go into cell j in the time interval $\Delta\tau$.

Concerning the last line of matrix P, if $p_{NN} = 1$ (the particle that leaves the system cannot come back), all other probabilities (p_{Nk} where k ≠ N) have to be considered as zero.

If the initial state of the system is E(0), then, by means of matrix P we can write:

$$E(1) = E(0) * P \tag{4.9}$$

By analogy:

$$E(2) = E(1) * P \; E(3) = E(2) * P...E(n+1) = E(n) * P \tag{4.10}$$

The last equations prove that the Markov chains [4.6] are able to predict the evolution of a system with only the data of the current state (without taking into account the system history). In this case, where the system presents perfect mixing cells, probabilities p_{ii} and p_{ij} are described with the same equations as those applied to describe a unique perfectly stirred cell. Here, the exponential function of the residence time distribution (p_{ii} in this case, see Section 3.3) defines the probability of exit from this cell. In addition, the computation of this probability is coupled with the knowledge of the flows conveyed between the cells. For the time interval $\Delta\tau$ and for $i = 1,2,3, ...N$ and $j = 1,2,3,......N - 1$ we can write:

$$p_{ii} = \exp\left(-\frac{\sum\limits_{i=1,i\neq j}^{N} Q_{ji}}{V_i} \Delta\tau\right) \tag{4.11}$$

$$p_{ij} = \frac{Q_{ji}}{\sum\limits_{i=1,i\neq j}^{N} Q_{ji}}\left[1 - \exp\left(-\frac{\sum\limits_{i=1,i\neq j}^{N} Q_{ji}}{V_i} \Delta\tau\right)\right] \tag{4.12}$$

When matrix P is filled, vector E(0) is known, the calculation for E(1), E(2), E(3)... E(n)... can be easily carried out. At this time, we can formulate the following question which is also valid in almost all chemical engineering cases: *What information is produced with the assistance of this stochastic model?* The answer to this question shows that the model is frequently used for:

1. calculating the system reaction to one disturbance impulse:

$$F(\tau) = e_N(n) = \sum_{n=0}^{\infty} e_{N-1}(n)\Delta n \quad , \quad \tau = n\Delta\tau \tag{4.13}$$

2. precisely estimating the mean residence time and the residence time variance around the mean residence time:

$$\tau_m = \sum_{n=0}^{\infty} (1 - e_N(\tau))\Delta\tau \tag{4.14}$$

$$\sigma^2 = 2\sum_{n=0}^{\infty} (1 - e_N(\tau))n\Delta\tau - \left|\sum_{n=0}^{\infty} (1 - e_N(n))\Delta\tau\right|^2 \tag{4.15}$$

3. appreciating the evolution with time of the function (λ) that shows the stirring intensity for our topological cell assembly:

$$\lambda(n) = \frac{e_{N-1}(n)}{1 - e_N(\tau)} \qquad (4.16)$$

We can observe that, with the help of simulation software, we can produce the numerical results which give the effect of the stirrer's number of turns, the position of the stirrer in the tank, the effect of the dimension of the stirring paddles, etc., on the model exits mentioned above.

4.1.2
Numerical Application

An elliptic-based cylindrical apparatus (D = 1 m, H = 1 m) contains a solution stirred with a 6-paddled stirrer (d = 0.4 m and b = 0.1 m). The stirrer is placed in the tank in such a position as to get the ratio h/H = 0.2 (see Fig. 4.1) and to work at n = 0.1, 0.3, 0.5, 0.7, 0.9, 1.1, 1.3, 1.5 revolutions/s. A compound with the same physical properties as the solution is fed close to the liquid surface. The obtained mixture is flushed out through a pipe placed near the base of the apparatus. The entry and the exit flows are identical (Q_{ex} = 0.0048 m³/s). Now the question is to obtain the dependences of the parameters characterizing this mixing case according to the number of revolutions of the stirrer.

Before developing the algorithm of calculation, we have to deduce the mixture topology. Then: $r = \dfrac{h/H}{1 - h/H} = 0.2/0.8 = 1/4$; because h/H<0.5, we can assert that the stirrer is placed in the lower part of the tank and then with n_{ch} = 1 we have n_i = 1, n_a = n_{ch} / r = 4. So the tank contains six elemental mixing cells: one in the lower region, one in the mixing region and four in the higher region; If Q_1 +Q_2 = Q and Q_2 / Q_1 = h/(H – h); then Q_1 = 4/5 Q and Q_2 = 1/5 Q. With these simple calculations, we can establish the flow topology shown in Fig. 4.2.

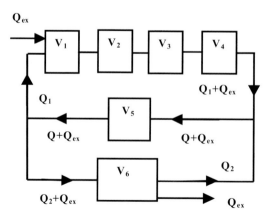

Figure 4.2 Topology of the numerical application 4.1.2.

According to Eqs. (4.11) and (4.12) and in agreement with Fig. 4.2 we can conclude that among the 49 probabilities only the following ones are not null:

$$p_{11} = \exp\left(-\frac{Q_1 + Q_{ex}}{V_1}\Delta\tau\right), \; p_{22} = \exp\left(-\frac{Q_1 + Q_{ex}}{V_2}\Delta\tau\right),$$

$$p_{33} = \exp\left(-\frac{Q_1 + Q_{ex}}{V_3}\Delta\tau\right), \; p_{44} = \exp\left(-\frac{Q_1 + Q_{ex}}{V_4}\Delta\tau\right),$$

$$p_{55} = \exp\left(-\frac{Q + Q_{ex}}{V_5}\Delta\tau\right), \; p_{66} = \exp\left(-\frac{Q_2 + Q_{ex}}{V_6}\Delta\tau\right),$$

$$p_{77} = 1, \; p_{12} = 1 - \exp\left(-\frac{Q_1 + Q_{ex}}{V_1}\Delta\tau\right), \; p_{23} = 1 - \exp\left(-\frac{Q_1 + Q_{ex}}{V_2}\Delta\tau\right),$$

$$p_{34} = 1 - \exp\left(-\frac{Q_1 + Q_{ex}}{V_3}\Delta\tau\right), \; p_{45} = 1 - \exp\left(-\frac{Q_1 + Q_{ex}}{V_4}\Delta\tau\right)$$

$$p_{51} = \frac{Q_1}{Q + Q_{ex}}\left(1 - \exp\left(-\frac{Q + Q_{ex}}{V_5}\Delta\tau\right)\right),$$

$$p_{56} = \frac{Q_2 + Q_{ex}}{Q + Q_{ex}}\left(1 - \exp\left(-\frac{Q + Q_{ex}}{V_5}\Delta\tau\right)\right)$$

$$p_{65} = \frac{Q_2}{Q_2 + Q_{ex}}\left(1 - \exp\left(-\frac{Q_2 + Q_{ex}}{V_6}\Delta\tau\right)\right),$$

$$p_{67} = \frac{Q_{ex}}{Q_2 + Q_{ex}}\left(1 - \exp\left(-\frac{Q_2 + Q_{ex}}{V_6}\Delta\tau\right)\right)$$

With these probabilities, the passing matrix can be written. For brevity, we use the following notations: $\alpha k = (Q_1 + Q_{ex})/V_k$ for $k = 1,....4$; $\alpha k = (Q + Q_{ex})/V_k$ for $k = 5$; $\alpha k = (Q_2 + Q_{ex})/V_k$ for $k = 6$; $\beta = Q_1/(Q + Q_{ex})$; $\gamma = Q_2/(Q + Q_{ex})$; $\delta = Q_2/(Q_2 + Q_{ex})$; $\varepsilon = Q_{ex}/(Q_2 + Q_{ex})$. The macro-relation (4.17) expresses our matrix of the transition probabilities:

$$P = \begin{vmatrix} e^{-\alpha 1\Delta\tau} & 1-e^{-\alpha 1\Delta\tau} & 0 & 0 & 0 & 0 \\ 0 & 0 & e^{-\alpha 2\Delta\tau} & 1-e^{-\alpha 2\Delta\tau} & 0 & 0 \\ 0 & 0 & 0 & 0 & e^{-\alpha 3\Delta\tau} & 1-e^{-\alpha 3\Delta\tau} \\ 0 & 0 & 0 & 0 & 0 & 0 \\ e^{-\alpha 4\Delta\tau} & 1 e^{-\alpha 4\Delta\tau} & 0 & 0 & \beta(1-e^{-\alpha 5\Delta\tau}) & 0 \\ 0 & 0 & e^{-\alpha 5\Delta\tau} & \gamma(1-e^{-\alpha 5\Delta\tau}) & 0 & 0 \\ 0 & 0 & 0 & \delta(1-e^{-\alpha 6\Delta\tau}) & e^{-\alpha 6\Delta\tau} & \varepsilon(1 e^{-\alpha 6\Delta\tau}) \\ 0 & 0 & 0 & 0 & 0 & 1 \end{vmatrix}$$

$$(4.17)$$

The numerical text of the calculation, shown in Fig. 4.3, leads to the program given in Fig. 4.4 and has the graphic interface (Fig. 4.5) associated with this program. A description of the graphic interface is given below:

- *the first window* (SMM1) is used for the introduction of the set of values which will be used for the simulation considered in the last window (SMM3) where the parameters are fixed in such a way as to have constant values by pair.
- in *the second window* (SMM2) with keys "∧" and "∨", the user moves among the values of vectors considered in the first window (Q_{ex}, d, h, n_i). Each press on the key leads to the calculation of the chosen situation. If the user supplements the fields with values that are not among those previously fixed, then pressing the button "Refresh" leads to the calculation of $F(\tau)$ vs τ, $\lambda(\tau)$ vs τ, $\sigma^2(\tau)$ vs τ, τ_m vs τ. The matrix of the passing probability is also established. All the charts considered in this window show the evolution of the mixture state towards the stationary state.
- *the third window* (SMM3) is used to show the effect of the stirring velocity and of the feed flow on the average residence time. It works with the values of the parameters selected in the first window.

1   **Initial data** /H=1,h=0.2 ,n_i=0.1 ,0.3,0.5,0.7,0.9,1.1,1.3,1.5 ,D=1 ,d=0.4,f(ρ,η)=1; b=0.1, etc

2   **Initial state**/ vector E(0)=[1 ,0,0,0,0,0,0] // Q_{ex}=0.0048//n_i=n_1//Vectors volumes..

3   ▶ **Preliminary computations**/V_m rel. (4.3) ,V_a rel (4.2) , V_i rel (4.1) // Q rel (4.4) //System for Q_1 and Q_2 by: Q_1+Q_2=Q and Q_1/Q_2=(H-h)/h // The cells volumes : V_5=V_m ,V_1=V_2=V_3=V_4=V_a/4 V_6=V_i//Values for : α_k=(Q_1+Q_{ex})/V_k with k=1,..4 ; α_k=(Q+Q_{ex})/V_k with k=5; α_k=(Q_2+Q_{ex})/V_k with k=6 ;β=Q_1/(Q+Q_{ex}) ; γ=Q_2/(Q+Q_{ex}) ; δ=Q_2/(Q_2+Q_{ex}) ;ε=Q_{ex}/(Q_2+Q_{ex})

4   **Matrix of transitions**// Matrix volume P: 7x7 / matrix elements with (4.17)

5   **Choose of $\Delta\tau$**// τ_{apr}=(0.786D*D*H)/Q_{ex} ;$\Delta\tau$= τ_{apr}/15   n=1

6   ▶**E(n)=P*E(n-1)**/ τ_m(n) rel (4.14) sum until n and with N=7 ,σ^2(n) rel. (4.15) sum until n and with N=7 , F(n) rel. (4.13) sum until n and with N=7 , λ(n) rel. (4.16) with N=7 / τ=n*$\Delta\tau$/ Extraction E(n)   **If τ_m(n)> τ_m(n-1)+error then**   n=n+1 , **Return to 6**   **If τ_m(n)<= τ_m(n)+error then** shows the figures F(n) and λ(n) and gives the values of σ^2(n) and τ_m(n)   **For n_i<=8 then n_i=n_{i+1}, Return** at the step **3**

7   **End**

Figure 4.3 Scheme of the computation algorithm.

```
unit d1;
interface
uses
  Windows, Messages, SysUtils, Classes, Graphics, Controls, Forms, Dialogs,
  StdCtrls;
type
  TForm1 = class(TForm) Button1: TButton;
    procedure Button1Click(Sender: TObject);
  private { Private declarations } public { Public declarations }
  end;
  const nc=8;error=1;
var e:array[0..10000,1..7] of real;
taum,lambda:array[0..10000] of real;
repeta:boolean;
V,alfa:array[1..6] of real;P:array[1..7,1..7] of real;
D,dm,b,h,hm,Q,Qex,Q1,Q2,Vm,Va,Vi,f,gamma,delta,epsilon:real;
tauapr,deltatau,tex,tau,suma,suma1,sigmap:real;
i,j,k,n,n1:integer;ni,dt:real;
var
  Form1: TForm1;
implementation
{$R *.DFM}
procedure TForm1.Button1Click(Sender: TObject);
begin      taum[0]:=0;taum[1]:=21;
for i:=1 to 7 do for j:=1 to 7 do P[i,j]:=0;P[7,7]:=1;
  n:=1; D:=1; dm:=0.4; b:=0.1; e[0,1]:=1;
for i:=2 to 7 do e[0,i]:=0;
h:=0.2;
HM:=1; Qex:=0.0048;
ni:=0.1;n:=1;
{3} repeat
n:=1;
Vm:=pi*b/60*(sqr(D)+D*dm/5+dm*dm/5); {rel 4.3} Va:=pi*D*h/4-Vm/2; {rel 4.3}
Vi:=pi*D*h/4*(HM-h); {rel. 4.1} f:=1;
Q:=10.5*dm*dm*b*ni*f; {f=1 } Q1:=Q/(1+1/((HM-h)*h)); Q2:=Q-Q1; Q1:=Q*4/5; Q2:=Q-Q1;
for i:=1 to 4 do V[i]:=Va/4; V[5]:=Vm; V[6]:=Vi;
for k:=1 to 4 do alfa[k]:=(Q1+qex)/V[k];
alfa[5]:=(Q+Qex)/V[5]; alfa[6]:=(Q2+Qex)/V[6]; gamma:=Q2/(Q+Qex); delta:=Q2/(Q+Qex);
epsilon:=Qex/(Q2+Qex); tauapr:=(0.768*D*D*HM)/Qex; deltatau:=tauapr/15;
for k:=1 to 4 do P[k,k]:=exp(-(Q1+Qex)/V[k]*deltatau);  P[5,5]:=exp(-(Q+Qex)/V[5]*deltatau);
P[6,6]:=exp(-(Q2+Qex)/V[6]*deltatau); {P[7,7]:=1; }
for k:=1 to 4 do P[k,k+1]:=1-exp(-(Q1+Qex)/V[k]*deltatau); tex:=1-exp(-(Q+Qex)/V[5]*deltatau);
P[5,1]:=Q1/(Q+Qex)*tex; P[5,6]:=(Q2+Qex)/(Q+Qex)*tex; tex:=1-exp(-(Q2+Qex)/V[6]*deltatau);
P[6,5]:=Q2/(Q2+Qex)*tex; P[6,7]:=Qex/(Q2+Qex)*tex; {n:=1;} tau:=n*deltatau; repeta:=true;
{6} while repeta do begin repeta:=false; for i:=1 to 7 do begin suma:=0;
for j:=1 to 7 do suma:=suma+P[i,j]*E[n-1,j]; E[n,i]:=suma; end;
suma:=0;suma1:=0; for i:=0 to n do begin suma:=suma+(1-e[i,7])*n*deltatau;
suma1:=suma1+(1-e[i,7])*deltatau; end;
sigmap:=2*suma-suma1*suma1; taum[n]:=suma/n; dt:=taum[n]-taum[n-1]; lambda[n]:=E[n,6]/(1-
E[n,7]);
suma:=0; {for i:=1 to n do suma:=suma+E[n,6]*deltatau; taum[n]:=suma;suma:=0;}
if taum[n]>taum[n-1]+error then begin repeta:=true;inc(n);end ;if n=1000 then repeta:=false;
end; beep; if taum[n]<=taum[n-1]+error then {grafice}; if ni<=1.5 then ni:=ni+0.2
until ni>1.5;
end; end.
```

Figure 4.4 Calculation program written in Matlab® language.

a)

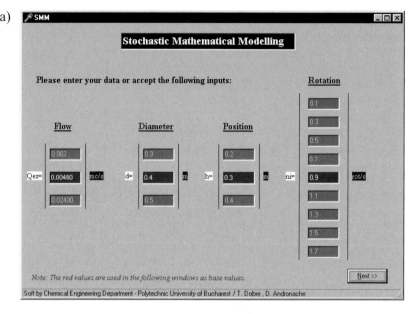

b)

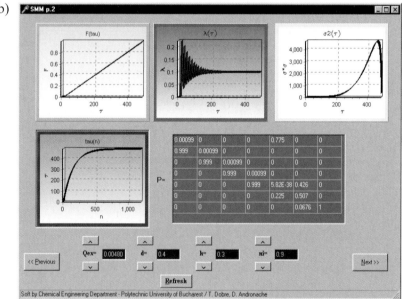

Figure 4.5 Graphic interface for application 4.1.2. (a) First window, (b) second window.

Figures 4.6 and 4.7 show other examples of simulations. For the case considered in the simulation given in Fig. 4.5(b), if, for example at time $\tau = 0$, we start the introduction of a constant signal into the system, the signal obtained at the exit after 500 s becomes stationary according to this value. It should be noticed

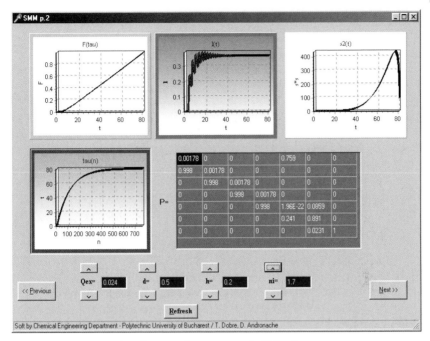

Figure 4.6 Mixing state when the external parameters take high values.

that the displacement towards the stationary state of mixing is dependent on all the external parameters that characterize this process of liquid mixing with a mechanical stirrer.

It is necessary to notice the very interesting aspect of the evolution of the function that characterizes the mixing intensity. As is observable when the stationary state of mixing is reached, the dispersion of the average residence time quickly moves towards a very small value, this behaviour is characteristic of the combination of the external parameters with the chosen topology.

In all situations, the estimated average residence time according to the ratio of the apparatus volume of liquid and of the flow passing through the apparatus gives smaller values compared to those calculated with the stochastic procedure.

The data presented in Figs. 4.5–4.7 show many interesting aspects with respect to the effect of the external parameters on the state of the mixing process with this type of mechanical stirrer. The most significant conclusions of this example are summarized below:

- The solution to the problem of the mechanical stirring of a liquid medium begins with the identification of the process components. This step is carried out using an identification particle which is placed in the elementary volumes V_1, V_2.........V_6 . The connection process is characterized by the probability showing that the identification particle moves from one volume of the topological space to another.

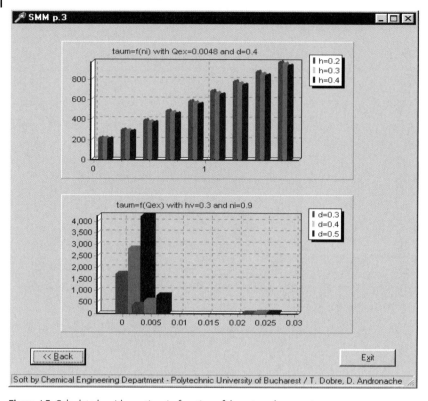

Figure 4.7 Calculated residence time in function of the external parameters.

- A calculation procedure has been used for the individual states of volumes and the flows circulating in the selected topology, in order to develop a simulation and calculation program of the mechanical stirring in a liquid medium.
- A graphic interface is suggested in order to make rapid simulations about a particular state of the system with the external parameters considered as significant by the researcher.
- The developed simulator allows the calculation of the displacement of the state of the mixing towards the stationary state as well as the characterization of this stationary state.

To end this section we can make some general observations about stochastic modelling:
- The jump from the description of the phenomenological process to its stochastic variant, which shows the process's elementary states and its connection procedure, is strongly dependent on the process cognition in terms of chemical engineering as well as on the researcher's ability and experience.

- The mathematical description of the modelled process uses a combination of one or more stochastic cores and phenomenological parts related to non-stochastic process components.
- The building of the mathematical model of a process with stochastic core and its transpositions as simulator, follows the steps considered in Fig. 3.4.
- The range of the values of the process factors considered by the simulator and the process exits considered by the graphic interface of the simulator requires very careful selection.

At this point in this chapter, it is easy to understand that, using the methodology above, the modelling of a chemical transformation presents no important difficulty if the chemical reaction is fitted in the general framework of the concepts of probability theory. Indeed, the discrete molecular population characterizing a chemical system can be described in terms of the joint probability of the random variables representing the groups of entities in the total population.

Until now, the use of stochastic mathematics to describe flow systems and, in particular, the residence time distribution, has been well developed. However, the models of processes based upon these principles have generally been less popular than those based upon the fundamental equations of motion and continuity (see Section 3.3). A random selection of 20 papers concerning the residence time distribution models shows that 12 of them are based on the stochastic motion of particles. Early stochastic modelling efforts in chemical engineering seem to be concentrated on a variety of generic systems with continuous flow, on processes with simultaneous chemical reaction and dispersion, on processes with internal reflux as well as on processes operating at unsteady state. So, in this domain, many papers and books aim at demonstrating the applicability of stochastic mathematics to the solution of fundamental chemical engineering problems, and in particular to the calculation of residence time and of the state of systems inside this residence time.

Stochastic models present a number of advantages over CFD models as far as the modelling of the residence time distribution of a complex flow system is concerned. These advantages are:

1. stochastic models are simple to develop,
2. they are computationally light,
3. they are simple to adapt to new systems, and
4. they are much simpler to solve than the full mathematical description.

The term "Markov chain" frequently appears in this chapter. This term is named after the Russian mathematician Andre Markov (1856–1922). The Markov theory is widely applied in many fields, including the analysis of stock-markets, traffic flows, queuing theories (e.g. modelling a telephone customer service hotline), reliability theories (e.g. modelling the time for a component to wear out) and many other systems involving random processes.

4.2
Stochastic Models by Probability Balance

The prediction of the results obtained with an industrial process is one of the fundamental objectives of modelling. This process prediction is necessary to obtain good information about the process management as well as a better knowledge of the process itself. If the process is rather complicated or if its laws of evolution are unknown, the application of a deterministic model is very difficult. However, if the elementary process components of the process are identified, then the application of a stochastic model can be realized, often with spectacular success. In this case, modelling begins with a complete descriptive model of the process where the identification of the participant elementary processes, their connections and the space topology where the process develops will be attentively examined. Thanks to this description, we can identify the factors that, all together, determine the process state.

The establishment of stochastic equations frequently results from the evolution of the analyzed process. In this case, it is necessary to make a local balance (space and time) for the probability of existence of a process state. This balance is similar to the balance of one property. It means that the probability that one event occurs can be considered as a kind of property. Some specific rules come from the fact that the field of existence, the domains of values and the calculation rules for the probability of the individual states of processes are placed together in one or more systems with complete connections or in Markov chains.

In the development of stochastic models, there are six successive steps:
1. The objective of the description of a process evolution, considering mainly the specific internal phenomena, is to precede the elementary processes (elementary states) components.
2. The identification of the elementary steps according to which the evolution of the investigated process (phenomena) is held.
3. The determination or the division of the transition probabilities from one state to another and the identification with respect to the connections if the stochastic process accepts a continuous or a discrete way.
4. The establishment of the balance equations of the probabilities. They show the probability of the process to exist in a given state, at a considered time in a formal geometry (system of selected coordinates).
5. The coupling of the univocity conditions to the problem established at the end of the probability balance.
6. The model resolution and its evaluation in order to give the models the requested exits in their relations with the entries.

In the following example, we show a more explicit explanation of the stochastic model genesis, particularization and evaluation.

4.2.1
Solid Motion in a Liquid Fluidized Bed

It is well known that fluidization with liquids is characterized by a very good homogeneity. However, when a liquid–solid suspension is mixed by fluidization and the size or density of the solid are not homogeneous, segregation is observed. To carry out the analysis of this problem, we can consider a two-dimensional fluidization system composed of pure water and 1 mm diameter glass balls. This system operates in the vicinity of a minimum fluidization state. One of these glass balls is coloured and thrown to the centre of the base of the bed [4.24]. The ball displacement will give qualitative and quantitative appreciation of the solid mixing during the fluidization process [4.5, 4.24]. The coloured ball displacement is recorded with a high-speed camera, which makes it possible to identify the trajectory and the displacement mechanism, allowing the identification of forward and backward displacements. The result of such an experiment is shown in Fig. 4.8. We can note that the global particle displacement results in a unique direction, in spite of the forward-and- backward displacements. If we decompose the particle displacement (different steps in Fig. 4.8), we can note that one "evolution of the state" is given by a forward or backward displacement.

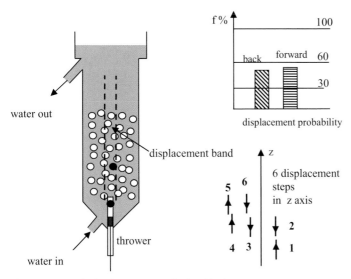

Figure 4.8 Particle displacement in a fluidized bed.

From the analysis of the decomposition of the images, we can observe that the movement ahead is dominant and controls the whole displacement. In the example given in Fig. 4.8, we can observe that the movement ahead (elementary state or process component) presents approximately the same frequency as the backward displacement. A radial movement is also possible but can be neglected if we consider only a very thin band in the centre of the bed. The same has been considered for stagnation. The description given above, shows that the first steps of the stochastic modelling are similar to those of the establishment of the model of transfer phenomena (stage of the descriptive model). The stochastic model here can be described as the mathematical solution given for a problem of forward-and- backward displacement [4.3, 4.4, 4.17] where the elementary processes considered are:

- elementary process of type I – displacement in the direction of the z-axis with velocity v_z.
- elementary process of type II – displacement in the opposite direction to axis z with a mean velocity $-v_z$.

In this case, the elements of the passage matrix of the particularized problem present very clear physical meanings:

- p_{11} is the probability for the coloured particle, which is displaced by a standard process I, to keep on being displaced by the same process (a positive displacement with speed v_z is followed by the same displacement).
- p_{22} is the probability for the coloured particle, which is displaced by a standard process II, to keep on being displaced by the same process (the same consideration as above but with speed $-v_z$).
- p_{12} is the probability for the coloured particle which is displaced by a standard process I, to skip to a standard process II (i.e. a negative displacement with respect to axis z occurs after a positive displacement with a speed v_z).
- p_{21} is the probability for the coloured particle which moves due to a standard process II, to skip to standard process I (i.e. a positive displacement with respect to axis z occurs after a negative displacement with speed $-v_z$).

If P_1 represents the probability for the process that evolved in state I to remains in this state after the interval of time $\Delta\tau$, then, because all the states are characterized by independent probabilities, we can write:

$$P_1 = 1 - \sum_{j=2}^{N} p_{1j} a \Delta\tau \tag{4.18}$$

Here N is the number of the independent states of the process (N = 2 in the analyzed case) and "a" is the frequency of exchange of an individual state. We can notice that the dimension for "a" is time^{-1} (T^{-1}). The same consideration as above can be used for the probability of the process evolving in state II. Then P_2 is written:

$$P_2 = 1 - \sum_{j=1, j \neq 2}^{N} p_{2j} a \Delta\tau \tag{4.19}$$

If we consider that the element is the coloured ball, the following relation writes the equations of the probability balance according to the model:

The probability that the element is in the z position at τ + Δτ and in state I	=	The probability that the element from z–Δz position at τ with state I evolves, in the next Δτ, with the same state	+	The probability that the element from z position at τ with state II evolves, in the following Δτ, towards state I	(4.20)

With relations (4.19) and (4.20), we can establish that $P_1(z, \tau + \Delta\tau)$ is given by Eq.(4.21). Equation (4.22), which gives $P_2(z, \tau + \Delta\tau)$, is written by the same procedure:

$$P_1(z, \tau + \Delta\tau) = P_1(z - v_z\Delta\tau)(1 - p_{12}a\Delta\tau) + P_2(z, \tau)p_{21}a\Delta\tau \qquad (4.21)$$

$$P_2(z, \tau + \Delta\tau) = P_2(z + v_z\Delta\tau)(1 - p_{21}a\Delta\tau) + P_1(z, \tau)p_{12}a\Delta\tau \qquad (4.22)$$

If we consider that $\Delta\tau \to 0$ in relations (4.21) and (4.22), we can write a two equation system with partial derivates in $P_1(z,\tau)$ and $P_2(z,\tau)$, as follows:

$$\lim_{\Delta\tau \to 0} \frac{P_1(z, \tau + \Delta\tau) - P_1(z - v_z\Delta\tau, \tau)}{\Delta\tau} = -p_{12}aP_1(z - v_z\Delta\tau, \tau) + p_{21}aP_2(z, \tau) \quad (4.23)$$

$$\lim_{\Delta\tau \to 0} \frac{P_2(z, \tau + \Delta\tau) - P_2(z + v_z\Delta\tau, \tau)}{\Delta\tau} = -p_{21}aP_2(z + v_z\Delta\tau, \tau) + p_{12}aP_1(z, \tau) \quad (4.24)$$

$$\begin{cases} \dfrac{\partial P_1(z, \tau)}{\partial \tau} + v_z\dfrac{\partial P_1(z, \tau)}{\partial z} = -ap_{12}P_1(z, \tau) + ap_{21}P_2(z, \tau) & (4.25) \\ \dfrac{\partial P_2(z, \tau)}{\partial \tau} - v_z\dfrac{\partial P_2(z, \tau)}{\partial z} = -ap_{21}P_2(z, \tau) + ap_{12}P_1(z, \tau) & (4.26) \end{cases}$$

In this system, we can take into account that $P_1(z,\tau)$ and $P_2(z,\tau)$ are the probabilities or probability densities, or can be considered as the concentrations which describe the type I or type II elementary action intensity.

To solve the model obtained, it is necessary to link it with the univocity conditions. They are obtained from the physical meanings of the problem:

1. The only way for the coloured ball to get into the layer is by a type I elementary action.
2. After the input, it is impossible for the coloured ball to exit the layer.
3. The only way for the coloured ball to exit the layer is by a type I elementary action (assuming that the marked particle has reached the end of the layer and cannot flow back)

These conditions can also be applied for cases where an impulse or signal is introduced in a continuous flow (for instance see Section 3.3):

$$z = 0 \quad \tau = 0 \quad P_1(0,0) = 1 \quad P_2(0.0) = 0 \tag{4.27}$$

$$z = H \quad \tau = 0 \quad P_1(H,0) = 0 \quad P_2(H.0) = 0 \tag{4.28}$$

$$z = 0 \quad \tau \succ 0 \quad P_1(0,\tau) = 0 \quad P_2(0.\tau) \neq 0 \tag{4.29}$$

$$z = H \quad \tau \succ 0 \quad P_1(H,\tau) \neq 0 \quad P_2(H.\tau) = 0 \tag{4.30}$$

In this example, two main situations can be considered:

Under the condition that $p_{12} = p_{21} = 1/2$, the system formed by Eqs. (4.25) and (4.26) takes the form:

$$\frac{\partial P_1(z,\tau)}{\partial \tau} + v_z \frac{\partial P_1(z,\tau)}{\partial z} = -\alpha P_1(z,\tau) + \alpha P_2(z,\tau) \tag{4.31}$$

$$\frac{\partial P_2(z,\tau)}{\partial \tau} - v_z \frac{\partial P_2(z,\tau)}{\partial z} = -\alpha P_2(z,\tau) + \alpha P_1(z,\tau) \tag{4.32}$$

where α corresponds to $a p_{12} = a p_{21} = a/2$.

From a practical point of view, the main interest may be given to the sum $P(z,\tau) = P_1(z,\tau) + P_2(z,\tau)$. It describes the density of probability when the particle reaches position z, at time τ, no matter what elementary action (type I or II), it has been subjected to. The derivation of Eqs.(4.31) and (4.32) with respect to τ and z, coupled to an algebraic calculation for $P_1(z,\tau)$ and with the elimination of $P_2(z,\tau)$ gives the following relation:

$$\frac{\partial P(z,\tau)}{\partial \tau} + \frac{1}{2\alpha} \frac{\partial^2 P(z,\tau)}{\partial \tau^2} = \frac{v_z^2}{2\alpha} \frac{\partial^2 P(z,\tau)}{\partial z^2} \tag{4.33}$$

In the resulting equation, we have the derivatives of the known transport equation as well as the second order derivative of the variable of the process with respect to the time. The type of model considered here is known as the hyperbolic *model*. Scheidegger [4.25] obtained a similar result and called it: *correlated random displacement*.

The hyperbolic model is easily reduced to a parabolic model if the value of the parameter α is large enough to reduce the expression $\dfrac{1}{2\alpha} \dfrac{\partial^2 P(z,\tau)}{\partial \tau^2}$ as much as possible. We have already noticed that "a" and then α correspond to the measurement of the passage frequency.

We can easily imagine the case of a group of very small particles (molecules for example), which quickly change positions; this produces the image describing the diffusion movement. Equation (4.34) describes the diffusion model or the model with a *parabolic* equation:

$$\frac{\partial P(z,\tau)}{\partial \tau} = \frac{v_z^2}{2\alpha} \frac{\partial^2 P(z,\tau)}{\partial z^2} \tag{4.34}$$

If we consider now that the condition (4.29) changes in order to obtain $P_2(0, \tau) = 0$ then according to the sum of Eqs.(4.27) and (4.29) we obtain the initial condition of a Dirac's pulse:

$$P(z, 0) = \delta(z) = \begin{vmatrix} 1 & z = 0 \\ 0 & z \neq 0 \end{vmatrix} \qquad (4.35)$$

The coloured particle is displaced in the fluidized bed according to the model and the laws of diffusion. Indeed, the solution to the diffusion model described by Eq. (4.34) and by the initial condition (4.35) is the following [4.26, 4.27]:

$$P(z, \tau) = \sqrt{\frac{\alpha}{2v_z^2\tau}} \exp\left(-\frac{\alpha z^2}{2v_z\tau}\right) \qquad (4.36)$$

The elliptic model given by Eq. (4.33) and the initial condition (4.35) gives the solution obtained with the relation (4.37) [4.5]:

$$P(z, \tau) = \frac{\alpha\exp(-\alpha\tau)}{2v_z}\left[I_0\left(\alpha\tau\sqrt{1 - \frac{z^2}{2v_z^2\tau^2}}\right) + \frac{1}{\sqrt{1 - \frac{z^2}{v_z^2\tau^2}}}I_1\left(\alpha\tau\sqrt{1 - \frac{z^2}{2v_z^2\tau^2}}\right) \right],$$

$$|z| \prec v_z\tau$$

$$P(z, \tau) = 0. \qquad |z| \succ v_z\tau \qquad (4.37)$$

$I_0(x)$ and $I_1(x)$, in Eq. (4.37), are the Bessel functions with imaginary arguments and they can be written as follows:

$$I_0(x) = \sum_{k=0}^{\infty} \frac{\left(\frac{x}{2}\right)^{2k}}{(k!)^2} \quad , \quad I_1(x) = \sum_{k=0}^{\infty} \frac{\left(\frac{x}{2}\right)^{2k+1}}{(k!)(k+1)!}. \qquad (4.38)$$

The graphical representations of solutions (4.36) and (4.38) are given in Fig. 4.9. The dimensionless variables $z_a = z/H$ and $t_a = \frac{v^2}{H^2}\left(\frac{\tau}{2a}\right)$ have been used here. The curves considered in this figure were drawn taking into account the values of $v_z = 0.1$ m/s, $\alpha = 10$ s^{-1} and $H = 0.2$ m. In Fig. 4.9, we can observe that:

1. Both models represent the same phenomenon because the curves $P(z,\tau)$ versus z_a and t_a (for the same conditions) are almost identical. The only difference is observed in the fields of very small times which are not of interest in this analysis.
2. The low values of $P(z, \tau)$, which are presented for $z_a > 0.15$, show that the marked particle has a strong conservative tendency because it keeps its position near the injection point where z_a is small.
3. In this particular case, the values considered for α and v_z are chosen without any deep experimental appreciation. How-

ever, in other situations, intensive experimental work would be necessary.

4. The mixing process is axial and the particle displacement is well represented by both models. However, the elliptic model seems to be a little more illustrative because it considers small values of z_a and t_a.

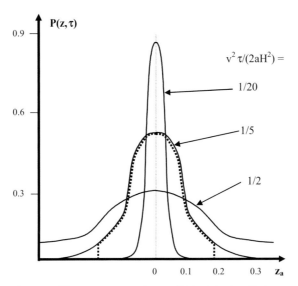

Figure 4.9 The space–time evolution of P (z, τ) by the elliptic model (dashed line) and hyperbolic model (continuous line).

In this problem of axial mixing, it should be specified that the calculated value of $\frac{v_z^2}{2\alpha}$ was approximately 10^{-3} m²/s; this is comparable with the typical experimental values of the axial dispersion coefficient in fluidized beds with liquids.

Another stochastic model (4.27)–(4.32) treatment can be made when the aim is to calculate the average time of residence and the axial dispersion coefficient. In this problem, we use the properties of the characteristic function, which is associated with the distribution function of the average time of residence [4.28, 4.29]. For this analysis we start with the Laplace transformation of the stochastic model when the system (4.31)–(4.32) is considered:

$$
\begin{cases}
sP_1(z,s) - P_1(z,0) + v_z\dfrac{dP_1(z,s)}{dz} = -p_{12}aP_1(z,s) + p_{21}aP_2(z,s) & (4.39) \\[2mm]
sP_2(z,s) - P_2(z,0) - v_z\dfrac{dP_2(z,s)}{dz} = -p_{21}aP_2(z,s) + p_{12}aP_1(z,s) & (4.40)
\end{cases}
$$

Taking into account the univocity conditions (for instance, look at relations (4.27)–(4.30)), $P_1(z, 0) = P_2(z,0) = 0$), the equations above can be written:

$$
\begin{cases}
v_z \dfrac{dP_1(z,s)}{dz} = -(s + p_{12}a)P_1(z,s) + p_{21}aP_2(z,s) & (4.41) \\[2mm]
v_z \dfrac{dP_2(z,s)}{dz} = (s + p_{21}a)P_2(z,s) - p_{12}aP_1(z,s) & (4.42)
\end{cases}
$$

Now, if Eq.(4.41) is derived according to s and in the obtained result, we make two replacements ($dP_2(z,s)/dz$ by its value given in Eq. (4.42) and $P_2(z,s)$ by its value from Eq. (4.41)), then both equations combine to give:

$$
v_z \frac{d^2P_1(z,s)}{dz^2} + a(p_{12} - p_{21})\frac{dP_1(z,s)}{dz} - \left[\frac{(ap_{12})^2 - (s + ap_{12})^2}{v_z}\right]P_1(z,s) = 0 \quad (4.43)
$$

After the group of relations (4.27) to (4.30) we can consider that the conditions that have to be coupled with Eq. (4.43) are:

$$
P_1(0,0) = 1 \quad , \quad P_1(0,s) = 0 \quad , \quad P_1(H,0) = 0 \qquad (4.44)
$$

These conditions introduce a complication with respect to the solution to the problem: $dP_1(0,0)/dz$ is absent and the top condition is opposed to the integration ($z = H$). This problem can be circumvented if we consider that the pulse at the input of the fluidized layer can be coupled or not with a particular condition on the output. To simplify the problem, we can choose in Eq.(4.43) $v_z = 1$ as the conventional unit/second. This selection implies that the z dimension (and thus H) would be measured in a conventional unit. When $p_{12} = p_{21} = p_{11} = p_{12} = 0.5$, Eq. (4.43) is simplified to Eq. (4.45), which has the general solution (4.46). Here, λ is given by relation (4.47):

$$
\frac{d^2P_1(z,s)}{dz^2} - \left[\left(\frac{2s + a}{2}\right)^2 - \left(\frac{a}{2}\right)^2\right]P_1(z,s) = 0 \qquad (4.45)
$$

$$
P_1(z,s) = C_1 e^{-\lambda z} + C_2 e^{+\lambda x} \qquad (4.46)
$$

$$
\lambda^2 = \left[\left(\frac{2s + a}{2}\right)^2 - \left(\frac{a}{2}\right)^2\right] \qquad (4.47)
$$

The solution of the system has to check the value of constants C_1 and C_2 (4.46). The relations (4.48) are thus obtained and lead to the solution (4.49):

$$
C_1 = \frac{(1+\lambda)^2}{(1+\lambda)^2 - (1-\lambda)^2 e^{-2\lambda H}} \qquad C_2 = \frac{(1-\lambda)^2 e^{-2\lambda H}}{(1+\lambda)^2 - (1-\lambda)^2 e^{-2\lambda H}} \qquad (4.48)
$$

$$
P_1(z,s) = \frac{(1+\lambda)^2 e^{-\lambda z} - (1-\lambda)^2 e^{-2\lambda(H-z)}}{(1+\lambda)^2 - (1-\lambda)^2 e^{-2\lambda H}} \qquad (4.49)
$$

Once Eq. (4.41) is adapted to the modifications carried out above, it can be used to give an expression for P_2 (z,s), then we can write:

$$P_2(z, s) = \frac{(1 - \lambda)^2 e^{-\lambda z} - (1 - \lambda)^2 e^{-2\lambda(H-z)}}{(1 + \lambda)^2 - (1 - \lambda)^2 e^{-2\lambda H}}$$

(4.50)

The function of the distribution of the residence time from 0 up to H can be obtained by the sum of the probabilities of the exit from the way. This is possible at z = H with an elementary action of type I and at z = 0 with a standard elementary action II. Thus, for the function of residence time distribution, the following equation can be written:

$$f(\tau) = P_1(H, \tau) + P_2(0, \tau)$$

(4.51)

The characteristic function for a distribution law of a random variable is the Laplace transform of the expression of the distribution law. For the analysis of the properties of the distribution of a random variable, the characteristic function is good for the rapid calculation of the centred or not, momentum of various orders. Here below, we have the definition of the characteristic function $\varphi_\tau(s)$ and its particularization with the case under discussion:

$$\varphi_\tau(s) = \int_0^\infty f(\tau)e^{-s\tau}d\tau = \int_0^\infty (P_1(H, \tau) + P_2(H, \tau))e^{-s\tau}d\tau = P_1(H, s) + P_2(0, s)$$ (4.52)

Relations (4.49) and (4.50) rapidly show what P_1 (H, s) and P_2 (0,s) are known and thus, in this case, Eq. (4.52) is written as follows:

$$\varphi_\tau(s) = \frac{Ch(\lambda Hs) + \dfrac{1 - \lambda^2}{2\lambda} Sh(\lambda H) - \dfrac{1 + \lambda^2}{2\lambda} Sh(\lambda Hs)}{\dfrac{1 + \lambda^2}{2\lambda} Sh(\lambda H) + Ch(\lambda H)}$$

Here, the sine and cosine hyperbolic functions (Sh and Ch) are well-known expressions. The average value of the residence time in the way 0–H can be described with the assistance of the characteristic function:

$$\tau_m = \int_0^\infty \tau f(\tau)d\tau = -\varphi_\tau{}'(0)$$

(4.53)

The calculation of an analytical derivative for $\varphi_\tau(s)$ by using relation (4.52) is very difficult and tedious. Here below, we make the numerical calculation for the derivative at s = 0. To do so, we use the relation that defines the derivative of a function in a point. We then obtain:

$$\varphi_\tau{}'(0) = \lim_{s \to 0} \frac{\varphi_\tau(s) - \varphi_\tau(0)}{s - 0} = \lim_{s \to 0} \frac{\varphi_\tau(s) - 1}{s}$$

$$= \lim_{s \to 0} \frac{Ch(\lambda Hs) - \lambda Sh(\lambda H) - Ch(\lambda H) - \dfrac{1 + \lambda^2}{2\lambda} Sh(\lambda Hs)}{\dfrac{1 + \lambda^2}{2\lambda} sSh(\lambda H) + sCh(\lambda H)}$$

$$= \frac{\lim\limits_{s \to 0}[(\lambda'Hs + \lambda H)Sh(\lambda Hs) - \lambda'HSh(\lambda H)] - \lim\limits_{s \to 0} \dfrac{\lambda^2 H}{s} \dfrac{Sh(\lambda H)}{\lambda H} - \lim\limits_{s \to 0} \dfrac{(1 + \lambda^2)H}{2} \dfrac{Sh(\lambda Hs)}{\lambda Hs}}{1 \lim\limits_{s \to 0}[\dfrac{1 + \lambda^2}{2\lambda} Sh(\lambda H) + Ch(\lambda H)]} \dots$$

$$= -\left[aH^2 + aH + \frac{H}{2}\right] \Big/ \left[\frac{H}{2} + 1\right] = -\frac{2aH^2 + 2aH + H)}{H + 2} \tag{4.54}$$

Consequently, the expressions of the mean residence time in the way 0–H and those of the linear distance traversed during motion can be written as follows:

$$\tau_m = \frac{2aH^2 + 2aH + H)}{H + 2} \quad , \quad l_m = v_z \tau_m = \frac{(2aH^2 + 2aH + H) * 1}{H + 2} \tag{4.55}$$

It is necessary to pay careful attention to these last two expressions where H is considered in conventional length units, which corresponds to a $v_z = 1$. For example if $v_z = 1$ cm/s, then the conventional unit (cu) is cm, therefore, in the relations, H would be expressed in cm. Another example shows that $v_z = 0.02$ m/s; so a value of the conventional length unit of 1 cu = 0.02 m is requested to make $v_z = 1$ cu/s. If, in this case, the trajectory is 0.2 m, for example, then, for H, H = 0.2/0.02 = 10 is used which corresponds to a dimensionless value. For very large H values, relation (4.55) can be simplified as follows: $\tau_m = 2a(H + 1) \approx 2aH$. This simplification can guide us towards various speculative conclusions with respect to the covered linear distance. Categorically, the result obtained can be explained by the perfect similarity of the final relationships with the well-known formulas used in mechanics.

The problem of theoretical calculation of an axial dispersion coefficient for this example of displacement of the coloured ball is solved in an way identical to the stochastic problem with three equal probable elementary actions (for instance look at the example of axial mixing in a mobile bed column).

If we consider that $p_{12} \neq p_{21}$. From a phenomenological point of view, it is easier to accept a difference between p_{12} and p_{21}. This is typical for a case where a directional internal force acts on a marked particle. As an example, we can consider a particle displacement given by a difference between the weight and the Archimedes force. In this case, the model to be analyzed is described by relations (4.43) and (4.44). For $v_z = 1$ u.c/s, Eq. (4.43) is written as below:

$$\frac{d^2 P_1(z, s)}{dz^2} + a(p_{12} - p_{21}) \frac{dP_1(z, s)}{dz} - \left[(ap_{12})^2 - (s + ap_{12})^2\right] P_1(z, s) = 0 \tag{4.56}$$

If $\alpha = ap_{12}, \beta = ap_{21}$ and $\lambda^2 = s(s + 2\alpha)$, then Eq. (4.56) becomes:

$$\frac{d^2 P_1(z,s)}{dz^2} + (\alpha - \beta)\frac{dP_1(z,s)}{dz} - -\lambda^2 P_1(z,s) = 0 \tag{4.57}$$

The discriminant associated with the characteristic equation connected to Eq. (4.57) is always positive ($\Delta = (\alpha - \beta)^2 + 4\lambda^2$) and the solution of the differential equation (4.57) is written like a sum of the exponential terms. In addition, the solution for $\alpha = \beta$ must find the former case presented. According to the example already discussed, we have new expressions for $P_1(z,s)$ and $P_2(z,s)$:

$$P_1(z,s) = \frac{(1+\lambda)^2 e^{-\frac{[(\alpha-\beta)+\sqrt{(\alpha-\beta)^2+4\lambda^2}]}{2}z} - (1-\lambda)^2 e^{-2[(\alpha-\beta)-\sqrt{(\alpha-\beta)^2+4\lambda^2}](H-z)}}{(1+\lambda)^2 - (1-\lambda)^2 e^{-2\lambda H}}$$

$$P_2(z,s) = \frac{(1-\lambda)^2 e^{-\frac{[(\alpha-\beta)+\sqrt{(\alpha-\beta)^2+4\lambda^2}]}{2}z} - (1-\lambda)^2 e^{-2[(\alpha-\beta)-\sqrt{(\alpha-\beta)^2+4\lambda^2}](H-z)}}{(1+\lambda)^2 - (1-\lambda)^2 e^{-2\lambda H}}$$

Now there is no obstacle to continuing with the estimation of the characteristic function, average residence time etc.

The last two applications, where the genesis, particularization and evaluation of a stochastic model were improved, undoubtedly show the capacity and the force of stochastic modelling.

4.3
Mathematical Models of Continuous and Discrete Polystochastic Processes

Polystochastic models are used to characterize processes with numerous elementary states. The examples mentioned in the previous section have already shown that, in the establishment of a stochastic model, the strategy starts with identifying the random chains (Markov chains) or the systems with complete connections which provide the necessary basis for the process to evolve. The mathematical description can be made in different forms such as: (i) a probability balance, (ii) by modelling the random evolution, (iii) by using models based on the stochastic differential equations, (iv) by deterministic models of the process where the parameters also come from a stochastic base because the random chains are present in the process evolution.

As was described in the section concerning modelling based on transfer phenomena, a general model can generate many particular cases. The same situation occurs in stochastic modelling processes. The particularization of some stochastic models results in a new image of chemical engineering processes. It is called the stochastic or polystochastic image. It is actually well accepted that almost all chemical engineering processes have a stochastic description [4.5–4.7, 4.30, 4.31].

Some ideas and rather simple concepts, which are fundamental for the alphabet of stochastic modelling, will be described here for some particular cases. It is obvious that knowledge of the alphabet of stochastic methods is only one area of knowledge necessary to become an expert in stochastic modelling. To this aim, a major study of the literature and especially a great personal experience in solving problems, together with a clear knowledge of the corresponding theory are necessary. Some of the aspects presented below will show how to apply polystochastic modelling in chemical engineering.

4.3.1
Polystochastic Chains and Their Models

In the problem of polystochastic chains, different situations can be considered. A first case is expressed by one or several stochastic chains, which keep their individual character. A second case can be defined when one or several random chains are complementary and form a completely connected system. In the first case, it is necessary to have a method for connecting the elementary states which define a chain.

4.3.1.1 Random Chains and Systems with Complete Connections

If we consider the example described at the beginning of this chapter, the element of study in stochastic modelling is the particle which moves in a trajectory where the local state of displacement is randomly chosen. The description for this discrete displacement and its associated general model, takes into consideration the fact that the particle can take one of the positions $i = 0, \pm 1, \pm 2, \pm 3$ ..... where i is a number contained in Z. The particle displacement is carried out step by step and randomly according to the type of component process (elementary state). The type of motion (of the process component) followed by the particle is denoted k. Here, $k \in K$, K is a field of the finite values and p_{ek} is the probability of passage from e to k. In addition, ε is a random variable, which gives the length of displacement for each process component; thus ε_k represents the length of displacement for the k-type motion. The distribution function of this random variable (ε) is written: $p_k(a)$, $a \in Z$. It represents the probability to have a step with length 'a' for a k-type displacement.

The process described above is thus repeated with constant time intervals. So, we have a discrete time $\tau = n\Delta\tau$ where n is the number of displacement steps. By the rules of probability balance and by the prescriptions of the Markov chain theory, the probability that shows a particle in position 'i' after n motion steps and having a k-type motion is written as follows:

$$P_k(n, i) = \sum_{e \in K} \sum_{a \in Z} P_k(n - 1, i - a) p_{ek} p_k(a) \tag{4.58}$$

In order to begin the calculations, we need to know some parameters such as the process components (k = 2 or k = 3, etc.), the trajectory matrix (p_{ek} in the model), and the equation that describes the distribution function of the path length for displacement k and for the initial state of the process

$P_k(0, i - a)$, $\forall i \in Z$ and $k \in K$. In our example, when the particle displacement is realized by unitary steps and in a positive direction (type I) or in a negative direction (type II) and where the path length distribution is uniform with a unitary value for both component processes, we have:

$$k = 1, 2 \ , \ p_1(a) = \begin{cases} 1 \ \text{ for } \ a = 1 \\ 0 \ \text{ for } \ a \neq 1 \end{cases}, \ p_2(a) = \begin{cases} 1 \ \text{ for } \ a = -1 \\ 0 \ \text{ for } \ a \neq -1 \end{cases}$$

For this case, the particularization of the relation (4.58) gives the following system:

$$\begin{cases} P_1(n, i) = P_1(n - 1, i - 1)p_{11} + P_2(n - 1, i - 1)p_{21} \\ P_{21}(n, i) = P_1(n - 1, i + 1)p_{12} + P_2(n - 1, i + 1)p_{22} \end{cases} \tag{4.59}$$

If, in addition to the standard process components (type I and II), we introduce a third one (position or displacement $k = 3$), which considers that the particle can keep a rest position, then the general model produces the following particularization:

$$\begin{cases} P_1(n, i) = P_1(n - 1, i - 1)p_{11} + P_2(n - 1, i - 1)p_{21} + P_3(n - 1, i - 1)p_{31} \\ P_2(n, i) = P_1(n - 1, i + 1)p_{12} + P_2(n - 1, i + 1)p_{22} + P_3(n - 1, i + 1)p_{32} \ \ (4.60) \\ P_3(n, i) = P_1(n - 1, i)p_{13} + P_2(n - 1, i)p_{23} + P_3(n - 1, i)p_{33} \end{cases}$$

$$p_k(a) = \begin{cases} 1 \ \ \text{ for } \ (k = 1, \ a = 1); (k = 2, \ a = -1); (k = 3, \ a = 0) \\ 0 \ \ \text{ for other cases} \end{cases} \tag{4.61}$$

Schmaltzer and Hoelscher [4.32] had suggested this model for the description of the axial mixing and the mass transfer in a packed column. Another particularization can be made in the case when the types of trajectory are chains corresponding to the completely random displacement (for example in the steps $k = 1$, which represent a displacement ahead, it is possible to have a small step towards the right or the left. In a k-type chain, the probability to realize a step towards the right is noted p_k whereas q_k represents the probability for the particle to realize a step towards the left (then, the probability $p_k(a)$ is expressed according to relation (4.63)).

The model which is obtained can be described by relations (4.62) and (4.63).

$$P_k(n, i) = \sum_{e \in K} \left[p_k P_e(n - 1, i - a_k) + q_k P_e(n - 1, i + a_K) \right] p_{ek} \tag{4.62}$$

$$p_k(a) = \begin{cases} p_k \ \text{ for } \ a = a_k \\ q_k \ \text{ for } \ a = -a_k \\ 0 \ \text{ for } \ a \neq a_k \text{and } a \neq -a_k \end{cases} \tag{4.63}$$

For the mathematical characterization of polystochastic chains, we often use the theory of systems with complete connections. According to the definition given in

Section 4.1, the group [(A,<u>A</u>) ,(B,<u>B</u>) ,u,P] defines a random system with complete connections. For each one of these systems two chains of the random variables are associated: S_n with values in A and E_n with values in B; the causal dependence between both chains is given by the function u as is shown here below:

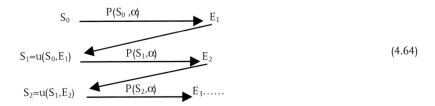

$$(4.64)$$

In system (4.64), $P(S_0,\alpha)$, $P(S_1,\alpha)$, etc., represent the conditioned probabilities of temporary transition from (A,<u>A</u>) to (B,<u>B</u>). The probability $P(S_n,\alpha)$ is given according to the following statement: "$P(S_n,\alpha)$ *is the probability that the phenomenon produced at time n+1 (E_{n+1}) belongs to α (with $\alpha \subset A$) with the condition that, at time n, the S_n state has already occurred*". The chain S_n with values in A, gives a Markov chain E_n (with values in B) which is in fact a complete connections chain. Before particularizing the model given in relation (4.58) into a case with two random variables, we need to explain the case of a particle displacement in a random system with complete connections [4.33]. As shown here, the jumps of a particle are randomly dimensioned by $a \in Z$; if we have a conditioned probability $p_k(a)$ in state $k \in K$, then the component elements characterized by $k \in K$ are different according to their nature (trajectory velocity, medium conditions, etc.). If the particle is positioned at $j \in Z$ and k is its temporary state, the passage probability to a new process $e \in K$ is $p_{ke}(j)$.

At the beginning of the process with $j \in Z$, where the state is a k-type process, the particle jumps distance a and, at the same time, evolves towards process e with the probability $p_e(a)p_{ke}(j)$. Consequently, it reaches j + a in a state e. Now, the new beginning is in j + a and in the e state and it jumps a' distance and realizes a state commutation towards e' ($e' \in K$) with $p_{e'}(a)p_{ee'}(j + a)$ probability.

The system keeps evolving considering that it is a random system with complete connections [(A,A) ,(B,<u>B</u>) ,u,P] and with the following particularizations: A = B = ZxK, $u(S_n,E_{n+1}) = u((j,k),(a,e)) = (j+a,e)$ and $P(S_n,...E_1S_1,S_0) = P((j,k);(a,e)) = p_e(a)p_{ke}(j)$ (please look at the definition of a random system with complete connections). The statement above is supported by the fact that $P(S_n,...E_1,S_1,S_0) = P((j,k);(a,e))$ with values in K is a real probability (we can observe that $\sum_{k \in K} p_e(a)p_{ke}(j) = 1$ because $\sum_{a \in Z} p_e(a) = 1$ and $\sum_{e \in Z} p_{ke}(j) = 1$). The particle displacement, described above, is represented by relation (4.64), which is particularized as follows:

$$(j,k) \xrightarrow{\quad p_k(a)p_{ke}(j) \quad} (a,e)$$

$$(j+a,e) \xleftarrow{\quad p_e(a)p_{ee}(j+a) \quad} (a',e')$$

$$(j+a+a',e')$$ ...... (4.65)

A complete particularization can be made in order to show more precisely that the case considered is a random system with complete connections. In this case, S_n is the first random vector:

$$S_n = \begin{vmatrix} S_n{}' \\ S_n{}'' \end{vmatrix} = \begin{vmatrix} \text{particle position at time } n \\ \text{process component at time } n \end{vmatrix}$$

whereas E_{n+1} is the second random vector:

$$E_{n+1} = \begin{vmatrix} E_{n+1}{}' \\ E_{n+1}{}'' \end{vmatrix} = \begin{vmatrix} \text{jump distance at time } n \\ \text{process component at time } n \end{vmatrix}$$

The function u is given by $S_{n+1} = u(S_n, E_{n+1})$ and the probability $P(S_n,....E_1,S_1,S_0)$ can be expressed as:

$$P\left(E_1 = \begin{vmatrix} a \\ e \end{vmatrix} / S_0 = \begin{vmatrix} j \\ k \end{vmatrix}\right) = P((j,k);(a,e))$$

$$P\left(E_{n+1} = \begin{vmatrix} a \\ e \end{vmatrix} / S_n, E_n, ...S_1, E_1; S_0\right) = P(S_n;(a,e)) \tag{4.66}$$

Obviously, S_n is a homogeneous Markov chain with a passage probability given by $P(S_{n+1} = (j,e)/S_n = (i,k)) = p_e(j-i)p_{ke}(i)$, whereas, E_n is a complete connection chain. In this case, the stochastic model (4.58) is known as the Chapman–Kolmogoraov model; it can be generalized by the Eq. (4.67):

$$P_k(n,i) = \sum_{e \in K} \sum_{a \in Z} P_k(n-1, i-a) p_{ek}(i-a) p_k(a) \tag{4.67}$$

In the case when the transition probabilities do not depend on the position, noted here by i, they are constant and therefore the chain E_n is a Markov chain.

4.3.2
Continuous Polystochastic Process

The stochastic models can present discrete or continuous forms. The former discussion was centred on discrete models. The continuous models are developed according to the same base as the discrete ones. Example 4.3.1 has already shown this method, which leads to a continuous stochastic model. This case can be generalized as follows: a particle in the z position moves in a medium with velocity v_k $k = 1,2,3,......n$, where n is a finite number; in the interval of time $\Delta\tau$, the probability to pass from speed v_k to v_j is $p_{kj} = \alpha_{kj}\,\Delta\tau$. This corresponds to a connection process of the Markov type. If $P_k(z,\tau + \Delta\tau)$ is the probability that the particle reaches z at $\tau + \Delta\tau$ with velocity v_k, then we can write:

$$P_k(z, \tau + \Delta\tau) = \sum_{j=1}^{n} P_j(x - v_k\Delta\tau, \tau)p_{jk}\; ; 1 < = j, k < = n \qquad (4.68)$$

We can notice that relation (4.68) describes the evolution of the particles having reached position z in time $\tau + \Delta\tau$ which were originally positioned among the particles at the distance $v_k\Delta\tau$ with respect to z. In the interval of time $\Delta\tau$, their velocity changes to v_k. In the majority of the displacement processes with v_k velocity, a complete system of events appears and, consequently, the matrix of passage from one velocity to another is of the stochastic type. This means that the addition of the probabilities according to the unit value limit is: $\sum_{j=1}^{n} p_{kj} = 1$, $\forall k = 1, n$. If, in relation (4.68), we replace $p_{kk} = 1 - \sum_{j=1, j\neq k}^{n} p_{kj}$ then, we can write:

$$P_k(z, \tau + \Delta\tau) = P_k(z - v_k\Delta\tau, \tau) - \left(\sum_{j=1, j\neq k}^{n} p_{kj} \right) P_k(z - v_k\Delta\tau, \tau)$$

$$+ \sum_{j=1}^{n} p_{jk}P_j(z - v_j\Delta\tau, \tau) \qquad (4.69)$$

In the equation system (4.69), the subscripts j, k are limited by the number of elementary states. Thus, we always have $0 < = j, k < = n$. Now, if we use a development around the point (x, τ) for the first term of Eq. (4.69) we have:

$$P_k(z - v_k\Delta\tau, \tau) = P_k(z, \tau) - v_k \frac{\Delta P_k(z, \tau)}{\Delta z} + ... \qquad (4.70)$$

The development described above transforms system (4.69) into the following system of n equations (k = 1, n) with partial derivatives:

$$\frac{\partial P_k(z, \tau)}{\partial \tau} + v_k \frac{\partial P_k(z, \tau)}{\partial z} = -\left(\sum_{j=,j\neq k}^{n} \alpha_{kj} \right) P_k(z, \tau) + \sum_{j=1, j\neq k}^{n} \alpha_{jk}P_j(z, \tau) \qquad (4.71)$$

If the parameters α_{kj} have constant values, then the model described by system (4.71) corresponds to a Markov connection linking the process components. In this case, as in general, the process components represent the individual displacements which can be characterized globally through the convective mixing of their spectra of speeds (v_k, k = 1,N).

The model developed in Section 4.3.1 is a particular case of the model (4.71) where $k = 2, v_1 = v_z, v_2 = -v_z$ (for instance see relations (4.25), (4.26)):

$$\frac{\partial P_1(z,\tau)}{\partial \tau} + v_z \frac{\partial P_1(z,\tau)}{\partial z} = -\alpha_{12}P_1(z,\tau) + \alpha_{21}P_2(z,\tau) \tag{4.72}$$

$$\frac{\partial P_2(z,\tau)}{\partial \tau} - v_z \frac{\partial P_{21}(z,\tau)}{\partial z} = -\alpha_{21}P_2(z,\tau) + \alpha_{12}P_2(z,\tau) \tag{4.73}$$

For $k = 3$, $v_1 = v_z$, $v_2 = -v_z$ $v_3 = 0$ (at $v_3 = 0$ the particle keeps a stationary position) we have the model (4.74)–(4.76) which has been successfully used in the analysis of axial mixing for a fluid that flows in a packed bed column [4.28]:

$$\frac{\partial P_1(z,\tau)}{\partial \tau} + v_z \frac{\partial P_1(z,\tau)}{\partial z} = -(\alpha_{12} + \alpha_{13})P_1(z,\tau) + \alpha_{21}P_2(z,\tau) + \alpha_{31}P_3(z,\tau) \tag{4.74}$$

$$\frac{\partial P_2(z,\tau)}{\partial \tau} - v_z \frac{\partial P_{21}(z,\tau)}{\partial z} = -(\alpha_{21} + \alpha_{23})P_2(z,\tau) + \alpha_{12}P_1(z,\tau) + \alpha_{32}P_3(z,\tau) \tag{4.75}$$

$$\frac{\partial P_1(z,\tau)}{\partial \tau} = -(\alpha_{31} + \alpha_{32})P_3(z,\tau) + \alpha_{13}P_1(z,\tau) + \alpha_{23}P_2(z,\tau) \tag{4.76}$$

A second continuous polystochastic model can be obtained from the transformation of the discrete model. As an example, we consider the case of the model described by Eqs. (4.62) and (4.63). If $P_k(z,\tau)$ is the probability (or, more correctly, the probability density which shows that the particle is in the z position at time τ with a k-type process) then, p_{kj} is the probability that measures the possibility for the process to swap, in the interval of time $\Delta\tau$, the elementary process k with a new elementary process (component) j. During the evolution with the k-type process state, the particle moves to the left with probability β_k and to the right with probability γ_k (it is evident that we take into account the fact that $\beta_k + \gamma_k = 1$). For this evolution, the balance of probabilities gives relation (4.77), which is written in a more general form in Eq. (4.78):

$$P_k(z,\tau+\Delta\tau) = \sum_{j=1}^{n} p_{jk}[\beta_k P_j(z-\Delta z,\tau) + \gamma_k P_j(z+\Delta z,\tau)] \; ; 1 <= j, k <= n \tag{4.77}$$

$$P_k(z,\tau+\Delta\tau) = \sum_{m}\sum_{j=1}^{n} P_k(z-\Delta z_m,\tau)p_{jk}(z-\Delta z_m)p_i(\Delta z_m) \tag{4.78}$$

Equation (4.78) is developed with the assistance of relation (4.77). To do so, it is necessary to consider the values of $p_{jk}(z - \Delta z_m) = p_{jk}$ as constant and the following relation for $p_i(\Delta z_m)$:

$$p_i(\Delta z_m) = \begin{cases} \beta_k & \text{for } \Delta z_m = \Delta z_k \\ \gamma_k & \text{for } \Delta z_m = -\Delta z_k \\ 0 & \text{for other cases} \end{cases}$$

Now, we have to take into account the following considerations for relation (4.77):
- the passage matrix is stochastic, it results in:

$$\sum_{j=1}^{n} p_{kj} = 1; \; .p_{kk} = 1 - \sum_{j=1,j\neq k}^{n} p_{kj}.$$

- the connection process is of Markov type and then: $p_{kj} = \alpha_{kj}\Delta\tau$.
- the probabilities $P_k(z - \Delta z, \tau)$ and $P_k(z + \Delta z, \tau)$ are continuous functions and can be developed around (z,τ):

$$P_k(z \mp \Delta z, \tau) = P_k(z, \tau) \mp \Delta z_i \frac{\Delta P_k(z, \tau)}{\Delta z_i} + \frac{\Delta z_i^2}{2} \frac{\Delta^2 P_k}{(\Delta z_i)^2} \mp .....$$

- the displacements to the left and to the right by means of an elementary k-process have the same probability and then
$\beta_k = \gamma_k = 1/2$.

These considerations result in the following equation:

$$\frac{\partial P_k(z, \tau)}{\partial \tau} = D_k \frac{\partial^2 P_k(z, \tau)}{\partial z^2} - \left(\sum_{j=1, j \neq k}^{n} \alpha_{kj} \right) P_k(z, \tau) + \sum_{j=1, j \neq k}^{n} \alpha_{jk} P_j(z, \tau) ;$$

$$1 <= j, k <= n \tag{4.79}$$

In Eq. (4.79), D_k represents the limit $D_k = \lim\limits_{\Delta\tau \to 0} \frac{\Delta z_k^2}{\Delta\tau}$, which has a finite value, and the dimension of a diffusion coefficient. It is called: *diffusion coefficient of the elementary k-process*.

If the displacement velocity of the particle is described by $+v_x$ and $-v_x$ in the x axis, $+v_y$ and $-v_y$ in the y axis and finally $+v_z$ and $-v_z$ in the z axis, we can consider the following diffusion coefficients:

$$D_{xx} = \lim_{\Delta\tau \to 0} \frac{\Delta x^2}{\Delta\tau}, D_{yy} = \lim_{\Delta\tau \to 0} \frac{\Delta y^2}{\Delta\tau}, D_{zz} = \lim_{\Delta\tau \to 0} \frac{\Delta z^2}{\Delta\tau}, D_{xy} = \lim_{\Delta\tau \to 0} \frac{\Delta x \Delta y}{\Delta\tau}, \text{etc.}$$

This definition of diffusion coefficients considers the non-isotropic diffusion behaviour in some materials. So, this stochastic modelling can easily be applied for the analysis of the oriented diffusion phenomena occurring in materials with designed properties for directional transport.

Model (4.79) describes an evolutionary process, which results from the coupling of a Markov chain assistance with some individual diffusion processes. This model is well known in the study of the coupling of a chemical reaction with diffusion phenomena [4.5, 4.6, 4.34, 4.35]. The models described by relations (4.63) and (4.79) can still be particularized or generalized. As an example, we can notice that other types of models can be suggested if we consider that the values of α_{kj} are functions of z or τ or $P_k(z,\tau)$ in Eq.(4.79). However, it is important to observe that the properties of the Markov type connections cannot be considered when $\alpha_{kj} = f(P_K(z, \tau))$.

Using *stochastic differential equations* can also represent the stochastic models. A stochastic differential equation keeps the deterministic mathematical model but accepts a random behaviour for the model coefficients. In these cases, the problems of integration are the main difficulties encountered. The integration of stochastic differential equations is known to be carried out through working methods that are completely different from those used for the normal differential equations

[4.36, 4.37]. We can overcome this difficulty if, instead of using the stochastic differential equations of the process, we use the analysis of the equations with partial derivatives that become characteristic for the passage probabilities (Kolmogorov-type equations).

The following practical example will illustrate this problem: a mobile device passes through an arbitrary space with a variable velocity. By means of the classic dynamics analysis, we can write that $\dfrac{dX(\tau)}{d\tau} = kv(\tau)$. In the stochastic language, this equation can be written as follows:

$$\frac{dX(\tau)}{d\tau} = F(X(\tau), v(\tau)), \quad X(0) = X_0 \tag{4.80}$$

where $F(X,v)$ is an operator ($F(X,v) = kv$ for example), defined for $R^n x R^m$ with values in R^n. It is able to be derivate in X, keeping the continuity in v.
- $v(\tau)$ is a Markov diffusion process (for instance, look at the model described by Eq. (4.79)). The following relations give the variances and the average (mean) values of this diffusion process:
- the variances:

$$\sigma_{ij} = \lim_{\Delta\tau \to 0} E\left\{ [v_i(\tau + \Delta\tau) - v_i(\tau)][v_j(\tau + \Delta\tau) - v_j(\tau)]/v(\tau) - v \right\} \tag{4.81}$$

- the mean values

$$m_j(v) = \lim_{\Delta\tau \to 0} E\left\{ [(v_j(\tau) - v(\tau)]/v(\tau) \right\} = v \tag{4.82}$$

Here, i, j = 1, 2,...m are subscripts which indicate the individual states of the device speed. The coupled process $(X(\tau),v(\tau))$ is a Markov process with values in R^{n+m} and with mean value and variances (X,v), (X,X) given by the following relations:

$$\lim_{\Delta\tau \to 0} \frac{1}{\Delta\tau} E\left\{ (X_i(\tau + \Delta\tau) - X(\tau))/(X(\tau), v(\tau)) \right\} = F_i(X(\tau), v(\tau)) \tag{4.83}$$

$$\lim_{\Delta\tau \to 0} \frac{1}{\Delta\tau} E\left\{ [X_i(\tau + \Delta\tau) - X(\tau)][v_j(\tau + \Delta\tau) - v_j(\tau)]/(X(\tau), v(\tau)) \right\} = 0 \tag{4.84}$$

$$\lim_{\Delta\tau \to 0} \frac{1}{\Delta\tau} E\left\{ [X_i(\tau + \Delta\tau) - X(\tau)][X_j(\tau + \Delta\tau) - X_j(\tau)]/(X(\tau), v(\tau)) \right\} = 0 \tag{4.85}$$

If $F(X,v) = v$ or if v is the device speed, then the stochastic differential equation (4.80) shows that the state of the device is a function which depends on position and speed. The device passes from one speed to another with the rules defined by a diffusion process and with an average value $m_j(v)$ and a variance σ_{ij}, $1 < = i, j < = n$. It is important to note that the passage probability densities of the coupled Markov process $(X(\tau),v(\tau))$ – written: $p = p(\tau,X,v,X_0,v_0)$ – should verify the following equation:

$$\frac{\partial p}{\partial \tau} = \frac{1}{2} \sum_{i,j=1}^{m} \frac{\partial^2 [\sigma_{ii}(v)p]}{\partial v_i \partial v_j} - \sum_{j=1}^{m} \frac{\partial [m_j(v)]}{\partial v_j} - \sum_{i=1}^{n} \frac{\partial [F_j p]}{\partial x_j} \tag{4.86}$$

The initial condition used with Eq. (4.86) shows that, at time $\tau = 0$, the stochastic evolution begins according to a signal impulse:

$$p(0, X, v, X_0, v_0) = \delta(X - X_0)\delta(v - v_0) \tag{4.87}$$

In Eq. (4.86), F_j is the average value for the coupled Markov process (see Eq. (4.83)). In Eq. (4.87) $p = p(\tau, X, v, X_0, v_0)$ corresponds to the probability density of the coupled process $(X(\tau), v(\tau))$. To calculate this density of probability at a predefined time $(p(\tau, X, v))$, we use the initial condition:

$$p(0, X, v) = \rho_v(0, v)\delta(X - X_0) \tag{4.88}$$

where $\rho_v(0, v)$ is the probability density of the process $v(\tau)$ at the start.

With solution $p(\tau, X, v)$, we can calculate the distribution process $X(\tau)$ after the integration for all the possible speeds:

$$p(\tau, X) = \int_{-\infty}^{\infty} p(\tau, X, v)dv \tag{4.89}$$

This method to solve stochastic differential equations has also been suggested to calculate the solutions of the stochastic models originated from the theory of random evolution [4.38, 4.39].

In the following paragraph, we will explain how this method is particularized for two examples. A random evolution is described with the assistance of a model, which is based on a dynamic system with operation equations called state equations, which have to undergo random variations. The first example is given by the evolution of a bacterial population that develops in a medium with a randomly changing chemical composition. A second example can be represented by the atmospheric distribution for polluting fumes produced by a power station when atmospheric turbulences change randomly. Many other examples illustrate these typical situations where a system in evolution changes its mode of evolution according to the random changes of the medium or according to the changing conditions of the process development. In these systems the process can evolve (move) into a stochastic or deterministic way at time "t" and, suddenly, at time "τ", the process undergoes another random descriptive evolution.

From a mathematical point of view, a random evolution is an operator $O(\tau, t)$ that is improved at both t and τ times. The linear differential equation is Eq. (4.90):

$$\frac{dO(\tau, t)}{d\tau} = -V(X(\tau))O(\tau, t) \quad \text{or} \quad \frac{dO(\tau, t)}{dt} = O(\tau, t)V(X(t)) \tag{4.90}$$

$V(X(\tau)$ (or $V(X(t))$ is the expression of an operator which depends on parameter X (or X(t)), which is the stochastic parameter characterizing the process. It is important to note that the correct expression of X(t) is X(t, e), where $e \in \Omega$. Here, Ω is the region where the elementary steps characterizing the process occur.

If we consider that V(X(t)) is a first order linear differential operator like V(X) = $v(X)\dfrac{d}{dz}$ with v(X) in R, then for each X value $(1, 2, \ldots n,)$ v(X) will be a constant that multiplies the operator $\dfrac{d}{dz}$ out. With this type of random evolution operator, we can describe the behaviour of a particle that at present moves in the z-axis with a random speed. This velocity is included in the speed spectrum of the integral process.

The concept of infinitesimal operator is frequently used when the random evolutions are the generators of stochastic models from a mathematical point of view. This operator can be defined with the help of a homogeneous Markov process X(t) where the random change occurs with the following transition probabilities:

$$p(t - \tau, X, A) = P(X(t) \in A / X(\tau) = X) \tag{4.91}$$

We have to notice that, for different X(t) values, we associate different values for the elements of the matrix of transition probabilities. When the movement randomly changes the value of X into a value around A, Eq. (4.91) is formulated with expressions giving the probability of process X(t) at different states. The infinitesimal operator [Qf] ([Qf] = Q by function f) is defined as the temporary derivative of the mean value of the stochastic process for the case when the process evolves randomly:

$$[Qf] = -\dfrac{d}{d\tau}\Big|_{\tau=t} (\smallint f(y)p(t - \tau, X)dy) = -\dfrac{d}{d\tau}\Big|_{\tau=t} E_{X/\tau}(f(X(t))) \tag{4.92}$$

If the Markov process, considered in Eq. (4.91), is characterized by n states, then the infinitesimal operator Q corresponds to a matrix (n, n) where the q_{ij} elements are:

$$q_{ij} = \lim_{t \to \tau} \dfrac{1}{t - \tau}[p_{ij}(t - \tau) - \delta_{ij}]$$

In the case where X(t) or X(t, e) corresponds to a diffusion process (the stochastic process is continuous), it can be demonstrated that Q is a second order elliptic operator [4.39– 4.42]. The solution of the equation, which defines the random evolution, is given by a formula that yields O(τ,t). In this case, if we can consider that $e_m(t, X)$ is the mean value of X(t) (which depends on the initial value of X_0), then, we can write the following equation:

$$e_m(t, X) = E_{X0}\{O(0, t)f(X(t))\} \tag{4.93}$$

Here, $e_m(t, X)$ gives the solution for Eq. (4.94) where the operators V(X) and Q work together:

$$\dfrac{de_m(t, X)}{dt} = V(X)e(t, X) + Qe_m(t, X) \quad \text{where} \quad e_m(0, X) = f(X) \tag{4.94}$$

The condition $e_m(0, X) = f(X)$ in the previous equation is a result of $O(t,t) = I$ where I is the identity operator.

Two examples, which show the methodology to be used in order to establish the random evolution operator, are developed below:

1. When $X(t)$ or, more correctly, $X(t, e)$ is a Brownian motion process (a displacement with multiple direction changes) and $V(X)$ is a function of real values, Eq. (4.90) gives the following solution:

$$O(\tau, t) = \exp \int_\tau^t V(X(\alpha) d\alpha) \tag{4.95}$$

Here, $e_m(t, X)$ is given to the computation with the relation (4.93). We obtain formula (4.96) where we can observe that $E_X = E_{X0}$ is a Wiener integral.

$$e_m(t, X) = E_{X0} \left\{ \exp \left(\int_0^t V(X(\alpha) d\alpha) f(X(t)) \right) \right\} \tag{4.96}$$

As far as the infinitesimal operator is elliptic $\left(Q = \frac{1}{2} \frac{d}{dX^2} \right)$, Eq. (4.94) gives, for $e_m(t, X)$, the following equation of partial derivatives:

$$\frac{\partial e_m}{\partial t} = V(X)e_m + \frac{1}{2} \frac{\partial^2 e_m}{\partial X^2} \quad \text{where} \quad e_m(0, X) = f(X) \tag{4.97}$$

2. When $X(t, e)$ is a "n states" process with the infinitesimal generator Q and when $V(X)$ with $X = 1,2,3,...n$ are first order differential generators, the particularization of relation (4.94) is given by a system of hyperbolic equations with constant coefficients. So when $V(X) = v(X)\frac{d}{dz}$ and $v(X)$ is in R, this system is described by Eq. (4.98). Here q_{xy} are the elements of the infinitesimal generator:

$$\frac{\partial e_m(t, X, z)}{\partial t} = v(X) \frac{\partial e_m(t, X, z)}{\partial z} + \sum_{y=1}^n q_{xy} e_m(t, X, z) \quad 1<X<n \tag{4.98}$$

For relation (4.98), the initial condition $e_m(0, X, z) = f(X, z)$ can be established according to the form considered for $v(X)$. This condition shows that, at the beginning of the random evolution and at each z position, we have different X states for the process. The examples described above show the difficulty of an analysis when the required process passes randomly from one stochastic evolution to another.

As was stated previously, the method of analysis for the stochastic differential equations, which gives the probability density $p(\tau,X,v)$ as a model solution, can be applied to build and analyze the models developed from the random evolution theory. At the same time, from the mathematical point of view, we have shown that a model solution for a process with random exchanges from one stochastic evolution to another can be carried out as $e_m(\tau, X, v)$ mean values. With reference to this model solution form, Gikham and Shorod [4.43] show that, in a stochastic process (X,v), the mean values for the process trajectories $(e_m(\tau, X, v))$ are given by:

$$e_m = e_m(\tau, X, v) = E_{Xv}\{f(X(\tau), v(\tau))\} \qquad (4.99)$$

otherwise these mean values satisfy the indirect equations of Kolmogorov:

$$\frac{\partial e_m}{\partial \tau} = \frac{1}{2}\sum_{i,j=1}^{m}\sigma_{ij}(v)\frac{\partial^2 e_m}{\partial v_j \partial v_i} + \sum_{j=1}^{m}m_j(v)\frac{\partial e_m}{\partial v_j} + \sum_{j=1}^{n}F_j(X, v)\frac{\partial e_m}{\partial X_j} \quad ;$$

$$e_m(0, X, v) = f(X, v) \qquad (4.100)$$

The practical example given below illustrates this type of process evolution and its solution. Here, we consider a displacement process such as diffusion with $v(\tau)$. The process presents a variance $\sigma(v)$ and a mean value $m(v)$ whereas $X(\tau)$ is an associated process which takes scalar values given by:

$$\frac{dX(\tau)}{d\tau} = h(v(\tau))X(\tau) , \quad X(0) = X \qquad (4.101)$$

It is evident that we must have real values for $h(v(\tau))$ so, $h(v)$: $R \rightarrow R$ where R is the domain of real numbers. The solution to Eq. (4.101) is:

$X(\tau) = X\exp[\int_0^\tau h(v(\alpha)d\alpha]$ and its average value, calculated by the Kolmogorov relation ((4.100)), corresponds to one of the possible solutions of Eq. (4.102). In this example relation (4.102) represents the particularization of Eq. (4.100).

$$\frac{\partial e_m}{\partial \tau} = \frac{1}{2}\sigma(v)\frac{\partial^2 e_m}{\partial v^2} + m(v)\frac{\partial e_m}{\partial v} + h(v)X\frac{\partial e_m}{\partial X} , \quad e_m(0, X, v) = f(X, v) \qquad (4.102)$$

If the function f(X,v), which gives the mean value, is particularized as $f(X,v) = Xg(v)$, where the derivative of g(v) can be calculated, then the expression for e_m becomes: $e_m = e_m(t, X, v) = E_{Xv}\{f(X(\tau), v(\tau))\} = XH(v,\tau)$. It is observable that it is easy to write that $H(v, \tau) = E_v[\exp[\int_0^\tau h(v(\alpha)d\alpha)]g(v(t))]$. At the same time, $H(v,\tau)$ verifies the partial derivative equation (4.103) which is developed from the replacement of the average value e_m and the function f(X,v) inside Eq. (4.102):

$$\frac{\partial H}{\partial \tau} = \frac{1}{2}\sigma(v)\frac{\partial^2 H}{\partial v^2} + m(v)\frac{\partial H}{\partial v} + h(v)V , \qquad H(0, v) = g(v) \qquad (4.103)$$

These equations and the example shown in Section 4.3.1 can be related if we consider that, when $v(\tau)$ is a Markov process with discrete valuesi = 1,.......n and with the infinitesimal generator Q, then, the e_{mi} mean values ($e_{mi} = e_{mi}(t, X) = E_{Xi}\{f_i(X(\tau))\}$) are the solutions to the following differential equation:

$$\frac{\partial e_{mi}}{\partial \tau} = \sum_{j=1}^{m} q_{ij} e_{mi} + \sum_{k=1}^{n} F_k(X, v_i) \frac{\partial e_{mi}}{\partial X_k} \ , \quad e_{mi}(0, X) = f_i(X) \tag{4.104}$$

A discussion concerning the equations assembly (4.104) can be carried out dividing it into its different component terms. If we consider the first term alone, we can observe that it represents a connection for the elementary processes with the passage matrix $e^{Q\tau}$. The second term corresponds to the transport or convection process at different speeds. Indeed, $v(\tau)$ is a two-states process with the infinitesimal generator Q and the function $F(X,v)$ given by the following formula:

$$Q = \begin{vmatrix} -\alpha & \alpha \\ \alpha & -\alpha \end{vmatrix} , \ F(X, \pm v) = \pm v \tag{4.105}$$

The particularization of the equations assembly (4.104) results in system (4.106):

$$\begin{cases} \dfrac{\partial e_{m1}}{\partial \tau} - v \dfrac{\partial e_{m1}}{\partial X} = -\alpha e_{m1} + \alpha e_{m2} \\[2mm] \dfrac{\partial e_{m2}}{\partial \tau} + v \dfrac{\partial e_{m2}}{\partial X} = -\alpha e_{m2} + \alpha e_{m1} \end{cases}$$

$$e_{m1}(0, X) = f_1(X) , \quad e_{m2}(0, X) = f_2(X) \tag{4.106}$$

If the process takes place along the z-axis, then we can write that X = z. Considering now that e_{m1} and e_{m2} are the average or mean probabilities for the process evolution with +v or –v states at the z position, we can observe a similitude between system (4.106) and Eqs. (4.31) and (4.32) that describe the model explained in the preceding paragraphs. The solution of the system (4.106) [4.5] is given in Eq. (4.107). It shows that the process evolution after a random movement depends not only on the system state when the change occurs but also on the movement dynamics:

$$e_{m1,2}(X, t) = E\left[f_{1,2}(X)(X + \int_0^t v(s)ds) \right] \tag{4.107}$$

4.3.3
The Similarity between the Fokker–Plank–Kolmogorov Equation and the Property Transport Equation

In Chapter 3, it was established that the local concentration Γ_A characterizes the state of one property (momentum, heat, mass of species, etc.) in a given system. In terms of the Γ_A concentration field, the differential form for the conservation of the property can be written as follows:

$$\frac{\partial \Gamma_A}{\partial \tau} + \text{div}(\overrightarrow{w}\Gamma_A) = \text{div}(D_{\Gamma A}\overrightarrow{\text{grad}}\Gamma_A) + \text{div}(\overrightarrow{J}_{SA}) + J_{v\Gamma}$$

This equation shows that the conservation of a property depends on the fortuitous or natural displacement of the property produced by vector $\vec{w}$, when that is generated through a volume (J_{VI}) or/and by a surface process (vector J_{SA}). The mentioned displacement is supplemented by a diffusion movement ($D_{\Gamma A}$ in the right part of the conservation equation). This movement is characterized by steps of small dimension occurring with a significant frequency in all directions. When the diffusion movement takes place against the vector $\vec{w}$ it is often called counter-diffusion. In the case of a medium, which does not generate the property, the relation can be written as follows:

$$\frac{\partial \Gamma_A(\tau, \bar{x})}{\partial \tau} + \frac{\partial}{\partial \bar{x}}(w(\tau, \bar{x})\Gamma_A(\tau, \bar{x})) = \frac{\partial}{\partial \bar{x}}\left(D_{\Gamma A}\frac{\partial}{\partial \bar{x}}(\Gamma_A(\tau, \bar{x}))\right) \tag{4.108}$$

where τ represents the time, $\bar{x}$ is a vector with n dimensions which represents the coordinates, $w(\tau, \bar{x})$ is also an n dimension vector and gives the speed which is bound to the position, $D_{\Gamma A}$ is an n × n matrix that contains the diffusion coefficients of the property with the local concentration Γ_A. If we assume that the component values of the diffusion matrix depend on the concentration values of the local property then Eq. (4.108), can be written as:

$$\frac{\partial \Gamma_A(\tau, \bar{x})}{\partial \tau} = -\frac{\partial}{\partial \bar{x}}\left(\left(w(\tau, \bar{x}) + \frac{\partial D_{\Gamma A}(\tau, \bar{x})}{\partial \Gamma_A}\frac{\partial \Gamma_A}{\partial \bar{x}}\right)\Gamma_A(\tau, \bar{x})\right) +$$
$$\frac{\partial}{\partial \bar{x}}\frac{\partial}{\partial \bar{x}}(D_{\Gamma A}(\tau, \bar{x})\Gamma_A(\tau, \bar{x})) \tag{4.109}$$

The vectors and the matrix described by relations (4.108) and (4.109) are given in Table 4.1. It should be specified that only the axial anisotropy in the case of an anisotropic medium was considered. However, if we want to take into account the anisotropy through the plane or the surface, we have to consider the terms of type $D_{\Gamma Axy}$ in the $D_{\Gamma A}(\tau, \bar{x})$ matrix.

Table 4.1 Vectors of Eqs. (4.108) and (4.109).

Case	Mono-dimensional	Tri-dimensional	n × n dimensions
vector			
$\bar{x}$	$\bar{x} = x$	$\bar{x} = \begin{pmatrix} x \\ y \\ z \end{pmatrix}$	$\bar{x} = \begin{pmatrix} x_1 \\ x_2 \\ x_3 \\ - \\ x_n \end{pmatrix}$
$w(\tau, \bar{x})$	$w(x, \tau)$	$\begin{pmatrix} w_x(x, y, z, \tau) \\ w_y(x, y, z, \tau) \\ w_z(x, y, z, \tau) \end{pmatrix}$	$\begin{pmatrix} w_{x1}(x_1, x_2, \; x_n, \tau) \\ w_{x2}(x_1, x_2, \; x_n, \tau) \\ w_{x3}(x_1, x_2, \; x_n, \tau) \\ - \\ w_{xn}(x_1, x_2, \; x_n, \tau) \end{pmatrix}$

Case	Mono-dimensional	Tri-dimensional	$n \times n$ dimensions
$D_{\Gamma A}(\tau, \bar{x})$	$D_{\Gamma A}$	$\begin{pmatrix} D_{\Gamma A} & 0 & 0 \\ 0 & D_{\Gamma A} & 0 \\ 0 & 0 & D_{\Gamma A} \end{pmatrix}$ isotropic medium	$\begin{pmatrix} D_{\Gamma A} & 0 & - & 0 \\ 0 & D_{\Gamma A} & - & 0 \\ - & - & - & 0 \\ 0 & 0 & - & D_{\Gamma A} \end{pmatrix}$ isotropic medium
		$\begin{pmatrix} D_{\Gamma x} & 0 & 0 \\ 0 & D_{\Gamma y} & 0 \\ 0 & 0 & D_{\Gamma z} \end{pmatrix}$ anisotropic medium	$\begin{pmatrix} D_{\Gamma x1} & 0 & - & 0 \\ 0 & D_{\Gamma x2} & - & 0 \\ - & - & - & 0 \\ 0 & 0 & - & D_{\Gamma xn} \end{pmatrix}$ anisotropic medium
$\dfrac{\partial D_{\Gamma A}}{\partial \Gamma_A} \dfrac{\partial \Gamma_A}{\partial \bar{x}}$	$\dfrac{\partial D_{\Gamma A}}{\partial \Gamma_A} \dfrac{\partial \Gamma_A}{\partial x}$	$\begin{pmatrix} \dfrac{\partial D_{\Gamma A}}{\partial \Gamma_A} \dfrac{\partial \Gamma_A}{\partial x} \\ \dfrac{\partial D_{\Gamma A}}{\partial \Gamma_A} \dfrac{\partial \Gamma_A}{\partial y} \\ \dfrac{\partial D_{\Gamma A}}{\partial \Gamma_A} \dfrac{\partial \Gamma_A}{\partial z} \end{pmatrix}$	$\begin{pmatrix} \dfrac{\partial D_{\Gamma A}}{\partial \Gamma_A} \dfrac{\partial \Gamma_A}{\partial x1} \\ \dfrac{\partial D_{\Gamma A}}{\partial \Gamma_A} \dfrac{\partial \Gamma_A}{\partial x2} \\ \dfrac{\partial D_{\Gamma A}}{\partial \Gamma_A} \dfrac{\partial \Gamma_A}{\partial xn} \end{pmatrix}$

If, in Eq. (4.109), we use the following notation:

$$A(\tau, \bar{x}) = \left(w(\tau, \bar{x}) + \frac{\partial D_{\Gamma A}}{\partial \Gamma_A} \frac{\partial \Gamma_A}{\partial \bar{x}} \right) \text{ then, we can write:}$$

$$\frac{\partial \Gamma_A(\tau, \bar{x})}{\partial \tau} = -\frac{\partial}{\partial \bar{x}}(A(\tau, \bar{x})\Gamma_A(\tau, \bar{x})) + \frac{\partial}{\partial \bar{x}}\frac{\partial}{\partial \bar{x}}(D_{\Gamma A}(\tau, \bar{x})\Gamma_A(\tau, \bar{x})) \qquad (4.110)$$

A careful observation of Eqs. (4.79), (4.80), (4.100) and their respective theoretical basis [4.44, 4.45], allows one to conclude that the probability density distribution that describes the fact that the particle is in position $\bar{x}$ at τ time, when the medium is moving according to one stochastic diffusion process (see relation (4.62) for the analogous discontinuous process), is given by Eq. (4.111). This relation is known as the Fokker–Planck–Kolmogorov equation.

$$\frac{\partial P(\tau, \bar{x})}{\partial \tau} = -\frac{\partial}{\partial \bar{x}}(A(\tau, \bar{x})P(\tau, \bar{x})) + \frac{\partial}{\partial \bar{x}}\frac{\partial}{\partial \bar{x}}(D(\tau, \bar{x})P(\tau, \bar{x})) \qquad (4.111)$$

There is an important analogy between the Fokker–Planck–Kolmogorov equation and the property transport equation. Indeed, the term which contains $A(\tau, \bar{x})$ describes the particle displacement by individual processes and the term which contains $D(\tau, \bar{x})$ describes the left and right movement in each individual displacement or diffusion. We can notice the very good similarity between the transport and the Kolmogorov equation. In addition, many scientific works show that both equations give the same result for a particular problem. However, large and important differences persist between both equations. The greatest difference is given by the presence of the speed vector in the $A(\tau, \bar{x})$ expression in Eq. (4.110).

Undeniably, the speed vector, by its size and directional character, masks the effect of small displacements of the particle. Another difference comes from the different definition of the diffusion coefficient, which, in the case of the property transport, is attached to a concentration gradient of the property; it means that there is a difference in speed between the mobile species of the medium. A second difference comes from the dimensional point of view because the property concentration is dimensional. When both equations are used in the investigation of a process, it is absolutely necessary to transform them into dimensionless forms [4.6, 4.7, 4.37, 4.44].

Both equations give good results for the description of mass and heat transport without forced flow. Here, it is important to notice that the Fokker–Plank–Kolmogorov equation corresponds to a Markov process for a stochastic connection. Consequently, it can be observed as a solution to the stochastic equations written below:

$$dX_\tau = A(\tau, X_\tau)d\tau + B(\tau, X_\tau)dW_\tau \tag{4.112}$$

Here, X_τ is the stochastic state vector, $B(\tau, X_\tau)$ is a vector describing the contribution of the diffusion to the stochastic process and W_τ is a vector with the same dimensions as X_τ and $B(\tau, X_\tau)$. After Eqs. (4.94) and (4.95), the W_τ vector is a Wiener process (we recall that this process is stochastic with a mean value equal to zero and a gaussian probability distribution) with the same dimensions as $D(\tau, X_\tau)$:

$$D(\tau, \bar{x}) = B(\tau, \bar{x})B^T(\tau, \bar{x}) \quad ; \quad D(\tau, X_\tau) = B(\tau, X_\tau)B^T(\tau, X_\tau) \tag{4.113}$$

By comparision with the property transport equation the advantage of a stochastic system of equations (SDE) is the capacity for a better adaptation for the numerical integration.

4.3.3.1 Stochastic Differential Equation Systems for Heat and Mass Molecular Transport

A good agreement is generally obtained between the models based on transport equations and the SDE for mass and heat molecular transport. However, as explained above, the SDE can only be applied when convective flow does not take place. This restrictive condition limits the application of SDE to the transport in a porous solid medium where there is no convective flow by a concentration gradient. The starting point for the transformation of a molecular transport equation into a SDE system is Eq. (4.108). Indeed, we can consider the absence of convective flow in a non-steady state one-directional transport, together with a diffusion coefficient depending on the concentration of the transported property:

$$\frac{\partial \Gamma_A}{\partial \tau} = \frac{\partial}{\partial x} \left(D_{\Gamma A}(\Gamma_A) \frac{\partial \Gamma_A}{\partial x} \right) \tag{4.114}$$

By introducing the stochastic Markov type connection process through the following equation:

$$\frac{\partial^2}{\partial x^2} (D_{\Gamma A}(\Gamma_A) \Gamma_A) = \frac{\partial}{\partial x} \left[D_{\Gamma A} \frac{\partial \Gamma_A}{\partial x} + \frac{dD_{\Gamma A}}{d\Gamma_A} \frac{\partial \Gamma_A}{\partial x} \Gamma_A \right] \tag{4.115}$$

the right term of Eq. (4.114) can be written as:

$$\frac{\partial}{\partial x} \left(D_{\Gamma A}(\Gamma_A) \frac{\partial \Gamma_A}{\partial x} \right) = -\frac{\partial}{\partial x} \left[\frac{dD_{\Gamma A}}{d\Gamma_A} \frac{\partial \Gamma_A}{\partial x} \Gamma_A \right] + \frac{\partial^2}{\partial x^2} (D_{\Gamma A}(\Gamma_A) \Gamma_A) \tag{4.116}$$

If we replace Eq. (4.116) by Eq. (4.114) we have:

$$\frac{\partial \Gamma_A}{\partial \tau} = -\frac{\partial}{\partial x} \left[\frac{dD_{\Gamma A}}{d\Gamma_A} \frac{\partial \Gamma_A}{\partial x} \Gamma_A \right] + \frac{\partial^2}{\partial x^2} (D_{\Gamma A}(\Gamma_A) \Gamma_A) \tag{4.117}$$

A simultaneous comparison between Eqs. (4.114), (4.117) and (4.113) results in the following identifications:

$$A(\tau, x) = \frac{dD_{\Gamma A}}{d\Gamma_A} \frac{\partial \Gamma_A}{\partial x}, \quad D(\tau, x) = D_{\Gamma A}(\Gamma_A), \quad B(\tau, x) = (D_{\Gamma A}(\Gamma_A))^{1/2}$$

Then, the SDE system can be written in the form:

$$dX(\tau) = \frac{dD_{\Gamma A}}{d\Gamma_A} \frac{\partial \Gamma_A}{\partial x} d\tau + (D_{\Gamma A}(\Gamma_A))^{1/2} dW(\tau) \tag{4.118}$$

The SDE and transport equation can be used with the same univocity conditions. For simple univocity conditions and functions such as $D_{\Gamma A}(\Gamma_A)$, the transport equations have analytical solutions. Comparison with the numerical solutions of stochastic models allows one to verify whether the stochastic model works properly. The numerical solution of SDE is carried out by space and time discretization into space subdivisions called bins. In the bins j of the space division i, the dimensionless concentration of the property ($\Gamma = \Gamma_A/\Gamma_{A0}$) takes the Γ_j value. Taking into consideration these previous statements allows one to write the numerical version of relation (4.118):

$$X_i(\tau + \Delta\tau) = X_i(\tau) + \left(\frac{\Gamma_{j+1} - \Gamma_{j-1}}{2\Delta x} \right) \left(\frac{dD_{\Gamma A}}{d\Gamma_j} \right)_{\Gamma_i} \Delta\tau + (2D_{\Gamma Aj}(\Gamma_j)\Delta\tau)^{1/2} \varphi_i \tag{4.119}$$

Here, φ_i is a random number for the calculation step "i". It is given by a standard procedure for the normal distribution values with a mean value of zero where $D_{\Gamma Aj}$ and Γ_j are the corresponding $D_{\Gamma A}$ and Γ values for the j bin and the particle position "i". The only limitation of the numerical method is concentrated in the fact that $\Delta\tau$ must have very small values in order to eliminate all the problems of non-convergence caused by the second term on the right half of the equation (4.119).

The approximated master equation, such as Eq. (4.111) with its associate (4.112), has computational advantages besides its obvious similarity to the convective-diffusion form. Even when this equation cannot be solved exactly, the numerical techniques for computing such equations are well established. More importantly, the derivation of this equation gives a clue to the identification of the terms of the vectors $A(\tau, X_\tau)$ and $B(\tau, X_\tau)$, which can be found independently without knowing the details of the transition probabilities required in the master equation (see assemblies (4.25)–(4.26) or (4.74)–(4.79)) and this is a great advantage. The set up of the Fokker–Planck–Kolmogorov equation into the form of Eq. (4.112) needs to take a time interval so small that X_τ does not change significantly but the Markovian assumption is still valid.

The Fokker–Planck–Kolmogorov approximation of the master equation is based on the assumption that all the terms greater than second order, which are extracted from the Taylor expansion of $P_k(z \pm \Delta z, \tau)$, vanish. This is rarely true in practice, however, and a more rational way of approximating the master equation is to systematically expand it in powers of a small parameter, which can be chosen approximately. This parameter is usually chosen in order to have the same size as the system.

4.4
Methods for Solving Stochastic Models

Once the stochastic model has been established, it is fed with data which characterize the inputs and consequently, if the model works correctly it produces data which represent the process output. The model solution is obtained:
- By an analytical solution given by a relation or by an assembly of relations and their exploitation algorithm showing how the output solutions are developed when the inputs are selected.
- By a numerical solution and the corresponding software.
- By another model, obtained by the transformation of the original model towards one of its boundaries and which can also be solved by an analytical or numerical solution. These models are called "limit stochastic models" or "asymptotic stochastic models".

The numerical as well as the asymptotic model solutions are estimated solutions, which often produce characteristic outputs of the model in different forms when compared to the natural state of the exits. Both stochastic and transfer phenomenon models present the same type of resolution process. The analysis developed in the paragraphs below can be applied equally to both types of models.

4.4.1
The Resolution of Stochastic Models by Means of Asymptotic Models

It is well-known that, from a practical view point, it is always interesting to be aware of the behaviour of a process near the boundaries of validity. The same statement can be applied to the stochastic model of a process for small stochastic disturbances which occur at large intervals of time. In this situation, we can expect the real process and its model not to be appreciably modified for a fixed time called "system answer time" or "constant time of the system". This statement can also be taken into account in the case of random disturbances with measurements realized at small intervals of time.

At the same time, it is known that, during exploitation of stochastic models, cases that show great difficulty concerning the selection and the choice of some parameters of the models frequently appear. As a consequence, the original models become unattractive for research by simulation. In these cases, the models can be transformed to equivalent models which are distorted but exploitable. The use of stochastic distorted models is also recommended for the models based on stochastic chains or polystocastic processes where an asymptotic behaviour is identified with respect to a process transition matrix of probabilities, process chains evolution, process states connection, etc. The distorted models are also of interest when the stochastic process is not time dependent, as, for example, in the stochastic movement of a marked particle occurring with a constant velocity vector, like in diffusion processes.

The diffusion model can usually be used for the description of many stochastic distorted models. The equivalent transformation of a stochastic model to its associated diffusion model is fashioned by means of some limit theorems. The first class of limit theorems show the asymptotic transformation of stochastic models based on polystochastic chains; the second class is oriented for the transformation of stochastic models based on a polystochastic process and the third class is carried out for models based on differential stochastic equations.

4.4.1.1 Stochastic Models Based on Asymptotic Polystochastic Chains
We begin the discussion by referring to the stochastic model given by relation (4.58), which is rewritten here as shown in relation (4.120). Here for a finite Markov connection process we must consider the constant time values for all the elements of the matrix $P = [p_{ik}]_{i,k \in K}$.

$$P_k(n, i) = \sum_{e \in K} \sum_{a \in Z} P_k(n - 1, i - a) p_{ek} p_k(a) \tag{4.120}$$

The interest is to produce a model for the computation of the probability to have the particle at the i position after n passages. This probability, which is denoted as $P(n, i)$, can be calculated by summing up all $P_k(n, i)$. So, $P(n, i) = \sum_k P_k(n, i)$.

Now, if we consider that the connection in our stochastic model is given by a Markov chain, which presents the quality to be a regular chain, then we can show that there exists a n_0 value, where the matrix P contains values which are constant and positive. Indeed we can write: $P^{n_0} \succ 0$. In this case, for the matrix connections, we reach the situation of $\lim_{n \to \infty} P^n = \Pi$, where Π is a stable stochastic matrix having identical lines. It is not difficult to observe that the elements of the stable matrix Π result from the product of the unity matrix I and the vector $V_\Pi{}'$, which is a transposition of the vector that contains the unchangeable transition probabilities from one state to another (V_Π). At the same time, vector V_Π has the quality to be the proper vector of the matrix of probabilities P and, consequently, its elements are the solution of the linear algebraic equation system: $\sum_{e \in K} \pi_e p_{ek} = \pi_k, k \in K$. The π_k substitution by p_{ek} in Eq. (4.120) gives the asymptotic model (4.121). Relation (4.122), where $P^{as}(n, i)$ is the result of the addition of probabilities $P_k^{(as)}(n, i)$, allows the calculation of the probability to have the particle in state i after n time sequences.

$$P_k^{as}(n, i) = \sum_{e \in K} \sum_{a \in Z} P_k^{as}(n - 1, i - a)\pi_k p_k(a) \tag{4.121}$$

$$P^{as}(n, i) = \sum_{k \in K} \pi_k \sum_{a \in Z} P^{as}(n - 1, i - a)p_{ek} p_k(a) \tag{4.122}$$

The model described by Eq. (4.122) is known as *the generalized random displacement or generalized random walk*.

Relation (4.123) is obtained when the model relation (4.122) is written for the case of a stochastic process with two states and constant length of the particle displacement (this model was previously introduced with relation (4.59)).

$$P^{as}(n, i) = \pi_1 P^{as}(n - 1, i - 1) + \pi_2 P^{as}(n - 1, i + 1) \tag{4.123}$$

With $\pi_1 = \pi_2 = 1/2$ we observe that relation (4.123) has the same form as the relation used for the numerical solving of the unsteady state diffusion of one species or the famous Schmidt relation. The model described by Eq. (4.123) is known as the *random walk with unitary time evolution*.

In order to identify the conditions that allow an asymptotic transformation, we show a short analysis particularized to the case of the model given by the assembly of relations (4.59). To this aim, we focus the observations on one property of a generator function which is defined as a function which gives the following equation for the probabilities of the distribution with the general discrete values $P_k(n, i)$:

$$G_k(n, z) = \sum_{i \in Z} P_k(n, i)z^i \quad , \quad k = 1, 2, \dots \quad , \quad z = e^i \tag{4.123}$$

If we particularize this last relation for the stochastic model given by the assembly of equations (4.59), we obtain the following relation for the vector $G(n.z)$:

$$G(n, z) = [G_1(n, z), G_2(n, z)] = G(0, z) \left[P \begin{pmatrix} z & 0 \\ 0 & 1/z \end{pmatrix} \right]^n \tag{4.123}$$

Here $P = [p_{ek}]_{e,k \in K=(1,2)}$ represents the matrix of the transition probabilities between both states of the process.

If we accept that the $\lim_{n \to \infty} P^n = \Pi$, then, using relation (4.123) we derive the equation of the asymptotic generator function:

$$G^{as}(n, z) = G(0, z) \left(\pi_1 z + \frac{\pi_2}{z} \right)^n \tag{4.124}$$

By computing the values of the generator function for $\theta \to 0$ (relations (4.123) and (4.124)), we can observe similarities (identities) between both relations. Indeed, we corroborate that these functions come from a process with identical behaviour and we have a correct asymptotic transformation of the original model. We can conclude that in the case when the transition matrix of probabilities has a regular state, the generator function of the polystochastic chain process when $n \to \infty$ goes from one generator function to a Markov chain related with the model that is, for the present discussion, characterized by relation (4.123)

All other discrete stochastic models, obtained from polystochastic chains, attached to an investigated process, present the capacity to be transformed into an asymptotic model. When the original and its asymptotic model are calculated numerically, we can rapidly observe if they converge by direct simulation. In this case, the comparison between the behaviour of the original model and the generator function of the asymptotic stochastic model is not necessary.

4.4.1.2 Stochastic Models Based on Asymptotic Polystochastic Processes

For the derivation of one asymptotic variant of a given polystochastic model of a process, we can use the perturbation method. For this transformation, a new time variable is introduced into the stochastic model and then we analyze its behaviour. The new time variable is $\tau' = \varepsilon^r t$, which includes the time evolution t and an arbitrary parameter ε, which allows the observation of the model behaviour when its values become very small ($\varepsilon \to 0$). Here, we study the changes in the operator $O(\tau, t)$ when $\varepsilon \to 0$ whilst paying attention to having stable values for t/ε or t/ε^2.

Two different types of asymptotic transformation methods can be used depending on the ratio of t/ε used: in the first type we operate with fixed values of t/ε whereas in the second type we consider t/ε^2.

As an example, we show the equation that characterizes a random evolution (see relation (4.90)) written without the arguments for the operator $O(\tau, t)$, but developed with the operator $V(X(\tau))$. We also consider that, when the random process changes, the operator $O(\tau, t)$ will be represented by an identity operator ($I = I(\tau, t)$):

$$\frac{dO}{d\tau} = VO = (\varepsilon V_1(\tau) + \varepsilon^2 V_2(\tau) + \varepsilon^3 V_3(X, \tau)O \quad ; \quad O(\tau, \tau) = I \tag{4.125}$$

Each operator considered in the total operator $V(X(\tau), \tau)$ keeps its own mean action when it is applied to one parameter (for example the mean action of operator V_1 on the parameter (function) f will be written as follows:

$$\overline{V}_1 f = \lim_{u \to \infty} \frac{1}{u} \int_t^{t+u} E\{V_1(f(t))dt\}.$$

The introduction of terms of higer order in Eq. (4.125) is not necessary as far as, in the characterization of chemical engineering processes, the differential equations are limited to equations of order two.

Some restrictions are imposed when we start the application of limit theorems to the transformation of a stochastic model into its asymptotic form. The most important restriction is given by the rule where the past and future of the stochastic processes are mixed. In this rule it is considered that the probability that a fact or event C occurs will depend on the difference between the current process $(P(C) = P(X(\tau) \in A/V(X(\tau)))$ and the preceding process $(P_\tau(C/e))$. Indeed, if, for the values of the group (τ, e), we compute $\pi_\tau = \max[P_\tau(C/e) - P(C)]$, then we have a measure of the influence of the process history on the future of the process evolution. Here, τ defines the beginning of a new random process evolution and π_τ gives the combination between the past and the future of the investigated process. If a Markov connection process is homogenous with respect to time, we have $\pi_\tau = 1$ or $\pi_\tau \to 0$ after an exponential evolution. If $\pi_\tau \to 0$ when τ increases, the influence of the history on the process evolution decreases rapidly and then we can apply the first type limit theorems to transform the model into an asymptotic model. On the contrary, if $I = \int_0^\infty \pi_\tau^{1/2} d\tau$, the asymptotic transformation of an original stochastic model can be carried out by a second-type limit theorem.

For the example considered above (Eq. (4.125)), the mean value of the random evolution at time t is $e_m = e_m(t, X) = E_X[O(0, X)]$ and this process parameter verifies Eq. (4.126). Here, Q is the infinitesimal generator that characterizes the connection processes of the stochastic model of the process. This property of $e_m = e_m(t, X) = E_X[O(0, X)]$ is a consequence of relation (4.94). So we can write:

$$\frac{de_m}{d\tau} = (\varepsilon V_1(t) + \varepsilon^2 V_2(t) + \varepsilon^3 V_3(t, X))e_m + Q e_m \tag{4.126}$$

In Eq. (4.126) we can change variable $\tau = t/\varepsilon$ in order to obtain a limit transformation after the first type theorem. The result is:

$$\frac{de_m}{d\tau} = (V_1(\tau/\varepsilon) + \varepsilon V_2(\tau/\varepsilon) + \varepsilon^2 V_3(\tau/\varepsilon, X))e_m + \frac{1}{\varepsilon} Q e_m \tag{4.127}$$

If the stochastic evolution $X(\tau, e)$ complies with the mixing condition $(\lim_{\tau \to \infty} \pi_\tau = 0)$ then, if $\varepsilon \to 0$; $O(0, \tau/\varepsilon)$ becomes $\tau \overline{V}_1$ through a probabilistic way. This shows that e_m, which is the solution of the differential equation (4.126), becomes e_m^{as} when $\varepsilon \to 0$ for a fixed τ/ε:

$$\frac{de_m^{as}}{d\tau} = \overline{V}_1 e_m^{as} ; \quad e_m^{as}(0, X) = I \tag{4.128}$$

Considering this last mathematical derivation, we observe that the stochastic process has been distorted by another one with a similar behaviour. In order to explain the meaning of $\overline{V}_1$ we consider the case of a connection between the two states of a stochastic process with the following infinitesimal generator:

$$Q = \begin{bmatrix} -q_1 & q_1 \\ q_2 & -q_2 \end{bmatrix}$$

The invariable measure for the infinitesimal generator is a stable matrix that complies with Eq. (4.17) and the following conditions: $PQ_s = Q_sP = O_s$, $OQ_s = Q_sO = 0$, $Q_s^2 = Q_s$. Here, P is the matrix of transition probabilities. For our considered case (stochastic process with two states) we obtain $Q_s = \begin{bmatrix} \dfrac{q_2}{q_1 + q_2} & \dfrac{q_1}{q_1 + q_2} \end{bmatrix}$; consequently, $\overline{V}_1$ can be written as: $\overline{V}_1 = \dfrac{q_2 V_1(1) + q_1 V_1(2)}{q_1 + q_2}$.

If we continue with the particularization of the two-state stochastic process, by considering that the first state is the diffusion type and the second state concerns convection (for instance see relations (4.72), (4.73), (4.79), (4.98) and (4.100)), then the equation system (4.127) can be written as follows:

$$\frac{de_{m1}}{d\tau} = \frac{d^2 e_{m1}}{dz^2} - \frac{q_1 e_{m1}}{\varepsilon} + \frac{q_1 e_{m2}}{\varepsilon}$$

$$\frac{de_{m1}}{d\tau} = \frac{de_{m1}}{dz} - \frac{q_2 e_{m1}}{\varepsilon} - \frac{q_2 e_{m2}}{\varepsilon}$$

$$e_{m1}(0, z) = e_{m2}(0, z) = f(z) \tag{4.129}$$

Looking at this assembly of equations and relation (4.127) simultaneously, we can easily identify that $V_1(1) = d^2/dz^2$ (i.e. is an elliptic operator), $V_1(2) = d/dz$, $V_2 = 0$, $V_3 = 0$. With these identifications and in accordance with the transformation theorem of the first type (Eq. (4.128)) when $\varepsilon \to 0$, e_{m1} and e_{m2} will be solution of following equation:

$$\frac{dv}{d\tau} = \frac{q_2}{q_1 + q_2} \frac{d^2 v}{dz^2} + \frac{q_1}{q_1 + q_2} \frac{dv}{dz} \quad ; \quad v(0, z) = f(z) \tag{4.130}$$

The condition $v(0, z) = f(z)$ (see relation (4.130)) corresponds to the situation when we have $e_{m1}(0, z) = e_{m2}(0, z) = f(z)$; otherwise, we use $v(0, z)$ as:

$$v(0, z) = \frac{q_2 f_1(z) + q_1 f_2(z)}{q_1 + q_2}.$$

This shows that the invariable measure determining the mixing procedures of stochastic process states is extended over the initial conditions of the process.

If we obtain $\overline{V}_1 = \dfrac{q_2 V_1(1) + q_1 V_1(2)}{q_1 + q_2} = 0$ for an experiment, we can conclude that the use of the theorems for the first type transformation is not satisfactory.

Then we have to apply the asymptotic model transformation by using the second type theorems. To do so, we choose the new time variable $\tau = \varepsilon^2 t$ and the relation (4.127) becomes:

$$\frac{de_m}{d\tau} = (\frac{1}{\varepsilon}V_1(\tau/\varepsilon^2) + V_2(\tau/\varepsilon^2) + \varepsilon V_3(\tau/\varepsilon^2, X))e_m + \frac{1}{\varepsilon^2}Qe_m \tag{4.131}$$

The theorems for the two-type transformation are based on the observation that, for $\varepsilon \to 0$ and fixed τ/ε^2, we have the operator $O\left(0, \frac{\tau}{\varepsilon^2}\right) \to \exp(\tau\overline{V})$, where $\overline{V} = \overline{V}_2 + \overline{V}_{11}$. Indeed, the mean value of the stochastic process from relation (4.131), noted as ε_m, becomes v, which is the solution of the differential equation $\frac{dv}{d\tau} = \overline{V}v$; $v(0) = I$. For a stochastic process with connections between two states the infinitesimal generator of connection is $Q_s = \begin{bmatrix} \frac{q_2}{q_1+q_2} & \frac{q_1}{q_1+q_2} \end{bmatrix}$, here $\overline{V}_2$ and $\overline{V}_1$ (and then $\overline{V}$) are given by the following equations: $\overline{V}_2 = \frac{q_2V_2(1) + q_1V_2(2)}{q_1+q_2}, \overline{V}_1 = \frac{-2V(1)V_2(2)_1}{q_1+q_2}$.

Now, we have to identify $\overline{V}_2$ and $\overline{V}_{11}$. To do so, we consider the case of two connected stochastic processes where each process is a diffusion type with two states. The example concerns one marked particle that is subjected to a two-state diffusion displacement. The particle can be considered as a molecular species (so the particle movement describes a mass transport process) and we can also take into account the total enthalpy of the process (heat transport process). This particular case of stochastic model, can be described with the assembly of relations (4.79). In the model, the mean probability of the existence of local species (e_{m1}) and the mean probability of the existence of local enthalpy (e_{m2}) are given by the assembly of relations (4.132):

$$\frac{\partial e_{m1}}{\partial \tau} = \frac{g(z)}{\varepsilon}\frac{\partial e_{m1}}{\partial z} + \alpha 1(z)\frac{\partial^2 e_{m1}}{\partial z^2} - \frac{q}{2}e_{m1} + \frac{q}{2}e_{m2}$$

$$\frac{\partial e_{m2}}{\partial \tau} = \frac{g(z)}{\varepsilon}\frac{\partial e_{m2}}{\partial z} + \alpha 2(z)\frac{\partial^2 e_{m2}}{\partial z^2} - \frac{q}{2}e_{m2} + \frac{q}{2}e_{m1}$$

$$e_{m1}(0, z) = f_1(z) \quad ; \quad e_{m2}(0, z) = f_2(z) \tag{4.132}$$

If, in the assembly of equations, we consider $q_1 = q_2 = q$ in the equation of the infinitesimal generator then, we can identify $V_1(1) = -V_1(2) = g(z)\frac{\partial}{\partial z}$, $V_2(1) = \alpha 1(z)\frac{\partial^2}{\partial z^2}$, $V_2(2) = \alpha 2(z)\frac{\partial^2}{\partial z^2}$, $V_3 = 0$. After the theorem of the two-type transformation, the solutions for e_{m1} and e_{m2} will tend towards the solution of the following particularization of the asymptotic model $\frac{dv}{d\tau} = \overline{V}v$; $v(0) = I$:

$$\frac{\partial v}{\partial \tau} = \frac{1}{2}\frac{\partial}{\partial z}\left(g(z)\frac{\partial v}{\partial z}\right) + \frac{a1(z) + a2(z)}{2}\frac{\partial^2 v}{\partial z^2} = \overline{V}_{11}v + \overline{V}_2 v \qquad (4.133)$$

$$v(0, z) = \frac{f_1(z)}{2} + \frac{f_2(z)}{2} \qquad (4.134)$$

To complete this short analysis, we can conclude that, for the asymptotic transformation of a stochastic model, we must identify: (i) the infinitesimal generator; (ii) what type of theorem will be used for the transformation procedure.

4.4.1.3 Asymptotic Models Derived from Stochastic Models with Differential Equations

Studies of the transformation of a stochastic model characterized by an assembly of differential equations to its corresponding asymptotic form, show that the use of a perturbation method, where we replace the variable t by: $t = \varepsilon^r \tau$, can be recommended without any restrictions [4.47, 4.48].

If we consider a process where the elementary states $v_1, v_2, ...v_N$ work with a Markov connection, this connection presents an associated generator of probability $(p_1, p_2, ...p_N)$ that verifies the invariable measure $\sum_{i=1}^{N} p_i v_i = 0$. Now, if the elementary states are represented by displacements with constant speed, then $X(\tau)$ can take scalar values and, consequently, $F(X, v_i) = v_i$. For this case, we consider that the mean values of $X(\tau)$, determined by their X_0 initial values and noted as $e_{mi}, i = 1, N$, verify relations (4.135) (see for instance Eq. (4.98)). The equations (4.135) consider that the displacement associated with space X occurs after the z direction; consequently we have:

$$\frac{\partial e_{mi}}{\partial \tau} = \sum_{j=1}^{N} q_{ij} e_{mj} + v_i \frac{\partial e_{mi}}{\partial z} \quad ; \quad e_{mi}(0, X) = f_i(z) \quad ; \quad i = 1...N \qquad (4.135)$$

Here, we use the classical perturbation procedure ($\varepsilon \to 0$ when $\tau \to \infty$) for the analysis of the asymptotic behaviour of mean values $e_{mi}(\tau, X) = e_{mi}(\tau, z)$. The following expression can be written for $e_{mi}(\tau, X)$, when we use the perturbation $t = \tau\varepsilon^2$:

$$e_{mi}^* = e_{mi}^*(\tau, X) = E_i\left\{ (f_v(\tau/\varepsilon^2)(X + \varepsilon \int_0^{\tau/\varepsilon^2} v(a)da) \right\} \qquad (4.136)$$

For the expected mean values $e_{mi}^*(\tau, X)$, as a result of the application of the time perturbation to the system (4.135), we derive the following differential equations system:

$$\frac{\partial e_{mi}^*}{\partial \tau} = \frac{1}{\varepsilon^2}\sum_{j=1}^{N} q_{ij} e_{mi}^* + \frac{1}{\varepsilon^2} v_i \frac{\partial e_{mi}^*}{\partial z} \quad ; \quad e_{mi}^*(0, X) = e_{mi}^*(0, z) = f_i(z) \quad ; \quad i = 1...N \quad (4.137)$$

Now, we can write Eq. (4.136) as:

$$e^*_{mi} = e^*_{mi}(\tau, X) = E_i\left\{(f_v(\tau/\varepsilon^2)(X + \sqrt{\tau}\frac{1}{\sqrt{\tau}}\int_0^\tau v(\alpha)d\alpha)\right\} \tag{4.138}$$

For our considered process, where the states $v_1, v_2, ...v_N$ are Markov connected, the variable term for this last relation $\left(\frac{1}{\sqrt{\tau}}\int_0^\tau v(\alpha)d\alpha\right)$ tends [4.5] towards a normal random variable with a zero mean value and variance $\sigma = \frac{1}{\tau}\int_0^\tau\int_0^\tau E\{v(\alpha)v'(\alpha')\}d\alpha d\alpha'$. Coupling this observation with relation (4.138) results in:

$$\lim_{\tau \to \infty} e^*_{mi}(\tau, X) = \int_{-\infty}^\infty \left\{\sum_{i=1}^N p_i f_i(z + \sqrt{\tau})\right\}\frac{e^{-\xi^2/2\tau^2}}{\sqrt{2\pi\sigma}}d\xi \tag{4.139}$$

Indeed, for $\varepsilon \to 0$ and $0 < \tau < \tau_0$, the solution of the system (4.137) with respect to $e^*_{mi}(\tau, X)$ will be uniformly displaced with respect to X (or z when the movement occurs along this direction) and the solution can be written as:

$$\frac{\partial v^0}{\partial \tau} = \frac{1}{2}\sigma^2\frac{\partial^2 v^0}{\partial z^2} \quad ; \quad v^0(0, X) = v^0(0, z) = \sum_{i=1}^N p_i f_i(z) \tag{4.140}$$

This last equation has a form similar to the famous equation of the single direction diffusion of a property in an unsteady state, the property here being the local concentration v^0. The diffusion coefficient is represented by the variance of the elementary speeds which are given by their individual states $v_1, v_2, ..., v_N$. It is important to notice the consistency of the definition of the diffusion coefficient.

Very difficult problems occur with the asymptotic transformation of original stochastic models based on stochastic differential equations where the elementary states are not Markov connected. This fact will be discussed later in this chapter (for instance see the discussion of Eq. (4.180)).

4.4.2
Numerical Methods for Solving Stochastic Models

In Section 4.2 we have shown that stochastic models present a good adaptability to numerical solving. In the opening line we asserted that it is not difficult to observe the simplicity of the numerical transposition of the models based on polystochastic chains (see Section 4.1.1). As far as recursion equations describe the model, the numerical transposition of these equations can be written directly, without any special preparatives.

When a stochastic model is described by a continuous polystochastic process, the numerical transposition can be derived by the classical procedure that change the derivates to their discrete numerical expressions related with a space discretisation of the variables. An indirect method can be used with the recursion equations, which give the links between the elementary states of the process.

The following examples detail the numerical transposition of some stochastic models. The numerical state of a stochastic model allows the process simulation.

Indeed, we can easily produce the evolution of the outputs of the process when the univocity conditions and parameters of the process are correctly chosen.

The first example concerns a stochastic model which is known as the Chapmann–Kolmogorov model and is mathematically characterized by Eq. (4.67). This model accepts a numerical solution that is developed using implicit methods and then computation begins with the corresponding initial and boundary conditions. If we particularize the Chapmann–Kolmogorov model to the situation where we have two elementary states of the process with constant displacement steps and without an i position dependence of the transition probabilities ($a = 1$, $p_k(a) = 1$ and $p_{ek}(i-a) = p_{ek}$) then, we obtain the model (4.59). Figure 4.10 presents its numerical structure. Many researchers have been using programs of this type to characterize the transport of species through various zeolites [4.49–4.52].

1	**Definition** :matrix $P_1(N,M)$,matrix $P_2(N,M)$/ N-passages ,M-positions/
2	**Data**: $p_{11}=$, $p_{21}=$ $p_{12}=$ $p_{22}=$ N= M=
3	**Univocity**: $P_1(1,1)=1$;$P_1(I+1,1)=0$ for I=1,N-1;$P_2(I,1)=0$ for I=1,N-1
4	**Start**
5	I=2
6	J=2
7	$P_1(I,J)=p_{11}P_1(I-1,J-1)+p_{21}P_2(I-1,J-1)$
8	$P_2(I,J)=p_{12}P_1(I-1,J-1)+p_{22}P_2(I-1,J-1)$
9	for J<M
10	**Write**:$P_1(I,J)$,$P_2(I,J)$
11	J=J+1
12	**Back to** 7
13	for J>M and I<N
14	**Write**:$P_1(I,J)$,$P_2(I,J)$
15	I=I+1
16	**Back to** 6
17	**Data treatment**: $P_1(I,J)$,$P_2(I,J)$/ Graphiques;other calculations..:
18	**Stop**

Figure 4.10 Numerical text of the stochastic model given in Eq. (4.59).

The second example discusses the numerical transposition of the asymptotic models based on polystochastic chains (see Section 4.4.1.1) where to compute the limit transition probabilities, we must solve the system $\sum_{e \in K} \pi_e p_{ek} = \pi_k$ $k = 0, 1, ..N$. If the number of process components, here noted as k, is greater than two, then we can use a successive approximation method for the estimation of the column vector Π. More precisely, we use the iteration chain $P\Pi^{(m+1)} = \Pi^{(m)}$ with the stop condition $\|\Pi^{(m)} - \Pi^{(m+1)}\| \leq M\lambda_2$. The determinant value λ_2 from the stop condition represents the second decreasing proper value of the transition probabilities matrix (P). In the stop condition, M is considered as an arbitrary constant value.

The third example presents the problem of numerical transposition of continuous stochastic models, which is introduced by the following general equation:

$$\frac{\partial P}{\partial \tau} + A\frac{\partial P}{\partial z} = BP + F \qquad (4.141)$$

where P is the vector that contains the probabilities $P_i(z, \tau)$ $i = 1, 2, ...N$;

A is the quadratic N×N matrix where the elements are constant numerical values; B the quadratic N×N matrix that contain functions of z and τ arguments; F the vector with elements defined by functions of z and τ arguments.

Before carrying out the discretization of the equations, we have to make a careful mathematical analysis of the problem in order to establish what its most convenient rewriting in order to facilitate the numerical solution. First we observe that between the matrix A, the proper values λ_j and the (left) proper vectors z_j, we have the equality $z_j A = \lambda_j z_j$. Consequently, as a result, the multiplication of Eq. (4.141) by z_j gives:

$$z_j \frac{\partial P}{\partial \tau} + z_j A \frac{\partial P}{\partial z} = z_j BP + z_j F \tag{4.142}$$

$$z_j \left(\frac{\partial P}{\partial \tau} + \lambda_j \frac{\partial P}{\partial z} - BP - F \right) = 0 \tag{4.143}$$

If the value of z_j is not zero, then relation (4.144) becomes:

$$\frac{\partial P}{\partial \tau} + \lambda_j \frac{\partial P}{\partial z} = BP + F \tag{4.144}$$

The left term of this last equation represents the differential state of vector P with respect to time for the family of curves $dz/d\tau = \lambda_j^{-1}, j = 1, 2, ...N$. So we can write relation (4.144) as:

$$\left(\frac{dP}{d\tau} \right)_{\lambda_j} = BP + F \tag{4.145}$$

where, for $\left(\frac{dP}{d\tau} \right)_{\lambda_j}$, we define the differential state of P after the normal curves $dz/d\tau = \lambda_j^{-1}$. The transformation given above, is still valid when all values $\lambda_j, j = 1, ...N$ are real and strictly different. However, if the A matrix gives complex values for some λ_j, then we can assert that our original model (described by Eq. (4.141)) is not a hyperbolic model. At the same time, the proper values of the matrix $A(\lambda_j, j = 1, ...N)$ give important information for fixing univocity conditions and solving the model. The following situations are frequently presented:

- when all λ_j verify that $\lambda_j \succ 0$, $j = 1, ...N$, we can specify the initial values $P(z)$, $z \succ 0$ and the boundary values at each time: $P(z_f, \tau)$, $\tau \succ 0$. The values $P(0, \tau)$ and $P(z_e, \tau)$ will be specified when the boundaries of the process are $z = 0$ and $z = z_e$;
- when all λ_j verify that $\lambda_j \prec 0$, $j = 1, ...N$ and when values $P(z_f, \tau)$ at the boundary line $z_f = 0$ are needed, we must specify the initial values of the probabilities for $z \le 0$;
- for positive and negative values λ_j, $j = 1, ...N$, we separate two domains with their respective univocity problems; when we have the following specifications $0 \prec z \prec z_e$, $\lambda_j \succ 0$, $j = 1, ...l$ with $1 \prec N$ and $P(z, 0) = f(z)$ simultaneously, we must complete the

univocity problem with functions or data for $P(z,0)$, $P(0,\tau)$ and $P(z_e,\tau)$, respectively.

After the establishment of the univocity conditions, we can begin the numerical treatment of the model. For this purpose, we can use the simplified model form (see relations (4.145)) or its original form (4.141)).

The change of a continuous polystochastic model into its numerical form is carried out using the model described by Eq. (4.71) rewritten in Eq. (4.146). The solution of this model must cover the variable domain $0 \prec z \prec z_e$, $0 \prec \tau \prec T$. In accordance with the previous discussion, the following univocity conditions must be attached to this stochastic model. Here, $f_k(z)$, $g_k(\tau)$ and $h(\tau)$ are functions that must be specified.

$$\frac{\partial P_k(z,\tau)}{\partial \tau} + v_k \frac{\partial P_k(z,\tau)}{\partial z} + \left(\sum_{j=1, j \neq k} \alpha_{kj} \right) P_k(z,\tau) - \sum_{j=1, j \neq k} \alpha_{jk} P_j(z,k) \quad ; \quad k = 1, N \tag{4.146}$$

$$\tau = 0 \quad , \quad 0 < z < z_e, \quad P_k(z,0) = f_k(z)$$

$$\tau > 0 \quad , \quad z = 0 \quad , \quad \begin{cases} P_k(0,\tau) = g_k(\tau) \text{ for } v_k(0,\tau) > 0 \\ \dfrac{dP_k(0,\tau)}{dz} = 0 \text{ for } v_k(0,\tau) < 0 \end{cases}$$

$$\tau > 0 \quad , \quad z = z_e \quad , \quad \begin{cases} P_k(z_e,\tau) = h_k(\tau) \text{ for } v_k(0,\tau) < 0 \\ \dfrac{dP_k(z_e,\tau)}{dz} = 0 \text{ for } v_k(0,\tau) > 0 \end{cases} \tag{4.147}$$

Concerning the boundary conditions of this problem, we can have various situations: (i) in the first situation, the probabilities are null but not the probability gradients at $z = 0$ zero. For example, for a negative speed $v_k(0,\tau)$, the particle is not in the stochastic space of displacement. However, at $z = 0$, we have a maximum probability for the output of the particle from the stochastic displacement space. Indeed, the flux of the characteristic probability must be a maximum and, consequently, $dP_k(0,\tau)/dz = 0$; (ii) we have a similar situation at $z = z_e$; (iii) in other situations we can have uniformly distributed probabilities at the input in the stochastic displacement space; then we can write the following expression:

$$g_k(\tau) = P_k(0,\tau) = p_k \text{ for } k = 1, N - 1 \;, \quad g_N(\tau) = P_N(0,\tau) = 1 - \sum_{k=1}^{N-1} p_k \;.$$

It is important to notice that the univocity conditions must adequately correspond to the process reality. Concerning the numerical discretisation of each variable space, model (4.146) gives the following assembly of numerical relations:

$$z = i * \Delta z \quad , \quad \tau = g * \Delta \tau \;, \quad i = 0, r \;; \quad g = 0, s \tag{4.148}$$

- For $v_k(z,\tau) > 0$

$$\frac{P_k(z,\tau + \Delta\tau) - P_k(z,\tau)}{\Delta\tau} + v_k(z,\tau)\frac{P_k(z + \Delta z,\tau) - P_k(z,\tau)}{\Delta z} +$$

$$\left(\sum_{j=1,j\neq k} \alpha_{kj}\right) P_k(z,\tau) - \sum_{j=1,j\neq n} \alpha_{jk} P_j(z,\tau) = 0$$

- For $v_k(z,\tau) < 0$

$$\frac{P_k(z,\tau + \Delta\tau) - P_k(z,\tau)}{\Delta\tau} + v_k(z,\tau)\frac{P_k(z,\tau) - P_k(z - \Delta z,\tau)}{\Delta z} +$$

$$\left(\sum_{j=1,j\neq k} \alpha_{kj}\right) P_k(z,\tau) - \sum_{j=1,j\neq n} \alpha_{jk} P_j(z,\tau) = 0$$

The balance between the unknown variables and the relations available for their estimation is given here: (i) for $T/\tau = s$ and $z_e/z = r$ we obtain $r * s * N$ unknowns (for each solving network point we must determine the values of $P_1(z,\tau)$, $P_2(z,\tau)....P_N(z,\tau)$); (ii) the system of equations to compute unknowns is made considering the particularization of:
- the relation (4.148) for all network points that are not in the boundaries; it gives a total of $(r - 2) * s * N$ equations.
- the second condition from the univocity problem of the model (4.146)–(4.147); this particularization gives $P_k(0,1),....P_k(0,s)$ so $s * N$ equations;
- the third condition from the univocity problem of the model (4.146)–(4.147); this particularization gives $s*N$ equations.

The algorithm to compute a stochastic model with two Markov connected elementary states is shown in Fig. 4.11. Here, the process state evolves with constant v_1 and v_2 speeds. This model is a particularization of the model commented above (see the assembly of relations (4.146)–(4.147)) and has the following mathematical expression:

$$\frac{\partial P_1(z,\tau)}{\partial\tau} + v_1\frac{\partial P_1(z,\tau)}{\partial z} + \alpha P_1(z,\tau) - \beta P_2(z,\tau) = 0$$

$$\frac{\partial P_2(z,\tau)}{\partial\tau} - v_2\frac{\partial P_1(z,\tau)}{\partial z} + \beta P_2(z,\tau) - \alpha P_1(z,\tau) = 0$$

$$P_1(z,0) = 0 \quad ; \quad P_2(z,0) = 0$$

$$P_1(0,0) = 1 \quad ; \quad P_1(0,\tau) = 0 \quad ; \quad P_2(0,\tau) = P_2(\Delta z,\tau)$$

$$P_2(z_e,\tau) = 0 \quad ; \quad P_1(z_e,\tau) = P_1(z_e - \Delta z,\tau) \tag{4.149}$$

It can easily be observed that the considered case (4.149) corresponds to the situation for $\tau = 0$: only one marked particle evolves through a stochastic trajectory (a type 1 displacement with v_1 speed). This example corresponds to a Dirac type input and the model output response or the sum $P_1(z_e, \tau) + P_2(0, \tau)$, represents the distribution function of the residence time during the trajectory (see also application 4.3.1).

1   **Definition** : Matrix $P_1(r,s)$; Matrix $P_2(r,s)$
2   **Data**: r= , s= , v_1= , v_2= , α= , β= , $\Delta\tau$= , Δz=
3   **Initial conditions** : $P_1(j,0)$=0 , $P_2(j,0)$=0 for j=1,r
4   **Boundary conditions** z=0 : $P_1(0,0)$=1, $P_1(0,g)$=0 for g=1,s
5   **Boundary conditions** z=z_e: $P_2(r,h)$=0 for h=0,s
6   i=1
7   **System solution**:

         J=1,r
         $P_2(0,i)$-$P_2(1,i)$=0
         $(P_1(j,i)-P_1(j,i-1))/\Delta\tau+v_1(P_1(j,i)-P_1(j-1,i))/\Delta z+\alpha P_1(j,i)-\beta P_2(j,i)$=0
         $(P_2(j,i)-P_2(j,i-1))/\Delta\tau-v_2(P_2(j,i)-P_2(j-1,i))/\Delta z-\alpha P_1(j,i)+\beta P_2(j,i)$=0
         $P_1(r,i)$-$P_1(r-1,i)$=0
8   **Write and transfer** through data processing: $P_1(j,i),P_2(j,i)$
9   For i<s
10 i=i+1
11 **Back** to 7
12 For i>s
13 **Stop**

Figure 4.11 Numerical text of the stochastic model given by Eq. (4.149).

4.4.3
The Solution of Stochastic Models with Analytical Methods

Examples 4.2 and 4.3 and the models from Section 4.4 show that the stochastic models can frequently be described mathematically by an assembly of differential partial equations.

The core of a continuous stochastic model can be written as Eq. (4.150). Here, $P(z,\tau)$ and a(z), b(z), c(z) are quadratic matrices and L is one linear operator with action on the matrix $P(z,\tau)$. In the mentioned equation, f(z) is a vector with a length equal to the matrix $P(z,\tau)$. In this model, z can be extended to a two- or a three-dimensional displacement:

$$\frac{\partial P(z,\tau)}{\partial \tau} + a(z)\frac{\partial P(z,\tau)}{\partial z} + b(z)\frac{\partial^2 P(z,\tau)}{\partial z^2} + c(z)L(P(z,\tau)) = f(z) \qquad (4.150)$$

This mathematical model has to be completed with realistic univocity conditions. In the literature, a large group of stochastic models derived from the model described above (4.150), have already been solved analytically. So, when we have a new model, we must first compare it to a known model with an analytical solution so as to identify it. If we cannot produce a correct identification, then we must analyze it so as to determine whether we can obtain an analytical solution. In both situations, we have to carry out different permissive but accepted manipulations of the original model:

- all algebraic transformations are accepted, especially the combination of model variables;
- model transformation with dimensionless variables and parameters;
- all integral transformations that produce a model and univocity conditions similar to those given in a problem with an analytical solution.

The analysis of the univocity conditions attached to the model shows that, here, we have an unsteady model where nonsymmetrical conditions are dominant.

The analytical solution of the model imposes the use of integral transformation methods [4.53]. With the kernel $K(z,\mu)$, the finite integral transformation of the function $P(z,\tau)$ is the function $P^1(\mu,\tau)$, which is defined with the following relation:

$$P^1(\mu,\tau) = \int_0^{z_e} K(\mu,z)P(z,\tau)dz \tag{4.151}$$

The Laplace integral transformation, used in Section 4.3.1, allows the indentification of its kernel as $K(z,\mu) = K(z,s) = e^{-s\tau}$. It corresponds to the case when we produce a transformation with time. So, for this case, we particularize the relation (4.151) as:

$$P^1(z,s) = \int_0^\infty P(z,\tau)e^{-s\tau}d\tau \tag{4.152}$$

So as to show how we use the integral transformation in an actual case, we simplify the general model relation (4.150) and its attached univocity conditions to the following particular expressions:

$$\frac{\partial P(z,\tau)}{\partial \tau} = a\left(\frac{\partial^2 P(z,\tau)}{\partial z^2}\right) + F(z,\tau) \tag{4.153}$$

$$z = 0 , \tau > 0 , c_1 a\frac{dP}{dz} + a_1 P = 0$$

$$z = z_e , \tau > 0 , c_2 a\frac{dP}{dz} + a_2 P = 0$$

$$z > 0 , \tau = 0 , P = f_0(z) \tag{4.154}$$

In Eqs. (4.153) and (4.154), we recognize the general case of an unsteady state diffusion displacement in a solid body.

The particularization of relation (4.151) to Eqs. (4.153) and (4.154) results in one image of the original model:

$$\frac{\partial P^1(\mu, \tau)}{\partial \tau} = a\left(\frac{\partial P^1(\mu, \tau)}{\partial \mu}\right) + F^1(\mu, \tau) \tag{4.155}$$

$$z = 0, \tau > 0, (c_1 a + \alpha_1) P^1(\mu, \tau) = f_1(\tau) \int_0^{z_e} K(\mu, z) dz$$

$$z = z_e, \tau > 0, (c_2 a + \alpha_2) P^1(\mu, \tau) = f_2(\tau) \int_0^{z_e} K(\mu, z) dz$$

$$z > 0, \tau = 0, P^1(\mu, \tau) = \int_0^{z_e} f_0(z) K(\mu, z) dz = f_0^1(\mu) \tag{4.156}$$

It is easy to observe that the conditions from Eq. (4.154) have been completed with the evolution with time of the stochastic trajectory at the start ($z = 0$) and at the end ($z = z_e$) (for instance, note the presence of $f_1(\tau)$ and $f_2(\tau)$ inside the assembly of conditions (4.156)). The image of the model has the analytical solution given by Eq. (4.157) [4.53]. The notations used here are specified thanks to relations (4.158) and (4.159):

$$P^1(\mu_n, \tau) = e^{-(a\mu_n^2/z_e^2)\tau}\left\{f_0^1(\mu_n) + \int_0^\tau e^{(a\mu_n^2/z_e^2)\tau} A(\mu_n, \tau) d\tau\right\} \tag{4.157}$$

$$A(\mu_n, \tau) = \frac{1}{c_1} F^1(\mu_n, \tau) + \left[\frac{K(\mu_n, z)}{c_1} f_1(\tau)\right]_{z=0} + \left[\frac{K(\mu_n, z)}{ac_2}\right]_{z=z_e} f_2(\tau) \tag{4.158}$$

$$F^1(\mu_n, \tau) = \int_0^{z_e} F(z, \tau) K(\mu_n, z/z_e) dz \tag{4.159}$$

The analytical solution for the original $P(z,\tau)$ is obtained with the inversion formula [4.53]. This solution is an infinite sum where the proper values μ_n, $n = 1,2,..\infty$ represent the summing parameters:

$$P(z, \tau) = \sum_{n=1}^\infty K(\mu_n, z/z_e) e^{-(a\mu_n^2/z_e^2)\tau}\left\{f_0^1(\mu_n) + \int_0^\tau e^{(a\mu_n^2/z_e^2)\tau} A(\mu_n, \tau) d\tau\right\} \tag{4.160}$$

Table 4.2 gives the kernels and characteristic equations for cases with various boundary conditions. Here h_1 and h_2 are defined by Fig. 4.12: $h_1 = \alpha_1/c_1 a$, $h_2 = \alpha_2/c_2 a$.

Table 4.2 The kernels and characteristic equations for the
stochastic model given by relations (4.153)–(4.154).

Univocity conditions		Kernel expression: $K(\mu_n, z/z_e)$	Equation for μ_n
$z = 0$	$z = z_e$		
$h_1 = \infty$	$h_2 = \infty$	$(\sqrt{2/z_e}) \sin \mu_n z/z_e$	$\sin(\mu) = 0$
$h_1 = \infty$	$h_2 = 0$	$(\sqrt{2/z_e}) \sin \mu_n z/z_e$	$\cos(\mu) = 0$
$h_1 = 0$	$h_2 = \infty$	$(\sqrt{2/z_e}) \cos \mu_n z/z_e$	$\sin(\mu) = 0$
$h_1 = 0$	$h_2 = 0$	$(\sqrt{2/z_e}) \cos \mu_n z/z_e$	$\sin(\mu) = 0$
$h_1 = 0$	$h_2 = ct$	$(\sqrt{2/z_e}) \left[\dfrac{\mu_n^2 + h_2^2 z_e^2}{\mu_n^2 + h_2^2 z_e^2 + h_2 z_e}\right]^{1/2} \cos(\mu_n z/z_e)$	$\mu \tan(\mu) = h_2 z_e$
$h_1 = ct$	$h_2 = \infty$	$(\sqrt{2/z_e}) \left[\dfrac{\mu_n^2 + h_1^2 z_e^2}{\mu_n^2 + h_1^2 z_e^2 + h_1 z_e}\right]^{1/2} \sin(\mu_n(1 - z/z_e))$	$\mu \cot(\mu) = -h_1 z_e$
$h_1 = ct$	$h_2 = 0$	$(\sqrt{2/z_e}) \left[\dfrac{\mu_n^2 + h_1^2 z_e^2}{\mu_n^2 + h_1^2 z_e^2 + h_1 z_e}\right]^{1/2} \cos(\mu_n(1 - z/z_e))$	$\mu \tan(\mu) = h_1 z_e$
$h_1 = ct$	$h_2 = ct$	$(\sqrt{2/z_e})*$ $\left[\dfrac{\mu_n \cos \mu_n z/z_e + h_1 z_e \sin \mu_n z/z_e}{[\mu_n^2 + h_1^2 z_e^2(1 + h_2 z_e/(\mu_n^2 + h_2^2 z_e^2) + h_1 z_e]^{1/2}}\right]$	$\tan(\mu) = \dfrac{\mu(h_1 + h_2)z_e}{\mu^2 - h_1 h_1 z_e^2}$

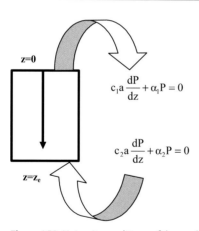

$$c_1 a \frac{dP}{dz} + \alpha_1 P = 0$$

$$c_2 a \frac{dP}{dz} + \alpha_2 P = 0$$

Figure 4.12 Univocity conditions of the model of diffusive and
unidirectional displacement (4.153).

We can illustrate this actual case by reaching an analytical solution if the following considerations are taken into account:
- $f_0(z) = 1$: this fact shows that we have uniformly distributed
  marked particles onto the displacement trajectory at the initial
  instant;

- $f_1(\tau) = f_2(\tau) = 0$ for $\tau \geq 0$ and $F(z, \tau) = 0$ for $0 \leq z \leq z_e$ and $\tau \geq 0$: we do not have any external intervention on boundaries, and it is not possible to have a generation of marked particles along the moving trajectory;
- $h_1 = 0$ and $h_2 = \infty$: the marked particle can leave the trajectory only at position $z = 0$.

In Table 4.2 our actual case can be identified to accept the kernel $K(\mu_n \, z/z_e) = (\sqrt{2/z_e}) \cos \mu_n z/z_e$ and $\sin(\mu) = 0$ as the characteristic equation for the proper values μ_n, $n = 1, \ldots \infty$. At the same time, it is identified that $A(\mu_n, \tau) = 0$ and, consequently, after a little modification, solution (4.160) becomes:

$$P(z, \tau) = \sum_{n=1}^{\infty} e^{-(a\mu_n^2/z_e^2)\tau} \cos(\mu_n z/z_e) \tag{4.161}$$

Figure 4.13 shows this dependence as a trend; here parameter Fo is recognized as the Fourier number (Fo $= a\tau/z_e^2$).

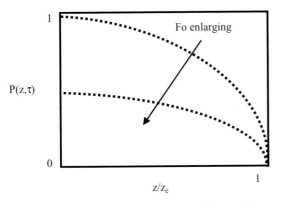

Figure 4.13 Time and space evolution of $P(z, \tau)$ (case of model (4.153)–(4.154)).

The model (4.140), which has been transformed, can easily accept this analytical solution. The desorption of one species from a saturated membrane, when the membrane surface respects the nonpermeable condition, can be described by this solution. Other cases seem to be more interesting, as for example, when the value of h_2 is constant and not null. In this last example, the conversion $P(z, \tau) = c(z, \tau)/c_0$ allows the calculation of the concentration field of mobile species through the membrane thickness.

The following section contains the particularization of the integral Laplace transformation for the case of the stochastic model given by the assembly of relations (4.146)–(4.147). This particularization illustrates how the Laplace transformation is used to solve partial differential equations. We start by applying the integral Laplace operator to all the terms of relation (4.146); the result is in:

$$\int_0^\infty e^{-s\tau}\left(\frac{\partial P_k(z,\tau)}{\partial\tau}+v_k\frac{\partial P_k(z,\tau)}{\partial\tau}+\left(\sum_{j=1,j\neq k}^N\alpha_{kj}\right)P_k(z,\tau)-\sum_{j=1,j\neq k}^N\alpha_{jk}P_j(z,\tau)\right)d\tau=0$$

$$(4.162)$$

The computing of the above integrals gives:

$$sP_k(z,s)-P_k(z,0)+v_k\frac{dP_k(z,s)}{dz}+\left(\sum_{j=1,j\neq k}^N\alpha_{kj}\right)P_k(z,s)-\sum_{j=1,j\neq k}^N\alpha_{jk}P_j(z,s)=0$$

$$(4.163)$$

This last result can be written as Eq. (4.164) and completed with the univocity conditions (4.165) resulting from the Laplace transformation of the original conditions written with relation (4.147):

$$v_k\frac{dP_k(z,s)}{dz}=-\left(s+\sum_{j=1,j\neq k}^N\alpha_{kj}\right)P_k(z,s)+\sum_{j=1,j\neq k}^N\alpha_{jk}P_j(z,s)+P_k(0)\qquad(4.164)$$

$$s\succ 0\;\;;\;\;z=0\;\;;\;\;\begin{cases}P_k(0,s)=\int_0^\infty e^{-s\tau}g_k(\tau)d\tau\;\text{ for }\;v_k\succ 0\\dP(0,s)/dz=0\;\text{ for }\;v_k\prec 0\end{cases}$$

$$s\succ 0\;\;;\;\;z=z_e\;\;;\;\;\begin{cases}P_k(z_e,s)=\int_0^\infty e^{-s\tau}h_k(\tau)d\tau\;\text{ for }\;v_k\prec 0\\dP(z_e,s)/dz=0\;\text{ for }\;v_k\succ 0\end{cases}\qquad(4.165)$$

From a mathematical view-point, this result is made up of a system of ordinary differential equations with its respective integration conditions. In many situations, similar systems for probabilities $P_k(z,s)$, $k=1,...N$ also have an analytical solution. Using the inverse transformation (Mellin–Fourier transformation) of each $P_k(z,s)$, $k=1,...N$, we obtain the originals $P_k(z,\tau)$, $k=1,...N$ as an analytical expression. We complete the problem of inverse transformation of each $P_k(z,s)$, $k=1,...N$ with two observations: (i) the original is frequently obtained by using a table of the Laplace transformed functions; in this table more associations for the image-original assembly can be tabulated; (ii) all the non-destroying algebraic manipulations of the Laplace image are accepted when we want an analytical expression for its original.

When system (4.164)–(4.165) does not have any analytical solution, we can use numerical integration coupled with interpolation for each function $P_k(z,s)$, $k=1,...N$; then we can obtain the originals $P_k(z,\tau)$, $k=1,...N$. However, this procedure gives an approximate result when compared to the direct numerical integration of the original model.

When we have discrete stochastic models, as those introduced through the poly-stochastic chains, we can obtain their image by using different methods: the Z transformation, the discrete Fourier transformation, the characteristic function of the process or the developing of the function of the process generator. The use of a characteristic or generator function has already been discussed in this book in some particular cases. Now we will focus on the use of the Z transformation to solve discrete stochastic models. For a function $u(\tau)$ and with a time network given as $\tau = n * \Delta\tau$, we can introduce the transformation Z by means of:

$$Z[u(\tau)] = Z[u(n)] = \sum_{n=0}^{\infty} u(n\Delta\tau)z^{-n} = \sum_{n=0}^{\infty} u(n)z^{-n} = F(z) \qquad (4.166)$$

It is not difficult to observe the recurrence property (4.167), which can be of interest for the F (z) construction:

$$Z[u(\tau - \Delta\tau)] = \frac{F(z)}{z} \qquad (4.167)$$

For the discrete stochastic model given by the group of relations (4.58), written considering a unitary and uniform displacement length for the step k:
$\left(p_k(a) = \begin{cases} 1 \text{ for } a = a_k \\ 0 \text{ for } a \neq a_k \end{cases} \right)$, the application of the Z transformation results in the following expression:

$$Z[P_k(n, i)] = \sum_{e \in K} Z[p_{ek}P_k(n - 1, i - a_k)] \qquad (4.168)$$

Considering the notation $F_k(z,j) = \sum_{n=0}^{\infty} P_k(n, j)z^{-n}$, we can rewrite this expression as shown in relation (4.169). If we particularize the model to address the simplification that considers non-fractionary values for the steps a_k, then we can easily solve the transformed model after $F_k(z,j)$. To do so, we must use a new discrete transformation where n in the $F_k(z,j)$ expression is replaced by j.

$$zF_k(z, j) = \sum_{e \in K} p_{ek}F_k(z, j - a_k) \qquad (4.169)$$

The Z transformation for a random and discrete variable results in facilitating the computation of the most important parameters used for a process characterization (mean values, momenta of various order, etc.). In example 4.3.1, we can use the obtained functions $F_k(z,j)$ to compute some parameters of this type because, in this case, we have a solution to the characteristic function of the stochastic model but not a complete and proper solution. The knowledge of the behaviour of the mean values of random variables is frequently enough to provide the stochastic model of the investigated process. For this purpose, the vector which contains the probability distributions of the process random variables $P_k(z,\tau)$, is used together with relation (4.170) in a finite space, to compute the mean values of the random variables of various order (non-centred moment of various order). The integration will carefully be corrected by bordering the space of the integral.

$$E_k^m(\tau) = \int_{-\infty}^{+\infty} z^m P_k(z, \tau)dz \quad , \quad k \geq 0 \qquad (4.170)$$

The derivate of the probabilities vector on the z axis results in:

$$\int_{-\infty}^{+\infty} \frac{dP_k(z,\tau)}{d\tau} z^m dz = z^m P_k(z,\tau)/_{-\infty}^{+\infty} - m \int_{-\infty}^{+\infty} z^{m-1} P_K(z,\tau) = -m E_k^{m-1} \tag{4.171}$$

Now we can particularize this transformation method (called method of momenta) for the model case given by the group of relations (4.146)–(4.147):

$$\frac{dE_k^m(\tau)}{d\tau} = -\left(\sum_{j=1,j\neq k}^{N} \alpha_{kj}\right) E_k^m(\tau) + m v_k E_k^{m-1}(\tau) + \sum_{j=1,j\neq k}^{N} \alpha_{jk} E_j^m(\tau) \tag{4.172}$$

$$\tau = 0 \ , \quad E_k^m(0) = \int_{-\infty}^{+\infty} z^m f_k(z) dz \tag{4.173}$$

If the obtained formulation $E_k^m(\tau)$ does not have any analytical solution, we can carry out its Laplace transformation. In this case, the images $E_k^m(s)$ can be written with the following recurrence relations:

$$\left(s + \left(\sum_{j=1,j\neq k}^{N} \alpha_{kj}\right)\right) E_k^m(s) - m v_k E_k^{m-1}(s) - \sum_{j=1,j\neq m}^{N} \alpha_{jk} E_j^m(s) - E_k^m(0) = 0 \ ;$$

$$k = 1, N \tag{4.174}$$

The solution of this equation system gives expressions $E_k^m(s)$, $k = 1, N$, which can be solved analytically by using an adequate inversion procedure. Indeed, the stochastic model has now an analytical solution but only with mean values. It is important to notice that when the analytical solution of a stochastic model pro-diuces only mean values it is important to make relationships between these results and the experimental work. This observation is significant because more of the experimental measurements allow the determination of the mean values of the variables of the process state, for the model validation or for the indentification of process parameters.

At the end of this short analysis about solving stochastic models using integral transformation, we can conclude that:

- by these methods we transform an original stochastic model into its image that is simpler and consequently more easily explored;
- we transform: (i) a problem with singular coefficients into a non-singular coefficient problem; (ii) a problem with a weak dependence on one parameter into an independent problem with respect to this parameter; (iii) an n order differential equation or a system with n differential equations into a system with n–1 order algebraic equations.
- by looking at the presented examples of transformation, it is not difficult to consider the problem of a model transformation as a general problem: indeed, the transformation presents a general form $v(t) = E(u(\tau(t)))$, where $\tau(t)$ is for each t a time randomly distributed with the law $h(s,t)$ and E is the mean value operator.

4.5
Use of Stochastic Algorithms to Solve Optimization Problems

In recent times, stochastic methods have become frequently used for solving different types of optimization problems [4.54–4.59]. If we consider here, for a steady state process analysis, the optimization problem given schematically in Fig. 4.14, we can wonder where the place of stochastic methods is in such a process. The answer to this question is limited to each particular case where we identify a normal type distribution for a fraction or for all the independent variables of the process ($X = [X_i]$). When we use a stochastic algorithm to solve an optimization problem, we note that stochastic involvement can be considered in [4.59]:

- the stochastic selection of the starting point of optimization. This shows that each starting point of optimization is selected by a stochastic procedure where all points have the same probability of being chosen; so we have here a multi-start problem for the objective function and the algorithm of optimization.
- the selection procedure for the establishment of a value for each independent variable of the process. Here, we use a random procedure in which a stochastic generator gives a value between the minimal and maximal accepted value for each variable of the process. We retain only the selected values producing a vector X that minimizes or maximizes the objective function of the process.

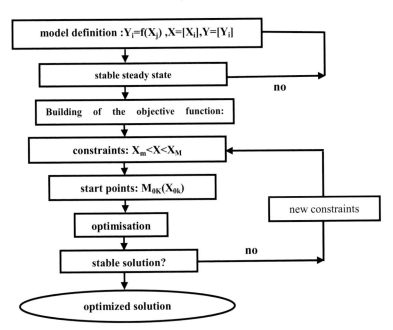

Figure 4.14 Description of the problem of optimization of a steady state process.

Figure 4.15 details this computation procedure, here s_{max} and i_{max} are respectively considered as the number of starting points and the number of acceptable iterations for one start.

1   Define the stationary mathematical model $Y=f(X,Y)$
2   Define an objective function $F=F(X,Y)$
3   Define the constraints $C=C(X,Y)$
4   Select a lower and upper bound for X   such as $X_{mi}<X_i<X_{Mi}$ for all i=1,N
5   Input s_{max} and i_{max} and start the selection of the state vectors $X_0(s)$, $s=1,s_{max}$
6   Random select $X=X_0=X_0(s)$◄
7   i=1
8   Solve the system $Y^0=f(X^0,Y^0)$◄
9   Compute $F^I=F(X^I,Y^I)$
10   Random select X $=X^I$ nearX^{I-1}◄
11   Solve the system $Y^I=f(X^I,Y^I)$ and compute F^I_{new}
12   Control of the constraints violations
13   For violations go back at 10
14   For $F^I_{new}<F^I$ go back at 10
15   For $F^I_{new}>F^I$ $F^I=F^I_{new}$,$X^I=X^0$; store and write F^I and X^I
16   For i<i_{max} go back at 8
17   For s<s_{max} go back at 6
18   End

Figure 4.15 The summary description of a stochastic procedure used for the maximization of the objective function.

The success of this computation method depends strongly on the dimension of the computation field which is considered here with the values of s_{max} and i_{max}. Indeed, when the values of i_{max} and s_{max} are greater than $2*10^4$ and 10 respectively, using this method can be problematic because of the size of the computation volume. It is important to notice that this method works without the preparations considered in the gradient optimizing procedures (see Section 3.5.5).

This procedure can easily be transformed to identify the parameters of a process as is shown in Fig. 4.15.

4.6
Stochastic Models for Chemical Engineering Processes

Stochastic modelling has been developing exponentially in all the domains of scientific research since 1950, when the initial efforts for the particularization of the stochastic theory in some practical domains were carried out. In 1960, James R. Newman, who was one of the first scientists in modern statistical theory, wrote the following about the stochastic theory particularization: *Currently in the period of dynamic indetermination in science, there is a serious piece of research that, if treated realistically, does not involve operations on stochastic processes* [4.8].

The stochastic process theory has been a major contribution to the opportune renewal of the basic stochastic theory resulting from some actual requirements and forced by the necessity of characterizing modern scientific processes. The scientific literature for the theory and practice of stochastic processes has been extensively scattered in many books and magazines. Many reviews and specialized books discuss the basic research lines in the theory of stochastic processes or present very interesting applications [4.8].

The practical applications of the stochastic process theory are multiple. This is a consequence of the capacity for this theory for predicting the future of a dynamic system by use of its history and its current state. Among the most famous applications we can note:

- The analysis of all types of movements, from atomic and molecular level [4.61–4.62] to the evolution of macroscopic systems such as atmospheric phenomena [4.63–4.64].
- The analysis of dynamic links for networks with locations where the time of service is stochastically distributed (computer networks, internet networks, etc.).
- The analysis of virtual experiments given with a stochastic model [4.65].
- The analysis of capital and fund movement [4.66, 4.67].
- The analysis and development of all types of games [4.68].
- The optimization and the control of all types of dynamic systems [4.68].

The applications of the stochastic theory in chemical engineering have been very large and significant [4.5–4.7, 4.49–4.59, 4.69–4.78]. Generally speaking, we can assert that each chemical engineering operation can be characterized with stochastic models. If we observe the property transport equation, we can notice that the convection and diffusion terms practically correspond with the movement and diffusion terms of the Fokker–Plank–Kolmogorov equation (see for instance Section 4.5) [4.79].

The following sections describe applications where stochastic models are used for the characterization of some momentum, heat and mass transport examples. For the beginner in stochastic modelling, these applications are relevant, firstly as practical examples, secondly as an explanation of the procedures and methodology for the creation of stochastic models and thirdly as examples of the use of stochastic models to obtain computation formula or algorithms for one or more investigated parameters of chemical engineering processes.

4.6.1
Liquid and Gas Flow in a Column with a Mobile Packed Bed

Columns packed with a moving bed are highly efficient for mass transfer in gas–liquid or vapour–liquid systems. This high transfer efficiency is a consequence of the rapid interface renewal brought about by the rapid movements of the packing particles [4.80–4.82].

In this example, we consider two types of operation carried out in the same type of packed column (shown in Fig. 4.16). The operating conditions depend on the values of the gas and liquid flow rates. In the former mode of operation, the column works in a wetted state whereas in the latter, we have a flooded packed state. The mobile packing bed is composed of spherical spheres with diameter 1–3 cm and density no greater than 500 kg/m³. For this type of device, recognized as mobile wetted packed bed (MWPB), the liquid and gas flow are usually characterized either by processing the parameters' relationships or through models that show:

- The state of gas and liquid hold-up for specified values of factors and parameters affecting the operation of the mobile packed bed;
- The state of phase mixing at each level of the working factors of the MWPB.

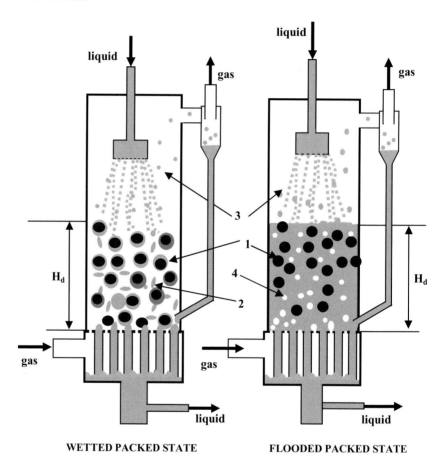

Figure 4.16 Regimes of work of the moving bed column. 1 – solid bed particle, 2 – drops of liquid in the bed, 3 – small drops, 4 – gas bubbles.

To deal with this problem, and more specifically with the working state of the MWPB, we will use stochastic modelling of the liquid and gas flow. When the MWPB operates with small liquid retention (wetted packed state) the liquid and gas hold-up are described by the concept of mean residence time (τ_{ml}, τ_{mg}) and the flow rate density (q_{vl}, q_{vg}) as follows:

$$\varepsilon_l = q_{vl}\tau_{ml}/H_d \tag{4.175}$$

$$\varepsilon_g = q_{vg}\tau_{mg}/H_d \tag{4.176}$$

The residence time for a liquid element flowing in a MWPB can be described by Eq. (4.176). Here τ_{ml0} is the liquid mean residence time for a standard mobile packed bed with a d_{p0}-diameter solid packing and a q_{vl0}-density liquid flow rate:

$$\varepsilon_l = \frac{q_{vl}\left(\dfrac{d_{p0}}{d_p}\right)^\alpha \left(\dfrac{q_{vl0}}{q_{vl}}\right)^\beta \tau_{ml0}}{H_d} \tag{4.177}$$

A stable hydrodynamic state (where the particles move about, in all directions, without any preferences) occurs in working states with a small liquid hold-up (mobile and wetted packed bed or MWPB). This displacement is the driving force of the liquid flow and we can then characterize the liquid flow by means of one stochastic model with three evolution states (for instance, see Fig. 4.17). Indeed, after this model, we accept that a liquid element is in motion with three independent evolution states:
- The liquid element moves with a flow rate $+v_x$ towards the positive direction of x;
- the liquid element moves with a flow rate $-v_x$ along the negative direction of x;
- the liquid element moves through the normal plane to x, or keeps its position.

After this description, we can appreciate the evolution of the liquid element in a MWPB through a continuous stochastic process. So, when the liquid element evolves through an i state, the probability of skipping to the j type evolution is written as $p_{ij}a\Delta\tau$. Consequently, we express the probability describing the possibility for the liquid element to keep a type I evolution as:

$$P_i = 1 - \sum_{j \neq i} p_{ij}a\Delta\tau \tag{4.178}$$

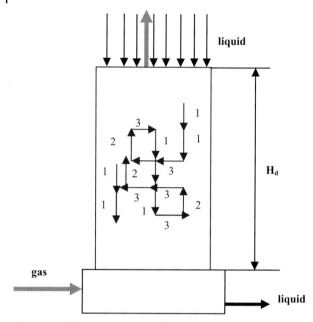

Figure 4.17 Elementary processes for the evolution of a liquid element in a MWPB.

If we consider the evolution of the liquid element together with the state of probabilities of elementary evolutions, we can observe that we have a continuous Markov stochastic process. If we apply the model given in Eq. (4.68), $P_1(z, \tau)$ is the probability of having the liquid element at position x and time τ evolving by means of a type 1 elementary process (displacement with a $+v_x$ flow rate along a positive direction of x). This probability can be described through three independent events:

- The liquid element which evolves at time τ, with the rate of evolution $+v_x$ to the position $x - \Delta x$, keeps the same evolution for the interval of time $\Delta\tau$; the probability of this occurrence is mathematically written as: $(1 - p_{12}a\Delta\tau - p_{13}a\Delta\tau)P_1(x - v_x\Delta\tau, \tau)$;
- the liquid element, which has evolved with rate $-v_x$ to position x in a time τ, changes to evolution rate $+v_x$ at the interval of time $\Delta\tau$; $p_{31}a\Delta\tau P_3(x, \tau)$ describes the probability of this occurrence;
- the last event is represented by the possibility for the liquid element that evolves at a rate 0 to a position x in a time τ, to change its evolution rate to $+v_x$ in the interval of time $\Delta\tau$; $p_{21}a\Delta\tau P_2(x, \tau)$, describes the probability of this occurrence.

Considering the type 1 evolution, we notice that $P_1(z, \tau)$ is obtained by the sum of the probabilities of independent events; so we obtain:

$$P_1(x, \tau + \Delta\tau) = [1 - (p_{12} + p_{13})a\Delta\tau]P_1(x - v_x\Delta\tau, \tau) + \\ p_{21}a\Delta\tau P_2(x, \tau) + p_{31}a\Delta\tau P_3(x, \tau) \tag{4.179}$$

By the same procedure, we obtain the probabilities of having the liquid element at x position in a time $\tau + \Delta\tau$ with a type 2 or type 3 evolution. The relations below describe these probabilities:

$$P_2(x, \tau + \Delta\tau) = [1 - (p_{21} + p_{23})a\Delta\tau]P_1(x - v_x\Delta\tau, \tau) + \\ p_{12}a\Delta\tau P_1(x, \tau) + p_{32}a\Delta\tau P_3(x, \tau) \tag{4.180}$$

$$P_3(x, \tau + \Delta\tau) = [1 - (p_{31} + p_{32})a\Delta\tau]P_3(x - v_x\Delta\tau, \tau) + \\ p_{13}a\Delta\tau P_1(x, \tau) + p_{23}a\Delta\tau P_2(x, \tau) \tag{4.181}$$

The probabilities $P_1(H_d, \tau)$ and $P_3(0, \tau)$ show the possibilities for the liquid element to leave the MWPB; these will consequently be used to compute the residence time of the liquid. When the time increases, $\Delta\tau$ is very small (near zero) and relations (4.180)–(4.181) become a particularization of the model (4.74)–(4.76), (for instance see Section 4.4.2):

$$\frac{\partial P_1(x, \tau)}{\partial \tau} + v_x\frac{\partial P_1(x, \tau)}{\partial x} = -(p_{12} + p_{13})aP_1(x, \tau) + p_{21}aP_2(x, \tau) + p_{31}aP_3(x, \tau) \tag{4.182}$$

$$\frac{\partial P_3(x, \tau)}{\partial \tau} - v_x\frac{\partial P_3(x, \tau)}{\partial x} = -(p_{31} + p_{31})aP_3(x, \tau) + p_{13}aP_1(x, \tau) + p_{23}aP_2(x, \tau) \tag{4.183}$$

$$\frac{\partial P_2(x, \tau)}{\partial \tau} = -(p_{21} + p_{23})aP_2(x, \tau) + p_{12}aP_1(x, \tau) + p_{32}aP_3(x, \tau) \tag{4.184}$$

The univocity conditions, necessary to solve the model, are established by the following considerations;
- when the liquid element gets into WPB at $\tau = 0$, it will not be present on the points where $x > 0$:

$$\tau = 0 \quad, \quad x > 0 \quad, \quad P_1(x, 0) = P_2(x, 0) = P_3(x, 0) = 0 \tag{4.185}$$

- when the liquid element gets into WPB at $x = 0$ with only an elementary type 1 process, we have:

$$\tau = 0 \quad, \quad x = 0 \quad, \quad P_1(0, 0) = 1 \tag{4.186}$$

- for all $\tau > 0$ at $x = 0$ and $x = H_d$, the liquid element cannot evolve with a type 1 or a type 3 process:

$$\tau > 0 \ , \quad x = 0 \ , \quad x = H_d \ , \quad P_1(0, \tau) = 0 \ , \quad P_3(H_d, \tau) = 0 \tag{4.187}$$

The process model expressed as a Laplace image is given by the following system of differential equations:

$$v_x \frac{dP_1(x, s)}{dx} = \left[-(s + ap_{12} + ap_{13}) + \frac{p_{21}ap_{12}a}{s + p_{21}a + p_{23}a} \right] P_1(x, s) +$$
$$\left[p_{31}a + \frac{p_{21}ap_{23}a}{s + p_{21}a + p_{23}a} \right] P_3(x, s) \tag{4.188}$$

$$v_x \frac{dP_3(x, s)}{dx} = \left[+(s + ap_{12} + ap_{13}) - \frac{p_{23}ap_{32}a}{s + p_{21}a + p_{23}a} \right] P_3(x, s) +$$
$$\left[p_{13}a + \frac{p_{12}ap_{23}a}{s + p_{21}a + p_{23}a} \right] P_1(x, s) \tag{4.189}$$

Here, the image $P_2(x.s)$ has been eliminated from the expressions that resulted from the first form of the Laplace model transformation. The observations on the packed particles evolving in WPB show that we do not have any preferential motion directions. We can extend this observation to the motion of the liquid element. Indeed, we can accept the equality of its transition probabilities: $p_{ij} = 1/3$; $i = 1, 3$; $j = 1, 3$. If we take into account this last consideration, together with a unitary value for the velocity evolution ($v_x = 1$ length units/s or $v_x = 1$ dm/s), we can assert that the model of the evolution of the liquid element flowing inside the MWPB is fully characterized. The solution of this model is carried out with its Laplace transformation of the differential equations of the model and by considering the corresponding univocity conditions. The result is given by Eqs. (4.190) and (4.191). As explained at the beginning of this section, $P_2(x, s)$ is missing here as a consequence of its elimination from the first state of the Laplace transformation of model differential equations:

$$\frac{dP_1(x, s)}{dx} = -\frac{(s + a)(3s + a)}{3s + 2a} P_1(x, s) + \frac{a(s + a)}{3s + 2a} P_3(x, s) \tag{4.190}$$

$$\frac{dP_3(x, s)}{dx} = \frac{(s + a)(3s + a)}{3s + 2a} P_3(x, s) + \frac{a(s + a)}{3s + 2a} P_1(x, s) \tag{4.191}$$

If we remove $P_3(x, s)$ from the assembly of relations (4.190) and (4.191), the result is:

$$\frac{d^2 P_1(x, s)}{dx^2} = \frac{3as(s + a)^2}{3s + 2a} P_1(x, s) \tag{4.192}$$

The general solution of this last differential equation is given by relation (4.193), where C_1 and C_2 are integration constants and λ^2 is expressed by relation (4.194):

$$P_1(x, s) = C_1 e^{\lambda(s+a)} + C_2 e^{-\lambda(s+a)} \tag{4.193}$$

$$\lambda^2 = \frac{3as}{3s + 2a} \tag{4.194}$$

Constants C_1 and C_2 are obtained from the univocity problem adapted to the Laplace transformation. The solution thus obtained is given below. In this relation, Λ is given by Eq. (4.196):

$$P_1(x, s) = \frac{1}{\Lambda}[(1 + \lambda)^2 e^{-\lambda x(s+a)} - (1 - \lambda)^2 e^{-2\lambda H_d + \lambda x(s+a)}] \tag{4.195}$$

$$\Lambda = (1 + \lambda)^2 - (1 - \lambda)^2 e^{-2\lambda H_d} \tag{4.196}$$

If we take into consideration the procedure that we used above to eliminate $P_3(x, s)$ from the system (4.190)–(4.191), we can obtain the following expression for $P_3(x, s)$:

$$P_3(x, s) = \frac{1}{\Lambda}[(1 - \lambda^2)e^{-\lambda x(s+a)} - (1 - \lambda^2)e^{-2\lambda H_d + \lambda x(s+a)}] \tag{4.197}$$

The residence time distribution function is found as a result of the addition of the probabilities showing the possibility for a liquid element to leave the MWPB (see also Section 4.3.1):

$$f(\tau, H_d) = P_1(H_d, \tau) + P_3(0, \tau) \tag{4.198}$$

The mean residence time for the liquid element evolution in a MWPB, can easily be obtained from the first derivative of the characteristic function of the residence time distribution:

$$\tau_{m0}(H_d) = -\varphi(0, H_d) \tag{4.199}$$

The analytical expression of our $\varphi(s, H_d)$ is obtained by coupling the basic formula of the characteristic residence time distribution function with the solutions of $P_1(x, s)$ and $P_3(x, s)$:

$$\varphi(s, H_d) = \int_0^\infty f(\tau, H_d)e^{-s\tau}d\tau = P_1(H_d, s) + P_3(0, s) \tag{4.200}$$

$$\varphi(s, H_d) = \frac{Ch(\lambda H_d s) - \dfrac{1 - \lambda^2}{2\lambda}Sh(\lambda H_d) - \dfrac{1 + \lambda^2}{2\lambda}Sh(\lambda H_d s)}{\dfrac{1 + \lambda^2}{2\lambda}Sh(\lambda H_d) + Ch(\lambda H_d)} \tag{4.201}$$

A similar result for $\varphi(s, H_d)$ but with a different relation for parameter λ was obtained in application 4.3.1. The value of the derivative $\varphi'(0, H_d)$ is then obtained using the definition of the derivative in a point:

$$\varphi'(0, H_d) = \lim_{s \to 0} \frac{\varphi(s, H_d) - \varphi(0, H_d)}{s - 0} = \lim_{s \to 0} \frac{\varphi(s, H_d) - 1}{s} =$$

$$\lim_{s \to 0} \frac{Ch(\lambda H_d s) - \lambda Sh(\lambda H_d) - Ch(\lambda H_d) - \dfrac{1 + \lambda^2}{2\lambda} Sh(\lambda H_d s)}{\dfrac{1 + \lambda^2}{2\lambda} s Sh(\lambda H_d) + s Ch(\lambda H_d)} \tag{4.202}$$

$$= \ldots = -\left(\frac{3}{2} H_d + \frac{H_d}{H_d + 2} \right)$$

It is easy to observe that the intensity of transition from one state to another (parameter a from relations (4.179)–(4.181)) does not influence the mean residence time characterizing the liquid evolution in the MWPB:

$$\tau_{mo}(H_d) = -\varphi'(0, H_d) = -\left(\frac{3}{2} H_d + \frac{H_d}{H_d + 2} \right).$$

Coming back to the problem of liquid fraction in the MWPB, we observe that the replacement of Eq. (4.177) by (4.202) imposes homogenizing units because we previously established that the mean residence time of the liquid was calculated considering $v_x = 1$ in dm/s and, consequently, H_d was used in decimeters. Now considering H_d in meters, relation (4.177) becomes:

$$\varepsilon_l = \left(\frac{30}{2} + \frac{10}{H_d + 2} \right) q_{vl} \left[\frac{d_{p0}}{d_p} \right]^\alpha \left[\frac{q_{vl0}}{q_{vl}} \right]^\beta \tag{4.203}$$

We complete the expression of MWPB liquid hold-up by considering $\alpha = 0.5$ and $\beta = 0.4$. These values are also used to hold-up the liquid in a countercurrent gas–liquid flow in a fixed packed bed. To complete the building of the MWPB liquid hold-up expression, we show the identified conditions corresponding to a standard defined MWPB. These conditions correspond to a spherical-shaped packing with $d_{p0} = 0.025$ m, $\rho_p = 300$ kg/m³, fluidized by air and wetted by water with a flowing density of $q_{vl0} = 2.5{*}10^{-3}$ m³/m²s. Indeed, the final expression is:

$$\varepsilon_l = 0.216 \left(\frac{30 H_d + 8}{30 H_d + 6} \right) q_{vl}^{0.6} d_p^{-0.5} \tag{4.204}$$

For other shapes of packing, we have to correct the previous relation with a coefficient ψ (shape factor). Relation (4.205) takes into account this correction:

$$\varepsilon_l = 0.216 \left(\frac{30 H_d + 8}{30 H_d + 6} \right) q_{vl}^{0.6} d_p^{-0.5} \psi^{-0.5} \tag{4.205}$$

The final expression giving the mean residence time for the liquid evolution in a MWPB of spherical particles is written as follows:

$$\tau_{ml} = 0.0143 \left(\frac{30 H_d}{2} + \frac{10 H_d}{10 H_d + 2} \right) d_p^{-0.5} q_{vl}^{-0.4} \tag{4.206}$$

Experimental validation. When a model cannot be verified by experiments, it can be considered as an excellent exercise which however does not have any practical significance. So as to validate the results of our stochastic model for a liquid flow in a MWPB, we will use data previously published [4.80, 4.82] for a model contacting bed of spherical particles, in which the gas and liquid fluids were respectively air and water. In these studies, the liquid residence time was estimated by measuring the response of a signal injected into the bed. The MWPB liquid hold-up was obtained by the procedure of instantly stopping the water and air at the bed input. The data obtained are reported in Figs. 4.18–4.20.

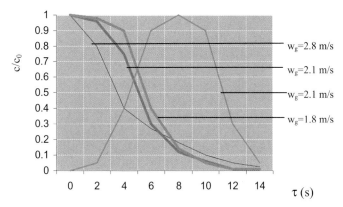

Figure 4.18 Response to signals in liquid for a WPB operated with air and water (packing of spherical particles with $d_p = 0.0275$ m and $\rho_p = 330$ kg/m³, $q_{vl} = 12$ m³/m²h , $H_0 = 0.18$ m).

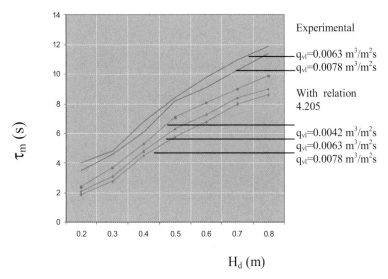

Figure 4.19 Liquid residence time for a MWPB operated with air and water (Spherical particles with $d_p = 0.0275$ m and $\rho_p = 330$ kg/m³).

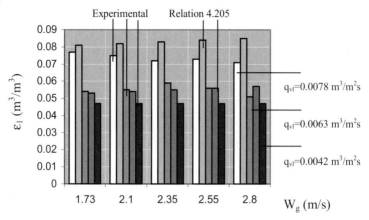

Figure 4.20 Liquid hold-up state for air and water operated MWPB.
(Spherical particles with $d_p = 0.0275$ m and $\rho_p = 330$ kg/m², $H_0 = 0.18$ m).

Each point on the curves in Fig. 4.19 corresponds to the mean value of various experimental results. We can notice that, even if we have good trends, the experimental and calculated values do not match well. This can be ascribed to model inadequacies, especially with respect to the liquid exit conditions; in that case, we considered that the MWPB output had occurred at $x = 0$ and at $x = H_d$ when it was experimentally observed that the liquid exit dominantly occurs at $x = H_d$. This results in a decrease in the mean residence time computed values. If we look at Figs. 4.20 to 4.22, which have been obtained at different operating conditions, we can conclude that we do not have major differences between the computed and experimental values of liquid MWPB hold up; then we can consider the equality of the transition probabilities between the individual states of the stochastic model to be realistic.

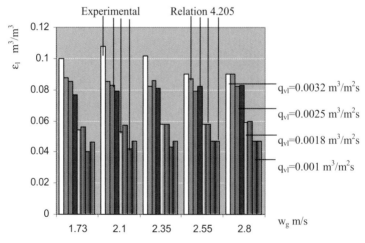

Figure 4.21 Liquid hold-up state for a MWPB operated with air and water. Packing particles of cylindrical shape with $d_p = 0.012$ m, $\rho_p = 430$ kg/m³, $H_0 = 0.18$ m.

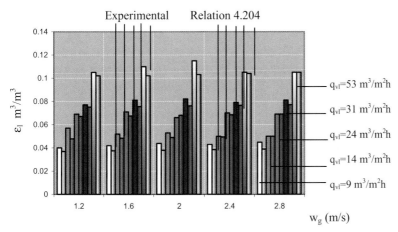

Figure 4.22 Liquid hold-up state for a WPB operated with air and water. Case of spherical particles with $d_p = 0.02$ m and $\rho_p = 220$ kg/m³; $H_0 = 0.22$ m.

According to Fig. 4.21, where the particles used for the bed are cylindrical, as well as to Fig. 4.22 obtained from the experimental data published by Chen [4.83], we observe that the MWPB liquid hold-up does not depend on gas velocity as was found in Fig. 4.20. In order to justify the value of the evolution velocity of a liquid element ($v_x = 1$ dm/s) used in the simulations, we drew Figs. 4.23 and 4.24. Figure 4.23 has been drawn after the data published by Cains and Prausnitz [4.84]. It shows the evolution of the liquid velocity when water fluidizes spherical particles of glass.

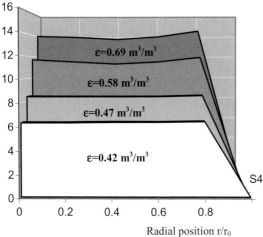

Figure 4.23 Liquid flow state for the fluidization of a bed of glass spheres with $d_p = 0.0032$ m.

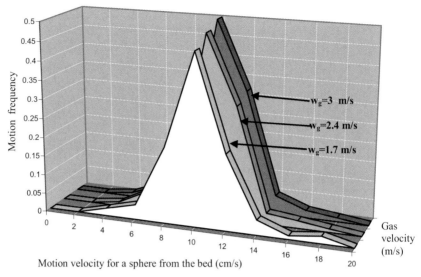

Figure 4.24 Velocity frequencies for the motion of a mobile
sphere inside of a MWPB. Case: Air–water operated, spheres
with $d_p = 0.0275$ m and $H_0 = 0.18$ m, $q_{vl} = 0.004$ m^3/m^{2}s.

In Fig. 4.24 we show the experimental data that present the frequency of the
velocity states for the motion of some marked particles inside MWPB. For this
purpose some particles of the packing have been coloured and their motion has
been recorded by means of a high speed video-recorder [4.81]. From both figures
it is obvious that the velocity range obtained is very near to the liquid element velocity considered for simulations. Indeed, if in the MWPB the most important
quantity of the liquid covers the fluidized particles, the evolution of the liquid velocity has to be identical to the fluidized particle velocity. However, it is obvious
that the physical properties of the liquid have to affect the residence time of the
liquid element that evolves inside the MWPB. Indeed, if we have a liquid different
from water we have to introduce the influence of its own physical properties on
the system response. The solution to this problem can be obtained by derivation
of the expression of the characteristic function of the liquid evolution as shown in
Eq. (4.206). Here H_d^e is considered as an equivalent height of the MWPB corresponding to a unitary value of the velocity of displacement of the liquid element
($v_x = 1$ conventional length unit/s).

$$\varphi'(0, H_d^e) = -\left(\frac{3}{2} H_d^e + \frac{H_d^e}{H_d^e + 2}\right) \tag{4.207}$$

The transposition of the equivalent height of the mobile packed bed (H_d^e) to a normal working unit is carried out through a correction function, which is applied to
a bed height corresponding to a MWPB operated with air and water. If we consider that the major contributions to the correction function expression are given

by liquid density, viscosity and superficial tension, we can complete the relations (4.205) and (4.206), which are then rewritten as:

$$\varepsilon_l = 0.216 \left(\frac{30H_d + 8}{30H_d + 6} \right) q_{vl}^{0.6} d_p^{-0.5} \psi^{-0.5} f(\eta_l, \sigma_l, \rho_l) \tag{4.208}$$

$$\tau_{ml} = 0.0143 \left(\frac{30H_d}{2} + \frac{10H_d}{10H_d + 2} \right) d_p^{-0.5} q_{vl}^{-0.4} f(\eta_l, \sigma_l, \rho_l) \tag{4.209}$$

Some computed values of the function $f(\eta_l, \sigma_l, \rho_l)$ obtained by using the data reported by Masao et al. [4.85], are given in Table 4.3. The analysis of the data of Table 4.3 shows that the expression of the function $f(\eta_l, \sigma_l, \rho_l)$ can be written as follows:

$$f(\eta_l, \sigma_l, \rho_l) = \eta_{rl}^{\alpha} \sigma_{rl}^{\beta} \rho_{rl}^{\gamma}) \tag{4.210}$$

Table 4.3 Some values of the function $f(\eta_l, \sigma_l, \rho_l)$ when the MWPB is operated with air and various liquids.

$f(\eta_l, \sigma_l, \rho_l)$	Water $\rho_l = 1$ g/cm³, $\eta_l = 1$ CP, $\sigma_l = 72.8$ d/cm	Ethanol $\rho_l = 0.8$ g/cm³, $\eta_l = 1.38$ CP, $\sigma_l = 22.5$ d/cm	Glycerol 25% $\rho_l = 1.07$ g/cm³, $\eta_l = 1.33$ CP, $\sigma_l = 70.8$ d/cm	Glycerol 65% $\rho_l = 1.16$ g/cm³, $\eta_l = 14.5$ CP, $\sigma_l = 67.5$ d/cm
Spheres of 170 kg/m³ $d_p = 0.02$ m, $w_g = 2$ m/s $q_{vl} = 0.0025$ m³/m²s	1	0.759	0.955	1.36
Spheres of 590 kg/m³ $d_p = 0.028$ m, $w_g = 3$ m/s $q_{vl} = 0.006$ m³/m²s	1	0.801	0.983	1.31
	$\rho_{rl} = 1$ $\eta_{rl} = 1$ $\sigma_{rl} = 1$	$\rho_{rl} = 0.8$ $\eta_{rl} = 1.33$ $\sigma_{rl} = 0.303$	$\rho_{rl} = 1.07$ $\eta_{rl} = 1.33$ σ_r	$\rho_{rl} = 1.16$ $\eta_{rl} = 14.45$ $\sigma_{rl} = 0.981$

The data of this table also allow the identification of parameters $\alpha = 0.158$, $\beta = 0.484$ and $\gamma = -1.65$. Based on these values, we can then write slightly more complex relations for the liquid hold-up and liquid residence time in the MWPB:

$$\varepsilon_l = 0.216 \left(\frac{30H_d + 8}{30H_d + 6} \right) q_{vl}^{0.6} d_p^{-0.5} \psi^{-0.5} \eta_{rl}^{0.158} \sigma_{rl}^{0.484} \rho_{rl}^{-1.65} \tag{4.211}$$

$$\tau_{ml} = 0.0143 \left(\frac{30H_d}{2} + \frac{10H_d}{10H_d + 2} \right) d_p^{-0.5} q_{vl}^{-0.4} \eta_{rl}^{0.158} \sigma_{rl}^{0.484} \rho_{rl}^{-1.65} \tag{4.212}$$

4.6.1.1 Gas Hold-up in a MWPB

In Fig. 4.16 the gas is shown to be the continuous phase of the MWPB. In this configuration, many direction changes in the flow of gas elements originate from the presence and moving of the packing spheres. If we consider that these changes take place randomly, then we can carry out the particularization of the stochastic flow description of the gas elements. This description has been successfully used for the case of gas–liquid flow in a fixed and in a mobile packed bed [4.28, 4.81].

The gas element evolves inside the MWPB in 3 states: displacement in the $+x$ direction with velocity $+v_x$; displacement in the $-x$ direction with velocity $-v_x$; non-motion or displacement in the horizontal plane. The mathematical writing of the stochastic model is given by relations (4.182)–(4.187). In order to particularize this model to the case of gas element evolution inside the MWPB, we take into account the following considerations:

1. The sense of the x-axis is determined by the global gas flow direction and is inverse with respect to the liquid flow.
2. It is difficult to select the values for the probabilities of passage between process states.
3. The gas flowing element rapidly passes from one actual process state to another and tends to follow an elementary process type 1state.
4. Concerning the evolution of the velocity of the gas flowing element, the skip velocity is added or subtracted from the local gas flow velocity in the bed, when the gas flowing element skips in the $+x$ and $-x$ directions respectively.

With these considerations, the model that describes the gas element motion inside the MWPB can be described by the following assembly of relations:

$$\frac{\partial P_1(x,\tau)}{\partial \tau} + (w_g + v_x)\frac{\partial P_1(x,\tau)}{\partial x} = -(p_{12}+p_{13})aP_1(x,\tau) + p_{21}aP_2(x,\tau)$$
$$+ p_{31}aP_3(x,\tau)$$
(4.213)

$$\frac{\partial P_3(x,\tau)}{\partial \tau} + (w_g - v_x)\frac{\partial P_3(x,\tau)}{\partial x} = -(p_{31}+p_{32})aP_3(x,\tau) + p_{13}aP_1(x,\tau)$$
$$+ p_{23}aP_2(x,\tau)$$
(4.214)

$$\frac{\partial P_2(x,\tau)}{\partial \tau} + w_g\frac{\partial P_2(x,\tau)}{\partial x} = -(p_{21}+p_{23})aP_2(x,\tau) + p_{12}aP_1(x,\tau)$$
$$+ p_{32}aP_3(x,\tau)$$
(4.215)

$$\tau = 0 \ , \quad x = 0 \ , \quad P_1(x,0) = P_2(x,0) = P_3(x,0) = 0$$
(4.216)

$$\tau = 0 \ , \quad x = 0 \ , \quad P_1(0,0) = 1 - u \quad P_2(0,0) = P_3(0,0) = u/2$$
(4.217)

$$\tau > 0 \;, \quad x = 0 \;, \quad P_1(0, \tau) = 0 \;, \quad P_2(0, \tau) = 0 \;, \quad P_3(0, \tau) = 0 \qquad (4.218)$$

This stochastic model is in fact one type of turbulent motion model. For the univocity problem, we consider that a gas element can be influenced by any type of elementary process after its insertion into the MWPB at $x = 0$. The permanent velocity w_g, pushes the gas element outside the bed at $x = H_d$ and through any of the elementary processes. The presented model can be completed by considering the different frequencies induced by passing from one elementary process to another: $p_{11}a_{11} = a_{11}$, $p_{12}a_{12} = a_{12}$, $p_{13}a_{13} = a_{13}$, etc.

The distribution of the residence time for the gas evolution inside the bed takes into account the statement above, which concerns the possibility for the gas to exit the bed through any of the elementary processes:

$$f(\tau, H_d) = P_1(\tau, H_d) + P_2(\tau, H_d) + P_3(\tau, H_d) \qquad (4.219)$$

Theoretical and experimental results of the gas hold-up inside a MWPB show that the data converge only when the p_{11} values are greater than 0.7. Figure 4.25 presents a simulation of the presented model, which intends to fit some experimental data [4.82]. In the presented simulation, the initial values of $P_1(0, 0)$, $P_2(0, 0)$ and $P_3(0, 0)$ injected into the model give an idea about the values of the transition probabilities; these are: $p_{11} = p_{21} = p_{31} = 0.7$, $p_{12} = p_{13} = p_{12} = p_{23} = p_{22} = p_{33} = p_{32} = 0.15$. In Fig. 4.25 we can see that we have all the necessary data to begin the computation of the mean residence time of a gas element evolving inside the MWPB. Indeed, relation (4.176) can now be used to calculate the gas hold-up in the bed.

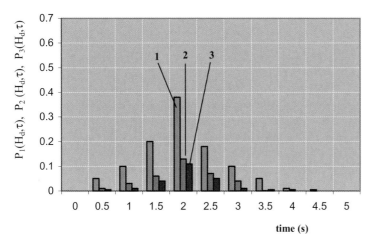

Figure 4.25 The gas exit from the MWPB in terms of probability of the elementary motion processes ($w_g = 2$ m/s, $v_x = 0.1$ m/s, $H_d = 0.9$ m, $1 - P_1(0,0) = 0.7$, $2 - P_2(0,0) = 0.15$, $3 - P_3(0,0) = 0.15$).

4.6.1.2 Axial Mixing of Liquid in a MWPB

The liquid flowing inside a MWPB can be described with a one-parameter dispersion flow model. As we show in Section 3.3, the axial mixing coefficient or, more correctly, the axial dispersion coefficient is the specific parameter for this model. Relation (3.112) contains the link between the variance of the residence time of liquid elements and the Peclet number. We can rewrite this relation so as to particularize it to the case of a MWPB. Here, we have the possibility to compute the variance of the residence time of the liquid through the stochastic model for the liquid flow developed previously in order to obtain the value of the axial dispersion coefficient:

$$\sigma^2 = \frac{2}{Pe} - \frac{2}{Pe^2}(1 - e^{-Pe}) = \frac{\varphi''(0, H_d)}{[\varphi'(0, H_d)]^2} - 1 \tag{4.220}$$

$$Pe = \frac{w_l H_d}{D_l} = \frac{H_d^2}{\tau_{ml} D_l} = -\frac{H_d^2}{\varphi'(0, H_d) D_l} \tag{4.221}$$

If we carefully observe the expression of the characteristic function of the residence time distribution for the evolution of a liquid element ($\varphi(s, H_d)$, relation (4.201)), we can notice that it is difficult to compute the expressions of the derivatives $\varphi'(0, H_d)$ and $\varphi''(0, H_d)$. Using the expansion of the hyperbolic sine and cosine respectively as multiplication series, we obtain the following simplified expression for the characteristic function:

$$Ch(z) = \left(1 + \frac{4z^2}{\pi^2}\right)\left(1 + \frac{4z^2}{3^2\pi^2}\right)\left(1 + \frac{4z^2}{5^2\pi^2}\right)......\left(1 + \frac{4z^2}{(2n+1)^2\pi^2}\right)... \tag{4.222}$$

$$Sh(z) = z\left(1 + \frac{z^2}{\pi^2}\right)\left(1 + \frac{z^2}{2^2\pi^2}\right)\left(1 + \frac{z^2}{3^2\pi^2}\right)......\left(1 + \frac{z^2}{(2n+1)^2\pi^2}\right)... \tag{4.223}$$

$$\varphi(s, H_d) = \frac{(2 + H_d) + (3 - H_d)s - 3H_d s^2 + 1.2H_d s^3}{(2 + H_d) + (3 + 3H_d + 1.2H_d^2)s} \tag{4.224}$$

Table 4.4 compares the residence time results obtained with the characteristic function given by the original equation (4.201) and with the simplified form (4.224). We can notice that the values obtained with the simplified form are good enough.

$$\varphi'(0, H_d) = \lim_{s \to 0}\frac{\varphi(s, H_d) - \varphi(0, H_d)}{s - 0} =$$
$$\lim_{s \to 0}\frac{1.2H_d s^3 - 3H_d s^2 - (1.2H_d^2 + 4H_d)s}{(1.2H_d^2 + 3H_d + 3)s^2 + 5(H_d + 2)s} = \tag{4.225}$$
$$\lim_{s \to 0}\frac{3.6H_d^2 s^2 - 6H_d s - (1.2H_d^2 + 4H_d)}{(2.4H_d^2 + 6H_d + 6)s + (H_d + 2)} = -\frac{1.2H_d^2 + 4H_d}{H_d + 2}$$

Table 4.4 Mean residence time of the liquid in a MWPB as a function of the bed height. First row – simplified characteristic function, second row – original characteristic function.

H_d (dm)			0.5	1	2	3	4
$-\varphi'(0, H_d) = \tau_{ml} = \dfrac{1.2H_d^2 + 4H_d}{H_d + 2}$	(s)		0.91	1.73	3.2	4.57	5.87
$-\varphi'(0, H_d) = \tau_{ml} = \dfrac{3}{2}H_d + \dfrac{H_d}{H_d + 2}$	(s)		0.95	1.83	3.5	5.1	6.67

The analytical computation for the first derivative of the characteristic function gives relation (4.226) where the functions $\alpha_i(H_d)$, $i = 1, 6$ are written with the relations (4.227)–(4.232):

$$\varphi'(s, H_d) = \frac{\alpha_1(H_d)s^3 + \alpha_2(H_d)s^2 + \alpha_3(H_d)s + \alpha_4(H_d)}{[\alpha_5(H_d) + \alpha_6(H_d)s]^2} \tag{4.226}$$

$$\alpha_1(H_d) = 7.2H_d^2 + 7.2H_d^3 + 2.88H_d^4 \tag{4.227}$$

$$\alpha_2(H_d) = -(9H_d + 1.8H_d^2) \tag{4.228}$$

$$\alpha_3(H_d) = -(12H_d + 6H_d^2) \tag{4.229}$$

$$\alpha_4(H_d) = -(8H_d + 6.4H_d^2 + 1.2H_d^3) \tag{4.230}$$

$$\alpha_5(H_d) = 2 + H_d \tag{4.231}$$

$$\alpha_6(H_d) = 3 + 3H_d + 1.2H_d^2 \tag{4.232}$$

For the second derivative $\varphi''(0, H_d)$ at point zero, we use the definition formula coupled with the l'Hospital rule for the elimination of the non-determination of $0/0$ type. The result is:

$$\varphi''(0; Hd) = \frac{\alpha_3(H_d)\alpha_5(H_d) - 2\alpha_4(H_d)\alpha_6(H_d)}{[\alpha_5(H_d)]^2} \tag{4.233}$$

Considering relations (4.220) and (4.221), we can observe that we have all the required elements to compute the axial dispersion coefficient. The theoretical computed values for the axial mixing coefficient for the case where the bed height has a practical importance are shown in Table 4.5. For the cases when the selection $v_x = 1$ dm/s is not justified by the operational conditions, we replace H_d by H_d^e. We can introduce H_d^e through equation (4.210):

$$H_d^e = \eta_{rl}^{0.158}\sigma_{rl}^{0.484}\rho_{rl}^{-1.65}H_d \tag{4.234}$$

Table 4.5 Evolution of the Peclet number and of the axial dispersion coefficient with the height of the MWPB (theoretical computation).

H_d (dm)	$-\varphi'(0, H_d)$ (s)	$\varphi''(0, H_d)$ (s²)	σ^2	Pe	$D_l * 10^2$ (m²/s)
0.5	0.95	2.33	1.33	–	–
1	1.83	6.32	0.887	0.5	1.07
2	3.50	19.08	0.557	2.1	0.54
3	5.10	37.98	0.46	2.8	0.63
4	6.67	64.83	0.43	3.3	0.73
5	8.21	93.67	0.388	3.7	0.823
6	9.75	130.32	0.371	3.9	0.946
7	11.28	172.79	0.36	4.1	1.05
8	12.80	221.05	0.35	4.4	1.14
9	14.31	275.16	0.343	4.6	1.23
10	15.93	335.04	0.0336	4.7	1.34

Experimental testing. The experimental work for the determination of the liquid axial mixing inside the packed bed has been carried out with the introduction of an impulse of NaCl (12 g/l) on the input of a MWPB which operates with air and water [4.80, 4.81] and with the recording of the evolution with time of the signal state at the liquid output. We assume that the system can be described by the dispersion model and that its analytical solution is given by Eqs. (3.106) and (3.107). Using these equations as well as the experimental results, we can then calculate the Pe number and the D_l coefficient values. Figure 4.26 indirectly shows the evolution of the signal at the output through the electric tension of a Wheatstone bridge having the resistive detector placed in the output flow of liquid.

The experimental data processing has previously been presented in example 3.3.5.3. Equations (3.106) and (3.107) can be simplified in order to allow a rapid identification of Pe and D_l as follows:

$$\ln\frac{c(\theta)}{c_0} = -\left(\frac{Pe}{4} + \frac{4}{Pe}\lambda_1^2\right)\theta + \frac{Pe}{2} +$$

$$\ln\frac{2\lambda_1}{\left(1 + \frac{Pe}{2}\right)\lambda_1 \sin(2\lambda_1) - \left[\frac{Pe}{2} + \left(\frac{Pe}{2}\right)^2 - \lambda_1^2\right]\cos(2\lambda_1)} = m\theta + n$$

$$(4.235)$$

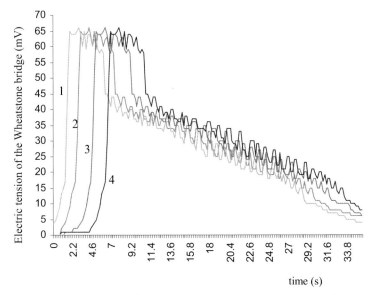

Figure 4.26 The output signal of the inverse unitary impulse of a solution of 12 g/l NaCl in the liquid of the MWPB.
$(1 - w_g = 1.31$ m/s, $q_{vl} = 0.0062$ m³/m²s, $H_0 = 0.18$ m , $d_p = 0.0275$ m , $\rho_p = 330$ kg/m³;
$2 - w_g = 1.31$ m/s, $q_{vl} = 0.0031$ m³/m² s, $H_0 = 0.18$ m , $d_p = 0.0275$ m , $\rho_p = 330$ kg/m³;
$3 - w_g = 2.4$ m/s, $q_{vl} = 0.0062$ m³/m² s, $H_0 = 0.18$ m , $d_p = 0.0275$ m , $\rho_p = 330$ kg/m³ ;
$4 - w_g = 2.4$ m/s, $q_{vl} = 0.0031$ m³/m² s, $H_0 = 0.18$ m , $d_p = 0.0275$ m , $\rho_p = 330$ kg/m³).

$$tg(2\lambda_1) = \frac{\frac{Pe}{2}\lambda_1}{\lambda_1^2 - \left(\frac{Pe}{4}\right)^2} \tag{4.236}$$

It is easy to notice that the values of the parameters m and n, from Eq. (4.235), will be estimated after the particularization of the least squares method for the dependence of $\ln\left(\frac{c(\theta)}{c_0}\right)$ vs θ

$$m = \frac{Pe}{4} + \frac{4}{Pe}\lambda_1^2 \tag{4.237}$$

$$n = \frac{Pe}{2} + \ln\frac{2\lambda_1}{\left(1 + \frac{Pe}{2}\right)\lambda_1 \sin(2\lambda_1) - \left[\frac{Pe}{2} + \left(\frac{Pe}{2}\right)^2 - \lambda_1^2\right]\cos(2\lambda_1)} \tag{4.238}$$

The value of the Pe number is obtained by solving the system formed by Eqs. (4.237) and (4.238). Two examples of data processing [4.81] are given in Table 4.6. Figure 4.27 presents the evolution of the Peclet number with the gas velocity [4.81–4.83, 4.86]. A comparison between the published and computed stochastic dependence of Pe vs w_g is given for cases 1 and 2, [4.81] ($H_0 = 0.36$ m, $q_{vl} = 0.0031$ and 0.0071 m^3/m^2 s, $d_p = 0.0275$ m , $\rho_p = 330$ kg/m^3).

Table 4.6 Experimental data processing for the estimation of Pe and D_l.

	Experimental conditions $w_g = 2$ m/s, $q_{vl} = 0.0047$ m^3/m^2 s, $H_0 = 1.8$ dm, $H_d = 2.1$ dm, $d_p = 0.0275$ m, $\rho_p = 330$ kg/m^3				**Experimental conditions** $w_g = 2.4$ m/s, $q_{vl} = 0.0047$ m^3/m^2 s, $H_0 = 1.8$ dm, $H_d = 2.1$ dm, $d_p = 0.0275$ m, $\rho_p = 330$ kg/m^3		
N°	$c(\theta)/c_0$	θ	$\ln c(\theta)/c_0$		$c(\theta)/c_0$	θ	$\ln c(\theta)/c_0$
	0	0	*	m = 1.29	0	0	*	m = 1.31
2	1	0.36	0	Solutions:	1	0.34	0	Solutions:
3	0.52	0.69	−0.69	$\lambda_1 = 0.608$	0.53	0.69	−0.65	$\lambda_1 = 0.618$
4	0.375	1.03	−0.98	Pe = 1.75	0.35	1.02	−1.04	Pe = 1.81
5	0.276	1.37	−1.35		0.26	1.38	−1.347	
6	0.23	1.71	−1.47	Stochastic	0.23	1.7	−1.47	Stochastic
7	0.11	2.05	−2.2	Pe = 2.4	0.12	2.04	−2.12	Pe = 2.55
8	0.05	2.4	−2.99		0.04	2.4	−3.2	
9	0.013	2.76			0.028	2.72		

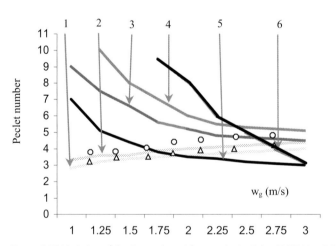

Figure 4.27 Variation of the Pe number with gas velocity. 1,2 – [4.81] $H_0 = 0.36$ m, $q_{vl} = 0.0031$ and 0.0071 m^3/m^2 s, $d_p = 0.0275$ m , $\rho_p = 330$ kg/m^3 ; 3,4 – [4.83] $H_0 = 0.16$ m, $q_{vl} = 0.0038$ and 0.016 m^3/m^2 s, $d_p = 0.016$ m , $\rho_p = 283$ kg/m^3; 5,6 – [4.86] $H_0 = 0.30$ m , $q_{vl} = 0.0025$ and 0.015 m^3/m^2 s $d_p = 0.0255$ m , $\rho_p = 173$ kg/m^3. ○ and △: calculated results for the cases 1 and 2.

Figure 4.28 illustrates the dependence of the axial mixing coefficient with the gas velocity for the examples given above. This graphic representation is derived form Fig. 4.27 according to the relation: $Pe = H_d^2/(\tau_{ml}D_l))$. Apparently, the stochastic model predictions of the liquid axial mixing versus gas velocity are in contradiction with the published data, especially at low gas velocity. This discordance can be explained if we consider that, at low gas velocity, we only have an incipient motion of the packing. Indeed, at low velocity, we can consider that we are near the conditions of a fixed packed bed, and the stochastic model which considers three states with the same probability: $p_{ij} = 1/3, \forall i = 1, 3; j = 1, 3$ is no longer sustained. At the same time, the calculated Pe values are slightly less than the experimental ones and then the experimentally obtained mixing is greater than when calculated by the model. However, it is important to notice the good trend observed for the axial mixing coefficient, which increases with the gas velocity as shown in Fig. 4.28.

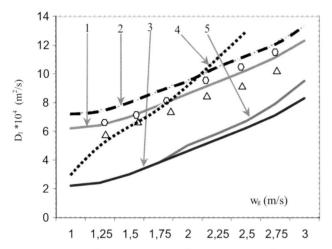

Figure 4.28 State of the axial mixing coefficient versus gas velocity.
$1,2 - [4.81]$ $H_0 = 0.36$ m, $q_{vl} = 0.0031$ and 0.0071 m^3/m^2 s, $d_p = 0.0275$ m ,
$\rho_p = 330$ kg/m^3 ; $3,4 - [4.83]$ $H_0 = 0.16$ m, $q_{vl} = 0.0038$ and 0.016 m^3/m^2 s,
$d_p = 0.016$ m , $\rho_p = 283$ kg/m^3; $5 - [4.86]$ $H_0 = 0.30$ m , $q_{vl} = 0.0025$ and
0.015 m^3/m^2 s $d_p = 0.0255$ m , $\rho_p = 173$ kg/m^3. $\bigcirc$ and $\triangle$: calculated results for the cases 1 and 2.

Before closing this discussion about the axial mixing of liquid inside the MWPB, it is important to note the significance of the result generated by the stochastic model. Concretely, we can compute the values of the axial mixing parameters for liquid flow inside the mobile wetted bed by a procedure (assembly of relations (4.202), (4.220), (4.221), (4.229)–(4.233)) that requires only the computation of the bed height (H_d). This last parameter strongly depends on all the factors that characterize fluidization: the density and diameter of the spheres, liquid flow density, gas velocity, all momentum transport properties of gas and liquid (gas and liquid density, viscosity, liquid superficial tension, etc.). Therefore, the axial mixing parameters for the liquid flow in the MWPB are influenced by all these factors. For the computation of the height of the MWPB, we can use previously published relations [4.80–4.81]. In these relations, the minimum fluidization velocity appears as an important variable. The liquid flow density and the minimum gas flow velocity for the fluidization, determine the value of this variable.

4.6.1.3 The Gas Fraction in a Mobile Flooded Packed Bed

Figure 4.16 shows that, in a mobile flooded packed bed (MFPB), the gas flow bubbles through the bed which is composed of the assembly of liquid and solid packing. The spheres that represent the mobile packing are frequently fluidized in the liquid and are predominantly near the surface. In this MFPB, the bed's gas fraction strongly depends on the feed gas flow rate and consequently depends on the apparent gas velocity. As far as many other factors can influence the bed's gas fraction, a stochastic approach to this problem can be convenient. Therefore, this section will be devoted to establishing one stochastic model for the gas movement inside the bed and to its use in solving the dependence between the gas fraction and the various factors of the process. Experimental observations indicate that, in a liquid–solid system, the gas bubbles have a non-organized motion, which can be associated with a stochastic process. In addition, the phenomenon of bubble association is frequently observed in all cases of gas bubbling in a liquid or suspension or in a liquid with large suspended solids. Looking at the motion of a single bubble in this type of system, we can consider that the bubbles will change their velocity in response to: (i) an interaction with the liquid or with the solid; (ii) an interaction with the bubbles in their vicinity. When d_k-diameter bubbles interact with the surrounding liquid, their velocities v_k can be calculated by the following relation

$$v_k = \sqrt{\frac{4gd_k(\rho_l - \rho_g)}{3\zeta_k\rho_g}}.$$ Here, ζ_k shows the hydrodynamic friction-resistance coefficient related with the movement of the bubble k. With $f_k^0(x, v_k, \tau)$, we note the function of the velocity distribution with respect to the individual k-type bubbles. We define the multiplication $f_k^0(x, v_k, \tau)dx$ as the fraction of bubbles having velocity v_k that are positioned between x and $x + dx$ at the time τ. For an intense bubbling situation, the interaction of the bubbles with the liquid or with the solid coexists with the bubble–bubble interactions. For this motion type, the function of distribution of velocities is noted as $f(x, v_k, \tau)$. Indeed, from the given motion description we can identify two basic motion processes of bubbles (two elementary evolution states): bubbles interacting with only the neighborhood liquid, when their velocities remain unchanged; bubbles interacting with other bubbles, when we have an evolution of the velocities (the velocity of one bubble can skip from one state to other m possible states).

If the probability for a motion caused only by bubbles–liquid interactions is $p_a = \alpha\Delta\tau$, then the following: $1 - p_a = 1 - \alpha\Delta\tau$, gives the probability for a displacement due to the bubble–bubble interactions. When we have the last type of motion in the interval of time $\Delta\tau$, one bubble changes its velocity v_e to velocity v_k

with the probability p_{ek}. Based on this description, the probability balance (please see the model relation (4.20)) gives the following equation:

$$f_k(x, v_k, \tau + \Delta\tau) = (1 - \alpha\Delta\tau) \sum_{e=1}^{m} p_{ek} f_e(x - v_k\Delta\tau, v_e, \tau) + \alpha\Delta\tau f_k^0(x - v_k\Delta\tau, v_k, \tau)$$

(4.239)

Relation (4.239) shows that "k" bubbles (bubbles having velocity v_k) reach point x at time $\tau + \Delta\tau$ because of the interaction with the other types of bubbles (the probability for this event is $1 - \alpha\Delta\tau$) or because of the interaction with the composite liquid–solid medium (the probability for this event is $\alpha\Delta\tau$). At the same time, the bubbles that originate from the position $x - v_k\Delta\tau$ without interaction with the nearly bubbles keep their velocity; so the local distribution function of these individuals velocities is $f_k^0(x, v_k, \tau)$. Due to the stochastic character of the described process, the transition probabilities from the state "e" to all "k" states verify the unification condition. Consequently, the probability p_{kk} will be written as

$p_{kk} = 1 - \sum_{e=1, e \neq k}^{m} p_{ke}$ and relation (4.239) will be rewritten as follows:

$$f_k(x, v_k, \tau + \Delta\tau) = (1 - \alpha\Delta\tau) \sum_{e=1, k \neq e}^{m} p_{ek} f_e(x - v_k\Delta\tau, v_e, \tau) +$$

$$\left(1 - \sum_{e=1, e \neq k}^{m} p_{ke}\right) f_k(x - v_k\Delta\tau, v_k, \tau) +$$

(4.240)

$$\alpha\Delta\tau f_k^0(x - v_k\Delta\tau, v_k, \tau)$$

The determination of the transition probabilities (p_{ek}, $\forall\, k, e = 1, ...m$) is a problem that requires careful analysis. Experimental observations show that, for bubbles moving in a liquid, two interaction rules can be accepted:

1. The assembly resulting from a bubble–bubble interaction, which takes the velocity of the bubble having the higher velocity:

$$Int(v_e, v_k) = max(v_e, v_k) \quad \forall\, e, k = 1, ...m$$

(4.241)

2. The assembly resulting from a bubble–bubble interaction, which takes a velocity higher than any of the individual velocities of the bubbles:

$$Int(v_e, v_k) = sup[\,max(v_e, v_k)\,] \quad \forall\, e, k = 1, ...m$$

(4.242)

Expressions (4.241) and (4.242) describe the well known observed phenomena of accelerated bubbling.

To estimate the probabilities p_{ek} or p_{ke}, we consider the behaviour of one individual bubble having velocity v_k at position x. Their interactions with the bubbles having velocities v_e with $v_e \prec v_k$ are described with the term $-\sum\limits_{e=1,e\neq k}^{m} p_{ke}f_e(....)$ in relation (4.240). Because our "k" bubble has the highest velocity, we derive that its relative velocity with respect to the type "e" bubbles is $v_k - v_e$. For the period of time $\Delta\tau$, the covered space is $(v_k - v_e)\Delta\tau$ and the number of type "e" bubbles met by our "k" bubble is $f_e(x, v_e, \tau)(v_k - v_e)\Delta\tau$. At the same time, the probability for our bubbles to realize a linear velocity change depends on the interactions number $f_e(x, v_e, \tau)(v_k - v_e)\Delta\tau$. So, for the transition probability p_{ke}, we can establish:

$$p_{ke} = \beta f_e(x, v_e, \tau)(v_k - v_e)\Delta\tau \quad ; \; k, e = 1; 2, ...m \tag{4.243}$$

When $v_e > v_k$, the interactions of the bubbles having a v_k velocity with the bubbles having a v_e velocity are described in relation (4.240) by the term $\sum\limits_{e=1,e\neq k}^{m} p_{ek}f_e(....)$.

After the analysis, updated here for the case of $v_e \prec v_k$, we obtain the following relation for the transition probabilities p_{ek}:

$$p_{ek} = \beta f_k(x, v_k, \tau)(v_e - v_k)\Delta\tau \quad ; \; e, k = 1, 2, ...m \tag{4.244}$$

These two last relations respect the following interaction rule: (1) the assembly resulting from a bubble–bubble interaction takes the higher velocity higher of any of the individual velocities of the bubbles.

For the interaction rule of type (2), relations (4.243) and (4.244) become respectively (4.245) and (4.246) where $\sup(v_k)$ and $\sup(v_e)$ are velocities which are higher than v_k and v_e respectively:

$$p_{ke} = \beta f_e(x, v_e, \tau)(\sup(v_k) - v_e)\Delta\tau \quad ; \; k, e = 1; 2, ...m \tag{4.245}$$

$$p_{ek} = \beta f_k(x, v_k, \tau)(\sup(v_e) - v_k)\Delta\tau \quad ; \; e, k = 1, 2, ...m \tag{4.246}$$

In order to transform Eq. (4.240) into a form that can be computed, we introduce the following considerations and definitions:
- the real effect of the bubbles interaction is:

$$\lim_{\Delta\tau \to 0} \frac{1}{\Delta\tau} \left[\sum_{e=1}^{m} p_{ek}f_e(x - v_k\Delta\tau, v_e, \tau) - \sum_{e=1}^{m} p_{ke}f_k(x - v_k\Delta\tau, v_k, \tau) \right]$$

$$= \beta \left[\sum_{e=1,v_e > v_k}^{m} f_e(...)(v_e - v_k)f_k(..) - \sum_{e=1,v_e < v_k} f_e(..)(v_k - v_e)f_k(...) \right] \tag{4.247}$$

$$= \beta f_k(..) \sum_{e=1}^{m} f_e(...)(v_e - v_k)$$

- the concentration of bubbles is linear along their trajectory:

$$b(x, \tau) = \sum_{e=1}^{m} f_e(x, v_e, \tau) = \sum_{e=1}^{m} f_k^0(x, v_e, \tau) \tag{4.248}$$

- the definition of the mean velocity ($\bar{v}(x, \tau)$) of bubbles along their trajectory is:

$$\bar{v}(x, \tau) = \frac{\sum\limits_{e=1}^{m} v_e f_e(x, v_e \tau)}{\sum\limits_{e=1}^{m} f(x, v_e, \tau)} \tag{4.249}$$

We can notice that relation (4.248) is a norma condition for the distribution functions $f_k^0(x, v_k, \tau)$ and $f_k(x, v_k, \tau)$. The probability balance of the motion of bubbles is developed by replacing relation (4.247) in Eq. (4.240) and by rearranging some terms:

$$\frac{f_k(x, v_k, \tau + \Delta\tau) - f_k(x - v_k\Delta\tau, v_k, \tau)}{\Delta\tau} =$$

$$- \alpha\left[f_k(x - v_k\Delta\tau, v_k, \tau - f_k^0(x - v_k\Delta\tau, v_k, \tau\right]$$

$$+ \beta f_k(x, v_k, \tau) \sum_{e=1}^{m} f_e(x, v_e, \tau)(v_e - v_k) + O_1(\Delta\tau) \tag{4.250}$$

$$+ \beta f_k(x - v_k\Delta\tau, v_k, \tau) \sum_{e=1}^{m} f_e(x - v_k\Delta\tau, v_e, \tau)(v_e - v_k) + O_2(\Delta\tau)$$

Here $O_1(\Delta\tau)$ and $O_2(\Delta\tau)$ have negligible values. By introducing the Taylor expansion of functions $f_k(x - v_k\Delta\tau, v_k, \tau)$ into relation (4.250), we obtain the following assembly of relations:

$$f_k(x - v_k\Delta\tau, v_k, \tau) = f_k(x, v_k, \tau) - v_k\Delta\tau \frac{\Delta f_k(x, v_k, \tau)}{\Delta x} + \ldots\ldots \tag{4.251}$$

$$\frac{\partial f_k}{\partial \tau} + v_k \frac{\partial f_k}{\partial x} = -\alpha(f_k - f_k^0) + \beta f_k \sum_{e=1}^{m} f_e(v_e - v_k) \tag{4.252}$$

The final form of the stochastic model of the gas bubbling in the liquid–solid system is written by coupling Eq. (4.252) with Eqs. (4.248) and (4.249):

$$\frac{\partial f_k(x, v_k, \tau)}{\partial \tau} + v_k \frac{\partial f_k(x, v_k, \tau)}{\partial x} = - \alpha[f_k(x, v_k, \tau) - f_k^0(x, v_k, \tau)] +$$

$$\beta f_k(x, v_k, \tau)b(x, \tau)[\bar{v}_k - v(x, \tau)] \quad k = 1, ..m \tag{4.253}$$

When the number of the elementary states of the process (m) is important, the discrete model (4.253) can be written in a continuous form:

$$\frac{\partial f(x, v, \tau)}{\partial \tau} + v \frac{\partial f(x, v, \tau)}{\partial x} = -\alpha[f(x, v, \tau) - f^0(x, v, \tau)] + \beta f(x, v, \tau)b(x, \tau)[v - \bar{v}(x, \tau)]$$

$$\tag{4.253'}$$

Here we have:

$$b(x, \tau) = \int_0^\infty f(x, v, \tau)dv = \int_0^\infty f^0(x, v, \tau)dv \tag{4.254}$$

$$\bar{v}(x, \tau) = \frac{\int_0^\infty vf(x, v, \tau)dv}{\int_0^\infty f(x, v, \tau)dv} \tag{4.255}$$

Taking into consideration the physical meaning of $f_k(x, v_k, \tau)$ and $f(x, v, \tau)$ and deriving $f(x, v, \tau)dxdv$, we define the number of the bubbles positioned at time τ between x and $x + dx$ and that gives the velocities in the interval $(v, v + dv)$. Relation (4.253) shows that: (i) the number of bubbles with velocity v decreases with the fraction $\alpha f(x, v, \tau)$ due to their interaction with the neighbouring medium; (ii) the number of bubbles with velocity v increases with the fraction $\alpha f^0(x, v, \tau)$ due to their interaction with the neighbouring medium; (iii) for $v > \bar{v}$ the number of the bubbles with velocity v increases respectively for $v < \bar{v}$ decreases due to the interactions with other bubbles.

The complete unsteady state stochastic model of the bubbling process is given coupling the assembly of relations (4.253)–(4.255) with the univocity conditions. The numerical analysis (checking) of this model can easily produce interesting data for the cases of bubbles coalescence and bubbles breaking. One interesting solution of this model corresponds to the case of a homogenous steady state bubbling which can be obtained with relation (4.253) and considering $\partial/\partial\tau = \partial/\partial x = 0$ and $1 + \dfrac{\beta}{\alpha}(\bar{v} - v) > 0$. Here $f(x, v, \tau)$ becomes $f(v)$:

$$f(v) = \frac{f^0(v)}{1 + \dfrac{\beta}{\alpha}(\bar{v} - v)b} \tag{4.256}$$

With the same conditions but considering $1 + \dfrac{\beta}{\alpha}(\bar{v} - v) = 0$, the solution for $f(v)$ is given by the new relation below written; this fact is equivalent to $v = \bar{v} + \alpha/(b\beta)$ which is the univocity condition for the model given by relations (4.253)–(4.255):

$$f(v) = \frac{f^0(v)}{1 + \dfrac{\beta}{\alpha}(\bar{v} - v)b} + \gamma b\delta\left[1 + \frac{\beta}{\alpha}(\bar{v} - v)b\right] \tag{4.256'}$$

This expression (bubble velocities distribution) shows that two types of flows participate in the bubbling process: the first type, introduced by the first term Eq. (4.256'), is the regular flow; the second flow type is called singular flow and is contained in the term where the Dirac function $\delta[1 + \frac{\beta}{\alpha}(\bar{v} - v)b]$ appears. The singular flow becomes unimportant when we have: (i) a velocity distribution in a restricted domain around $\bar{v}$; (ii) a slow concentration of bubbles. Both cases are coupled when we can consider that $\gamma \to 0$ in relation (4.256'). If we multiply the left and right terms of equation (4.256') by v and then integrate it for all velocities of the bubbles, we obtain relations (4.257)–(4.258). Here we used $\bar{v} = \int_0^\infty vf(v)dv / \int_0^\infty f(v)dv$ for the identification of the mean bubble velocity and for the variance of the velocities around the mean velocity:

$$\sigma^2 = \int\limits_0^\infty (\bar{v} - v)vf(v)dv \Big/ \int\limits_0^\infty f(v)dv = \int\limits_0^\infty v\left[1 + \frac{\beta}{\alpha}(\bar{v} - v)b\right]f(v)dv = \int\limits_0^\infty vf^0(v)dv \quad (4.257)$$

$$\bar{v} - \frac{\beta}{\alpha}\sigma^2 b = \bar{v}^0 \quad (4.258)$$

Equation (4.258) gives the mean bubble velocity which is described through the apparent gas velocity $\bar{v} = w_g/\varepsilon_g$, whereas the linear bubble concentration is given through the bubble diameter and the gas fraction $b = \left(\dfrac{6}{\pi}\right)^{1/3}\dfrac{\varepsilon_g^{1/3}}{d_{mb}}$. With these two last expressions we can write that:

$$w_g = \frac{\bar{v}^0\varepsilon_g}{1 - k\varepsilon_g^{1/3}} \quad (4.259)$$

where $k = \left(\dfrac{6}{\pi}\right)^{1/3}\dfrac{\beta\sigma^2}{\alpha\bar{v}d_{mb}}$ contains all the unknown factors introduced during model building. Here, σ^2 depends on $\bar{v}$. Indeed, it appears of interest to describe a case where σ^2 becomes independent with respect to $\bar{v}$. In relation (4.259), when $\varepsilon_g \to 1$, we have a limit case where $k = 1$ and the relation between the gas fraction and the gas apparent velocity becomes:

$$w_g = \frac{\bar{v}^0\varepsilon_g}{1 - \varepsilon_g^{1/3}} \quad (4.260)$$

To compute the mean bubble velocity ($\bar{v}^0$), it is necessary to know the mean bubble diameter and some physical properties of the medium in which the bubbles evolve.

Experimental checking. Figure 4.29 compares the theoretical calculations and experimental results for the evolution of $\varepsilon_g/(1 - \varepsilon_g^{1/3})$ with $w_g/\bar{v}^0$ for a MFPB. The bed height has been fixed at 0.25 m. Two liquid–solid systems with solid fractions of 0.1 and 0.3 m³/m³ have been chosen. A 0.0275 m diameter spheres of density 980–1030 kg/m³ were selected as mobile packing. Air and water are the working fluids. The gas fraction results from bed expansion when gas flows through the liquid–solid system.

The computation of the mean bubble velocity was based on the bubble diameter resulting from the bubble forming at each submerged orifice from the gas bubbling arrangement. In Fig. 4.29, we can observe that the stochastic model gives a good trend with respect to the experimental results, even if some discordance appears, especially at small values of the gas fraction. In all cases, the model underestimates the experimental results, the underestimation is between 5 and 15% for values of $w_g/\bar{v}^0$ greater than 0.3.

The problem of bubble motion in a liquid is fundamental in chemical reaction engineering because about 25% of all chemical reactions occur in bi-phase systems. As we have shown, the gas–liquid two-phase flow prevailing in a bubble column is extremely complex. It is dominated by a rich variety of logical configurations and exhibits inherent unsteadiness. As a consequence, the modelling of this flow is an attractive subject and constitutes an excellent subject for stochastic modelling.

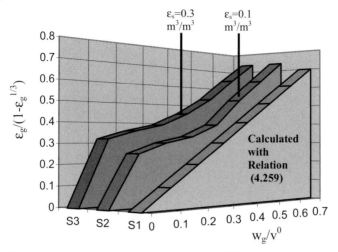

Figure 4.29 Gas hold-up versus gas velocity for a MFPB.

4.6.2
Species Movement and Transfer in a Porous Medium

Frequently we define a porous medium as a solid material that contains voids and pores. The notion of "pore" requires some observations for an accurate description and characterization. If we consider the connection between two faces of a porous body we can have opened and closed or blind pores; between these two faces we can have pores which are not interconnected or with simple or multiple connections with respect to other pores placed in their neighborhood. In terms of manufacturing a porous solid, certain pores can be obtained without special preparation of the raw materials whereas designed pores require special material synthesis and processing technology. We frequently characterize a porous structure by simplified models (Darcy's law model for example) where parameters such as volumetric pore fraction, mean pore size or distribution of pore radius are obtained experimentally. Some porous synthetic structures such as zeolites have an apparently random internal arrangement where we can easily identify one or more cavities; the connection between these cavities gives a trajectory for the flow inside the porous body (see Fig. 4.30).

The diameter or the radius of the pores is one of the most important geometric characteristic of porous solids. In terms of IUPAC nomenclature, we can have macropores (mean pore size greater than 5×10^{-8} m), mesopores (between 5×10^{-8} and 2×10^{-9} m) and micropores (less than 2×10^{-9} m). The analysis of species transport inside the porous structure is very important for the detailed description of many unit operations or applications; among them we can mention: suspension filtration, solid drying and humidification, membrane processes (dialysis, osmosis, gaseous permeation ......), flow in catalytic beds, ion exchange, adsorp-

(a)

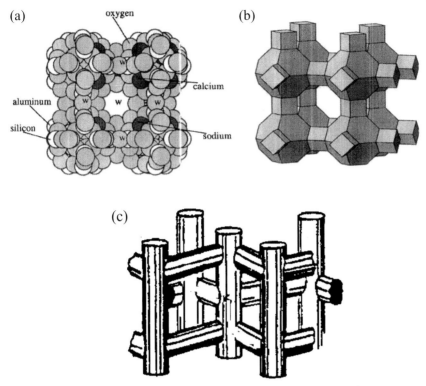

(b)

(c)

Figure 4.30 Structure of a zeolite material. (a) Internal view – atoms, (b) internal view of windows, (c) central-cavities and windows.

tion, solid–liquid extraction, the dispersion of therapeutic species inside an animal or human body, species penetration in porous soils etc.

The most used methods for the characterization of flow and species transport inside a porous body include the identification of the characteristics of the pores of the porous structure and the particularisation of classic transport equations to this case. These equations are generally associated with equations describing the solid–fluid interaction, adsorption, capillary condensation and flow due to the capillary forces etc. Concerning the species displacement (flow) problem inside a porous structure, we can consider the following classification:

- For pores with a radius between 10^{-3} and 10^{-7} m, the theory of Poiseuille flow is valid; so the mean force for fluid flow between two planes is expressed by the pressure difference; this can be a consequence of differential actions of external and capillary or/ and gravitational forces.
- For pores having a mean radius between 10^{-8} and 10^{-9} m, we explain the porous body flow by the Knudsen theory; here the diffusion coefficient and, consequently, the flow, strongly depend on

the molecular weight of the species (two or more species moving inside the porous body will present different displacement velocities);

- For pores smaller than 10^{-9} m, a molecular sieving effect can be present and the movement of one or more species inside the porous solid occurs due to the molecular interactions between the species and the network of the porous body; here, for the description of species displacement, the theory of molecular dynamics is frequently used. The affinity between the network and the species is the force that controls the molecular motion; at the same time, the affinity particularities, which appear when two or more species are in motion inside the porous structure, explain the separation capacity of those solids. We can use a diffusive characterisation of species motion inside a porous solid by using the notion of conformational diffusion.

Porous solids generally have a pores size distribution and in many cases this results in a complicated transport mechanism which is a combination of the different mechanisms described above. This is also the case when the pores size are ranged near the boundaries between these different mechanisms.

All these different mechanisms of mass transport through a porous medium can be studied experimentally and theoretically through classical models (Darcy's law, Knudsen diffusion, molecular dynamics, Stefan–Maxwell equations, dusty-gas model etc.) which can be coupled or not with the interactions or even reactions between the solid structure and the fluid elements. Another method for the analysis of the species motion inside a porous structure can be based on the observation that the motion occurs as a result of two or more elementary evolutions that are randomly connected. This is the stochastic way for the analysis of species motion inside a porous body. Some examples that will be analysed here by the stochastic method are the result of the particularisations of the cases presented with the development of stochastic models in Sections 4.4 and 4.5.

4.6.2.1 Liquid Motion Inside a Porous Medium

The classic and stochastic methods used for the analysis of liquid flow inside a porous medium are strongly related. These interactions are given by the relationships between the parameters of both types of models. We show here that the analysis of the flow of a liquid through a porous medium, using a stochastic model, can describe some of the parameters used in deterministic models such as:

- parameters from Darcy's law;
- parameters that appear in the equation of flow continuity in a porous medium;
- parameters used by the models explaining the flow mechanism inside a porous solid.

Apparently the parameters of stochastic models are quite different from those of classic (deterministic) models where the permeability, the porosity, the pore radius, the tortuosity coefficient, the specific surface, and the coefficient of the effective diffusion of species represent the most used parameters for porous media characterization. Here, we will present the correspondence between the stochastic and deterministic parameters of a specified process, which has been modelled with a stochastic and deterministic model in some specific situations.

4.6.2.1.1 Stochastic Modelling of Dispersion of a Liquid in a Porous Body

The dispersion of a liquid that flows inside a porous medium is the macroscopic result of some individual motions of the liquid determined by the pore network of the solid structure. These motions are characterised by the local variations of the velocity magnitude and direction. Accepting the simplified structure of a porous structure shown in Fig. 4.31, the liquid movement can be described by the motion of a liquid element in a +x direction (occurring with the probability p) compared to the opposite motion or −x displacement (here q gives the probability of evolution and Δx represents the length portion of the pore which is not in contact with the nearby pores). Indeed, the balance of probability that shows the chances for the liquid element to be at time τ in x position can be written as follows:

| The probability to have the fluid element at time τ in position x | = | The probability to have the fluid element at time $\tau - \Delta\tau$ in position $x - \Delta x$ with an evolution along $+x$ for the next $\Delta\tau$ time | + | The probability to have the fluid element at time $\tau - \Delta\tau$ in position $x + \Delta x$ with an evolution along $-x$ for the next $\Delta\tau$ time | (4.261) |

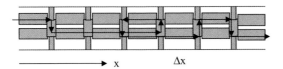

x Δx

Figure 4.31 Fluid movement inside a uniform porous body.

When we express relation (4.261) mathematically we have:

$$P(x, \tau) = pP(x - \Delta x, \tau - \Delta\tau) + qP(x + \Delta x, \tau - \Delta\tau) \tag{4.262}$$

The Taylor expansions of $P(x + \Delta x, \tau - \Delta\tau)$ and $P(x - \Delta x, \tau - \Delta\tau)$ are used as their right term:

$$\frac{\Delta P(x, \tau)}{\Delta\tau} + (p - q)\frac{\Delta x}{\Delta\tau}\frac{\Delta P(x, \tau)}{\Delta x} = \frac{\Delta x^2}{2\Delta\tau}\frac{\Delta^2 P(x, \tau)}{\Delta x^2} \tag{4.263}$$

The term $(p - q)\dfrac{\Delta x}{\Delta \tau}$ has a velocity dimension ($L\,T^{-1}$) and physically represents the net velocity of the liquid moving in the flow direction (w). The ratio $\dfrac{\Delta x^2}{2\Delta \tau}$ has the dimension of a diffusion coefficient ($L^2\,T^{-1}$) and is recognized as the dispersion coefficient (D). With these observations we can rewrite relation (4.263) in order to obtain Eq. (4.264) which is the equation that characterizes the dispersive flow in one dimension (for instance see Section 3.35).

$$\frac{\partial P(x, \tau)}{\partial \tau} + w\frac{\partial P(x, \tau)}{\partial x} = D\frac{\partial^2 P(x, \tau)}{\partial x^2} \tag{4.264}$$

It is not difficult to observe that, using this simple stochastic model of liquid flow inside the porosity, we obtain that the parameters of the model, such as the net flow velocity (w) and the dispersion coefficient (D), are determined by the porous structure. This last parameter is considered here through the value of Δx (length of one pore which is not in contact with nearby pores).

In order to solve the model equation, we must complete it with the univocity conditions. In some cases, relations (3.100)–(3.107) can be used as solutions for the model particularized for the process. The equivalence between both expressions is that $c(x, \tau)/c_0$ appears here as $P(x, \tau)$. Extending the equivalence, we can establish that $P(x, \tau)$ is in fact the density of probability associated with the repartition function of the residence time of the liquid element that evolves inside a uniform porous structure.

In the scientific literature, we can find a large quantity of experimental results where the flow characterization inside a porous medium has shown that the value of the dispersion coefficient is not constant. Indeed, for the majority of porous structures the diffusion is frequently a function of the time or of the concentration of the diffusing species. As far as simple stochastic models cannot cover these situations, more complex models have been built to characterize these dependences. One of the first models that gives a response to this problem is recognized as the *model of motion with states having multiple velocities*.

With this model, the liquid element evolves inside a porous solid with random motions having the velocities $v_i, i = 1, ..m$. These random skips of velocity from one state to another can be explained by random changes in pore sections and pore interconnections. This description can be completed with the consideration that here the elementary connection between the states (from one velocity or flow to another) becomes a Markov type connection: $p_{ij} = p_{ij}^* a\Delta \tau = \alpha_{ij}\Delta \tau$. We can observe that, for a randomly chosen length of time, the component of the process remains unchanged (the motion of the liquid with velocity v_i); after this length of time, the liquid element changes its velocity by skipping to another elementary state of process and again it keeps this new value (v_j) constant during this new length of time. For this stochastic description, the balance of probabilities gives relation (4.265). Here $P_i(x, \tau + \Delta \tau)$ represents the density of the probability that shows the possibility of the existence of the liquid element at time $\tau + \Delta \tau$ in the position x with the evolution $v_i, i = 1, ...m$.

$$P_i(x, \tau + \Delta\tau) = \sum_{j=1}^{m} p_{ji} P_j(x - v_i \Delta\tau) \quad, \quad i = 1, \ldots m \tag{4.265}$$

If we go back to Section 4.4, then we discover that the model presented herein is identical to the model presented at the beginning of Section 4.4.2. Based on the analogy principle, we can extend the treatment of this previous model (Section 4.4.3) to the model in progress. Consequently, the assembly of relations (4.265) takes the following form:

$$\frac{\partial P_i(x, \tau)}{\partial \tau} = -v_i \frac{\partial P_i(x, \tau)}{\partial x} - \left(\sum_{j=1, j \neq i}^{m} a_{ij} \right) P_i(x, \tau) + \sum_{j=1, j \neq i}^{m} a_{ji} P_j(x, \tau) \tag{4.266}$$

Equation (4.266) shows that the time evolution of the fraction of the fluid (or fluid elementary particles) that reaches position x with velocity v_i at time τ is determined by the following types of particles (i) particles having velocity v_i and leaving position x; (ii) particles having velocity v_i and reaching position x ; (iii) particles reaching position x and changing their velocity from v_j to v_i. For the particular case where we have two evolution states for the fluid velocity ($v_1 = +v$, $v_2 = -v$) the general model (4.266) is written as the set of relations (4.267). Here, the consideration of $a_{12} = a_{21} = a$ shows that we have a case of isotropic porous solid:

$$\frac{\partial P_1(x, \tau)}{\partial \tau} = -v \frac{\partial P_1(x, \tau)}{\partial x} - aP_1(x, \tau) + aP_2(x, \tau)$$
$$\frac{\partial P_2(x, \tau)}{\partial \tau} = v \frac{\partial P_2(x, \tau)}{\partial x} - aP_2(x, \tau) + aP_1(x, \tau) \tag{4.267}$$

This model is of interest because it can be easily reduced to a hyperbolic form of the transport model of one property. With some particular univocity conditions, this hyperbolic model accepts analytical solutions, which are similar to those of an equivalent parabolic model. The hyperbolic model for the transport of a property is obtained by coupling the equation $P(x, \tau) = P_1(x, \tau) + P_2(x, \tau)$ to relations (4.267) and then eliminating the terms $P_1(x, \tau)$ and $P_1(x, \tau)$. The result can be written as:

$$\frac{\partial P(x, \tau)}{\partial \tau} + \frac{1}{2a} \frac{\partial^2 P(x, \tau)}{\partial \tau^2} = \frac{v^2}{2a} \frac{\partial^2 P(x, \tau)}{\partial x^2} \tag{4.268}$$

$P(x, \tau)$ gives the probability of having the liquid element flowing inside the porous solid, in position x at time τ. By a simple analysis of the hyperbolic model for the property transport, (relation (4.268)) we can conclude that, in the case when parameter α has a high value, the term $\dfrac{1}{2a} \dfrac{\partial^2 P(x, \tau)}{\partial \tau^2}$ can be negligible with respect to other terms. The result is the conversion of the hyperbolic model into a parabolic model. For the transport in one dimension, this model is given by the partial differential equation:

$$\frac{\partial P(x, \tau)}{\partial \tau} = \frac{v^2}{2a} \frac{\partial^2 P(x, \tau)}{\partial x^2} \tag{4.269}$$

Now, we can have a special univocity case that considers the following unitary impulse as presented in the example of Section 4.2.1 as signal to the flow input inside the porous solid:

$$P(x, 0_-) = \begin{cases} 0 & \text{for } x \leq 0 \\ 0 & \text{for } x \succ 0 \end{cases}, P(x, 0_+) = \begin{cases} 1 & \text{for } x \leq 0 \\ 0 & \text{for } x \succ 0 \end{cases} \tag{4.270}$$

The solution for the coupling of the model equation (4.269) with the above speci-fied conditions (relations (4.270)) can be reached using the solution given by Crank [4.87] for the response given by a similar model to a unitary impulse input:

$$P(x, \tau) = \int_{-\infty}^{0} P^{imp}(|x - \xi|, \tau)d\xi = -\int_{\infty}^{x} P^{imp}(\eta, \tau)d\eta = \int_{x}^{\infty} P^{imp}(\eta, \tau)d\eta \tag{4.271}$$

For the particularization of this last equation to our problem, we have to take into account the following observations: (i) $P(x, \tau)$ is normalized (its values are included in the interval [0, 1]); (ii) $P(x, \tau)$ is symmetric with respect to the plane $x = 0$. So we can write:

$$P(x, \tau) = \frac{1}{2} - \int_{0}^{x} P^{imp}(\eta, \tau)d\eta \quad \text{for } x > 0$$

$$P(x, \tau) = \frac{1}{2} + \int_{x}^{0} P^{imp}(\eta, \tau)d\eta \quad \text{for } x < 0 \tag{4.272}$$

The same particularization procedure is used to establish a solution for the hyper-bolic model (4.268) coupled to conditions (4.270). The solutions for $P^{imp}(x, \tau)$, which correspond to the parabolic and hyperbolic models, are presented in Sec-tion 4.2.1 (for instance see relations (4.36) and (4.37)). The results for the probabil-ities $P(x, \tau)$ are given by the following relations:

$$x > 0 \ ; \ P(x, \tau) = \frac{1}{2} - \int_{0}^{x} \sqrt{\frac{\alpha}{2v^2\tau}} \exp\left(-\frac{\alpha\eta^2}{2v^2\tau}\right)d\eta$$

$$= \frac{1}{2} - \int_{0}^{\frac{x}{\sqrt{2v^2\frac{\tau}{\alpha}}}} \frac{1}{\sqrt{\pi}} e^{-z^2}dz = \frac{1}{2} - \frac{1}{2}\text{erf}\left(\frac{x}{\sqrt{2v^2\frac{\tau}{\alpha}}}\right) \tag{4.273}$$

$$x < 0 \ ; \ P(x, \tau) = \frac{1}{2} + \frac{1}{2}\text{erf}\left(\frac{x}{\sqrt{2v^2\frac{\tau}{\alpha}}}\right)$$

The solution for the hyperbolic model is given by Eqs. (4.274). For negative values of x (x < 0) the computation for the model solutions is developed by the same pro-cedure given above:

$$x > 0 \; ; \; P(x,\tau) = \frac{1}{2} - \int_0^x \frac{\alpha e^{-\alpha \tau}}{2v} \left[I_0 \left(\alpha\tau\sqrt{1 - \frac{\eta^2}{v^2\tau^2}} \right) + \frac{1}{\sqrt{1 - \frac{\eta^2}{v^2\tau^2}}} I_1 \left(\alpha\tau\sqrt{1 - \frac{\eta^2}{v^2\tau^2}} \right) \right] d\eta$$

$$= \frac{1}{2} e^{-\alpha\tau} \left[I_0 \left(\frac{\alpha}{v}\sqrt{v^2\tau^2 - x^2} \right) + 2 \sum_{n=1}^{\infty} \left(\frac{v\tau - x}{v\tau + x} \right)^{\frac{n}{2}} I_n \left(\frac{\alpha}{v}\sqrt{v^2\tau^2 - x^2} \right) \right] \quad \text{for} \; 0 \prec x \prec v\tau$$

$$P(x,\tau) = 0 \; \text{for} \; x \succ v\tau \tag{4.274}$$

With the hyperbolic and parabolic models, we can describe the evolution for the existence probability $P(x,\tau)$, which is shown in Fig. 4.32. Major differences between both models can be observed at small values of time.

The hyperbolic model shows a fast evolution of the probability $P(x.\tau)$ at the spatial distance $x = y\tau$ with respect to $x = 0$ or more precisely at $x/[v^2\tau/\alpha]^{0.5} = 1/2\exp(-\alpha\tau)$. At moderate or large time, we cannot observe a difference between the predicted values of $P(x.\tau)$ from the models. This is due to the rapid decrease with time of the magnitude of the rapid evolution of the predicted probability $P(x.\tau)$ in the hyperbolic model. It is important to specify that the hyperbolic model keeps a fast evolving probability $P(x.\tau)$ for all possible univocity conditions at small time. It is difficult to demonstrate experimentally the prediction of the stochastic hyperbolic model for the liquid dispersion inside a porous solid because the predicted skip is very fast $P(x.\tau)$ and not easily measurable.

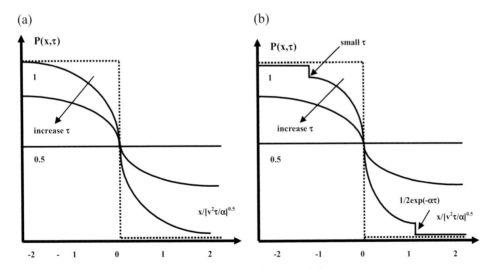

Figure 4.32 Differences between the parabolic and hyperbolic models for the calculation of the evolution of $P(x,\tau)$. (a) Parabolic model, (b) hyperbolic model.

In the characterization of porous membranes by liquid or gaseous permeation methods, the interpretation of data by the hyperbolic model can be of interest even if the parabolic model is accepted to yield excellent results for the estimation of the diffusion coefficients in most experiments. This type of model is currently applied for the time-lag method, which is mostly used to estimate the diffusion coefficients of dense polymer membranes; in this case, the porosity definition can be compared to an equivalent free volume of the polymer [4.88, 4.89].

Coming back to the stochastic analysis of the elementary particle motion inside the porous solid, we can notice that this analysis introduces a consistent explanation of the parameters participating in the coefficient of diffusion or dispersion, which is written as $D = \lim\limits_{\Delta x \to 0, \Delta\tau \to 0} \dfrac{(\Delta x)^2}{2\Delta\tau} = \dfrac{v^2}{2\alpha}$. Indeed, it is determined by the motion velocity of the species and by the frequency of the changes of direction of the velocity. Because v and α have specific values for each individual species–porous structure couple, then the diffusion is the mechanism which allows separating out such species when they permeate through a porous membrane.

The particularization of the limit theorem of the second type to model (4.267) (for instance see also Section 4.5.1.2, relations (4.132)–(4.134)) shows that the stochastic model of the process becomes asymptotic with the parabolic model. Indeed, we can identify the expressions $Q = \begin{pmatrix} -\alpha & \alpha \\ \alpha & -\alpha \end{pmatrix}$, $\overline{V}_1 = 0$, $\overline{V}_{11} = \dfrac{v^2}{2\alpha}\dfrac{\partial^2}{\partial x^2}$ that transform model (4.267) into model (4.269). The deviation of the original stochastic model to a parabolic one is not a definitive argument to eliminate the use of the hyperbolic model for the practical interpretation of some experimental data on membrane permeation.

In porous solids made of larger elements such as fixed packed beds, where the characteristic dimension of the packing is d (for example the diameter of a packed solid), the frequency of the velocity change is $\alpha = v/d$ (after each flow through an element of the packed bed, the local fluid velocity v changes its direction). Now if we use this value of α in the dispersion coefficient, we obtain the famous relation $Pe = (vd)/D = 2$, which gives the value of the dispersion coefficient when a fluid flows through a packed bed [4.90].

This stochastic model of the flow with multiple velocity states cannot be solved with a parabolic model where the diffusion of species cannot depend on the species concentration as has been frequently reported in experimental studies. Indeed, for these more complicated situations, we need a much more complete model for which the evolution of flow inside of system accepts a dependency not only on the actual process state. So we must have a stochastic process with more complex relationships between the elementary states of the investigated process. This is the *stochastic model of motion with complete connections*. This stochastic model can be explained through the following example: we need to design some flowing liquid trajectories inside a regular porous structure as is shown in Fig. 4.33. The porous structure is initially filled with a fluid, which is non-miscible with a second fluid, itself in contact with one surface of the porous body. At the start of the process, the second fluid begins to flow inside the well-structured porosity by means of a process with the following characteristics:

- in a given time only a small portion of the liquid inside the porosity is in a moving state;
- the points where the liquid gets into and out of the porous solid are randomly distributed with time;
- we cannot exclude the possibility for the liquid element to come back to a previous position;
- the present motion of the liquid element depends on its previous state;
- at each spatial position and timing for the liquid motion, we can identify four elementary states of the motion: k = 1 forward lead; k = 2 backward lead; k = 3 left lateral movement; k = 4 right lateral movement;
- the change in liquid velocity occurs not only when the flowing liquid changes direction;
- the length of the step of the liquid movement is randomly distributed.

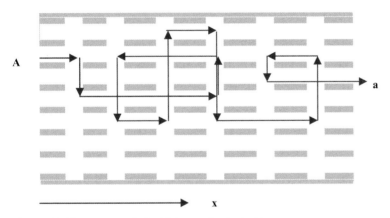

Figure 4.33 The trajectory of a liquid element flowing inside a regular porous structure.

If we combine all the aspects above with the descriptions of basic stochastic processes, then we can conclude that we have the case of a stochastic process with complete and random connections (see Section 4.4.1.1).

If a liquid element is initially in an i position moving with a k type of motion, the probability that shows its coming to a j position as a result of n motion steps is given by:

$$P_k^*(n, i, j) = \sum_{e \in K} \sum_{a \in Z} p_{ke}(i) p_k(a) P_k^*(n-1, i+a, j) \qquad (4.275)$$

In relation (4.275), we recognize $p_{ke}(i)$, which represents the transition probability from a type k motion into a type e motion at position i. By $p_k(a)$, we express the distribution of the length of steps related with type k motion.

For this case, the random system with complete connections $[(A,\mathbf{A}),(B,\mathbf{B}),u,P]$ presents the following particularisations: $A = B = ZxK$, $u(S_n,E_{n+1}) = u((i,k),(a,e)) = (i+a,e)$ and $P(S_n...E_1,S_1, S_0) = P((i,k);(a,e)) = p_e(a)p_{ke}(i)$.

The probability of the type k motion for the liquid element which gets to the j position, in a period of time given by the n evolution steps, is given by:

$$P_k(n,j) = \sum_{e \in K} \sum_{a \in Z} p_{ek}(j-a)p_k(a)P_k(n-1,j-a) \qquad (4.276)$$

For a practical computation, these two last equations need: (i) one procedure that gives the transition probabilities from e to k state at each $j - a$ position ($p_{ek}(j - a)$); (ii) some practical relations that express, for $k = 1$, $k = 2$, etc., the distributions of the step lengths ($p_k(a)$). It is not difficult to establish that the transition probabilities $p_{ek}(j - a)$ depend strongly on the totality of the previous trajectory. As a consequence, the fluid flowing trajectory is continuously updated step by step.

4.6.2.1.2 Stochastic Models for Deep Bed Filtration

Deep bed filtration is used to clarify suspensions with a small content of solids. This process, which is usually applied for water treatment, is based on the flowing of a fluid through a deep bed of granular solids such as sand. During the flowing inside the granular bed, interaction forces between one particle from the suspension and one particle from the granular bed occur. This interaction allows the particle from the suspension to latch onto the particle of the bed. This elementary process occurs in many points placed through all the granular bed. The quantity of the solid retained in one element of the system cannot exceed the quantity held-up by the open spaces of the granular bed. When this retained quantity approaches the quantity determined by the bed porosity we can assert that the bed is clogged. After clogging, we can regenerate the granular bed by a counter current liquid fluidization but, depending on the type of bed filtration, other regeneration processes can be used. For this filtration case, the quantity of particles retained by the bed increases with time and, consequently, the filtrate flow rate decreases if we do not increase the filtration pressure difference.

In addition to the traditional deep bed filtration, other interesting examples of different processes and techniques can be described by the same basic principle: (i) the tangential micro-filtration and ultra-filtration where a slow deep filtration produces the clogging of the membrane surface; (ii) some processes of impregnation of porous supports with a sol in order to form a gel which, after precipitation, will form a membrane layer. Here the sol penetration inside the support is fundamental for the membrane quality.

The modelling of the deep bed filtration based on the description of the trajectory of a suspension particle and on the deposition on a bed of solid has been extensively published. However, until now, the majority of the results generated by these models were not satisfactory, because the models generally consider simplifications with respect to the action of the forces that cause the deep bed filtration. While reviewing the forces occurring in the deep bed filtration, the complexity of this operation can easily be noticed. Figure 4.34, which aims to indirectly present those forces, shows the movement of one particle of the suspension around one particle of the granular bed. Among the most important forces considered in deep bed filtration we have:

- The inertial force, that expresses the tendency of the microparticle to keep moving when it is under the influence of the hydrodynamic trajectory imposed by the flow around one element of the fixed bed. The action of this force is given by the number $In = (\rho_p d^2 w_f)/(18\eta d_s)$; here d is the microparticle diameter, $d_s = D/2$ represents the diameter of the element of the granular bed, ρ_p gives the density of the microparticle which is expected to deposit, η is the viscosity of the flowing liquid and w_f measures the real local liquid velocity.
- The gravitational force characterizing the settling capacity of the microparticle from the suspension flowing inside the porosity. This is given by the Stokes number, $St = [g(\rho_p - \rho)d^2]/(18\eta w_f)$, which is a ratio between the Stokes settling velocity and the local velocity for the flowing suspension.
- The diffusion force, giving the local action of the Brownian motion on the deposition of the microparticle. A modified Peclet

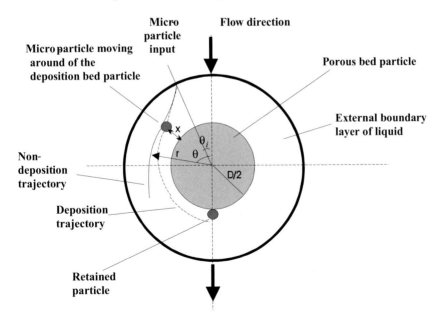

Figure 4.34 Scheme of the microparticle retention by one particle of a fixed bed.

number, $Pe = (3\pi\eta d d_s w_f)/(kT)$, is then considered. It is the ratio between the Stokes and Brownian forces, which together influence the microparticle movement;

- the laminar flow force characterizing the action of the flowing liquid on the microparticle; when the flowing field around the element of the porous structure is not uniform, an undesired rotation movement will be induced for the microparticle. The effect of the laminar flow force can be considered by means of the Reynolds number ($Re = (w_f d_s \rho)/\eta$).

The common action of these four forces results in the global mechanism that produces the approach of microparticles to deposition elements of the porous solid. At distances shorter than 1 µm, other forces come into play and produce the fixation of microparticles onto elements of the porous structure. Among these forces we have:

- The electrostatic force that appears when microparticles and the deposition element of the bed have electric charges; when the electric charges of both entities have identical signs, we have a repulsive force, the value of which is predicted by relation (4.277). The suspensions including ionized substances contain an excellent source of charged microparticles.

$$F_R^{(x)} = \frac{\exp\left[-kd\left(\frac{2x}{d} - 2\right)\right]}{1 + \exp\left[-kd\left(\frac{2x}{d} - 2\right)\right]} \tag{4.277}$$

- The Van der Waals force that is caused by the molecular vibrations of the material composing microparticles and deposition elements. This is an attraction force that strongly depends on the interparticles distance and on the wavelength ($\bar{\lambda}$) that characterizes the assembly microparticle–deposition element. Relation (4.278) gives a qualitative indication of the value of this force. In this relation, $F(u)$ gives a function which decreases rapidly with the distance (x).

$$F_{vw}^{(x)} = \frac{1}{\left(\frac{2x}{d} - 2\right)^2} F\left(\frac{\frac{2x}{d} - 2}{\bar{\lambda}}\right) \tag{4.278}$$

- The hydrodynamic adhesion force that expresses the resistance occurring when microparticles latch onto deposition elements. It is caused by the liquid that must be extracted out of the space between two particles when both microparticles adhere. This force allows the slowing down of microparticles adhesion and offers a possibility for drowning it in the flowing suspension.
- The detachment force that realizes the detachment of the assembly microparticle–deposition element; when the number of the

retained microparticles on one deposition element is important, this force is very active. This force is not a short distance action force.

As far as these forces, which present various origins and specificities, determine an assembly of very complex interactions between the suspension of microparticles and the deposition elements of the porous solid it is impossible to build a completely phenomenological model for deep bed filtration. Nevertheless, various empirical models have been developed by simplifying the assumptions concerning the description of interactions. Among these models, we have the famous filtration coefficient model or the Mint model [4.81]. This filtration coefficient noted as $\lambda(\lambda_0, c_{ss})$ depends on its initial value (λ_0) and on the local concentration of the retained solid around the bed deposition elements (c_{ss}). It is defined as the fraction of the solid retained from the suspension in an elementary length of the granular bed:

$$\lambda(\lambda_0, c_{ss}) = -\frac{dc_{vs}}{c_{vs}} \frac{1}{dx} \qquad (4.279)$$

In Fig. 4.35, the mass balance of solid in an elementary volume is given when the suspension flow is considered as a plug flow. This figure allows the establishment of relationships between the local concentration of the solid in suspension and the retained solid in the bed.

The mass balance of the retained solid is described by Eq. (4.280). The Mint deterministic model results from the coupling of this relation with the definition of the filtration coefficient. The result is written in Eq. (4.281) for the start time of the filtration and in Eq. (4.282) for the remaining filtration time. Here α is the detachment coefficient of the retained particle; its dimension is T^{-1}.

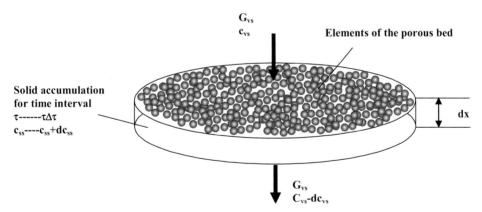

Figure 4.35 Scheme for the mass balance of the retained solid in deep bed filtration.

$$-\frac{dc_{vs}}{dx} = \frac{A}{G_{vs}} \frac{\partial c_{ss}}{\partial \tau} = \frac{1}{w_f} \frac{\partial c_{ss}}{\partial \tau} \qquad (4.280)$$

$$\frac{\partial c_{ss}}{\partial \tau} = w_f \lambda_0 c_{vs} \quad ; \quad \tau = 0 \qquad (4.281)$$

$$\frac{\partial c_{ss}}{\partial \tau} = w_f \lambda_0 c_{vs} - \alpha c_{ss} \quad ; \quad \tau \succ 0 \qquad (4.282)$$

Relation (4.283), obtained by coupling Eq. (4.282) and (4.280), presents the time derivation results in (4.284). Replacing the term $\partial c_{ss}/\partial \tau$ in (4.284) by (4.280) results in the famous Mint model equation (4.285). Relations (4.286) and (4.287) are the most commonly used univocity conditions of this model: (i) before starting filtration, the bed does not contain any retained solid; (ii) during filtration, the bed is fed with a constant flow rate of suspension, which has a constant concentration of solid.

$$\frac{\partial c_{vs}}{\partial x} = \lambda_0 c_{vs} - \frac{\alpha}{w_f} c_{ss} \qquad (4.283)$$

$$\frac{\partial^2 c_{vs}}{\partial x \partial \tau} = \lambda_0 \frac{\partial c_{vs}}{\partial \tau} - \frac{\alpha}{w_f} \frac{\partial c_{ss}}{\partial \tau} = 0 \qquad (4.284)$$

$$\frac{\partial^2 c_{vs}}{\partial x \partial \tau} + \lambda_0 \frac{\partial c_{vs}}{\partial \tau} + \alpha \frac{\partial c_{vs}}{\partial x} = 0 \qquad (4.285)$$

$$\tau = 0 \qquad x \geq 0 \qquad c_{vs} = 0 \qquad (4.286)$$

$$\tau \geq 0 \qquad x = 0 \qquad c_{vs} = c_{v0} \qquad (4.287)$$

Relations (4.288)–(4.290) give one solution for the Mint model. It is not difficult to verify that this solution cannot cover the requirement of relation (4.281). The given solution is a series with a rapid convergence due to the strong evolution of the chain T_n. A good result will consequently be obtained by limiting the sum of Eq. (4.288) to four or five terms:

$$\frac{c_{vs}}{c_{v0}} = \sum_{n=1}^{\infty} \exp(-\lambda_0 x) \frac{(\lambda_0 x)^{n-1}}{(n-1)!} T_n \exp(-\alpha \tau) \qquad (4.288)$$

$$T_n = T_{n-1} - \frac{(\alpha \tau)^{n-2}}{(n-2)!} \qquad (4.289)$$

$$T_1 = \exp(\alpha \tau) \qquad (4.290)$$

A second solution to this model is given by Eq. (4.291), which is an assembly of i-order Bessel functions with real argument $I_i((\lambda_0 x \alpha \tau)^{1/2})$:

$$\frac{c_{vs}}{c_{v0}} = \exp(-(\lambda_0 x + \alpha\tau)) \sum_{i=1}^{\infty} \left(\frac{\alpha\tau}{\lambda_0 x}\right)^{i/2} I_i\left[(\lambda_0 x \alpha\tau)^{1/2}\right] \tag{4.291}$$

In the Mint model, we have to take into account the following considerations: (i) the initial filtration coefficient λ_0, which is a parameter, presents a constant value after time and position; (ii) the detachment coefficient, which is another constant parameter; (iii) the quantity of the suspension treated by deep filtration depends on the quantity of the deposited solid in the bed; this dependency is the result of the definition of the filtration coefficient; (iv) the start of the deep bed filtration is not accompanied by an increase in the filtration efficiency. These considerations stress the inconsistencies of the Mint model: 1. valid especially when the saturation with retained microparticles of the fixed bed is slow; 2. unfeasible to explain the situations where the detachment depends on the retained solid concentration and /or on the flowing velocity; 3. unfeasible when the velocity of the mobile phase inside the filtration bed, varies with time; this occurrence is due to the solid deposition in the bed or to an increasing pressure when the filtration occurs with constant flow rate. Here below we come back to the development of the stochastic model for the deep filtration process.

A *stochastic model of deep bed filtration* [4.5] identifies two elementary processes for the evolution of the micro-particle in the filtration bed:

1. A type I process that considers the motion of microparticles occurring with a velocity $v_1 = v$; this velocity is induced by the surrounding flowing fluid (physically this type of process corresponds to the non-deposition of the microparticle);
2. A type II process that shows the possibility for the microparticle to deposit; from the viewpoint of the motion, the velocity of this process is $v_2 = 0$.

The stochastic model accepts a Markov type connection between both elementary states. So, with $\alpha_{12}\Delta\tau$, we define the transition probability from type I to type II, whereas the transition probability from type II to a type I is $\alpha_{21}\Delta\tau$. By $P_1(x, \tau)$ and $P_2(x, \tau)$ we note the probability of locating the microparticle at position x and time τ with a type I or respectively a type II evolution. With these introductions and notations, the general stochastic model (4.71) gives the particularization written here by the following differential equation system:

$$\begin{cases} \dfrac{\partial P_1(x, \tau)}{\partial \tau} = -v\dfrac{\partial P_1(x, \tau)}{\partial x} - \alpha_{12}P_1(x, \tau) + \alpha_{21}P_2(x, \tau) \\[2mm] \dfrac{\partial P_2(x, \tau)}{\partial \tau} = -\alpha_{21}P_2(x, \tau) + \alpha_{12}P_1(x, \tau) \end{cases} \tag{4.292}$$

For the transformation of the stochastic model into a form, such as the Mint model, that allows the computation of $c_{vs}(x, \tau)/c_{v0}$, we consider that this ratio gives a measure of the probability to locate the microparticle in the specified position: $P(x, \tau) = P_1(x, \tau) + P_2(x, \tau)$. We can simplify our equations by eliminating probabilities $P_1(x, \tau)$ and $P_2(x, \tau)$ with the use of this last definition and the relations of system (4.292). The result is the following interesting partial differential equation:

$$\frac{\partial^2 P(x, \tau)}{\partial \tau^2} + v \frac{\partial^2 P(x, \tau)}{\partial x \partial \tau} + v \alpha_{21} \frac{\partial P(x, \tau)}{\partial x} + (\alpha_{21} + \alpha_{12}) \frac{\partial P(x, \tau)}{\partial \tau} = 0 \tag{4.293}$$

By considering the combined variable $z = x - vt/2$, we remove the mixed partial differential term from Eq. (4.293). The transformation obtained is the hyperbolic partial differential equation (4.294). This equation represents a new form of the stochastic model of the deep bed filtration and has the characteristic univocity conditions given by relations (4.295) and (4.296). The univocity conditions show that the suspension is only fed at times higher than zero. Indeed, here, we have a constant probability for the input of the microparticles:

$$\frac{\partial^2 P(z + v\tau/2, \tau)}{\partial \tau^2} - \frac{v^2}{4} \frac{\partial^2 P(z + v\tau/2, \tau)}{\partial z^2} + \frac{v}{2}(\alpha_{21} - \alpha_{12}) \frac{\partial P(z + v\tau/2, \tau)}{\partial z} +$$
$$(\alpha_{21} + \alpha_{12}) \frac{\partial P(z + v\tau/2, \tau)}{\partial \tau} = 0 \tag{4.294}$$

$$\tau = 0 \ , \quad x > 0 \ , \quad z = x \quad P(z, \tau) = 0 \tag{4.295}$$

$$\tau > 0 \ , \quad x = 0 \ , \quad z > 0 \quad P(z, \tau) = P_0 \tag{4.296}$$

In this stochastic model, the values of the frequencies skipping from one state to another characterize the common deep bed filtration. This observation allows the transformation of the above-presented hyperbolic model into the parabolic model, given by the partial differential equation (4.297). With the univocity conditions (4.295) and (4.296) this model [4.5] agrees with the analytical solution described by relations (4.298) and (4.299):

$$\frac{v^2}{4(\alpha_{21} + \alpha_{12})} \frac{\partial^2 P\left(z + \frac{v\tau}{2}, \tau\right)}{\partial z^2} = \frac{v}{2}\left(\frac{\alpha_{21} - \alpha_{12}}{\alpha_{21} + \alpha_{12}}\right) \frac{\partial P\left(z + \frac{v\tau}{2}, \tau\right)}{\partial z} + \frac{\partial P\left(z + \frac{v\tau}{2}, \tau\right)}{\partial \tau} = 0 \tag{4.297}$$

$$x - \frac{v\alpha_{12}}{\alpha_{21} + \alpha_{12}} < 0 \ , \quad \frac{P(x, \tau)}{P_0} = \frac{1}{2}\left\{1 + \mathrm{erf}\left[\frac{x - \dfrac{v\alpha_{12}}{\alpha_{21} + \alpha_{12}}\tau}{\sqrt{\dfrac{v^2}{\alpha_{21} + \alpha_{12}}}}\right]\right\} \tag{4.298}$$

$$x - \frac{v\alpha_{12}}{\alpha_{21} + \alpha_{12}} > 0 \ , \quad \frac{P(x, \tau)}{P_0} = \frac{1}{2}\left\{1 - \mathrm{erf}\left[\frac{x - \dfrac{v\alpha_{12}}{\alpha_{21} + \alpha_{12}}\tau}{\sqrt{\dfrac{v^2}{\alpha_{21} + \alpha_{12}}}}\right]\right\} \tag{4.299}$$

It is well known that only experimental investigation can validate or invalidate a model of a process. For the validation of the model developed above, we use the experimental data of the filtration of a dilute $Fe(OH)_3$ suspension (the concentration is lower than 0.1 g $Fe(OH)_3$ /l) in a sand bed with various heights and particle diameters. The experiments report the measurements at constant filtrate flow rate and give the evolution with time of the concentration of $Fe(OH)_3$ at the bed output when we use a constant solid concentration at the feed. Figure 4.36 shows the form of the time response when deep bed filtration occurs. The concentration of the solid at the exit of the bed is measured by the relative turbidity (exit turbidity/ input turbidity*100). The small skips around the mean dependence, which appear when the clogging bed becomes important, characterize the duality between the retention and dislocation of the bed-retained solid. This dislocation shows that the Mint model consideration with respect to the detachment coefficient is not acceptable, especially when the concentration of the bed-retained solid is high.

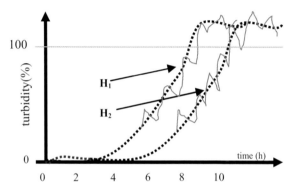

Figure 4.36 Response curves for the deep bed filtration of a suspension of $Fe(OH)_3$ in water.

The data from Fig. 4.36, that show the evolution of $c_{vs}(H, \tau)/c_{vo}$ versus time, have been used to identify the model parameters α_{12} and α_{21}. Here, H is the height of the fine sand granular bed used as porous filter. We have also selected the following process factors: the porous bed height (H), the mean diameter of the particles in the sand granular bed (d_g), the filtrate flow rate (G_v), the content of $Fe(OH)_3$ in the water (noted here as C_0 and c_{vo} in the model) and the fluid temperature as an indirect consideration of the liquid viscosity (t). Table 4.7 shows the results of these computations.

We can immediately observe that the assumption of the height values for the parameters α_{12} and α_{21} is excellently covered by the experimental starting data. Secondly, we find that all the process factors influence all the values of the parameters of the stochastic model.

Table 4.7 Influence of the factors of the process on the parameters of the stochastic model.

Deep bed filtration factor	Factor value	Stochastic model parameters	
		α_{12}	α_{21}
H [cm]			
$t = 20\,°C$	2	1.14	326
$G_v = 20$ cc/min	3	0.592	458.9
$d_g = 0.5–0.3$ mm	5	0.262	420.22
$C_0 = 6.75$ mg/l	6	17.66	9611.41
t [°C]			
H = 6 cm	20	17.95	9805.1
$G_v = 20$ cc/min	30	2.6	2748.14
$d_g = 0.5–0.31$ mm	35	22.71	1347.5
$C_0 = 6.75$ mg/l	40	4.337	3028
G_v [cm³/min]			
H = 6 cm	20	3.054	3020.66
$t = 30\,°C$	30	13.34	6698.96
$d_g = 0.5–0.31$ mm	40	62.08	22605.24
$C_0 = 6.75$ mg/l	50	118.2	42908.68
C_0[mg/l] Fe(OH)₃			
H = 6 cm			
$t = 30\,°C$	6.75	111.7	40493
$d_g = 0.5–0.31$ mm	13.49	82.7	28999.5
$G_v = 50$ cm³/min	26.98	34.18	10294.41
d_g [mm]			
H = 6 cm	0.31–0.2	754.98	355456.89
$C_0 = 6.75$ mg/lFe(OH)₃	0.5–031	110.65	39773
$T_f = 30\,°C$	0.63–0.5	22.409	8795.98
$G_v = 50$ cm³/min	0.85–0.63	23.82	6449.82

Using a regression analysis, the following dependences have been obtained:

$$a_{12} = 67.75 + 3.003 Gv - 2.872 C_0 - 256.28 dg$$

$$a_{21} = 3.081 \cdot 10^4 + 1111 Gv - 1268 C_0 - 1.023 \cdot 10^5 dg$$

It is important to notice that these relations show the independence of the parameters of the stochastic model with respect to the height of the porous bed. With the identified values of a_{12} and a_{21}, we can now simulate the deep bed filtration process by computing Eqs. (4.298) and (4.299), which show how the dimensionless $c_{vs}(H, \tau)/c_{vo} = P(H, \tau)/P(0, \tau)$ evolve with time.

Figure 4.37 gives the combination of the simulation results obtained with the model and with an assembly of experimental data. We have to notice that the values of the factors for the relations that give the transition frequencies must respect the dimensional units from Table 4.7. These relations make it possible to formulate the optimisation of the filtration problem and then to establish the combination of factors allowing deep bed filtration at minimum financial cost.

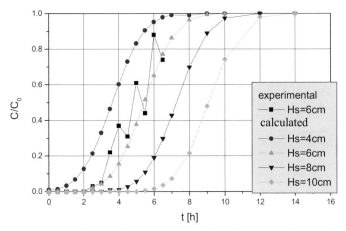

Figure 4.37 Simulated and experimental time dependence of the dimensionless solid concentration in the suspension at the bed output.
($G_v = 50$ cm^3/min, $C_0 = 6.75$ mg Fe(OH)$_3$/l, T = 30 °C, $d_g = 0.4$ mm.)

If we want to make a more complete stochastic model, it is recommended to consider a process with three elementary states which are: the microparticles motion in the direction of the global flow, the microparticles fixation by the collector elements of the porous structure and the washing of the fixed microparticles. In this case, we obtain a model with six parameters: $a_{12}, a_{13}, a_{21}, a_{23}, a_{31}, a_{32}$. This is a rather complicated computation.

The discussed stochastic model presents the capacity to be converted into a steady state model; in addition, an interesting asymptotic transformation can also be carried out. For the conversion of the model into a steady state one, we consider a time interval $\Delta\tau$, where the probabilities for the system to change by means of a type 1 or type 2 evolution process are given by $\alpha_1\Delta\tau$ and $\alpha_2\Delta\tau$ respectively and the transition probabilities of the process are described by $p_{12} = \alpha_{12}\Delta\tau$ and $p_{21} = \alpha_{21}\Delta\tau$. Indeed, the model (4.292) will be expressed as follows:

$$\begin{cases} \dfrac{\partial P_1(x,\tau)}{\partial\tau} = -v\dfrac{\partial P_1(x,\tau)}{\partial x} - \alpha_1 P_1(x,\tau) + \alpha_{21}P_2(x,\tau) \\ \dfrac{\partial P_2(x,\tau)}{\partial\tau} = -\alpha_2 P_2(x,\tau) + \alpha_{12}P_1 \leq (x,\tau) \end{cases} \tag{4.300}$$

With $P_1(x,\tau) + P_2(x,\tau) = P(x,\tau)$, we obtain relation (4.301) and its corresponding steady state (4.302):

$$\frac{\partial^2 P(x,\tau)}{\partial\tau^2} + v\frac{\partial^2 P(x,\tau)}{\partial x\partial\tau} + v\alpha_2\frac{\partial P(x,\tau)}{\partial x} + (\alpha_2 + \alpha_1)\frac{\partial P(x,\tau)}{\partial\tau} +$$
$$(\alpha_1\alpha_2 - \alpha_{12}\alpha_{21})P(x,\tau) = 0 \tag{4.301}$$

$$v\alpha_2\frac{dP(x)}{dx} + (\alpha_1\alpha_2 - \alpha_{12}\alpha_{21})P(x) = 0 \tag{4.302}$$

When the particularization condition $x = 0, P(x) = P_0 = 1$ is used for the differential equation (4.302), its solution respects relation (4.303). The correspondence with concentrations $c_{vs}(x)$ and c_{v0} is presented by means of relation (4.304).

$$P(x) = P_0\exp\left(-\frac{\alpha_1\alpha_2 - \alpha_{12}\alpha_{21}}{v\alpha_2}\right) = P_0\exp\left(-\frac{\alpha x}{v}\right) = \exp\left(-\frac{\alpha x}{v}\right) \tag{4.303}$$

$$c_{vs}(x) = c_{v0}\exp\left(-\frac{\alpha x}{v}\right) \tag{4.304}$$

It is important to notice the didactic importance of this last relation, because the deep bed filtration process cannot operate at steady state.

The asymptotic transformation of the discussed stochastic model (see relation 4.292 and Section 4.5.1.2) is carried out with the identification of the operators:

$$V_1(1) = -v\partial/\partial x \;,\;\; V_1(2) = 0 \;,\;\; Q = \begin{pmatrix} -\alpha_{12} & \alpha_{12} \\ \alpha_{21} & -\alpha_{21} \end{pmatrix} \;,\;\; \overline{V}_1 = -\frac{v\alpha_{21}}{\alpha_{12} + \alpha_{21}}\frac{\partial}{\partial x}.$$

The resulting asymptotic model is described by the following equation and the univocity conditions given by relations (4.295) and (4.296):

$$\frac{\partial P(x,\tau)}{\partial\tau} + \frac{v\alpha_{21}}{\alpha_{12} + \alpha_{21}}\frac{\partial P(x,\tau)}{\partial x} = 0 \tag{4.305}$$

We can observe that the asymptotic model of the deep bed filtration has no term concerning the dispersion of the flowing fluid. At the same time, it is important to emphasize the fact that this model, considered as its deterministic equivalent, is frequently used for the characterization of fluids seeping into the soil.

4.6.2.2 Molecular Species Transfer in a Porous Solid

When a fluid flows through a porous solid or a porous bed, the species forming the fluid can present some affinities with the solid particles. The affinity of species with respect to the contacting solid, which here form what is called a stationary phase, can be the result of different phenomena such as adsorption, ion exchange, steric exclusion and absorption.

The separation of species by affinity is the principle of chromatographic processes, as shown schematically in Fig. 4.38. At time $\tau = 0$, we introduce a small quantity of A and B species mixture (probe injection) at the column input in a carrier fluid flowing inside a fine granular bed of porous medium. The motion of species A and B caused by the flowing carrier creates the conditions necessary for their separation. If species A and B present a sorption phenomenon on the granular solid, the separation will take place as a consequence of a specific adsorption–desorption process repeated along the porous bed. On the contrary, the carrier must be inert with respect to the interactions with the granular bed. Figure 4.38 shows that the separation of A and B is not complete at all the local points placed in the first part of the length of the granular bed. So, it is important to emphasize that complete separation of A and B is attained only if the combination of the bed

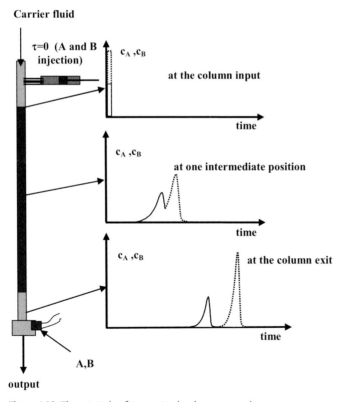

Figure 4.38 The principle of separation by chromatography.

length and the carrier velocity results in a good residence time. This residence time must comply with the time needed for the separation of A and B, which depends on the individual adsorption–adsorption process.

If we introduce a discrete feed of A and B, characterized by a reasonable interval of time between two inputs, and a discrete collection of A-carrier and B-carrier outputs into Fig. 4.38, then we will have a chromatographic separator.

When the carrier is a liquid, the instrumentation includes a pump, an injector, a column, a detector and a recorder or a data acquisition system, connected to a computer. The heart of the system is the column where the separation occurs. Since the stationary phase is composed of micrometric porous particles, a high-pressure pump is required to move the mobile phase through the column. The chromatographic process begins by injecting the solute into the top of the column by an impulse type injection. The separation of the components occurs during the elution of the mobile phase through the column.

The majority of chromatographic separations as well as the theory assume that each component elutes out of the column as a narrow band or a Gaussian peak. Using the position of the maximum of the peak as a measure of retention time, the peak shape conforms closely to the equation: $C = C_{max} \exp[-(t - t_R)^2 / 2\sigma^2]$. The modelling of this process, by traditional descriptive models, has been extensively reported in the literature.

As has been explained previously in this chapter, the building of a stochastic model starts with the identification of the individual states of the process.

For a chromatographic separation, each i species has three individual elementary evolutions (here we consider $i = 2,...N$ because $i = 1$ corresponds to the carrier which is not retained by the granular bed):

1. motion with velocity $+v$ in the sense of carrier flow (type 1 process);
2. adsorption on the solid (a type 2 process; the fixation on the solid stops the species motion);
3. motion with velocity $-v$.

If we consider that the connecting process is Markovian, then we can write the balance of the probabilities for $P_1^{(j)}(x, \tau)$, $P_2^{(j)}(x, \tau)$ and $P_3^{(j)}(x, \tau)$. Here $a_i^{(j)}\Delta\tau$ gives the probabilities for j species to change their i evolution state. Through $a_{ik}^{(j)}\Delta\tau$ ($i = 1, 3; k = 1, 3$) we consider the transition probabilities of species j between states i and k. With the statements above, we can write the following balance relations:

$$P_1^{(j)}(x, \tau) = (1 - a_1^{(j)}\Delta\tau)P_1^{(j)}(x - \Delta x, \tau - \Delta\tau) + a_{21}^{(j)}\Delta\tau P_2^{(j)}(x, \tau - \Delta\tau) + a_{31}^{(j)}\Delta\tau P_3^{(j)}(x + \Delta x, \tau - \Delta\tau)$$

(4.306)

$$P_2^{(j)}(x, \tau) = (1 - a_1^{(j)}\Delta\tau)P_2^{(j)}(x, \tau - \Delta\tau) + a_{12}^{(j)}\Delta\tau P_1^{(j)}(x, \tau - \Delta\tau) + a_{32}^{(j)}\Delta\tau P_3^{(j)}(x, \tau - \Delta\tau)$$

(4.307)

$$P_3^{(i)}(x,\tau) = (1 - \alpha_3^{(i)}\Delta\tau)P_3^{(i)}(x + \Delta x, \tau - \Delta\tau) + \alpha_{13}^{(i)}\Delta\tau P_1^{(i)}(x + \Delta x, \tau - \Delta\tau) +$$
$$\alpha_{23}^{(i)}\Delta\tau P_2^{(i)}(x, \tau - \Delta\tau) \tag{4.308}$$

Using the Taylor expansion of the probabilities $P_i^{(i)}(x \pm \Delta x, \tau - \Delta\tau)$ at Δx and $\Delta\tau \to 0$ for processing these balances, the results on the stochastic differential model are given by the relations of the assembly (4.309)

$$\begin{cases} \dfrac{\partial P_1^{(i)}(x,\tau)}{\partial\tau} = -v\dfrac{\partial P_1^{(i)}}{\partial x} - \alpha_1^{(i)}P_1^{(i)} + \alpha_{21}^{(i)}P_2^{(i)}(x,\tau) + \alpha_{31}^{(i)}P_3^{(j)} \\[3mm] \dfrac{\partial P_2^{(i)}(x,\tau)}{\partial\tau} = -\alpha_2^{(i)}P_2^{(i)} + \alpha_{12}^{(i)}P_1^{(i)}(x,\tau) + \alpha_{32}^{(i)}P_3^{(j)} \\[3mm] \dfrac{\partial P_3^{(i)}(x,\tau)}{\partial\tau} = v\dfrac{\partial P_1^{(i)}}{\partial x} - \alpha_3^{(i)}P_3^{(i)} + \alpha_{13}^{(i)}P_1^{(i)}(x,\tau) + \alpha_{23}^{(i)}P_2^{(j)} \end{cases} \tag{4.309}$$

Since these equations do not have an acceptable form for the description of the chromatographic separation, we have used them to build up the Lapidus model [4.92] by considering only one positive motion for the carrier fluid. Indeed, we will introduce $P_3^{(i)}(x,\tau) = 0$ into the general model. The result is given by the assembly of equations (4.310):

$$\begin{cases} \dfrac{\partial P_1^{(i)}(x,\tau)}{\partial\tau} = -v\dfrac{\partial P_1^{(i)}}{\partial x} - \alpha_1^{(i)}P_1^{(i)} + \alpha_{21}^{(i)}P_2^{(i)}(x,\tau) \\[3mm] \dfrac{\partial P_2^{(i)}(x,\tau)}{\partial\tau} = -\alpha_2^{(i)}P_2^{(i)} + \alpha_{12}^{(i)}P_1^{(i)}(x,\tau) \end{cases} \tag{4.310}$$

Now we have to particularize the obtained model by considering that the retention of species j occurs by one adsorption–desorption process. So, if the j species desorbs from the solid, it has to appear in the mobile phase. We can express this consideration mathematically with $\alpha_1^{(i)} = \alpha_{12}^{(i)}$ and respectively $\alpha_2^{(i)} = \alpha_{21}^{(i)}$. Now the model can be written as follows:

$$\begin{cases} \dfrac{\partial P_1^{(i)}(x,\tau)}{\partial\tau} = -v\dfrac{\partial P_1^{(i)}}{\partial x} - \alpha_{12}^{(i)}P_1^{(i)}(x,\tau) + \alpha_{21}^{(i)}P_2^{(i)}(x,\tau) \\[3mm] \dfrac{\partial P_2^{(i)}(x,\tau)}{\partial\tau} = -\alpha_{21}^{(i)}P_2^{(i)}(x,\tau) + \alpha_{12}^{(i)}P_1^{(i)}(x,\tau) \end{cases} \tag{4.311}$$

$$\tau = 0 \ , \quad x > 0 \ , \quad P_1^{(i)}(x,\tau) = P_2^{(i)}(x,\tau) = 0 \tag{4.312}$$

$$\tau = 0_+ \ , \quad x = 0 \ , \quad P_1^{(i)}(x,\tau) = P_{10}^{(i)} \ , \quad P_2^{(i)}(x,\tau) = 0 \ , \quad \sum_j P_{10}^{(i)} = 1 \tag{4.313}$$

With the univocity conditions given in relations (4.312) and (4.313), the stochastic model becomes ready to be used in simulation.

If we agree with the absorption–desorption equilibrium of the j species at each point of the bed and considering that the absorption–desorption process obeys the well known Langmuir isotherm, then we can write: $\alpha_{12}^{(i)} = \beta_{12}^{(i)}(1 - P_2^{(i)}(x, \tau)) = \beta_{12}^{(i)} P_1^{(i)}(x, \tau)$ and the following local relation is obtained:

$$P_2^{(i)}(x, \tau) = \frac{\beta_{12}^{(i)} P_1^{(i)}(x, \tau)}{\alpha_{21}^{(i)} + \beta_{12}^{(i)} P_1^{(i)}(x, \tau)} \tag{4.314}$$

With these considerations, we can transform Eq. (4.311) from the stochastic model (4.311)–(4.313):

$$\begin{cases} \dfrac{\partial P_1^{(i)}(x, \tau)}{\partial \tau} = -v \dfrac{\partial P_1^{(i)} x, \tau)}{\partial x} - \beta_{12}^{(i)} P_1^{(i)}(x, \tau)(1 - P_2^{(i)}(x, \tau) + \alpha_{21}^{(i)} P_2^{(i)}(x, \tau) \\[3mm] \dfrac{\partial P_2^{(i)}(x, \tau)}{\partial \tau} = -\alpha_{21}^{(i)} P_2^{(i)} + \beta_{12}^{(i)} P_1^{(i)}(x, \tau)(1 - P_2^{(i)}(x, \tau) \quad j = 1, \dots N_c \end{cases} \tag{4.315}$$

This model given by the system of equations (4.315) together with the conditions (4.312) and (4.313) can easily generate the chromatographic curves that are presented in Fig. 4.38. For this purpose, we simulate the state of existence probability of each species along the chromatographic bed. For two species, the sums $P^{(1)}(x, \tau) = P_1^{(1)}(x, \tau) + P_2^{(1)}(x.\tau)$ and $P^{(2)}(x, \tau) = P_1^{(2)}(x, \tau) + P_2^{(2)}(x.\tau)$ respectively show the state of species 1 and species 2 along the chromatographic bed. Indeed, we can identify the parameters of the stochastic model if we consider here the conventional identification $P^{(1)}(x, \tau) = c^{(1)}(x, \tau)/c^{(1)}(0, \tau)$, where $c^{(1)}(x, \tau)/c^{(1)}(0, \tau)$ has been established experimentally.

An interesting transformation of the stochastic model can be carried out when the derivate $\dfrac{\partial P_1^{(i)}(x, \tau)}{\partial \tau}$ is smaller than $\dfrac{\partial P_2^{(i)}(x, \tau)}{\partial \tau}$. This situation corresponds to the case when the variation of a fraction of the j species in the mobile phase is smaller than the fraction of the j species in the solid phase. To obtain this transformation, we operate in two steps: (i) we derivate again the first equation from system (4.311) with respect to time; (ii) with this derivate and the equation remaining from system (4.311), we eliminate probability $P_2^{(i)}(x, \tau)$. The result is:

$$\frac{\partial P_1^{(i)} x, \tau)}{\partial x \partial \tau} + \frac{\alpha_{12}}{v} \frac{\partial P_1^{(i)}(x, \tau)}{\partial \tau} + \alpha_{21} \frac{\partial P_1^{(i)}(x, \tau)}{\partial x} = 0 \tag{4.316}$$

The univocity conditions can be obtained from Eq.(4.311) which, at $\tau = 0$, results in the problem described by Eq. (4.317), which presents solution (4.318). This last relation represents the initial condition from the univocity problem of model (4.316):

$$v \frac{\partial P_1^{(i)}(x, 0)}{\partial x} - \alpha_{21}^{(i)} P_1^{(i)}(x, 0) \quad , \quad P_1^{(i)}(0, 0) = P_{10}^{(i)} \tag{4.317}$$

$$P_1^{(i)}(x, 0) = P_{10}^{(i)} e^{-\left(\frac{\alpha_{12} x}{v}\right)} \tag{4.318}$$

Other conditions of the univocity problem, give the probabilities to have j species at the bed input. For an impulse at the input we have $P_1^{(i)}(0,0) = P_{10}^{(i)}$ and $P_1^{(i)}(0,\tau) = 0$. With all these conditions we can build the relation (4.319), which gives the explicit solution to this transformed model [4.33]. Here, the Bessel function $I_0(y)$ is introduced with relation (4.320):

$$P_1^{(i)}(x,\tau) = P_{10}^{(i)} e^{-\left(\frac{a_{12}^{(i)}x}{v}\right)} \left[e^{-\left(\frac{\left(a_{12}^{(i)}\right)^2 \tau}{a_{21}}\right)} I_0\left(2\sqrt{\frac{\left(a_{12}^{(i)}\right)^2 x\tau}{a_{21}^{(i)}}}\right) + \frac{v}{a_{12}^{(i)}x} \int_0^{\frac{\left(a_{12}^{(i)}\right)^2 x\tau}{a_{31}^{(i)}}} e^{-\frac{v\tau}{a_{12}^{(i)}x}} I_0(2\sqrt{\tau d\tau}) \right]$$

(4.319)

$$I_0(y) = \sum_{k=0}^{\infty} \frac{(y/2)^{2k}}{(k!)^2}$$

(4.320)

This solution describes the evolution of the concentration of the j species in the carrier fluid from the input to the output of the chromatographic bed. Once the parameters are estimated, this solution can be used for the evaluation of the bed height (length of chromatographic column) needed for one actual separation (given values for $c_{10}^{(i)}$).

4.6.3
Stochastic Models for Processes with Discrete Displacement

In this type of process, the flow pattern inside a device is considered to occur in separated compartments. Each compartment is characterized by its own volume, input and exit flow rates. The circulation between all compartments is given in the scheme showing the flow topology of application 4.1.2 (Fig. 4.2). The system studied here corresponds in detail to the scheme shown in Fig. 4.39. Here we have species in motion inside a porous medium with active sites (as for example catalytic sites); species skip randomly from one site (or agglomeration) to another and inside the site they randomly interact with the site components. One agglomeration can support one or more visits of the flowing species and the time needed for one skip is low in comparison with the residence time inside one agglomeration. The model presented can easily be used to describe the visit of a very important person to a reception where the participants are distributed in various groups; considering that our VIP agrees with the protocol, he must visit each group; the time spent by the VIP with each group is established proportionally with the number of members of the group.

This type of model with compartment flow pattern can easily be applied in many chemical engineering devices such as chemical reactors, mechanical stirrers, absorption, rectification and liquid-liquid extraction columns [4.18, 4.19, 4.94]. Nevertheless, the practical applications of these models present some difficulties because of their high number of parameters. For example, in the application of Section 4.12 (the numerical application of the mechanical stirring of a liq-

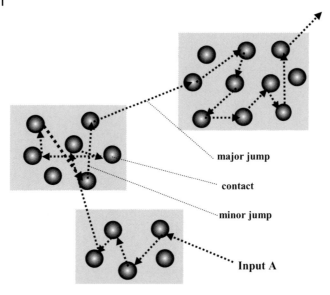

Figure 4.39 Scheme showing the motion of species A in a randomly organized porous body.

uid) the volume of cells, the flow rates and the fluid current topology are some of the parameters necessary for the model translation as simulator. Despite this major difficulty, these models remain very prolific for the production of theoretical data. In addition, these models can also be easily modified when we introduce some new conditions or when we change one or more of the existing conditions.

If we observe this type of modelling from the point of view of the general theory of the stochastic models, we can presume that it is not very simple. Indeed, the specific process which takes place in one compartment k = 1, 2, 3...., N, defines the possible states of a fluid element (the elementary processes of the global stochastic process) and the transition describing the fluid element flowing from one compartment to another represents the stochastic connections. Consequently, p_{ik} i = 1, 2....., N are the transition probabilities from the i to the k compartment and $P_k(\tau)$ is the probability of having, at time τ, the fluid element inside the k compartment. With these notations, the probability balances for $P_k(\tau + \Delta\tau)$ can be written as follows:

$$P_k(\tau + \Delta\tau) = \sum_{i=1}^{N} p_{ik} P_i(\tau) \tag{4.321}$$

Because, when $\tau = n\Delta\tau$, the discrete case is usually applied, relation (4.321) becomes:

$$P_k(n+1) = \sum_{i=1}^{N} p_{ik} P_i(n) \tag{4.322}$$

During the time interval $\Delta\tau$, the fluid element exits compartment i and flows into compartment k, which it cannot leave. This is the condition for the selection of one realistic value of this time interval. So, for one cellular topology with $N - 1$ cells, the number N defines the output from the system. It is not difficult to observe then that we have:

$$\sum_{j=1}^{N} P_j(N) = 1 \tag{4.323}$$

The multiplication $P_k(n)\Delta\tau$ gives the existence probability or the probability to have the fluid element in compartment k in the interval of time defined by $n\Delta\tau$ and $(n + 1)\Delta\tau$. In other words, it is the response of compartment k to an impulse signal. For $k = N$, we can observe that the probability $P_N(n)\Delta\tau$ makes it possible for the fluid element to leave the cells assembly in the same interval of time $\tau = n\Delta\tau$ and $\tau + \Delta\tau = (n + 1)\Delta\tau$. Furthermore, because $P_{N-1}(n)$ gives the distribution of the residence time for our assembly of compartments, then we can conclude that the response to one step impulse can be written as:

$$F(\tau) = F(n\Delta\tau) = P_N(n) = \sum_{n=0}^{n} P_{N-1}(n)\Delta\tau \tag{4.324}$$

With this response, it is easy to obtain some important parameters characterizing the flow in the cellular assembly: the mean residence time (τ_m), the variance around the mean residence time (σ^2) and the flow intensity function $(\lambda(n))$:

$$\tau_m = \frac{\sum_{n=1}^{\infty} nP_{N-1}(n)\Delta\tau}{\sum_{n=1}^{\infty} P_{N-1}(n)} \tag{4.325}$$

$$\sigma^2 = \sum_{n=1}^{\infty} (n\Delta\tau - \tau_m)^2 P_{N-1}(n) \tag{4.326}$$

$$\lambda(n) = \frac{P_{N-1}(n)}{1 - P_N(n)} \tag{4.327}$$

The basic relation of our stochastic model (relation (4.322)) can be written as the vectorial equation (4.328), where $E(n)$ gives the vector of the system state (relation (4.329)) and the matrix P (relation (4.330)) contains the transition probabilities:

$$E(n + 1) = P * E(n) \tag{4.328}$$

$$E(n) = [P_1(n), P_2(n), P_3(n), ...P_{N-1}(n), P_N(n)] \tag{4.329}$$

$$P = \begin{bmatrix} P_{11} & P_{12} & - & P_{1N-1} & P_{1N} \\ P_{21} & P_{22} & - & - & P_{2N} \\ - & - & - & - & - \\ P_{N-11} & P_{N-12} & - & P_{N-1N-1} & - \\ P_{N1} & P_{N2} & - & P_{N-1N} & P_{NN} \end{bmatrix} \tag{4.330}$$

In actual applications, the vector of the system state is used to observe the system evolution through characteristic parameters such as species concentrations, temperature, pressure, etc.

4.6.3.1 The Computation of the Temperature State of a Heat Exchanger

In this example, we can use a deterministic model based on the particularization of the unsteady state heat balance and transfer equations. The particularization can be carried out considering either the whole exchanger or a part of it. The model that can present different degrees of complication is determined by the heat exchanger construction and by the models of flow used for the hot and cold fluids.

If we consider plug flow models for both fluids, the heat exchanger dynamics can be described using the following model:

$$\frac{\partial t_1}{\partial \tau} + w_1 \frac{\partial t_1}{\partial x} = -\frac{4k}{d\rho_1 c_{p1}}(t_1 - t_2) \tag{4.331}$$

$$\frac{\partial t_2}{\partial \tau} + w_2 \frac{\partial t_2}{\partial x} = -\frac{4k}{d\rho_2 c_{p2}}(t_1 - t_2) - \frac{4k_e}{D\rho_2 c_{p2}}(t_2 - t_e) \tag{4.332}$$

$$\tau = 0 \ , \quad x > 0 \ , \quad t_1 = f_1(x) \ , \quad t_2 = f_2(x) \tag{4.333}$$

$$\tau > 0 \ , \quad x = 0 \ , \quad t_1 = g_1(\tau) \ , \quad t_2 = g_2(\tau) \tag{4.334}$$

$$\frac{1}{k} = \frac{1}{a_1} + \frac{\delta_p}{\lambda_p} + \frac{1}{a_2} \ , \quad \frac{1}{k_e} = \frac{1}{a_e} + \frac{\delta_{pe}}{\lambda_{pe}} + \frac{1}{a_2} \ ,$$

$$a_1 = h(w_1, \rho_1, c_{p1}, \lambda_1) \ , \quad a_2 = h(w_2, \rho_2, c_{p2}, \lambda_2) \tag{4.335}$$

The nomenclature of the equations above is: t_1 and t_2 – temperature of fluid 1 and fluid 2 respectively, w_1 and w_2 – mean velocities for hot and cold fluid, ρ_1 and ρ_2 – fluid densities, c_{p1} and c_{p2} – fluid sensible heats, d and D – specific diameters of the basic pipe and mantle of the heat exchanger, a_1 and a_2 – partial heat transfer coefficients around the basic pipe, δ_p and δ_{pe} – thickness of the basic pipe and the mantle, λ_p and λ_{pe} – thermal conductivities of the basic pipe and mantle walls, k and k_e – total heat transfer coefficients, t_e – external temperature of the heat exchanger.

This model has the remarkable characteristic of considering the heat loss in the external media with the term $\dfrac{4k_e}{D\rho_2 c_{p2}}(t_2 - t_e)$. The evolutions $t_1(x, \tau)$ and $t_2(x, \tau)$ result from the numerical integration of the model; for this purpose we need the analytical or discrete expressions for the functions $f_1(x), f_2(x), g_1(\tau)$ and $g_2(\tau)$.

The stochastic model of this problem is obtained after introducing a cellular structure and a flow topology. The partition of the heat exchanger into individual cells is carried out as follows:

1. The exchanger contains N_C perfectly mixed cells exposed to the hot fluid flow and N_r cells of the same type where a cold fluid exists; frequently $N_c = N_r = N$ but this fact is not obligatory;

2. The inter-fluid walls can be divided into N_p cells where a cell separates one or more hot cells from one or more cold cells;

3. The thermal capacity is symbolized as $C_{Cj}, j = 1, N_c$, $C_{ri}, i = 1, N_r$ and $C_{pk}, k = 1, N_p$, for the hot, cold and inter-fluid wall cells respectively whereas, the heat flows corresponding to a cold or hot J cell are: $(G_{cij}c_{cj}(t_i' - t_r)$, $G_{rij}c_{cj}(t_i'' - t_r)$; $\alpha_{cj}A_j(t_j' - t_{pj}')$ and $\alpha_{rj}A_j(t_{pj}'' - t_j'')$; those are considered as q_{ij}^c, q_{ij}^r when the temperature differences are unitary.

4. If, for the interval of time $\Delta\tau$, one or many i cells coupled to a j cell change their temperature, then the temperature of the j cell will change too.

5. The heat transfer from a cell i to a cell j occurs in a time interval $\Delta\tau$ with the probability p_{ij}.

As far as the explanation above allows one to express the studied system with the necessary objects of a cellular stochastic model, we can now describe the temperature changes inside the exchanger with a discrete Markov evolution that starts with an input cell of the hot or cold fluid. Indeed, relations (4.328) or (4.329) can now be particularized giving the expressions below whereas the matrix of the transition probabilities is described with relation (4.338).

$$t_j(n+1) = \sum_{i=1}^{N} t_i(n)p_{ij} \tag{4.336}$$

$$T(n+1) = T(n)P \tag{4.337}$$

As explained above in our actual application, we have to begin by identifying the cellular structure and flow topology, consequently we have first carefully established the cellular structure after the heat capacities of different fluids or materials: $C_{ci} = m_{ci}c_{ri}, i = 1, N_c$; $C_{rj} = m_{rj}c_{rj}, j = 1, N_r$; $C_{pk} = m_{pk}c_{pk}, k = 1, N_p$.

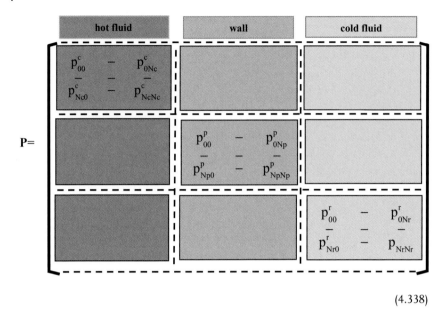

$$(4.338)$$

The elements of p_{jj} type, which characterize the capacity of the heat carrier to be in a j cell during a time $\Delta\tau$, are obtained by considering the perfect mixing inside the cell. With this consideration, we introduce the fact that the carrier residence time follows a Poisson distribution:

$$p_{jj} = 1 - \frac{\sum\limits_{i=1,i\neq j}^{N} q_{ij}}{C_j}\Delta\tau \qquad (4.339)$$

For our stochastic process, the probabilities $p_{ij}, i \neq j$ result from the Markov connections which are described as follows:

$$p_{ij} = \frac{q_{ij}}{C_j}\Delta\tau \qquad (4.340)$$

The probabilities characterizing the hot or cold fluid input into the cellular system, or p_{j0}, $j \neq 0$, are calculated with relation (4.340) but considering $C_0 \to \infty$. So, all $p_{j0}, j \neq 0$ are null and, consequently $p_{00} = 1$. The transition probabilities obey the norm conditions, which require the verification of the equality:

$$\sum_{i=0}^{N} p_{ij} = 1, \forall j = 1, N.$$

As we have shown at the beginning of this section, the application purpose consists in the establishment of a procedure for the thermal dynamics of the hot or cold fluid, when we have a rapid temperature change at the heat exchanger input. The topology of the heat exchanger of this example is shown in Fig. 4.40. If we consider, as an initial condition, that both fluids have the same temperature, we will not have a heat flow between the cells of the cellular assembly. Now, if we take into account a particular operation case where $V_1 = V_2 = 0.5 * V$, $C_1 = C_2 = 0.5C$, $G_{v1}\rho_1 c_{p1} = G_{v2}\rho_2 c_{p2}$, then we can compare the stochastic solution to an analytical solution of the deterministic model. The relation (4.341), which indicates the heat flow rate between both fluids, has been written with the intention of presenting the physical meaning of q_{12} and q_{21}. Indeed, when $t_1'' = t_2'' = t_r$, we do not have any heat flow inside the exchanger and the system state for $\tau = 0$ is represented by $\Delta T_1 = \Delta T_2 = 0$.

$$Q_{12} = kA(t_1'' - t_2'') = kA(t_1' - t_r)\frac{t_1'' - t_r}{t_1' - t_r} - kA(t_1' - t_r)\frac{t_2'' - t_r}{t_1' - t_r}$$

$$= q_{12}\Delta T_1 - q_{21}\Delta T_2$$

(4.341)

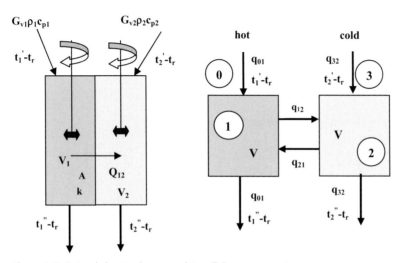

Figure 4.40 A simple heat exchanger and its cellular representation.

With the topology shown in Fig. (4.40) the matrix of the transition probabilities can be written as shown below:

$$P = \begin{bmatrix} p_{00} & p_{01} & p_{02} & p_{03} \\ p_{10} & p_{11} & p_{12} & p_{13} \\ p_{20} & p_{21} & p_{22} & p_{23} \\ p_{30} & p_{31} & p_{32} & p_{33} \end{bmatrix}$$

(4.342)

If we consider the relations (4.339) and (4.340) respectively for p_{ij}, we obtain:

$p_{00} = 1, p_{10} = q_{10}/C_0 = 0, p_{20} = q_{20}/C_0 = 0, p_{30} = q_{30}/C_0 = 0,$
$p_{01} = q_{01}/C_1 == (G_{v1}\rho_1 c_{p1})/(V_1\rho_1 c_{p1}) * \Delta\tau == (G_{v1}/V) * \Delta\tau = \alpha,$
$p_{02} = 0, p_{03} = 0, p_{11} = 1 - [(q_{01} + q_{21})/C_1]\Delta\tau =$
$p_{21} = p_{12} = \alpha, p_{22} = 1 - (q_{32} + q_{12}) = 1 - 2\alpha,$
$p_{23} = 0, p_{13} = 0, p_{32} = (q_{32}/C_2) * \Delta\tau = \alpha, p_{33} = 1.$

And the matrix giving the temperature change of the cells is written as follows:

$$T(1) = T(0)P = \begin{bmatrix} 1 & 0 & 0 & 0 \end{bmatrix} * \begin{bmatrix} 1 & \alpha & 0 & 0 \\ 0 & 1-2\alpha & \alpha & 0 \\ 0 & \alpha & 1-2\alpha & 0 \\ 0 & 0 & 0 & 1 \end{bmatrix} = \begin{bmatrix} 1 & \alpha & 0 & 0 \end{bmatrix},$$

$$T(2) = T(1)P = \begin{bmatrix} 1 & \alpha & 0 & 0 \end{bmatrix} * \begin{bmatrix} 1 & \alpha & 0 & 0 \\ 0 & 1-2\alpha & \alpha & 0 \\ 0 & \alpha & 1-2\alpha & 0 \\ 0 & 0 & 0 & 1 \end{bmatrix} = \begin{bmatrix} 1 & 2\alpha-2\alpha^2 & 0\alpha & 0 \end{bmatrix} \dots$$

The natural temperature values are:

$$(t''_1(1) - t_r) = (t_1''(0) - t_r) + \alpha(t_1' - t_r);$$
$$(t_2''(1) - t_r) = (t_2''(0) - t_r) + 0(t_1' - t_r);$$
$$(t_1''(2) - t_r) = (t_1''(1) - t_r) + (2\alpha - 2\alpha^2)(t_1' - t_r);$$
$$(t_2''(2) - t_r) = (t_2''(1) - t_r) + \alpha(t_1' - t_r), \text{etc} \dots$$

The results above are obtained from:

$$T_1(1) = T_1(0)(1+\alpha) \Leftrightarrow \frac{t_1''(1) - t_r}{t_1' - t_r} = (1+\alpha)\frac{t_1''(0) - t_r}{t_1' - t_r}, \text{etc} \dots$$

It is not difficult to show that the analytical solution of the deterministic model is given in relations (4.343) and (4.344) [4.94], where the parameter λ is

$$\lambda = \frac{KA}{G_{v1}\rho_1 c_{p1}} = \frac{KA}{G_{v2}\rho_2 c_{p2}}$$

$$\Delta T_1 = \frac{1}{1+2\lambda}\left\{1 + \lambda - \frac{1}{2}(1+2\lambda)\exp(-\tau) - \frac{1}{2}\exp[-(1+2\lambda)\tau]\right\} \tag{4.343}$$

$$\Delta T_2 = \frac{1}{1+2\lambda}\left\{\lambda - \frac{1}{2}(1+2\lambda)\exp(-\tau) + \frac{1}{2}\exp[-(1+2\lambda)\tau]\right\} \tag{4.344}$$

Figure 4.41 compares the data predicted by the deterministic model with the stochastic model. In this figure, we have to specify that: (i) for $\lambda = 0$, the value of the corresponding α in the transition matrix of the probabilities results from the simplification $q_{12} = q_{21} = 0$; (ii) the case being analyzed corresponds to a rapid increase in the temperature of fluid 1 in the exchanger input; so t_1 is the highest temperature.

The extension of the stochastic method for actual exchangers depends strongly on the correctness of the projected cellular topology and on the reality of the estimated transition probabilities. Figure 4.42 shows an example of an actual heat

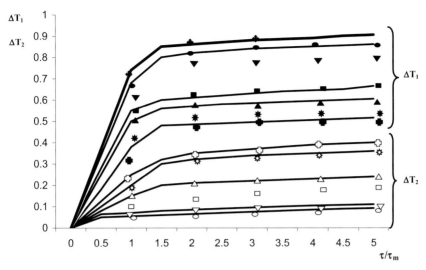

Figure 4.41 Dynamic evolution of the heat exchanger from figure 4.39.
Analytical solution – continuous lines; Stochastic solution – discrete points:
$\lambda = 0$ (✛), $\lambda = 0.1$ (○,●), $\lambda = 0.2$(▽,▼), $\lambda = 0.4$ (□,■), $\lambda = 1$ (△,▲), $\lambda = 0$(✳,✚); $\lambda = 5$(◌,✿).

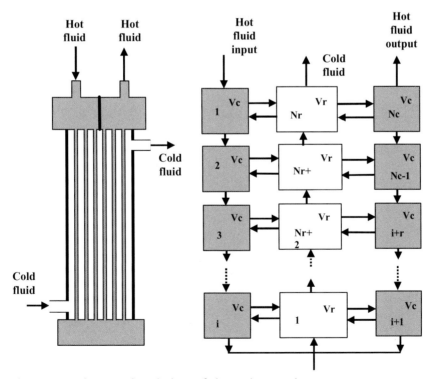

Figure 4.42 Stochastic topological scheme of a heat exchanger with two
configurations for the flow pattern of the hot fluid (countercurrent and co-current).

exchanger and its division into cells. It is the case of a tubular heat exchanger, where the hot fluid passes through the exchanger twice (in countercurrent and in co-current). The walls separating the fluids have not been divided into cells; because we considered that the heat accumulated by the walls was insignificant with respect to the heat transferred between the hot and cold fluids. In such example, the respective volumes of the heat exchanger cells are *a priori* different and result in a much more complex situation when compared with the previous example discussed in this chapter.

4.6.3.2 Cellular Stochastic Model for a Countercurrent Flow with Recycling

The example presented in this section is a system where two countercurrent fluids flow through N identical cells; Fig. 4.43 describes this system schematically. In this simplified case, we consider that, at each cell level, we have a perfect mixing flow and that for a "k" cell, the actual transition probabilities are p_{kk}, p_{kk-1} and p_{kk+1}. Indeed, these probabilities are expressed as:

$$p_{kk-1} = \frac{aG_v\Delta\tau}{V} \quad, \quad p_{kk+1} = \frac{(1+a)G_v\Delta\tau}{V} \quad, \quad p_{kk} = 1 - (p_{kk-1} + p_{kk+1}) \qquad (4.345)$$

When we have the same fraction of recycling in the system and when the cells have the same volume, we can rewrite relation (4.345) as:

$$p_{kk-1} = \frac{a_kG_v\Delta\tau}{V_{k-1}} \quad, \quad p_{kk+1} = \frac{(1+a_k)G_v\Delta\tau}{V_k} \quad, \quad p_{kk} = 1 - (p_{kk-1} + p_{kk+1}) \qquad (4.345)$$

We can observe that the first and the last cell of the system are in contact with only one cell: cell number 2 and number N − 1 respectively. So, in the matrix of the transition probabilities, the values p_{13} and p_{N-2N} will be zero. It is easily noticed that, if we have a complete matrix of the transition probabilities, then we can compute the mean residence time, the dispersion around the mean residence time and the mixing intensity for our cells assembly. The relations (4.324)–(4.326) are used for this purpose.

$$P = \begin{bmatrix} p_{11} & p_{12} & 0 & 0 & - & 0 & 0 \\ p_{21} & p_{22} & p_{23} & 0 & - & 0 & 0 \\ 0 & p_{32} & p_{33} & p_{34} & - & 0 & 0 \\ 0 & 0 & p_{43} & p_{44} & p_{45} & 0 & 0 \\ 0 & 0 & 0 & 0 & 0 & p_{N-1N} & p_{NN} \end{bmatrix} \qquad (4.346)$$

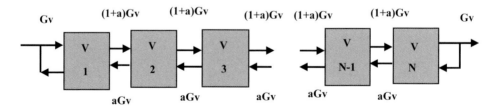

Figure 4.43 Countercurrent and recycle flow model.

If, besides hydrodynamics and mixing, we want to consider other phenomena, such as a chemical reaction, we have to separate the probabilities characterizing each particular phenomenon:

- $p_{jk} = \alpha_{jk}\Delta\tau$ – gives the probability for the species to skip from cell j to cell k in the time $\Delta\tau$;
- $p_{ke} = \alpha_{ke}\Delta\tau$ – gives the probability for the species to quit the system during the time interval $\Delta\tau$ (this probability exists when the k cell presents an open output);
- $p_{kr} = \alpha_{kr}\Delta\tau$ – is the probability that quantifies the species transformation as a result of its chemical interaction inside the k cell at $\Delta\tau$.

With this notation and considering the definition of $P_k(\tau)$ already given, the probability balance, with respect to the k cell for the time interval between τ and $\tau + \Delta\tau$ can be written as:

$$P_k(\tau + \Delta\tau) = \left[1 - \left(\sum_{j,j \neq k} \alpha_{kj}\Delta\tau\right)\right][1 - (\alpha_{ke} + \alpha_{kr})\Delta\tau]P_k(\tau) + \left(\sum_{j,j \neq k} \alpha_{jk}\Delta\tau\right)P_j(\tau)$$

(4.347)

If we consider that $\Delta\tau$ is very small, we obtain the concluding form of our cellular stochastic model. It describes the cellular countercurrent flow with recycling and chemical reaction:

$$\frac{dP_k(\tau)}{d\tau} = -(\alpha_{ke} + \alpha_{kr})P_k(\tau) - \left(\sum_{j,j \neq k} \alpha_{kj}\right)P_k(\tau) + \left(\sum_{j,j \neq k} \alpha_{jk}\right)P_j(\tau) \ ;$$

$$k = 1, N \ ; \ j = 1, N$$

(4.348)

Each term from the right side of this representative equation of the model has a particular meaning. The first term shows that the number of the reactant species molecules in the k cell decreases as a result of the consumption of species by the chemical reaction and the output of species from the cell. The second term describes the reduction of the number of molecules as a result of the transport to other compartments. The last term gives the increase in the number of the species in the k compartment because of the inputs from the other cells of the assembly. With reference to the mathematical formalism, our model is described by an ordinary system of differential equations. Indeed, for calculations we must specify the initial state of the probabilities. So, the vector $P_k(0), k = 1, N$ must be a known vector. The frequencies $\alpha_{ke}, \alpha_{kr}, \alpha_{kj}, \alpha_{jk}$ will be established by means of the cellular assembly topology and kinetic data. It is evident that the frequency α_{kr} will be related to the reaction process taking place in the cell.

References

4.1 R. Higbie, *Trans. A Inst. Chem. Engrs.* **1935**, *31*, 365.

4.2 P. J. Dankwertsz, *A.I.Ch.E. J.* **1955**, *1*, 456.

4.3 F. Feller, *Stochastic Processes*, John Wiley, New York, 1953.

4.4 V. V. Kafarov, *Cybernetic Methods for Technologic Chemistry*, Mir, Moscow, 1969.

4.5 O. Iordache, *Compound Stochastic Processes Applied in Transport Phenomena*, Romanian Academy, Bucharest, 1981.

4.6 A. Tamir, *Applications of Markov Chains in Chemical Engineering*, Elsevier, Amsterdam, 1998.

4.7 N. G. VanKampen, *Stochastic Processes in Physics and Chemistry*, Elsevier, North-Holland, Amsterdam, 2001.

4.8 D. N. Shanbhag (Ed.), *Handbook of Statistics: Stochastic Processes: Theory and Method*, Elsevier, North Holland, Amsterdam, 2000.

4.9 R. B. Bird, W. E. Steward, E. N. Lightfoot, *Transport Phenomena*, John Wiley, New York, 1960.

4.10 Em. A. Bratu, *Processes and Installations for Industrial Chemistry*, Bucharest Polytechnic Institute, Bucharest, 1959.

4.11 Em. A. Bratu, *Processes and Apparatus for Industrial Chemistry I, II*, Technical Book, Bucharest, 1970.

4.12 O. Onicescu, G. Mihoc, *Comp. Rend. Acad. Sci.* **1935**, *200*, 174.

4.13 O. Onicescu, G. Mihoc, *Bull. Sci. Math.* **1935**, *59*, 174.

4.14 O. Onicescu, G. Mihoc, *Romanian Academy, Studies and Researches*, Bucharest, 1943.

4.15 O. Onicescu, *Probability and Random Processes*, Scientific and Encyclopedic Book, Bucharest, 1977.

4.16 J. R. Doob, *Stochastic Processes Theory*, John Wiley, New York, 1953.

4.17 M. Iosifescu, *Finite Markov Chains and Their Applications*, Technical Book, Bucharest, 1977.

4.18 J.Y. Oldshue, *Fluid Mixing Technology*, McGraw-Hill, New York, 1983.

4.19 H. Holland, J. Chapman, *Liquid Mixing and Processing in Stirred Tanks*, Reinhold, New York, 1966.

4.20 Em. A. Bratu, *Unit Operations for Chemical Engineering, I*, Technical Book, Bucharest, 1983.

4.21 L. N. Braginsky, V. I. Begatcev, G. Z. Kofman, *Teor. Osn. Him. Technol.* **1968**, *2*(1), 128.

4.22 L. N. Braginsky, V. I. Begatcev, V. M. Barabash, *Mixing of Liquids. Physical Foundations and Methods of Calculation*, Himya Publishers, Leningrad, 1984.

4.23 L. N. Braginsky, Y. N. Kokotov, *J. Disp. Sci. Tech.* **1993**, 14, 3.

4.24 M. Filipescu, *Modeling of Homogeny Fluidization: Thesis*, Polytechnic Institute, Bucharest, 1981.

4.25 A.E. Scheidegger, *Can. J. Phys.* **1958** , *36*, 649.

4.26 J. Crank, *Mathematics of Diffusion*, Clarendon Press, Oxford, 1956.

4.27 H. S. Carslow, J. C. Jaeger, *Conduction of Heat in Solids*, Clarendon Press, Oxford, 1959.

4.28 S. K. Scrinivasan, M. K. Mehata, *A.I.Ch.E.J.* **1972**, *18*(3), 650.

4.29 T. Dobre, Highly Efficient Mass Transfer Apparatus- Mobile Packed Bed Column: Thesis, Polytechnic Institute of Bucharest, 1985.

4.30 D. K. Pickard, E. M. Tory, *Can. J. Chem. Eng.* **1977**, *55*, 655.

4.31 E.M. Tory, *Chem. Eng. J.* **2000**, *80*, 81.

4.32 D. K. Schmalzer, H. E. Hoelscher, *A.I.Ch.E.J.* **1971**, *17*, 104.

4.33 O. Iordache, M. A. Iosifescu, *Papers of 6th Romanian Conference on Probability*, pp.217–225, Brasov, 1979.

4.34 D. Revuz, *Markov Chains*, Elsevier, North-Holland, Amsterdam, 1984.

4.35 I. J. Gikham, A. N. Shorod, *Introduction to the Theory of Random Processes*, Saunders, Philadelphia, 1969.

4.36 I. J. Gikham, A. N. Shorod, *Stochastic Differential Equations*, Springer-Verlag, Berlin,1972.

4.37 P. E. Kloeden, E. Platen, *Numerical Solution of Stochastic Differential Equations*, Springer-Verlag, Berlin, 1992.

4.38 M. Pinsky, *Probabilistic Methods in Differential Equations*, Springer-Verlag, Berlin,1975.

4.39 P. R. Iranpour, P. Chacon, *Basic Stochastic Processes*, McMillan, New York,1988.

4.40 K. Yoshida, *Functional Analysis*, Springer-Verlag, Berlin,1965.

4.41 E. B. Dynkin, *Markov Processes*, Academic Press, New York, 1965.

4.42 K. Burdzy, M. D. Frankel, A. Pauzner, On the Time and Direction of Stochastic Bifurcation, in *Asymptotic Methods in Probability and Statistics*, B. Szyskowics, (Ed.), Elsevier, North-Holland, Amsterdam,1998.

4.43 R. A. Dabrowski, H. Dehling, Jump Diffusion Approximation for a Markovian Transport Model, in *Asymptotic Methods in Probability and Statistics*, B. Szyskowics, (Ed.), Elsevier, North-Holland, Amsterdam, 1998.

4.44 M. Laso, *A.I.Ch.E.J.* **1994**, 40, 1297.

4.45 D.K. Pickard, M.E. Tory, Dispersion Behavior-A Stochastic Approach, in *Advances in the Statistical Science*, B. I. MacNeil, J. G. Umphrey (Eds.), Vol. 4, Ch. 1, D. Reidel, Dordrecht, 1987.

4.46 A. T. Bharuca Reid, *Elements of Theory of Markov Processes and Their Applications*, McGraw-Hill, New York, 1960.

4.47 R. L. Stratanovich, *Conditional Markov Processes and Their Applications*, Elsevier, Amsterdam, 1968.

4.48 L. Berkes, Results and Problems Related to the Point wise Central Limit Theorem, in *Asymptotic Methods in Probability and Statistics*, B. Szyskowics (Ed.), Elsevier, North-Holland, Amsterdam,1998.

4.49 J. Karger, M. D. Ruthven, *Diffusion in Zeolites and Other Micoporous Solids*, John Wiley , New York, 1992.

4.50 Y. N. Chen, M. D. Ruthven, *Molecular Transport and Reaction in Zeolites*, VCH Publishers, New York, 1994.

4.51 J. Weitkamp, Separation and Catalysis by Zeolites, in *Catalysis and Adsorption*, J. Vedrine, A. Jacobs (Eds.), Elsevier, Amsterdam, 1991.

4.52 R. Haberlandt, J. Kager, *Chem.Eng. J.* **1999**, 74, 15.

4.53 A. Luikov, *Heat and Mass Transfer*, Mir Publishers, Kiev, 1980.

4.54 P. B. Fernando, N. P. Efstraitos, M. S. Pedro, *Ind. Eng. Chem. Res.* **1999**, 38(8), 3056.

4.55 R. J. Banga, *Ind. Eng. Chem. Res.* **1997**, 36(6), 2252.

4.56 L. C. Cheng, Y. S. Daim, *Ind. Eng. Chem. Res.* **2000**, 39(7), 2305.

4.57 Z. Novak, Z. Kravanja, *Ind. Eng. Chem. Res.* **1999**, 38(7), 2680.

4.58 N. P. Efstratios, *Ind. Eng. Chem. Res.* **1999**, 36(6), 2262.

4.59 R. A. Felipe, M. R. Carlos, V. N. Torres, *J. Biotechnol.* **1999**, 68, 15.

4.60 A. V. Ermoshin, V. Engel, *Chem. Phys. Lett.* **2000**, 332, 162, 110.

4.61 A. Gaizauskas, S. A. Berzanskas, H. K. Feller, *Chem. Phys.* **1998**, 235, 1-3, 123.

4.62 J. Christina, N. Breton, P. Daegelen, *J.Chem.Phys.* **1997**, 107(8), 2903.

4.63 R. Bocatti, *Wave Mechanics for Oceanic Engineering*, Elsevier, Amsterdam, 1998.

4.64 C.V. Singh, *Int. J. Clim.* **1998**, 18(14), 1611.

4.65 R. J. Koehler, B. A. Owen, Computer Experiments, in *Handbook of Statistics 13: Design and Analysis of Experiments*, S. Gosh, R. Rao (Eds.), Elsevier, North-Holland, Amsterdam, 1996.

4.66 G. A. Mallinaris, A.W. Brock, *Stochastic Methods in Economics and Finance*, 7th edn., Elsevier, North-Holland, Amsterdam,1996.

4.67 P. Embrechts, R. Frey, H. Furror, Stochastic Processes in Insurance and Finance, in *Handbook of Statistics 19: Stochastic Processes: Theory and Method*, N. E. Shanbhag, R. C. Rao (Eds.) Elsevier, North-Holland, Amsterdam, 2000.

4.68 V. Zaharov, Games and Stochastic Control, in *Control Applications of Optimization 2000*, Vol. 2, V. Zaharv (Ed.), Elsevier Science, Amsterdam, 2000.

4.69 B. S. Poppe, *Turbulent Flows*, Cambridge University Press, Cambridge, 2000.

4.70 V. Kudrna, P. Hsal, L. Vejmola, *Collect. Czech. Chem. Commun.* **1994**, 59(2), 345.

4.71 A. G. Maria, J. Colussi, *J. Phys. Chem.***1996**, 100(46), 18214.

4.72 B. K. Mishra, *Powder Technol.*, **2000**, 110(3), 246.

4.73 M. S. Cannon, S. B. Brewster, D. L. Smoot, *Combust. Flame*, **1998**, 113, 135.

4.74 H. Bertiaux, *Powder Technol.* **1999**, 105(1-3), 266.

4.75 A.W. Curtin, Curr. Opin. Solid State Mater. Sci. **1996**, 1(5), 674.

4.76 H. Bertiaux, *Chem. Eng. Sci.* **2000**, 55(19), 4117.

4.77 Z. Liang, A. M. Ioanidis, I. Chatzis, *Chem. Eng. Sci.* **2000** , 55(22), 5247.

4.78 N. Scheerlinck, P. Verboven, D. J. Stigter, D. J. Baerdemaeker, J. van Impe, M. B. Nicolai, *Int. J. Num. Methods Eng.* **2001**, 51(8), 961.

4.79 E. A. Stillman, H. J. Freed, *J. Chem. Phys.* **1980**, 72(1), 550.

4.80 T. Dobre, O. Floarea, *Rev. Chim*, **1984**, 35(9), 867.

4.81 T. Dobre, O. Floarea, *Rev. Chim*, **1985**, 36(11), 1021.

4.82 T. Dobre, *Rev. Chim.* **1992**, 37(8), 879.

4.83 B. N. Chen, W. J. Douglas, *Can. J. Chem. Eng.* **1968**, 47(2),113.

4.84 J. E. Cains, M. I. Prausnitz, *Ind. Eng. Chem.* **1959**, 5(12), 441.

4.85 K. Masao, K. Tabei, K. Murata, *Ind. Eng. Chem. Proc. Des. Dev.* **1978**, 17(4), 568.

4.86 A. I. Kovali, V. A. Bespalov, O. G. Kulesov, P. A. Jukov, *Teor. Osn. Him. Technol.* **1975**, 9(6), 887.

4.87 W. S. Rutherford, D. D. Do, *Adsorption*, **1997**, 3, 283.

4.88 W. S. Rutherford, D. D. Do, *Chem. Eng. J.* **1999**, 74, 155.

4.89 O. Levenspiel, *Chemical Reaction Engineering*, John Wiley, New York, 1999.

4.90 R.C. Darton, (Ed.) *Modeling of Solid-Fluid Separation*, NATO Series, Ninjof, New York, 1985.

4.91 P. M. Hertjes, P. M. Lerk, *Trans. Inst. Chem. Eng.* **1967**, 45, T138.

4.92 R. Arris, N. R. Amundson, *Mathematical Method in Chemical Engineering*, Vol. 2, Prentice Hall, New Jersey, 1973.

4.93 L. T. Fan, M. S. K. Chan, Y. K. Ahn, W. Y. Wen, *Can. J. Chem. Eng.* **1969**, 47, 141.

4.94 V. V. Kafarov, P. V. Voroviev, A. N. Klipinitzer, *Teor. Osn. Him. Technol.* **1972**, 6(4), 113.

5
Statistical Models in Chemical Engineering

The models based on the equations of transport phenomena and on stochastic models contain an appreciable quantity of mathematics, software creation, computer programming and data processing.

In many countries, a high level in mathematics is not a requirement for achieving a good knowledge in theoretical and practical chemistry or in chemical engineering, so, chemists or chemical engineers do not often have a deep knowledge of mathematics even though most areas of chemistry are often based upon quantitative measurements and computation. For example, the statistical validation of the techniques currently used in laboratories specializing in chemical analysis may be necessary to maintain the laboratory accreditation and/or for legal reasons In this case, the chemists or chemical engineers, who may have left formal training in mathematics 10 or 20 years before, could suddenly be faced with the need to brush up on statistics.

An important number of reference books on chemistry and chemical engineering statistics [5.1–5.11] have been published by specialists. The chemists and chemical engineers who intend to attend programs on statistical modelling of processes, must have a good basic knowledge in descriptive statistics, distribution of random variables and statistics hypotheses, and be able to carry out the experiments connecting the various measurements. These basic notions are therefore introduced in the following examples and discussions:

Descriptive statistics. A series of physical measurements can be described numerically. If, for example, we have recorded the concentration of 1000 different samples in a research problem, it is not possible to provide the user with a table giving all 1000 results. In this case, it is normal to summarize the main trends. This can be done not only graphically, but also by considering the overall parameters such as mean and standard deviation, skewness etc. Specific values can be used to give an overall picture of a set of data.

Distribution for random variables. The concept of distribution is fundamental to statistics. If a series of measurements is extracted from a great number of similar non-produced measurements (called population), we obtain a population sample. However, it is not possible to have the same mean characteristics for all the sam-

Chemical Engineering. Tanase G. Dobre and José G. Sanchez Marcano
Copyright © 2007 WILEY-VCH Verlag GmbH & Co. KGaA, Weinheim
ISBN: 978-3-527-30607-7

ples, because errors and noise influence the characterization properties of each sample. In fact, it is impossible for each sample to be identical. The distribution of measurements is often approximated by a normal distribution, although, in some cases, this does not represent an accurate model. If a sufficient number of measurements is taken during the analysis of samples, it is possible to see whether they fit into such a distribution.

If the number of samples with characteristics presenting a normal distribution is not significant, then we can have an error structure. This situation can also be due to outliers, i.e. samples that are atypical of the population or that might have been incorrectly labeled or grouped.

Statistics hypotheses and their testing. In many cases, the measurements are used to answer qualitative questions. For example, for the quality control of a batch of liquid products, a concentration analysis is carried out. If the analysis of a sample from the batch results in a higher concentration with respect to a reference value then we can reject the batch. In this case, we can use different tests to validate the rejection or acceptance of the batch. One example of such tests is the comparison of the mean values. Concerning the example described above, the measurements are realized by two groups of researchers, A and B. Group A has recorded twenty concentrations in a series of samples and has obtained a mean concentration value of 10 g/l and a deviation of 0.5 g/l. Group B, who monitors the same series of samples, has obtained a mean concentration value of 9.7 g/l and a deviation of 0.4 g/l. Then both mean values and deviations must be compared so as to answer the following questions: Are they actually different? What is the probability for both groups to have measured the same fundamental parameters? Is this difference in mean values simply caused by a different sampling or a variation in the measurement technique?

Relating measurements. Evaluating the relationships between the different types of measurements of the variables that are coupled or not to a process is fundamental in statistics. In the case of variables coupled to a process, the separation in the class of independent variables (x_i, $I = 1$, n) and dependent variables (y_j, $j = 1$, p) must be established based on the schematic representation of the process (see Fig. 1.1 in Chapter 1). The statistical models will be built based on experimental measurements. However, good models can be developed only if experimental results are obtained and processed from a statistical analysis. The analysis of neural networks processes, which are also statistical models, represents a modern and efficient research technique based on the experimental measurement of one actual process.

The first step for the analysis of a statistical modelling problem concerns the definition of the concept of statistical models. This definition is based on the diagram shown in Fig. 5.1 (which is a variation of Fig. 1.1 in Chapter 1). Statistical modelling contains all the statistical and mathematical procedures that use measured data of y_i ($i = 1$,P) and x_j ($j = 1$,N) simultaneously in order to obtain the multiple inter-dependences between dependent and independent variables. The relation (5.1) obtained on this basis represents the statistical model of a process:

$$y_i = f_i(x_1, x_2 .... x_n), i = 1, p \tag{5.1}$$

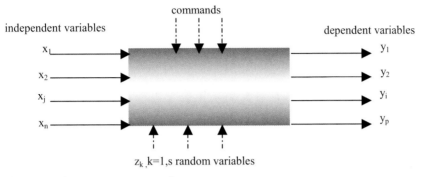

Figure 5.1 Schematic representation of a process.

5.1
Basic Statistical Modelling

The statistical modelling of a process can be applied in three different situations: (i) the information about the investigated process is not complete and it is then not possible to produce a deterministic model (model based on transfer equations); (ii) the investigated process shows multiple and complex states and consequently the derived deterministic or stochastic model will be very complex; (iii) the researcher's ability to develop a deterministic or stochastic model is limited.

The statistical modelling of a process presents the main advantage of requiring nothing but the inputs and outputs of the process (the internal process phenomena are then considered as hidden in a black box). We give some of the important properties of a statistical model here below:

1. As far as a statistical model has an experimental origin, it presents the property to be a model which could be verified (verified model).
2. Statistical models are strongly recommended for process optimization because of their mathematical expression and their being considered as verified models.
3. Classic statistical models cannot be recommended for the analysis of a dynamic process because they are too simple. Dynamic processes are better described by using the artificial neural network.

Two types of experiments can produce the data needed to establish statistical models. *Passive experiments* refer to the classical analysis of an experimental process investigation. They occur when the sets of experiments have been produced (in an industrial or in a pilot unit) either by changing the values of independent process variables one by one or by collecting the statistical materials obtained with respect to the evolution of the investigated process. *Active experiments* will be produced after the establishment of a working plan. In this case, the values of each of the independent variables of the process used for each planned experiment are obtained by specific fixed procedures.

To start the procedure of the statistical modelling of a process, we have to produce some initial experiments. These experiments will allow us:

1. to identify the domain of the value for each independent variable.
2. to identify the state of the dependent variables when the independent variables of the process increase.
3. to determine whether the state of the dependent variables of the process is affected by the interaction of the independent variables.

Dispersion and correlation analyses are used to process the data obtained in the preliminary experiments. The goal of these statistical analyses is to have qualitative or quantitative answers to points 2 and 3 mentioned above. Finally, when all the statistical data have been collected, a correlation and regression analysis will be used to obtain the inter-dependence relationships between the dependent and the independent variables of the process (see relation (5.1)).

In a process, when the value domain of each of the independent variables is the same in the passive and in the active experiments simultaneously, two identical statistical models are expected. The model is thus obtained from a statistical selection and its different states are represented by the response curves, which combine the input parameters for each of the output parameters.

Now, if the obtained model is used to produce output data and these are compared with the corresponding experimental results, some differences can be observed. This behavior is expected because the model has been extended outside the selection of its bases and this extension is only permissible if it is possible to take into account the confidence limits of the model.

Each of the independent variables x_1, x_2, x_3, ... x_N is frequently called a *factor* whereas the N-dimensional space containing the coordinates x_1, x_2, x_3, ... x_N is called *factorial space*; the *response surface* is the representation of one response function into N-dimensional space. A statistical model with a unique response surface would characterize the process that shows only one output (Fig. 5.1).

In a model, the number of response surfaces and the number of process outputs are the same. Figure 5.2 shows the surface response for a chemical reaction where the degree of transformation of the reactive species (dependent variables) in an expected product is controlled by their concentration and temperature (two independent variables). When we look at this figure, it is not difficult to observe that the maximum degree of transformation can easily be established; so, ignoring the economic aspects of the process, the optimal states of the temperature and concentration are automatically given by the maximum conversion.

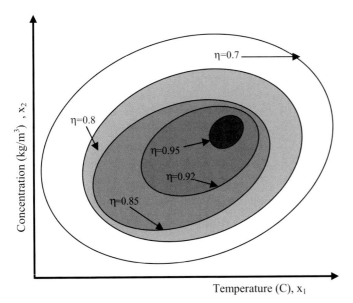

Figure 5.2 Reaction efficiency ($y = \eta$) versus temperature (x_1) and limiting reactant concentration (x_2).

The basis of the statistical model is given by the Taylor expansion of relation (5.1). It is established for the vicinity of the factors of the process where a fixed/an established value is given to the dependent variable (y_{i0}). In this expansion, y_{i0} results in the y_i value when the factors take the corresponding $x_{10}, x_{20},...x_{N0}$ values:

$$
\begin{aligned}
y_i = y_{i0} &+ \sum_{j=1}^{N}\left(\frac{\partial f_i}{\partial x_j}\right)_0 (x_j - x_{j0}) - \frac{1}{2!}\sum_{j=1}^{N}\sum_{k\neq J}^{N}\left(\frac{\partial^2 f_i}{\partial x_j \partial x_k}\right)_0 (x_j - x_{j0})(x_k - x_{k0}) - \\
&\frac{1}{2!}\sum_{j=1}^{N}\left(\frac{\partial^2 f_i}{\partial x_j^2}\right)_0 (x_j - x_{j0})^2 + \\
&\frac{1}{3!}\sum_{j=1}^{N}\sum_{k=1, k\neq j, k\neq l}^{N}\sum_{l=1; l\neq j, l\neq k}^{N}\left(\frac{\partial^3 f_i}{\partial x_j \partial x_k \partial x_l}\right)(x_j - x_{j0})(x_k - x_{k0})(x_l - x_{l0}) + \\
&\frac{1}{3!}\sum_{j=1}^{N}\sum_{k=1, k\neq j}^{N}\left(\frac{\partial^3 f_i}{\partial^2 x_j^2 \partial x_k}\right)(x_j - x_{j0})^2(x_k - x_{k0}) + \frac{1}{3!}\sum_{j=1}^{N}\left(\frac{\partial^3 f_i}{\partial x_j^3}\right)(x_j - x_{j0})^3 - \\
&\frac{1}{4!}\sum_{j=1}^{N}\sum_{k=1, k\neq j, j, m}^{N}\sum_{l=1, l\neq j, k, m}^{N}\sum_{m=1, m\neq j, k, l}^{N}\left(\frac{\partial^4 f_i}{\partial x_j \partial x_k \partial x_l \partial x_m}\right) \\
&(x_j - x_{j0})(x_k - x_{k0})(x_l - x_{l0})(x_m - x_{m0}) + ...
\end{aligned}
\tag{5.2}
$$

It is not difficult to observe that the Taylor expression can be transposed as Eq. (5.3) where the index "i" has been extracted because it stays unchanged along the relation:

$$
\begin{aligned}
y^{(i)} = \beta_0^{(i)} &+ \sum_{j=1}^{N} \beta_j^{(i)} x_j + \sum_{j=1}^{N} \sum_{k=1,k\neq j}^{N} \beta_{jk}^{(i)} x_j x_k + \sum_{j=1}^{N} \beta_{jj}^{(i)} x_j^2 + \\
&\sum_{j=1}^{N} \sum_{k=1,k\neq j,l}^{N} \sum_{l=1,l\neq j,k}^{N} \beta_{jkl}^{(i)} x_j x_k x_l + \sum_{j=1}^{N} \sum_{k=1,k\neq j}^{N} \beta_{jjk}^{(i)} x_j^2 x_k + \sum_{j=1}^{N} \beta_{jjj}^{(i)} x_j^3 + \\
&\sum_{j=1}^{N} \sum_{k=1,k\neq j,j,m}^{N} \sum_{l=1,l\neq j,k,m}^{N} \sum_{m=1,m\neq j,k,l}^{N} \beta_{jklm}^{(i)} x_j x_k x_l x_m + \dots\dots
\end{aligned}
\tag{5.3}
$$

From the analysis of Eqs. (5.2) and (5.3) we can observe that each β coefficient has a specific expression. As an example, relation (5.4) shows the definition expression for $\beta_0^{(i)}$:

$$
\beta_0^{(i)} = y_{i0} - \sum_{j=1}^{N} \left(\frac{\partial f_i}{\partial x_j} \right)_0 x_{j0} + \frac{1}{2!} \sum_{j=1}^{N} \sum_{k=1,k\neq j}^{N} \left(\frac{\partial^2 f_i}{\partial x_j \partial x_k} \right)_0 x_{j0} x_{k0} + \frac{1}{2!} \sum_{j=1}^{N} \left(\frac{\partial^2 f_i}{\partial x_j^2} \right)_0 x_{j0}^2 - \dots
\tag{5.4}
$$

In Eq. (5.4) we can note that, in fact, the model describes the relationship between the process variables. Nevertheless, coefficients $\beta_0^{(i)}, \beta_j^{(i)}$, etc are still unknown because the functions $f_i(x_1,x_2,x_3,\dots x_N)$ are also unknown.

A real process is frequently influenced by non-commanded and non-controlled small variations of the factors and also by the action of other random variables (Fig. 5.1). Consequently, when the experiments are planned so as to identify coefficients $\beta_0^{(i)}, \beta_j^{(i)}$, etc, they will apparently show different collected data. So, each experiment will have its own $\beta_0^{(i)}, \beta_j^{(i)}$, etc. coefficients. In other words, each coefficient is a characteristic random variable, which is observable by its mean value and dispersion.

Coefficients $\beta_0^{(i)}, \beta_j^{(i)}$, etc. (called *regression coefficients*) can be identified by means of an organised experiment. Since they have the quality to be the estimators of the real coefficients defined by relation (5.4), two questions can be formulated:

1. What is the importance of each coefficient in the obtained model?
2. What confidence can be given to each value of $\beta_0^{(i)}, \beta_j^{(i)}$, etc. when they are established as the result of programmed experiments?

The aim of statistical modelling is certainly not to characterize the relationship in a sample (experiment). So, after the identification of $\beta_0^{(i)}, \beta_j^{(i)}$.... etc, it is important to know what *confidence limits* can be given to the obtained model.

Each $\beta_0^{(i)}, \beta_j^{(i)}$..., etc. coefficient signification, is formally estimated. For instance, in this example, $\beta_0^{(i)}$ is the constant term for the regression relationship, $\beta_j^{(i)}$ corresponds to the linear effects of the factors, $\beta_{jk}^{(i)}$ gives the effect of the interaction of x_j and x_k factors on the regression relationship, etc.

In relation (5.3), where $\beta_0^{(i)}, \beta_j^{(i)}$... etc. are the unknown parameters, we can observe that among the different methods to identify these parameters, the method of least-squares can be used without any restriction. So, the identification of $\beta_0^{(i)}, \beta_j^{(i)}$... etc. coefficients has been reduced to the functional minimisation shown in relation (5.5):

$$\Phi^{(i)}(\beta_0^{(i)}, \beta_j^{(i)}, \beta_{jk}^{(i)}, \ldots) = \sum_{i=1}^{Ne} \left(y_i^{(i),ex} - y_i^{(i),th} \right)^2 \tag{5.5}$$

where "Ne" gives the dimension of the experimental sample produced for the identification of the parameters; $y_i^{(i),ex}$ is the "i" experimental value of the output (i) and $y_i^{(i),th}$ is the "i" model-computed value of the output (i). This $y_i^{(i),th}$ is obtained using relation (5.3) and the numerical values of $x_{ji}, j = 1, N$. The dimension of the model (for the identification of the parameters) depends on the number of terms considered in relation (5.3). Table 5.1 gives the number of coefficients to be identified when the number of the factors of the process and the statistical model degree are fixed at the same time.

Table 5.1 Number of coefficients to be identified for the polynomial state of a statistical model.

Number of factors of the process	Statistical model with polynomial state (polynomial degree)			
	Number of identifiable coefficients			
	First degree	Second degree	Third degree	Fourth degree
2	3	6	10	15
3	4	10	20	35
4	5	15	35	70
5	6	21	56	126

In Table 5.1, where the statistical model is presented in a polynomial state, a rapid increase in the number of identifiable coefficients can be observed as the number of factors and the degree of the polynomial also increase. Each process output results in a new identification problem of the parameters because the complete model process must contain a relationship of the type shown in Eq. (5.3) for each output (dependent variable). Therefore, selecting the "Ne" volume and particularizing relation (5.5), allows one to rapidly identify the regression coefficients. When Eq. (5.5) is particularized to a single algebraic system we take only one input and one output into consideration. With such a condition, relations (5.3) and (5.5) can be written as:

$$y^{(1)} = y^{(1),\text{th}} = y = f_1(x_1, \beta_0^{(1)}, \beta_1^{(1)}, \beta_{11}^{(1)}; \beta_{111}^{(1)}....) = f(x, \beta_0, \beta_1, \beta_2, ...) \tag{5.6}$$

$$\Phi^{(1)}(\beta_0^{(1)}, \beta_1^{(1)}, \beta_{11}^{(1)}; \beta_{111}^{(1)}..) = \Phi(\beta_0, \beta_1, \beta_2, ..) = \sum_{i=1}^{Ne}(y_i - f(x_i, \beta_0, \beta_1, \beta_2.))^2 = \min \tag{5.7}$$

Now, developing the condition of the minimum of relation (5.7) we can derive relation (5.8). It then corresponds to the following algebraic system:

$$\frac{\partial\Phi(\beta_0, \beta_1, \beta_2, ...)}{\partial\beta_0} = \frac{\partial\Phi(\beta_0, \beta_1, \beta_2, ...)}{\partial\beta_1} = \frac{\partial\Phi(\beta_0, \beta_1, \beta_2, ...)}{\partial\beta_2} = ..$$
$$= \frac{\partial\Phi(\beta_0, \beta_1, \beta_2, ...)}{\partial\beta_n} = 0 \tag{5.8}$$

Here β_n is the last coefficient from the function $f(x, \beta_0, \beta_1, \beta_2, ...)$. The different coefficients of this function multiply the x^n monomial, and "n" gives the degree of the polynomial that establishes the y–x relationship.

The computing of the derivates of relation (5.8) results in the following system of equations:

$$\begin{cases}
\displaystyle\sum_{i=1}^{Ne} y_i \frac{\partial f(x_i, \beta_0, ..\beta_n)}{\partial\beta_0} - \sum_{i=1}^{Ne} f(x_i, \beta_0, \beta_1...\beta_n) \frac{\partial f(x_i, \beta_0, \beta_1, ...\beta_n)}{\partial\beta_0} = 0 \\[3mm]
\displaystyle\sum_{i=1}^{Ne} y_i \frac{\partial f(x_i, \beta_0, ..\beta_n)}{\partial\beta_1} - \sum_{i=1}^{Ne} f(x_i, \beta_0, \beta_1...\beta_n) \frac{\partial f(x_i, \beta_0, \beta_1, ...\beta_n)}{\partial\beta_1} = 0 \\[3mm]
\text{---} \\[1mm]
\displaystyle\sum_{i=1}^{Ne} y_i \frac{\partial f(x_i, \beta_0, ..\beta_n)}{\partial\beta_n} - \sum_{i=1}^{Ne} f(x_i, \beta_0, \beta_1...\beta_n) \frac{\partial f(x_i, \beta_0, \beta_1, ...\beta_n)}{\partial\beta_n} = 0
\end{cases} \tag{5.9}$$

The system above contains N equations and consequently it will produce a single real solution for $\beta_0, \beta_1, ..., \beta_n$ (n unknowns). It is necessary to specify that the size of the statistical selection, here represented by Ne, must be appreciable. Moreover, whenever the regression coefficients have to be identified, Ne must be greater than n. This system (5.9) is frequently called: *system of normal equations* [5.4, 5.12–5.14].

In relation (5.5) we can see that $\Phi(\beta_0, \beta_1, \beta_2, ...)$ can be positive or null for all sorts of real $\beta_0, \beta_1, \beta_2, ...\beta_n$. As a consequence, it will show a minimal value for the identified $\beta_0, \beta_1, \beta_2, ...\beta_n$. Thus, the description of function $f(x, \beta_0, \beta_1, \beta_2, ...)$ results in a particularization of system (5.9).

If the number of independent variables is increased in the process, then the regression function will contain all the independent variables as well as their simple or multiple interactions. At the same time, the number of dependent variables also increases, and, for each of the new dependent variables, we have to consider the problem of identifying the parameters.

Coefficients $\beta_0, \beta_1, \beta_2, ...\beta_n$ or $\beta_0^{(i)}, \beta_j^{(i)}, \beta_{jk}^{(i)}, ...$ can be considered as the estimators of the real coefficients of the Taylor expansion in the relationship between the variables of model (5.1). These coefficients are estimators of maximum confidence because their identification starts with minimizing the function which contains the square deviation between the observed and computed values of the output variables. The quality of the identified coefficient and, indirectly, the quality of the regression model depend firstly on the proposed regression function. Then, the quality of the regression function imposes the volume of experiments needed to produce the statistical model. Indeed, with a small number of experiments we cannot suggest a good regression function. However, in the case of a simple process, the regression function can be rapidly determined with only a few experiments. It is important to note that, after the identification of the coefficients, the regression model must be improved with a signification test. Only the coefficients that have a noticeable influence on the process will be retained and the model that contains the established coefficients will be accepted.

In Fig. 5.3 the different steps of the statistical modelling of a process are shown. These steps include the analysis of the variables, the planning and developing of experimental research and the processing of the experimental data needed to establish the model. We can observe that the *production of the statistical model of a process is time consuming and that the effort to bypass experimentation is considerable*. With respect to this experimental effort, it is important to specify that it is sometimes difficult to measure the variables involved in a chemical process. They include concentrations, pressures, temperatures and masses or flow rates. In addition, during the measurement of each factor or dependent variable, we must determine the procedure, as well as the precision, corresponding to the requirements imposed by the experimental plan [5.4]. When the investigated process shows only a few independent variables, Fig. 5.3 can be simplified. The case of a process with one independent and one dependent variable has a didactic importance, especially when the regression function is not linear [5.15].

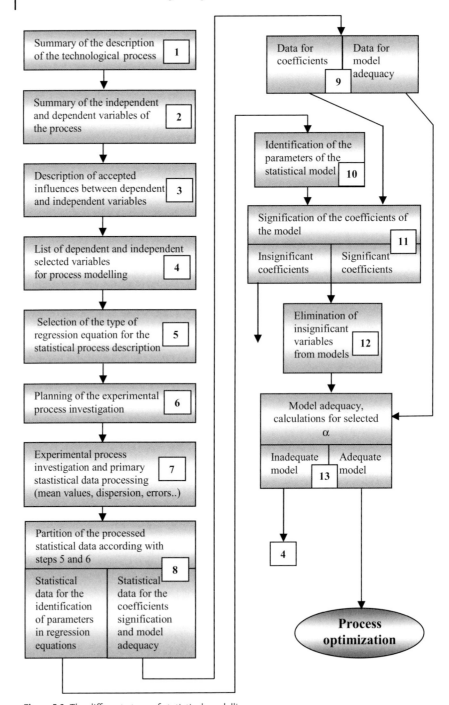

Figure 5.3 The different steps of statistical modelling.

5.2
Characteristics of the Statistical Selection

When we consider a process with only one input and one output variable, the experimental analysis of the process must contain enough data to describe the relationship between the dependent variable "y" and the independent variable "x". This relation can be obtained only if the data collected result from the evolution of one stationary process, and then supplementary experimental data can be necessary to demonstrate that the process is really in a stationary state.

As an actual process, we can consider the case of an isothermal and isobaric reactor working at steady state, where the input variable is the reactant's concentration and the output process variable (dependent variable) is the transformation degree. In this case, the values of the data collected are reported in Table 5.2. We can observe that we have the proposed input values (a prefixed set-point of the measurements) and the measured input values.

Table 5.2 Data for the characterization of y vs. x.

Current number for input	Proposed input value of x (set point)	Measured x value		Measured y value
		i	x_i	y_i
1	13.5	1	14.2	0.81
		2	13.5	0.75
		3	13.8	0.77
		4	14.3	0.75
		5	13.4	
2	20	1	20.5	0.66
		2	21.2	0.64
		3	19.8	0.63
		4	19.8	0.68
		5	19.5	0.65
		6		0.67
3	27	1	27.0	0.61
		2	27.4	0.59
		3	26.9	0.58

Table 5.2 Continued.

Current number for input	Proposed input value of x (set point)	Measured x value		Measured y value
		i	x_i	y_i
4	34	1	35.2	0.52
		2	34.7	0.49
		3	34.3	0.48
		4	35.1	0.55
		5	34.5	0.53
5	41	1	42.3	0.47
		2	42.6	0.43
		3	42.9	0.39
		4	41.8	0.46

In experimental research, each studied case is generally characterized by the measurement of x (x_i values) and y (y_i values). Each chain of x and each chain of y represents a statistical selection because these chains must be extracted from a very large number of possibilities (which can be defined as populations). However, for simplification purposes in the example above (Table 5.2), we have limited the input and output variables to only 5 selections. To begin the analysis, the researcher has to answer to this first question: "what values must be used for x (and corresponding y) when we start analysing of the identification of the coefficients by a regression function?" Because the normal equation system (5.9) requires the same number of x and y values, we can observe that the data from Table 5.2 cannot be used as presented for this purpose. To prepare these data for the mentioned scope, we observe that, for each proposed x value (x = 13.5 g/l, x = 20 g/l, x = 27 g/l, x = 34 g/l, x = 41 g/l), several measurements are available; these values can be summed into one by means of the corresponding mean values. So, for each type of x_i data, we use a mean value, where, for example, i = 5 for the first case (proposed x = 13.5 g/l), i = 3 for the third case, etc. The same procedure will be applied for y_i where, for example, i = 4 for the first case, i = 6 for the second case, etc.

With this method, we can create such couples as $(\bar{x}_1, \bar{y}_1), (\bar{x}_2, \bar{y}_2), \ldots (\bar{x}_5, \bar{y}_5)$ characterizing each case presented in Table 5.2. Thus, they can be used without any problem to solve the system of normal equations. Each class of finite data x_i or y_i with i ≻ 1 represents a statistical selection.

The most frequently used statistical measure for a selection is the mean value. For a selection x_i with i = 1, n, the mean value ($\bar{x}$) will be computed by the following relation:

$$\bar{x} = \frac{1}{n}(x_1 + x_2 + x_3 + \ldots + x_{n-1} + x_n) = \frac{1}{2}\sum_{i=1}^{n} x_i \tag{5.10}$$

In order to complete the selection characterization, we can use the variance or dispersion that shows the displacement of the selection values with respect to the mean value. Relations (5.11) and (5.12) give the definition of the dispersion:

$$s^2 = \frac{1}{n-1}\sum_{i=1}^{n}(x_i - \bar{x})^2 = \frac{1}{n-1}\left[\sum_{i=1}^{n} x_i^2 - \frac{1}{N}\left[\sum_{i=1}^{n} x_i\right]^2\right] \tag{5.11}$$

$$s^2 = \frac{n\sum_{i=1}^{n} x_i^2 - \left(\sum_{i=1}^{n} x_i\right)^2}{n(n-1)} \tag{5.12}$$

It is often necessary to simplify the calculations by replacing the initial selection by another one, which presents the same mean value and dispersion [5.8, 5.9]. Therefore, if, for each value x_i, $i = 1,n$ of the selection, we subtract the x_0 value, we obtain a new selection u_i, $i = 1,n$

$$u_i = x_i - x_0 \tag{5.13}$$

computing the mean value and the dispersion for this new selection we have:

$$\bar{u} = \frac{1}{n}\sum_{i=1}^{n}(x_i - x_0) = \bar{x} - x_0 \tag{5.14}$$

$$s_u^2 = \frac{1}{n-1}\sum_{i=1}^{n}(u_i - \bar{u})^2 = \frac{1}{n-1}\sum_{i=1}^{n}(x_i - \bar{x})^2 = s^2 \tag{5.15}$$

Table 5.3 shows the values obtained after the calculation of the mean values and the dispersions respect to the statistical data presented in Table 5.2.

It is very important to pay attention to two important aspects: (i) the selection is a sample drawn from a population; (ii) the scope of the statistical analysis is to characterize the population by using one or more selections.

It is easily observable that each selection x_i and its associated y_i shown in Tables 5.2 or 5.3 correspond to a sample extracted from each type of population. In the current example we have 5 populations, which give the input reactant concentration, and 5 populations for the transformation degree of the reactant. In the tables, the first population associated to the input concentration corresponds to the experiment where the proposed concentration has the value 13.5 g/l.

During the experiment, the numerical characterization of the population is given by the concentration of the reactant associated to the flow of the material fed into the reactor. Therefore, this reactant's concentration and transformation degree are random variables. As has been explained above (for instance see Chapters 3 and 4), the characterization of random variables can be realized taking into account the mean value, the dispersion (variance) and the centred or non-centred momentum of various degrees. Indeed, the variables can be characterized by the following functions, which describe the density of the probability attached

Table 5.3 Mean values and dispersions for the statistical data given by Table 5.2.

Current number for input	Proposed input value for x	Measured x value				Measured y value		
		i	x_i	$\bar{x}$	s^2_x	y_i	$\bar{y}$	s^2_y
1	13.5	1	14.2	13.86	$(14.2-13.86)^2 +$ $(13.5-13.86)^2 +$ $(13.8-13.86)^2 +$ $(14.3-13.86)^2 +$ $(13.4-13.86)^2$ $= 0.654$ $s^2 = 0.654/4$ $= 0.1635$	0.81	0.77	$(0.81-0.77)^2 +$ $(0.75-0.77)^2 +$ $(0.77-0.77)^2 +$ $(0.75-0.77)^2$ $= 0.0024$ $s^2 = 0.0024/3$ $= 0.0008$
		2	13.5			0.75		
		3	13.8			0.77		
		4	14.3			0.75		
		5	13.4					
2	20	1	20.5	20.16	0.473	0.66	0.655	0.00035
		2	21.2			0.64		
		3	19.8			0.63		
		4	19.8			0.68		
		5	19.5			0.65		
		6				0.67		
3	27	1	27.0	27.1	0.07	0.61	0.593	0.0001015
		2	27.4			0.59		
		3	26.9			0.58		
4	34	1	35.2	34.76	0.148	0.52	0.514	0.00083
		2	34.7			0.49		
		3	34.3			0.48		
		4	35.1			0.55		
		5	34.5			0.53		
5	41	1	42.3	42.4	0.22	0.47	0.438	0.001291

to the continuous random variable: repartition (5.16); mean value (5.17); variance (dispersion) (5.18); non-centred momentum of 'i' order (5.19); centred momentum of 'i' order (5.20):

$$F(x) = P(X \leq x) = \int_{-\infty}^{x} f(x)dx \tag{5.16}$$

$$\mu = E(X) = \int_{-\infty}^{+\infty} xf(x)dx \tag{5.17}$$

$$\sigma^2 = E[(X - \mu)^2] = \int_{-\infty}^{+\infty} (x - \mu)^2 f(x)dx \tag{5.18}$$

$$m_i = E(X^i) = \int_{-\infty}^{+\infty} x^i f(x)dx \tag{5.19}$$

$$M_i = E[(X - \mu)^2] = \int_{-\infty}^{+\infty} (x - \mu)^i f(x)dx \tag{5.20}$$

The transposition from a selection to a population raises the following fundamental questions: *When a selection characterizes its original population? What is its procedure?* Until now, there has been no existing procedure able to prove whether or not a selection reproduces its original population identically. However, this fact can be improved if it is assumed that $\mu = \bar{x}$ and $\sigma^2 = s^2$. Nevertheless, we have to verify whether these identities are realistic using an acceptable confidence degree.

5.2.1
The Distribution of Frequently Used Random Variables

The distribution of a population's property can be introduced mathematically by the repartition function of a random variable. It is well known that the repartition function of a random variable X gives the probability of a property or event when it is smaller than or equal to the current value x. Indeed, the function that characterizes the density of probability of a random variable (X) gives current values between x and x + dx. This function is, in fact, the derivative of the repartition function (as indirectly shown here above by relation (5.16)). It is important to make sure that, for the characterization of a continuous random variable, the distribution function meets all the requirements. Among the numerous existing distribution functions, the normal distribution (N), the chi distribution (χ^2), the Student distribution (t) and the Fischer distribution are the most frequently used for statistical calculations. These different functions will be explained in the paragraphs below.

The famous normal distribution can be described with the following example: a chemist carries out the daily analysis of a compound concentration. The samples studied are extracted from a unique process and the analyses are made with identical analytical procedures. Our chemist observes that some of the results are scarcely repeated whereas others are more frequently obtained. In addition, the concentration values are always found in a determined range (between a maximum and a minimum experimental result). By computing the apparition frequency of the results as a function of the observed apparition number and the total number of analysed samples, the chemist begins to produce graphic relationships between the apparition frequency of one result and the numerical value of the experiment.

The graphic construction of this computation is given in Fig. 5.4. Two examples are given: the first concerns the processing of 50 samples and the second the processing of 100 samples. When the mean value of the processed measurements has been computed, we can observe that it corresponds to the measurement that has the maximum value of apparition frequency. The differences observed between the two measurements are the consequence of experimental errors [5.16, 5.17]. Therefore, all the measurement errors have a normal distribution written as a density function by the following relation:

$$f(x) = \frac{1}{\sigma\sqrt{2\pi}} e^{-\frac{(x-\mu)^2}{2\sigma^2}} \qquad (5.21)$$

Here μ and σ^2 are, respectively, the mean value and the dispersion (variance) with respect to a population. These characteristics establish all the integral properties of the normal random variable that is represented in our example by the value expected for the species concentration in identical samples. It is not feasible to calculate the exact values of μ and σ^2 because it is impossible to analyse the population of an infinite volume according to a single property. It is important to say that μ and σ^2 show physical dimensions, which are determined by the physical dimension of the random variable associated to the population. The dimension of a normal distribution is frequently transposed to a dimensionless state by using a new random variable. In this case, the current value is given by relation (5.21). Relations (5.22) and (5.23) represent the distribution and repartition of this dimensionless random variable. Relation (5.22) shows that this new variable takes the numerical value of "x" when the mean value and the dispersion are, respectively, $\mu = 0$ and $\sigma^2 = 1$.

$$u = \frac{x - \mu}{\sigma} \qquad (5.22)$$

$$f(u) = \frac{1}{\sqrt{2\pi}} e^{-\frac{u^2}{2}} \qquad (5.23)$$

$$F(u) = \frac{1}{\sqrt{2\pi}} \int_{-\infty}^{u} e^{-\frac{u^2}{2}} du = erf(u) \qquad (5.24)$$

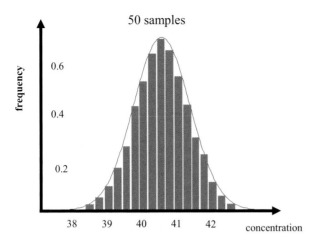

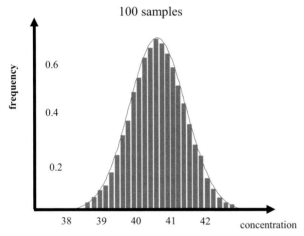

Figure 5.4 Graphic introduction of normal distribution.

Before presenting some properties of normal distribution, we have to present the relation (5.25) that gives the probability for which one random variable is fixed between a and b values (a<b), with the repartition function:

$$P(a < X < b) = F(b) - F(a) \tag{5.25}$$

The particularization of this general relation (5.25) to the dimensionless normal distribution results in the following observations:

1. the current value of the random variable is positioned within the interval $[\mu-\sigma, \mu+\sigma]$ with a probability equal to 0.684
2. the current value of the random variable is positioned within the interval $[\mu-2\sigma, \mu+2\sigma]$ with a probability equal to 0.955

3. the current value of the random variable is positioned within
   the interval [μ−3σ, μ+3σ] with a probability equal to 0.9975.

The observations mentioned above, which are graphically represented in Fig. 5.4, can also be demonstrated mathematically. For example, for the observation which gives $P(\mu - \sigma \prec x \prec \mu + \sigma) = 0.683$, we consider $a = (x - \mu)/\sigma = [(\mu - \sigma) - \mu]/\sigma = -1$ and $b = (x - \mu)/\sigma = [(\mu + \sigma) - \mu]/\sigma = +1$; with Eqs. (5.25) and (5.24) we can derive $P(-1 < u < +1) = \mathrm{erf}(1) - \mathrm{erf}(-1) = 0.8413 - 0.1586 = 0.6823$.

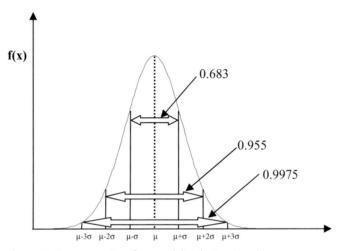

Figure 5.5 Some properties of a normal distribution (population (σ) or sample extracted (s)).

By using normal distribution, we can introduce other random variables, which are very important for testing the significance of $\beta_0, \beta_1, \beta_2, \beta_{12}...$ coefficients as well as for testing the model confidence (see Fig. 5.3).

The first of these random variables is the chi distribution (χ^2). It is derived from relation (5.26), which defines the expression of the current random variable. Here μ and σ are the characteristics of a normal distribution; x_i is the current i value for the same normal distribution. It is easy to observe that a χ^2 distribution adds positive values, consequently $\chi^2 \in (0, \infty)$ and χ^2 is a dimensionless random variable. Relation (5.27) expresses the density of the χ^2 random variable. Here $\upsilon = n - 1$ represents the degrees of freedom of the χ^2 variable:

$$u_i = \frac{x_i - \mu}{\sigma} \approx \frac{x_i - \overline{x}}{\sigma}, \quad \chi^2 = \sum_{i=1}^{n} u_i^2 \tag{5.26}$$

$$f_\upsilon(\chi^2) = \frac{1}{2^{\frac{\upsilon}{2}} \sigma^\upsilon \Gamma\left(\frac{\upsilon}{2}\right)} (\chi^2)^{\frac{\upsilon}{2} - 1} e^{-\frac{(\chi^2)}{2\sigma^2}} \tag{5.27}$$

For a rapid calculation, we can use the tabulated data values for the repartition function of the χ^2 variable $F_\upsilon(\chi^2)$. These tabulated data are obtained with Eq. (5.28):

$$F_\upsilon(\chi^2) = \int_0^{\chi_a^2} f_\upsilon(\chi^2) d\chi^2 = 1 - \alpha \tag{5.28}$$

The second important random variable for statistical modelling is the Student (t) variable. It is derived from a normal variable, which is associated with "u" and χ^2 dimensionless random variables. Relation (5.29) introduces the current value of the Student (t) random variable:

$$t = \frac{u}{\sqrt{\dfrac{\chi^2}{\upsilon}}} \tag{5.29}$$

Equation (5.30), where $\Gamma(\upsilon)$ is given by relation (5.31) shows the probability to have a Student random variable with values between t and t + dt; so this relation gives the density function of the Student variable distribution:

$$f_\upsilon(t) = \frac{\Gamma\left(\dfrac{\upsilon+1}{2}\right)}{\sqrt{(\upsilon+1)\pi}\,\Gamma\left(\dfrac{\upsilon}{2}\right)} \left(1 + \frac{t^2}{\upsilon}\right)^{-\frac{\upsilon+1}{2}} \tag{5.30}$$

$$\Gamma(\upsilon) = \int_0^\infty t^{\upsilon-1} e^{-t} dt \tag{5.31}$$

The third random variable is the Fischer variable. It is defined by the use of two normal variables, each of which is expressed by a χ^2 random variable. The current Fischer variable is given by Eq. (5.32) where $\upsilon_1 = n - 1$ and $\upsilon_2 = m - 1$ represent the degrees of freedom associated, respectively, to random variables χ_1^2 and χ_2^2.

$$x = \frac{\chi_1^2}{\chi_2^2} \cong \frac{\left[\displaystyle\sum_{i=1}^{n} \dfrac{(x_i - \bar{x})_1^2}{\sigma_1^2}\right]}{\left[\displaystyle\sum_{i=1}^{m} \dfrac{(x_i - \bar{x})_2^2}{\sigma_2^2}\right]} \cong \frac{\dfrac{\sigma_2^2}{\upsilon_2}}{\dfrac{\sigma_1^2}{\upsilon_1}} \tag{5.32}$$

The values of the Fischer variable are within the interval $(0, \infty)$. The density of probability for this variable is given by Eq. (5.33):

$$f_{\upsilon_1,\upsilon_2}(x) = f_{\nu_1}(\chi_1^2)/f_{\nu_2}(\chi_2^2) \tag{5.33}$$

For a rapid calculation of this variable, we can use the tabulated values for the Fischer repartition function $F_{\upsilon_1,\upsilon_2}(x)$ corresponding to the confidence limits $\alpha = 0.05$ and $\alpha = 0.01$.

5.2.2
Intervals and Limits of Confidence

The paragraphs above show that μ and σ^2 are the most important characteristics for a random variable attached to a given population. Nevertheless, from a practical point of view, the main characteristics of μ and σ^2 remain unknown. Therefore, we have the possibility to draw one or more statistical selection(s) concerning the property considered by the associated random variable from a population. However, with this procedure, we cannot estimate μ and σ^2 for the whole population directly from the mean value and dispersion of the selection. The acceptance of the statement, which considers that the population mean value μ is placed in an interval containing the selection mean value ($\bar{x}$), must be completed with the observation that the placement of μ near $\bar{x}$ is a probable event. The probability of this event is recognized as the "confidence", "probability level" or "confidence level". A similar processing is carried out for σ^2 and s^2. If we define the probability by α, which shows that μ or σ^2 are not placed in a confidence interval, then, $1 - \alpha$ is the probability level or confidence level. α is frequently called the "significance limit". Figure 5.6 gives the graphic interpretation for α in the case of a normal repartition with $\mu = 0$ and $\sigma^2 = 1$.

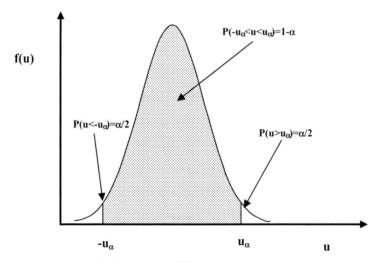

Figure 5.6 Diagram for the definition of the significance level.

Considering Fig. 5.6, we observe that, if we have a very high confidence level, then $1 - \alpha \rightarrow 1$ and the domain for the existence of parameters (μ, σ^2) is high. As far as our scope is to produce the relations between the population and the selection characteristics, i.e. between the couples (μ, σ^2) and ($\bar{x}, s^2$), we can write Eq. (5.17) in a state that introduces the mean value ($\bar{x}$) and volume (n) of the selection. In relation (5.34) the population mean value has been divided into n parts. Now, if for each interval $a_{i-1} - a_i$, the population mean value is compared with the mean value of the selection that has a similar volume, then relation (5.34) can be written as (5.35):

$$\mu = \int\limits_{-\infty}^{+\infty} xf(x)dx = \int\limits_{-\infty}^{a_1} xf(x)dx + \int\limits_{a_1}^{a_2} xf(x)dx + \dots + \int\limits_{a_n}^{+\infty} xf(x)dx = \mu_1 + \mu_2 + \dots + \mu_n$$

(5.34)

$$\mu = \bar{x}_1 + \bar{x}_2 + \bar{x}_3 + \dots + \bar{x}_n$$

(5.35)

Now if we consider the population variance (dispersion), each identical interval $a_{i-1} - a_i$ presents a dispersion which depends on the global σ^2, thus, we can write:

$$\sigma^2 = \sigma_1^2 + \sigma_2^2 + \dots + \sigma_n^2 = \left(\frac{\sigma^2}{n} + \dots + \frac{\sigma^2}{n} \right) = n \left(\frac{\sigma^2}{n} \right)$$

(5.36)

The above relation shows that each of the n divisions of the population has the σ^2/n dispersion. Now, considering that a division $\bar{x} - \mu$ is a normal random variable and that the mean value of this variable is zero, we can transform relation (5.22) into relation (5.37) where u keeps its initial properties (mean value is zero and dispersion equal to unity):

$$u = \frac{\bar{x} - \mu}{\sigma/\sqrt{n}}$$

(5.37)

It is known that $P(a \leq \chi^2 \leq b) = P \left(a \leq \frac{(n-1)s^2}{\sigma^2} \leq b \right) = (1 - \alpha)$ and then, with an accepted significance limit, we can derive the confidence interval considering that $a = \chi_{1-a/2}^2$ and $b = \chi_{a/2}^2$. Thus, we obtain the following results:

$$P \left(\chi_{1-a/2}^2 \leq \frac{(n-1)s^2}{\sigma^2} \leq \chi_{a/2}^2 \right) = 1 - \alpha$$

or:

$$\chi_{1-a/2}^2 \leq \frac{n-1}{\sigma^2} s^2 \quad \text{and} \quad \chi_{a/2}^2 \geq \frac{n-1}{\sigma^2} s^2$$

and:

$$\sigma^2 \leq \frac{n-1}{\chi_{1-a/2}^2} s^2 \quad \text{and} \quad \sigma^2 \geq \frac{n-1}{\chi_{a/2}^2} s^2$$

(5.38)

The intersection of the expressions contained in Eq. (5.38) gives the expression for the confidence interval $I = \left(\frac{n-1}{\chi_{a/2}^2} s^2 ; \frac{n-1}{\chi_{1-a/2}^2} s^2 \right)$. Here, for $\chi_{a/2}^2$ and $\chi_{1-a/2}^2$, we use tabulated or computed values which correspond to the degrees of freedom $v = (n-1)$ where n is the number of selected experiments.

When the selection contains a small number of measurements (for example n<25), the confidence interval for the mean value will be obtained by the use of the dimensionless Student variable given here by the current value (5.39):

$$t = \frac{\mu}{\sqrt{\frac{\chi^2}{\upsilon}}} = \frac{\frac{\bar{x} - \mu}{\sigma/\sqrt{n}}}{\sqrt{\frac{\upsilon s^2}{\upsilon \sigma^2}}} = \frac{\bar{x} - \mu}{s}\sqrt{n} \tag{5.39}$$

Because $t \in (-\infty, +\infty)$ for a fixed significance level, we can write $P(-t_\alpha \le t \le t_\alpha) = 1 - \alpha$. Now the substitution of Eq. (5.39) into $P(-t_\alpha \le t \le t_\alpha) = 1 - \alpha$ results in the following relations:

$$P\left(-t_\alpha \le \frac{\bar{x} - \mu}{s}\sqrt{n} \le t_\alpha\right) = 1 - \alpha \tag{5.40}$$

or:

$$-t_\alpha \le \frac{\bar{x} - \mu}{s}\sqrt{n} \quad \text{and} \quad t_\alpha \ge \frac{\bar{x} + \mu}{s}\sqrt{n} \tag{5.41}$$

and:

$$\mu \le \bar{x} + t_\alpha\frac{s}{\sqrt{n}} \quad \text{and} \quad \mu \ge \bar{x} - t_\alpha\frac{s}{\sqrt{n}} \tag{5.42}$$

The expressions from relation (5.42) show that the confidence interval for a mean value with a small number of measurements is:

$$I = \left(\bar{x} - t_\alpha\frac{s}{\sqrt{n}}; \bar{x} + t_\alpha\frac{s}{\sqrt{n}}\right).$$

5.2.2.1 A Particular Application of the Confidence Interval to a Mean Value

The scope of this section is to show a practical application of the confidence interval to a mean value. The example below concerns the data given in Table 5.2. In order to verify the correctness of the data obtained, the chemist has carried out new measurements in case the proposed x should be near 20 g/l. Table 5.4 gives the new results obtained for the concentration of the reactant in the reactor feed. Concerning these data two questions are raised:

1. What is the confidence interval for the mean value of the population from which the selection in Table 5.4 has been extracted?
2. What is the difference between these new data and those given in Table 5.2?

Table 5.4. New values of the limiting reactant concentration in the reactor feed (Data equivalent to column 2 in Table 5.2).

Sample number (i)	Concentration x_i, g/l	Sample number (i)	Concentration x_i, g/l	Sample number (i)	Concentration x_i, g/l	Sample number (i)	Concentration x_i, g/l
1	19.4	9	21.2	17	18.4	25	21.6
2	22.2	10	18.7	18	18.1	26	20.4
3	21.9	11	19.3	19	18.9	27	18.5
4	23.2	12	18.7	20	22.0	28	20.8
5	19.8	13	23.5	21	18.5	29	18.8
6	21.3	14	22.5	22	20.5	30	22.1
7	17.8	15	18.9	23	18.7	31	20.7
8	23.2	16	19.3	24	21.1	32	19.2

The answers to the questions above are obtained numerically with the following procedure and the corresponding algorithm:

1. We compute the selection mean value ($\bar{x}$) and the dispersion (s^2) with the data from Table 5.4 and with Eqs. (5.10) and (5.12).

   Result: $\bar{x} = 20.3$ g/l; $s^2 = 3.86$; $s = 1.92$ g/l

2. We accept the equality between the population and the selection dispersion, i.e. $\sigma^2 = s^2$

   Result: $\sigma^2 = 3.86$; $\sigma = 1.92$ g/l

3. We establish the probability significance level (α).

   Result: $\alpha = 0.05$

4. Equation $\dfrac{1}{\sqrt{2\pi}} \int_{-u_\alpha}^{u_\alpha} e^{-\frac{u^2}{2}} du = 1 - \alpha$ is resolved in order to estimate u_α.

   Result: $u_\alpha = 1.96$.

   Observation: For this purpose we must use a computer program. Alternatively, we can also use the tabulated data of the normal u_α at various fixed α.

5. We obtain the mean value confidence with relation

   $$I = \left(\bar{x} - u_\alpha \frac{\sigma}{\sqrt{n}}; \bar{x} + u_\alpha \frac{\sigma}{\sqrt{n}}\right)$$

   Result: $I = (19.5; 21)$

6. We calculate the selection mean value ($\bar{x}$) and the dispersion
   (s^2) with the data from Table 5.2 column 2 and with Eqs.
   (5.10) and (5.12).
   Results: $\bar{x} = 20.16$ g/l; $s^2 = 0.473$; $s = 0.687$ g/l).

7. According to point 2 of the present algorithm, we accept the
   equality between the population and the selection dispersion
   $\sigma^2 = s^2$.
   Results: $\sigma^2 = 0.473$; $\sigma = 0.687$ g/l

8. We observe that for χ^2 variable $\nu = n - 1$.
   Result: $\nu = 4$

9. Equation $\displaystyle\int_{-t_\alpha}^{t_\alpha} \frac{\Gamma\left(\dfrac{\upsilon+1}{2}\right)}{\sqrt{\upsilon\pi}\,\Gamma\left(\dfrac{\upsilon}{2}\right)} \left(1 + \frac{t^2}{\upsilon}\right)^{-\frac{\upsilon+1}{2}} dt = 1 - \alpha$ is solved for t_α
   unknown.
   Result: $t_\alpha = 2.776$

10. We obtain the mean value confidence according to relation

$$I = \left(\bar{x} - t_\alpha \frac{s}{\sqrt{n}}\,; \bar{x} - t_\alpha \frac{s}{\sqrt{n}}\right)$$

   Result: $I = (19.307;\ 21.013)$

11. Conclusion: The obtained results for the confidence intervals
   $I = (19.5;\ 21)$ and $I = (19.307;\ 21.013)$ show that the compared selections are almost the same or have a similar
   origin.

5.2.2.2 An Actual Example of the Calculation of the Confidence Interval for the Variance

The purpose of this section is to show the calculation of the confidence interval
for the variance in an actual example. The statistical data used for this example
are given in Table 5.3. In this table, the statistically measured real input concentrations and the associated output reactant transformation degrees are given for five
proposed concentrations of the limiting reactant in the reactor feed. Table 5.3 also
contains the values of the computed variances for each statistical selection. The
confidence interval for each mean value from Table 5.3 has to be calculated
according to the procedure established in steps 6–10 from the algorithm shown in
Section 5.2.2.1. In this example, the number of measurements for each experiment is small, thus the estimation of the mean value is difficult. Therefore, we
can compute the confidence interval for the dispersion $\left(I = \left(\dfrac{n-1}{\chi^2_{a/2}}s^2\,;\dfrac{n-1}{\chi^2_{1-a/2}}s^2\right)\right)$

for each experiment only if we establish the degrees of freedom ($v = n - 1$, where n is the number of experiments from each experimentation), and for a chosen α, we obtain the quintiles values $\chi^2_{\alpha/2}$ and $\chi^2_{1-\alpha/2}$. These are the solutions of the following system of equations:

$$
\begin{cases}
\displaystyle\int_{\chi^2_{1-\alpha/2}}^{\chi^2_{\alpha/2}} f_v(\chi^2)d\chi^2 = 1 - \alpha \\[4mm]
\displaystyle\int_{\chi^2_{1-\alpha/2}}^{\infty} f_v(\chi^2)d\chi^2 = 1 - \alpha/2
\end{cases}
\tag{5.43}
$$

Table 5.5 gives the results obtained for the mean value and dispersion intervals for a significance limit $\alpha = 0.05$.

Table 5.5 The confidence intervals of the mean value and dispersion for the data from Table 5.3.

Current number for input	n x/y	$\bar{x}$	t_α	I from Eq. (5.42)	s^2_x	$\chi^2_{1-\alpha/2}$	$\chi^2_{\alpha/2}$	I from Eq. (5.38)
1	5/4	13.86	2.571	13.67; 14.04	0.163	1.15	11.1	0.058; 0.265
2	5/6	20.16	2.571	20.43; 19.97	0.473	1.15	11.1	0.172; 1.641
3	3/3	27.10	3.182	27.23; 26.97	0.070	0.352	7.81	0.018; 0.398
4	5/5	34.76	2.571	34.58; 34.94	0.148	1.15	11.1	0.053; 0.514
5	4/4	42.40	2.776	42.09; 42.705	0.220	0.711	9.49	0.069; 0.928

Current number for input	$\bar{y}$	t_α	I from (5.48)	$s^2_y * 10^2$	$\chi^2_{1-\alpha/2}$	$\chi^2_{\alpha/2}$	I from (5.44)
1	0.77	2.776	0.772; 0.768	0.080	0.711	9.49	0.033; 0.331
2	0.655	2.447	0.654; 0.656	0.035	1.64	12.6	0.013; 0.106

Table 5.5 Continued.

Current number for input	$\bar{y}$	t_α	I from (5.48)	$s^2_y * 10^2$	$\chi^2_{1-\alpha/2}$	$\chi^2_{\alpha/2}$	I from (5.44)
3	0.593	3.182	0.592; 0.594	0.010	0.352	7.81	0.002; 0.036
4	0.514	2.571	0.513; 0.515	0.083	1.15	11.1	0.03; 0.288
5	0.438	2.776	0.434; 0.442	0.129	0.711	9.49	0.041; 0.544

5.2.3
Statistical Hypotheses and Their Checking

The introduction of the formulation of the statistical hypotheses and their checking have already been presented in Section 5.2.2.1 where we proposed the analysis
of the comparison between the mean values and dispersions of two selections
drawn from the same population. If we consider the mean values in our actual
example, the problem can be formulated as follows: if $\bar{x}_1$ is the mean value calculated with the values in Table 5.4 and $\bar{x}_2$ is the mean value for another selection
extracted from the same population (such as for example $\bar{x}_2$, which is the limiting
reactant concentration at the reactor input for Table 5.2, column 2) we must demonstrate whether $\bar{x}_1$ is significantly different from $\bar{x}_2$.

A similar formulation can be established in the case of two different dispersions
in two selections extracted from the same population. Therefore, this problem can
also be extended to the case of two populations with a similar behaviour, even
though, in this case, we have to verify the equality or difference between the mean
values μ_1 and μ_2 or between the variances σ_1^2 and σ_2^2. We frequently use three
major computing steps to resolve this problem and to check its hypotheses:

- First, we begin the problem with the acceptance of the zero or
  null hypothesis. Concerning two similar populations, the null
  hypothesis for a mean value shows that $\mu_1 = \mu_2$ or $\mu_1 - \mu_2 = 0$.
  Thus, we can write $\sigma_1^2 = \sigma_2^2$ or $\sigma_1^2 - \sigma_2^2 = 0$ for dispersion. We
  have $\bar{x}_1 = \bar{x}_2$ or $\bar{x}_1 - \bar{x}_2 = 0$ for both selections and $s_1^2 = s_2^2$ or
  $s_1^2 - s_2^2 = 0$ for the mean value and dispersion respectively.
- Then, we obtain the value of a random variable associated to the
  zero hypothesis and to the commonly used distributions, we
  establish the value of the correlated repartition function, which is
  in fact a probability of the hypothesis existence.

- Finally, we accept a confidence level and we compare this value with those given by the repartition function and we eventually accept or reject the null hypothesis according to this comparison.

Table 5.6 presents the statistical hypotheses frequently formulated and the tests used for their validation.

Table 5.6 Frequently formulated statistical hypotheses and their validation tests.

Current number for input	Comparison state	Zero hypothesis	Test used	Computed value for the random variable	Associated probability	Condition of rejection
1	Two populations and two selections. Parameters: μ_1, σ_1^2 population 1 μ_2, σ_2^2 population 2; $\bar{x}_1, s_1^2$ selection 1 $\bar{x}_2, s_1^2$ selection 2	$\mu_1 = \mu_2$ or $\bar{x}_1 = \bar{x}_2$	u	$u = \dfrac{\mu_1 - \mu_2}{\sigma_1/\sqrt{n}}$ or $u = \dfrac{\bar{x}_1 - \bar{x}_2}{\sigma_1/\sqrt{n}}$	$P(X \leq u) =$ $\int_{-\infty}^{u} \frac{1}{\sqrt{2\pi}} e^{-u^2} du$	$P(X \leq u) >$ $1 - \alpha$
2	Same as 1 but for selections with a small volume	$\mu_1 = \mu_2$ or $\bar{x}_1 = \bar{x}_2$	t $v = n-1$	$t = \dfrac{\mu_1 - \mu_2}{\sigma_1}\sqrt{n}$ or $t = \dfrac{\bar{x}_1 - \bar{x}_2}{\sigma_1}\sqrt{n}$	$P(X \leq t) =$ $\int_{-\infty}^{t} f_v(t)dt$	$P(X \leq t) >$ $1 - \alpha$
3	The n volume of selection and its population	$s_1^2 = \sigma^2$	χ^2 $v = n-1$	$\chi^2 = \dfrac{n-1}{\sigma^2}s^2$	$P(X \leq \chi^2) =$	$P(X \leq \chi^2) >$ $1 - \alpha$
4	Two selections of n_1 and n_2 volumes	$s_1^2 = s_2^2$ or $s_1^2 > s_2^2$	F $v_1 = n_1-1$ $v_1 = n_2-1$	$F = \dfrac{s_1^2}{s_2^2}$	$P(X \leq F) =$ $\int_{-\infty}^{F} f_{v1,v2}(F)dF$	$P(X \leq F) >$ $1 - \alpha$

In order to clarify this conceptual discussion we will use the actual example we have been working on in this chapter. First, it is required to verify whether dispersion s_1^2, which characterizes the selection given in Section 5.5.2.1, is similar to dispersion s_2^2, established in Table 5.5, column 2. Indeed, it is known that these selections have been extracted from the same original population. The response to this question is obtained with the calculation methodology described above. This computation is organized according to the algorithmic rule proposed at the beginning of this paragraph, so:

- We write the actual H_0 hypothesis: $H_0 : s_1^2 = s_2^2$
- We compute the current value of the Fischer random variable associated to the dispersions s_1^2 and s_2^2: $F = s_1^2/s_2^2$;
  Result: $F = 3.86/0.473 = 8.16$
- We establish the degrees of freedom for the Fischer variable:
  $\upsilon_1 = n_1 - 1, \upsilon_2 = n_2 - 1$;
  Results $v_1 = 3$, $v_2 = 4$
- We obtain the probability of the current Fischer variable by computing the value of the repartition function:

$$P(X \leq 8.16) = \int_0^{8.16} f_{\upsilon 1,\upsilon 2}(F)dF;$$

$$Result:\ P(X \leq 8.16) = \int_0^{8.16} f_{\upsilon 1,\upsilon 2}(F)dF = 0.97$$

- We accept the most used significance level $\alpha = 0.05$
- We observe that $P(X \leq 8.16) = 0.97 \succ 1 - \alpha = 0.95$ and, as a consequence, we reject the zero hypothesis.

5.3
Correlation Analysis

When the preliminary steps of the statistical model have been accomplished, the researchers must focus their attention on the problem of correlation between dependent and independent variables (see Fig. 5.1). At this stage, they must use the description and the statistical selections of the process, so as to propose a model state with a mathematical expression showing the relation between each of the dependent variables and all independent variables (relation (5.3)). During this selection, the researchers might erroneously use two restrictions: Firstly, they may tend to introduce a limitation concerning the degree of the polynomial that describes the relation between the dependent variable $y^{(i)}$ and the independent variables x_j, $j = 1,n$; Secondly, they may tend to extract some independent variables or terms which show the effect of the interactions between two or more independent variables on the dependent variable from the above mentioned relationship.

The problem of simplifying the regression relationship can be omitted if, before establishing those simplifications, the specific procedure that defines the type of the correlations between the dependent and independent variables of the process, is applied on the basis of a statistical process analysis.

Classical dispersion analyses, dispersion analyses with interaction effects and especially correlation analyses can be used successfully to obtain the information needed about the form of an actual regression expression. Working with the statistical data obtained by the process investigation, the dispersion and the correlation analyses, can establish the independent process variables and the interactions of independent variables that have to be considered in a regression expression [5.18, 5.19].

For a process with one dependent variable and one independent variable, the statistical process analysis gives one chain with values of y_i, $i = 1,n$ and another one with values of x_i, $i = 1,n$. Here, n is the number of the processed experiments. The correlation analysis shows that the process variables y and x are correlated if the indicator cov(x,y), given here by relation (5.44), presents a significant value:

$$cov(y, x) = \frac{\sum_{i=1}^{n} (x_i - \bar{x})(y_i - \bar{y})}{(n - 1)} \tag{5.44}$$

We observe that the covariance indicator (cov(x,y)) has an expression which is similar to the dispersion of a statistical selection datum near the mean value (Eq. (5.11)). It is important to specify that the notion of variance (or dispersion) differs completely from the notion of covariance.

If the multiplication $(x_i - \bar{x})(y_i - \bar{y})$ from the covariance definition (5.44) gives a positive number, then the figurative point (x_i, y_i) will be placed in the first or third quadrant of an x,y graphic representation, whereas, the figurative point (x_i, y_i) will be placed in the second or fourth quadrant. Now if the x and y variables are independent, then the placement probability of the figurative point is the same for all quadrants. So, in this case, we have the graphic representation from Fig. 5.7(a), and the sum $\sum_{i=1}^{n} (x_i - \bar{x})(y_i - \bar{y})$ tends to zero or to a very small number. For the case when x and y are dependent, then the placement probability is not the same for all four quadrants and consequently the sum $\sum_{i=1}^{n} (x_i - \bar{x})(y_i - \bar{y}) \neq 0$. This last situation is shown in Fig. 5.7(b).

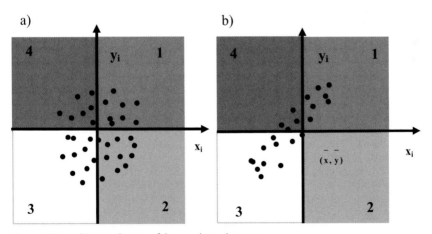

Figure 5.7 Graphic introduction of the correlation between statistical variables, (a) independent variables, (b) dependent variables.

The x and y covariance increases or decreases with the values of $(x_i - \bar{x})$ and $(y_i - \bar{y})$. Thus, if we repeat the statistical experiment in order to obtain the chains of values x_i, y_i $i = 1, n$ and if we compute again the cov(x,y), this new cov value can be different from the cov initially calculated. This distortion is eliminated if we replace the covariance by the correlation coefficient of the variables:

$$r_{yx} = \frac{\sum\limits_{i=1}^{n} (x_i - \bar{x})(y_i - \bar{y})}{\sqrt{\sum\limits_{i=1}^{n} (x_i - \bar{x})^2 (y_i - \bar{y})^2}} = \frac{\text{cov}(y, x)}{s_x s_y} \tag{5.45}$$

It is easy to observe that the domain of the values of the correlation coefficient is placed between –1 and +1 and that $r_{yx} : R \rightarrow [-1, 1]$.

The following observations can also be made with respect to the correlation coefficient:

- If the value of the correlation coefficient approaches zero, then we can accept x and y variables to be independent. So, the variations on the dependent variable do not affect the independent variable;
- When the correlation coefficient takes a positive value, the independent and dependent variables increase simultaneously. The opposite case corresponds to a negative value of the correlation coefficient;
- The extreme values ($r_{yx} = 1; r_{yx} = -1$) for the correlation coefficient show that a linear relationship exists between the dependent and independent variables.

The discussion presented above for the case when the process has only one input can easily be extended to a process with more than one independent variable (many inputs). For example, when we have one dependent and two independent variables, we can compute the $r_{yx_1}, r_{yx_2}, r_{yx_1x_2}$.. coefficients. All the observations concerning r_{yx} stay unchanged for each $r_{yx_1}, r_{yx_2}, r_{yx_1x_2}$.. When this process involves two inputs, if we obtain $r_{yx_1} = 1, r_{yx_2} = -1, r_{yx_1x_2} = 1$ and if the other possible correlation coefficients approach zero, then dependence $y = \beta_0 + \beta_1 x_1 - \beta_2 x_2 + \beta_{12} x_1 x_2$ is recommended to build the statistical model of the process.

If we once more consider the example studied throughout this chapter, we can use the statistical data presented in Table 5.3 in order to compute the value of the correlation coefficient. However, before carrying out this calculation, we can observe an important dependence between variables x and y due to the physical meaning of the results in this table. The value obtained for the correlation coefficient confirms our a priori assumption because the cov has a value near unity. It shows that a linear relationship can be established between process variables. The results of these calculations are shown in Table 5.7.

Table 5.7 Calculation of the correlation coefficient between the reactant conversion degree and the input concentration (the statistical data used are from Table 5.3)

Current number for input	x_i	$\bar{x}$	$(x_i - \bar{x})$	y_i	$\bar{y}$	$(y_i - \bar{y})$	$(x_i - \bar{x})$ $* (y_i - \bar{y})$	$(x_i - \bar{x})^2$	$(y_i - \bar{y})^2$
1	13.86		−13.91	0.770		0.1762	−2.45975	193.655	0.03104
2	20.16		−7.616	0.655		0.0612	−0.46609	58.003	0.00374
3	27.70		−0.076	0.593	0.594	−0.0008	0.00006	0.0057	0.00006
4	34.76	27.78	6.984	0.514		−0.0798	−0.55732	48.776	0.00636
5	42.40		14.62	0.437		−0.1568	−2.29304	213.861	0.02496
$\sum_{i=1}^{5}$	138.88			2.969			5.776144	514.30	0.070672

The value of the correlation coefficient is: $r_{yx} = (5.7761/(514.3*0.0707)^2) = 0.957$.

We can eliminate all the false dependent variables from the statistical model thanks to the correlation analysis. When we obtain $r_{y_1 y_2} = 1$ for a process with two dependent variables (y_1, y_2), we have a linear dependence between these variables. Then, in this case, both variables exceed the independence required by the output process variables. Therefore, y_1 or y_2 can be eliminated from the list of the dependent process variables.

5.4
Regression Analysis

Regression analysis is the statistical computing procedure that begins when the model regression equations have been established for an investigated process. The regression analysis includes [5.18, 5.19]:
- the system of normal equations for the particularizations to an actual case, in which the relationship between each dependent variable and the independent process variables is established on the basis of Eq. (5.3);
- the calculations of the values of all the coefficients contained in the mathematical model of the process;
- the validation of the model coefficients and of the final statistical model of the process.

The items described above have already been introduced in Fig. 5.3 where the steps of the development of the statistical model of a process are presented. It should be pointed out that throughout the regression analysis, attention is commonly concentrated on the first and second aspects, despite the fact that virgin statistical data are available for the third aspect. Normally, this new non-used data (see Fig. 5.3) allows the calculation of the reproducibility variance (s_{rp}^2) as well as the residual variance, which together give the model acceptance or rejection. In fact, this aspect contains the validation of the hypothesis considering that $s_{rp}^2 = s_{rz}^2$; it is clear that the use of the Fischer test (for instance, see Table 5.6) is crucial in this situation. The following paragraphs contain the particularization of the regression analysis to some common cases. It is important to note that these examples differ from each other by the number of independent variables and the form of their regression equations.

5.4.1
Linear Regression

A linear regression occurs when a process has only one input (x) and one output variable (y) and both variables are correlated by a linear relationship:

$$y^{th} = y = f(x, \beta_0, \beta_1) = \beta_0 + \beta_1 x \tag{5.46}$$

This relation is a particularization of the general relation (5.3). Indeed, polynomial regression presents the limitation of being first order. In accordance with Eq. (5.46), the system of equations (5.9) results in the following system for the identification of β_0 and β_1:

$$\begin{cases} \sum_{i=1}^{N} y_i - \sum_{i=1}^{N} (\beta_0 + \beta_1 x_i) = 0 \\ \sum_{i=1}^{N} y_i x_i - \sum_{i=1}^{N} (\beta_0 + \beta_1 x_i) x_i = 0 \end{cases} \tag{5.47}$$

which is equivalent to:

$$\begin{cases} N\beta_0 + \beta_1 \sum_{i=1}^{N} x_i = \sum_{i=1}^{N} y_i \\ \beta_0 \sum_{i=1}^{N} x_i + \beta_1 \sum_{i=1}^{N} (x_i)^2 = \sum_{i=1}^{N} y_i x_i \end{cases} \tag{5.48}$$

Now it is very simple to obtain coefficients β_0 and β_1 as the Cramer solution of system (5.39). The following expressions for β_0 and β_1 are thus obtained:

$$\beta_0 = \frac{\sum_{i=1}^{N} y_i \sum_{i=1}^{N} x_i^2 - \sum_{i=1}^{N} x_i \sum_{i=1}^{N} y_i x_i}{N \sum_{i=1}^{N} x_i^2 - \left(\sum_{i=1}^{N} x_i\right)^2} \tag{5.49}$$

$$\beta_1 = \frac{N \sum\limits_{i=1}^{N} y_i x_i - \sum\limits_{i=1}^{N} x_i \sum\limits_{i=1}^{N} y_i}{N \sum\limits_{i=1}^{N} x_i^2 - \left(\sum\limits_{i=1}^{N} x_i\right)^2} \qquad (5.50)$$

After the calculation of β_1 we can extract β_0 from relation (5.51) where $\bar{x}$ and $\bar{y}$ are the mean values of variables x and y respectively. Otherwise, this relation can also be used to verify whether β_0 and β_1 are correctly obtained by relations (5.50) and (5.51):

$$\beta_0 = \bar{y} - \beta_1 \bar{x} \qquad (5.51)$$

The next step in developing a statistical model is the verification of the significance of the coefficients by means of the Student distribution and the reproducibility variance.

The problem of the significance of the regression coefficients can be examined only if the statistical data take into consideration the following conditions [5.19]:

1. The error of the measured input parameter (x) must be minor. In this case, any error occurring when we obtain "y" will be the consequence of the non-explicit input variables. These non-explicit variables are input variables which have been rejected or not observed when the regression expression was proposed.
2. When the measurements are repeated, the results of the output variable must present random values with a normal distribution (such samples are shown in Table 5.2).
3. When we carry out an experimentation in which "N" is the dimension of each experiment and where each experiment is repeated "m" times, the variances $s_1^2, s_2^2, ...s_N^2$, which are associated to the output variable, should be homogeneous.

The testing of the homogeneity of variances concerns the process of primary preparation of the statistical data. It is important to note that this procedure of homogeneity testing of the output variances is in fact a problem which tests the zero hypothesis, i.e.: $H_0 : s_1^2 = s_2^2 = ... = s_N^2$. For this purpose, we comply with the following algorithm:

1. We compute the mean values of samples with respect to the output process variable:

$$\bar{y}_i = \sum_{k=1}^{m} y_{ik}/m \quad i = 1, 2, 3, ...N \qquad (5.52)$$

2. With the mean values and with each one of the experiments we establish the variables $s_1^2 = s_2^2 = ... = s_N^2$ as well as their maximum values:

$$s_i^2 = \frac{\sum\limits_{i=1}^{m} (y_{ik} - \bar{y}_i)^2}{m - 1} \tag{5.53}$$

3. We proceed with the calculation of the sum of the variances that give the value of the testing process associated to the Fischer random variable:

$$s^2 = \sum_{i=1}^{N} s_i^2 \qquad F = \frac{s_{max}^2}{s^2} \tag{5.54}$$

4. At this point we identify the values which have the same degrees of freedom as variable F and thus we obtain an existence probability of this random variable between 0 and the computed value of point c):

$$\upsilon_1 = N, \upsilon_2 = m - 1 \; P(X \leq F) = \int\limits_{-\infty}^{F} f_{\upsilon 1, \upsilon 2}(F) dF \tag{5.55}$$

5. For a fixed significance level α, all the variances $s_1^2 = s_2^2 = \;.... = s_N^2$ will be accepted as homogenous if we have:

$$P(X \leq F) \leq 1 - \alpha \tag{5.56}$$

6. When the homogeneity of the variances has been tested, we continue to compute the values of the reproducibility variance with relation (5.64):

$$s_{rp}^2 = s^2 / N \tag{5.57}$$

In statistics, the reproducibility variance is a random variable having a number of degrees of freedom equal to $\upsilon = N(m - 1)$. Without the reproducibility variances or any other equivalent variance, we cannot estimate the significance of the regression coefficients. It is important to remember that, for the calculation of this variance, we need to have new statistical data or, more precisely, statistical data not used in the procedures of the identification of the coefficients. This requirement explains the division of the statistical data of Fig. 5.3 into two parts: one significant part for the identification of the coefficients and one small part for the reproducibility variance calculation.

The significance estimation of β_0 and β_1 coefficients is, for each case, a real statistical hypothesis, the aim of which is to verify whether their values are null or not. Here, we can suggest two zero hypotheses ($H_{01} : \beta_0 = 0$ and $H_{02} : \beta_1 = 0$) and by using the Student test (see Table 5.6), we can find out whether these hypotheses are accepted or rejected.

In a more general case, we have to carry out the following calculations:

- firstly: the values of the t_j variable using relation (5.58) where β_j is the j regression coefficient, and s_{β_i} represents the corresponding β_j mean square root of variance $s_{\beta_j}^2$:

$$t_j = \frac{|\beta_j|}{s_{\beta_i}} \tag{5.58}$$

- secondly: the existence probability of the t_j value of the Student variable, where υ is the number of the degrees of freedom respect to the calculation of the t_j value:

$$P_j(X \le t_j) = \int_{-\infty}^{t_j} f_\upsilon(t)dt$$

- finally: if we have $P_j(X \le t_j) > 1 - \alpha$, the zero hypothesis for β_j so that $H_{0j} : \beta_j = 0$ will be rejected. In this case, β_j is an important coefficient in the relationship between the regression variables. The opposite case corresponds to the acceptance of the H_{0j} hypothesis.

It is then important to show that, in case of generalization, the mean square root of the variances with respect to the mean β_j value as well as its variances have the quality to respect the law of the accumulation of errors [5.13, 5.16, 5.19]. As a result, the mean square root of the variances will have a theoretical expression, which is given by:

$$s_{\beta_i} = \sqrt{\sum_{i=1}^{N} \left(\frac{\partial \beta_i}{\partial y_i}\right) s_i^2} \tag{5.59}$$

Because, in a normal case, we have the homogenous variances $s_1^2 = s_2^2 = ... = s_N^2 = s_{rp}^2$, then for the case of a linear regression, we can particularize relation (5.59) in order to obtain the following relations:

$$s_{\beta_0} = \sqrt{\frac{s_{rp}^2 \sum_{i=1}^{N} x_i^2}{N \sum_{i=1}^{N} x_i^2 - \left(\sum_{i=1}^{N} x_i\right)^2}} \tag{5.60}$$

$$s_{\beta_1} = \sqrt{\frac{s_{rp}^2 N}{N \sum_{i=1}^{N} x_i^2 - \left(\sum_{i=1}^{N} x_i\right)^2}} \tag{5.61}$$

After estimation of the significance of the coefficients, each non-significant coefficient will be excluded from the regression expression and a new identification can be made for all the remaining coefficients. This new calculation of the remaining regression coefficients is a consequence of the fact that these regression coefficients are in an active interrelated state. Before ending this problem, we must verify the model confidence, i.e. we must check whether the structure that remains after the testing of the significance coefficients, is adequate or not. For the example discussed above, the model is represented by the final expression of regression. Its confidence can thus be verified using the Fischer test the orientation of which is to verify the statistical hypothesis: $H_{0m} : s_{rz}^2 = s_{rp}^2$ suggesting the equality of the residual and reproducibility variances. The Fischer test begins with the calculation of the Fischer random variable value: $F = s_{rz}^2/s_{rp}^2$. Here the degrees of freedom have the values $\upsilon_1 = N - 1, \upsilon_2 = N - n_\beta$, where N is the number of statistical data used to calculate s_{rz}^2 in υ_1, as well as in υ_2. Here, n_β introduces the number of regression coefficients that remain in the final form of the regression expression. For a process with only one output (only one dependent variable) the residual variance measures the difference between the model computed and the mean value of the output:

$$s_{rz}^2 = \frac{\sum_{i=1}^{N} (\hat{y}_i - \bar{y})^2}{N - n_\beta} \tag{5.62}$$

$$\hat{y}_i = \beta_0 + \beta_1 x_i \quad i = 1, N \tag{5.63}$$

After calculating the value of the random variable F, we establish the reproducibility variances and carry out the test according to the procedure given in Table 5.6. Exceptionally, in cases when we do not have any experiment carried out in parallel, and when the statistical data have not been divided into two parts, we use the relative variance for the mean value (s_y^2) instead of the reproducibility variance. This relative variance can be computed with the statistical data used for the identification of the coefficients using the relation (5.64):

$$s_y^2 = \frac{\sum_{i=1}^{N} (y_i - \bar{y})^2}{N - 1} \tag{5.64}$$

In this case, the value of N for υ_1 and υ_2 is the same and it is equal to the number of experiments accepted for the statistical calculations. Coming back to the problem of the model adequacy, it is clear that the zero hypothesis has been transformed into the following expression: $H_{0m} : s_{rz}^2 = s_y^2$.

5.4.1.1 Application to the Relationship between the Reactant Conversion and the Input Concentration for a CSR

The statistical data shown in Table 5.2 were obtained for an isothermal continuously stirred reactor (CSR) with a spatial time of 1.5 h. With these experimental data, we can formulate a relationship between the reactant conversion (y) and the input concentration (x). For the establishment of a statistical model based on a linear regression, we have a coefficient of regression close to 1 (found in Table 5.7 which contains the values obtained with the same statistical data). However, we did not have any additional experiments carried out in parallel and consequently we cannot establish a real reproducibility variance. The correlation coefficient from Table 5.7, sustains the proposal of a linear dependence between the conversion (y) and the input concentration of the reactant (x): $y = \beta_0 + \beta_1 x$. Table 5.8 shows the statistical data and the results of some calculations needed for the determination of β_0 and β_1.

Table 5.8 The statistical data and calculated parameters for the estimation of β_0 and β_1.

i =	x_i	y_i	$(x_i)^2$	$(y_i x_i)$	$\bar{x}$	$\bar{y}$
1	13.86	0.77	194.8816	10.6722		
2	20.16	0.655	406.4256	13.2048		
3	27.70	0.593	767.29	16.4261	27.776	0.5938
4	34.76	0.514	1208.2576	17.86664		
5	42.40	0.437	1797.76	18.5288		
$\sum_{i=1}^{N}$	138.88	2.969	4374.6	76.67		

Thus, for β_0 and β_1 we obtain:

$$\beta_1 = \frac{N \sum_{i=1}^{N} y_i x_i - \sum_{i=1}^{N} x_i \sum_{i=1}^{N} y_i}{N \sum_{i=1}^{N} x_i^2 - \left(\sum_{i=1}^{N} x_i\right)^2} = \frac{5 * 76.67 - 138.88 * 2.969}{5 * 43754.6 - (138.88)^2} = -0.0112 \ ;$$

$$\beta_0 = 0.0112 * 27.76 + 0.5938 = 0.92692$$

The significance estimation of β_0 and β_1 is made by computing the residual variance and the variance relative to the mean value of the dependent variable. The results corresponding to these calculations are shown in Table 5.9.

Table 5.9 Computed values of the residual and relative variances.

Number (N)	1	2	3	4	5	$\sum\limits_{i=1}^{N}$
x_i	13.86	20.16	27.70	34.76	42.40	138.88
y_i	0.77	0.655	0.593	0.514	0.437	2.969
$y_i - \bar{y}$	0.1762	0.0612	−0.0008	−0.0798	−0.1568	
$\hat{y} = \beta_0 - \beta_1 x_i$	0.884	0.701	0.617	0.538	0.452	
$\hat{y} - \bar{y}$	0.2902	0.1072	0.0232	−0.0558	−0.1418	
Variance	$s_{rz}^2 = (0.2902^2 + 0.1072^2 + 0.0232^2 +$ $0.0558^2 + 0.1418^2)/4 = 0.02986675$			$s_y^2 = (0.1762^2 + 0.0612^2 + 0.0008^2 +$ $0.0798^2 + 0.1568^2)/4 = 0.0164367$		

Now, we can obtain the variances due to β_0 and β_1 by using relations (5.60) and (5.61):

$$s_{\beta_0} = \sqrt{\frac{s_{rp}^2 \sum\limits_{i=1}^{N} x_i^2}{N \sum\limits_{i=1}^{N} x_i^2 - \left(\sum\limits_{i=1}^{N} x_i\right)^2}} = \sqrt{\frac{0.01644 * 4374.6}{5 * 4374.6 - (138.88)^2}} = 0.1667$$

$$s_{\beta_1} = \sqrt{\frac{s_{rp}^2 N}{N \sum\limits_{i=1}^{N} x_i^2 - \left(\sum\limits_{i=1}^{N} x_i\right)^2}} = \sqrt{\frac{0.01644 * 5}{5 * 4374.6 - (138.88)^2}} = 0.0018$$

The estimations of the β_0 and β_1 significance are computed by the procedure given in Table 5.6. The results are shown in Table 5.10.

Table 5.10 The significance of β_0 and β_1 coefficients estimated by the Student test.

Hypothesis	v	T	P(X<t), relation (5.30)	1−α	Conclusion
$\beta_0 = 0$	4	0.92692/0.1667 = 5.5	0.87	0.95	β_0 important
$\beta_1 = 0$	4	0.0112/0.0018 = 6.2	0.91	0.95	β_1 important

At this point, we have to think about the problem of the model confidence. For this purpose we have to consider that:
- the value of the Fischer variable is F = 0.0298/0.0164 = 1.817;

- for $1 - \alpha = 0.95$, we obtain $F = F_{0.05} = 3.24$ by solving the equation
$$1 - \alpha = \int_0^F f_{4,4}(F)dF;$$
- we accept the zero hypothesis $H_{0m} : s_{rz}^2 = s_y^2$ because
$F_{0.05} = 3.24 \succ F = 1.817$.

In other words, the reactant transformation degree (η), depends on the input reactant concentration (c_0), according to the following relation: $\eta = 0.92692 - 0.0112c_0$. The results obtained here show that physical and chemical processes occurring in the reactor of this case under study are not simple. It is well known that for a reaction occurring in a CSR with a simple kinetics, the degree of transformation is not significantly dependent on the input reactant concentration. For example, if a first order reaction occurs in a CSR, η will depend only on the residence time and the kinetic reaction constant $\eta = k_r \tau_s / (k_r \tau_s + 1)$.

5.4.2
Parabolic Regression

If the regression expression is a polynomial, then, by applying the method of least squares to identify the coefficients and compute the values of the coefficients, we obtain a simple linear system. If we particularize the case for a regression expression given by a polynomial of second order, the general relation (5.3) is reduced to:

$$y^{th} = y = f(x, \beta_0, \beta_1) = \beta_0 + \beta_1 x + \beta_{11} x^2 \tag{5.65}$$

By computing the derivatives of the system of normal equations $\dfrac{\partial f(x, \beta_0, \beta_1, \beta_{11})}{\partial \beta_0} = 1, \dfrac{\partial f(x, \beta_0, \beta_1, \beta_{11})}{\partial \beta_1} = x, \dfrac{\partial f(x, \beta_0, \beta_1, \beta_{11})}{\partial \beta_{11}} = x^2$, we establish the system of equations which is necessary to calculate the values of $\beta_0, \beta_1, \beta_{11}$:

$$\begin{cases} \beta_0 N + \beta_1 \sum_{i=1}^{N} x_i + \beta_{11} \sum_{i=1}^{N} x_i^2 = \sum_{i=1}^{N} y_i \\ \beta_0 \sum_{i=1}^{N} x_i + \beta_1 \sum_{i=1}^{N} x_i^2 + \beta_{11} \sum_{i=1}^{N} x_i^3 = \sum_{i=1}^{N} y_i x_i \\ \beta_0 \sum_{i=1}^{N} x_i^2 + \beta_1 \sum_{i=1}^{N} x_i^3 + \beta_{11} \sum_{i=1}^{N} x_i^4 = \sum_{i=1}^{N} y_i x_i^2 \end{cases} \tag{5.66}$$

The same procedure is used if we increase the polynomial degree given by the regression equation. In this case, the tests of the coefficient significance and model confidence are implemented as shown in the example developed in Section 5.4.1.1. It is important to note that we must use relation (5.59) for the calculation of the variances around the mean value of β_j.

5.4.3
Transcendental Regression

For statistical samples of small volume, an increase in the order of the polynomial regression of variables can produce a serious increase in the residual variance. We can reduce the number of the coefficients from the model but then we must introduce a transcendental regression relationship for the variables of the process. From the general theory of statistical process modelling (relations (5.1)–(5.9)) we can claim that the use of these types of relationships between dependent and independent process variables is possible. However, when using these relationships between the variables of the process, it is important to obtain an excellent ensemble of statistical data (i.e. with small residual and relative variances).

It is well known that using an exponential or power function can also describe the portion of a polynomial curve. Indeed, these types of functions, which can represent the relationships between the process variables, accept to be developed into a Taylor expansion. This procedure can also be applied to the example of the statistical process modelling given by the general relation (5.3) [5.20].

In this case, the calculation of the coefficients for the transcendental regression expression can be complicated because, instead of a system of normal equations (5.9), we obtain a system of non-linear equations. However, we can simplify the calculation by changing the original variables of the regression relationship. In fact, changing the original variables results in the mathematical application of one operator to the expression of the transcendental regression. As an example, we can consider the relations (5.67)–(5.69) below, where the powers or an exponential transcendental regression are transformed into a linear regression:

$$y^{th} = y = f(x, \beta_0, \beta_1) = \beta_0 \beta_1^x \tag{5.67}$$

$$y^{th} = y = f(x, \beta_0, \beta_1) = \beta_0 x^{\beta_1} \tag{5.68}$$

$$\lg y = \lg \beta_0 + x \lg \beta_1 \ , \quad z = \lg y \ , \quad \beta_0' = \lg \beta_0 \ , \quad \beta_1' = \lg \beta_1 \ , \quad z = \beta_0' + \beta_1^x \tag{5.69}$$

Coefficients β_0', β_1' can easily be obtained by using the method of least squares. Nevertheless, the interest is to have the original coefficients of the transcendental regression. To do so, we apply an inverse operator transformation to β_0' and β_1'. Here, we can note that β_0' and β_1' are the bypassed estimations for their correspondents β_0 and β_1.

5.4.4
Multiple Linear Regression

When the studied case concerns obtaining a relationship for the characterization of a process with multiple independent variables and only one dependent variable, we can use a multiple linear regression:

$$y^{th} = y = f(x_1..x_k, \beta_0, \beta_1...\beta_k) = \beta_0 + \beta_1 x_1 + \beta_2 x_2 + ... + \beta_k x_k \tag{5.70}$$

It is clear that Eq. (5.70) results from the general relation (5.3). In this case, when k = 2, we have a regression surface whereas, when k>2, a hypersurface is obtained. For surface or hypersurface constructions, we have to represent the corresponding values of the process parameters (factors and one dependent variable) for each axis of the phase's space. The theoretical starting statistical material for a multiple regression problem is given in Table 5.11.

Table 5.11 The starting statistical material for a multiple regression.

i	x_1	x_2	x_3	....	x_k	y
1	x_{11}	x_{21}	x_{31}	....	x_{k1}	y_1
2	x_{12}	x_{22}	x_{32}	....	x_{k2}	y_2
3	x_{13}	x_{23}	x_{33}	....	x_{k3}	y_3
.	...	...	...	....	...	..
.	...	...	...	....	...	..
N	x_{1N}	x_{2N}	x_{3N}	....	x_{kN}	y_N

The starting data are frequently transformed into a dimensionless form by a *normalization* method in order to produce a rapid identification of the coefficients in the statistical model. The dimensionless values of the initial statistical data (y_i^0 and x_{ji}^0) are computed using Eqs. (5.71) and (5.72), where s_y, s_{xj} are the square roots of the correspondent variances:

$$y_i^0 = \frac{y_i - \bar{y}}{s_y} \quad , \quad x_{ji}^0 = \frac{x_{ji} - \bar{x}_j}{s_{xj}} \quad , \quad i = 1, N \; ; \; j = 1, k \tag{5.71}$$

$$s_y = \sqrt{\frac{\sum_{i=1}^{N} (y_i - \bar{y})^2}{N - 1}} , \; s_{xj} = \sqrt{\frac{\sum_{i=1}^{N} (x_{ji} - \bar{x}_j)^2}{N - 1}} \tag{5.72}$$

At this step of the data preparation, we can observe that each column of the transformed statistical data has a zero mean value and a dispersion equal to one. A proof of these properties has already been given in Section 5.2 concerning a case of normal random variable normalization.

Then, considering the statistical data from Tables 5.11 and 5.12 and using the statistical correlation aspects (see Section 5.3), we can observe that the correlation coefficients are the same for variables y, x_j and y_i^0, x_{ji}^0 (relation (5.73)). This observation remains valid for the correlations concerning x_j and x_l.

$$r_{yx_j} = r_{y^0x_j^0} = \frac{1}{N-1} \sum_{i=1}^{N} y_i^0 x_{ji}^0 \tag{5.73}$$

$$r_{x_jx_l} = r_{x_j^0x_l^0} = \frac{1}{N-1} \sum_{i=1}^{N} x_{ji}^0 x_{li}^0 , \quad j\neq l \quad j,l = 1,2,...k \tag{5.74}$$

Table 5.12 The dimensionless statistical data for a multiple regression.

i	x_1^0	x_2^0	x_3^0	....	x_k^0	Y^0
1	$x^0{}_{11}$	$x^0{}_{21}$	$x^0{}_{31}$	....	$x^0{}_{k1}$	y_1^0
2	$x^0{}_{12}$	$x^0{}_{22}$	$x^0{}_{32}$	....	$x^0{}_{k2}$	y_2^0
3	$x^0{}_{13}$	$x^0{}_{23}$	$x^0{}_{33}$	....	$x^0{}_{k3}$	y_3^0
.	...	...	...	....	...	..
.	...	...	...	....	...	..
N	$x^0{}_{1N}$	$x^0{}_{2N}$	$x^0{}_{3N}$	....	$x^0{}_{kN}$	y_N^0

The observations mentioned above are important because they will be used in the following calculations. As was explained above, the mean value of the dependent normalized variable is zero, consequently the regression expression with the normalized variables can be written as:

$$y^{0\,th} = f^0(x_1^0, ..x_k^0, a_1, ..a_k) = a_1 x_1^0 + a_2 x_2^0 + ... + a_k x_k^0 \tag{5.75}$$

It is evident that, for the identification of the a_j coefficients, we have to determine the minimum of the quadratic displacement function between the measured and computed values of the dependent variable:

$$\Phi(a_1, a_2, ...a_k) = \sum_{i=1}^{N} (y_i^0 - y_i^{0\,th})^2 = \text{min} \tag{5.76}$$

thus, we obtain the minimum value of function $\Phi(a_1, a_2...., a_k)$ when we have:

$$\frac{\partial\Phi(a_1..a_k)}{\partial a_1} = \frac{\partial\Phi(a_1..a_k)}{\partial a_2} = ... = \frac{\partial\Phi(a_1..a_k)}{\partial a_k} = 0 \tag{5.77}$$

the relation above can be developed as follows:

$$
\begin{cases}
\alpha_1 \sum_{i=1}^{N} (x_{1i}^0)^2 + \alpha_2 \sum_{i=1}^{N} (x_{2i}^0 x_{1i}^0) + \ldots\ldots + \sum_{i=1}^{N} (x_{Ni}^0 x_{1i}^0) = \sum_{i=1}^{N} (y_1^0 x_{1i}^0) \\[2mm]
\alpha_1 \sum_{i=1}^{N} (x_{1i}^0 x_{2i}^0) + \alpha_2 \sum_{i=1}^{N} (x_{2i}^0)^2 + \ldots\ldots + \sum_{i=1}^{N} (x_{Ni}^0 x_{2i}^0) = \sum_{i=1}^{N} (y_1^0 x_{2i}^0) \\[2mm]
\text{---} \\[2mm]
\alpha_1 \sum_{i=1}^{N} (x_{1i}^0 x_{Ni}^0) + \alpha_2 \sum_{i=1}^{N} (x_{2i}^0 x_{Ni}^0) + \ldots\ldots + \sum_{i=1}^{N} (x_{Ni}^0)^2 = \sum_{i=1}^{N} (y_1^0 x_{Ni}^0)
\end{cases} \tag{5.78}
$$

The system (5.80) for the identification of the α_j coefficients is obtained after multiplying each term of system (5.78) by $1/(N-1)$ and after coupling this system with relations (5.73), (5.74) and (5.79):

$$
\frac{1}{N-1} \sum_{i=1}^{N} (x_{ji}^0)^2 = s_{x_j^0}^2 = 1 \tag{5.79}
$$

$$
\begin{cases}
\alpha_1 + \alpha_2 r_{x_1 x_2} + \alpha_3 r_{x_1 x_3} + \ldots\ldots + \alpha_k r_{x_1 x_k} = r_{yx_1} \\[2mm]
\alpha_1 r_{x_2 x_1} + \alpha_2 + \alpha_3 r_{x_2 x_3} + \ldots\ldots + \alpha_k r_{x_2 x_k} = r_{yx_2} \\[2mm]
\text{---} \\[2mm]
\alpha_1 r_{x_k x_1} + \alpha_2 r_{x_k x_2} + \alpha_3 r_{x_k x_3} + \ldots\ldots + \alpha_k = r_{yx_k}
\end{cases} \tag{5.80}
$$

Considering the commutability property of the correlations of coefficients ($r_{x_i x_j} = r_{x_j x_i}$) we can solve the above system. After solving it with unknown $\alpha_1, \alpha_2 \ldots \alpha_k$, we can determine the value of the correlation between the coefficients of the process variables by using Eq. (5.81):

$$
R_{yx_i} = \sqrt{\alpha_1 r_{yx_1} + \alpha_2 r_{yx_2} + \ldots\ldots + \alpha_k r_{yx_k}} \tag{5.81}
$$

When the statistical sample is small, the multiple linear correlation coefficient must be corrected. The correction is imposed by the fact that, in this case, the small number of degrees of freedom ($\upsilon = N - n_\beta$ is small) adds errors systematically. Therefore, the most frequently used correction is given by:

$$
R_{yx_i}^c = \sqrt{1 - (1 - R_{yx_i}^2) \frac{N-1}{N - n_\beta}} \tag{5.82}
$$

At this point, we have to consider coefficients $\alpha_1, \alpha_2 \ldots \alpha_k$ according to the dimensional relationship between the process variables (5.70). For this purpose, we must transform α_j into β_j, and $j = 1,k$. Indeed, these changes can take place using the following relations: $\beta_j = \alpha_j s_y / s_{xj}$, $j = 1, 2, \ldots k$, $j \neq 0$, $\beta_0 = \bar{y} - \sum_{i+}^{N} \beta_j \bar{x}_j$.

Now we have to estimate the reproducibility of the variance, to carry out the confidence tests for the coefficients so as to establish the final model.

5.4.4.1 Multiple Linear Regressions in Matrix Forms

The regression analysis, when the relationship between the process variables is given by a matrix, is frequently used to solve the problems of identification and confidence of the coefficients as well as the problem of a model confidence. The matrix expression is used frequently in processes with more than two independent variables which present simultaneous interactive effects with a dependent variable. In this case, the formulation of the problem is similar to the formulation described in the previous section. Thus, we will use the statistical data from Table 5.11 again in order to identify the coefficients with the following relation:

$$y^{th} = y = f(x_1..x_k, \beta_0, \beta_1...\beta_k) = \beta_0 x_0 + \beta_1 x_1 + \beta_2 x_2 + ... + \beta_k x_k \tag{5.83}$$

The first step in this discussion concerns the presentation of the matrix of the independent variables (X), the experimental observation vector of the dependent variable (Y) and the column matrix of the coefficients (B) as well as the transposed matrix of the independent variables (X^T). All these terms are introduced by relation (5.84). A fictive variable x_0, which takes the permanent value of 1, has been considered in the matrix of the independent variables:

$$X = \begin{bmatrix} x_{01} & x_{11} & . & . & x_{k1} \\ x_{02} & x_{12} & . & . & x_{k2} \\ . & . & . & . & . \\ . & . & . & . & . \\ x_{0N} & x_{1N} & . & . & x_{NN} \end{bmatrix} \quad Y = \begin{bmatrix} y_1 \\ y_2 \\ . \\ . \\ y_N \end{bmatrix} \quad B = \begin{bmatrix} \beta_0 \\ \beta_1 \\ \beta_2 \\ . \\ \beta_k \end{bmatrix}$$

$$\tag{5.84}$$

$$X^T = \begin{bmatrix} x_{01} & x_{02} & . & . & x_{0N} \\ x_{11} & x_{12}^T & . & . & x_{1N} \\ . & . & . & . & . \\ . & . & . & . & . \\ x_{k1} & x_{k2} & . & . & x_{kN} \end{bmatrix}$$

The particularization of the system of the normal equations (5.9) into an equivalent form of the relationship between the process variables (5.83), results in the system of equations (5.85). In matrix forms, the system can be represented by relation (5.86), and the matrix of the coefficients is given by relation (5.87). According to the inversion formula for a matrix, we obtain the elements for the inverse matrix of the matrix multiplication (XX^T), where (X^T) is the transpose matrix of the matrix of independent variables.

Relation (5.89) gives the value of each element of matrix (XX^T), where the symbol d_{jk} represents a current element, as shown in relation (5.88),

$$
\begin{cases}
\beta_0 \sum_{i=1}^{N} (x_{0i})^2 + \beta_1 \sum_{i=1}^{N} (x_{0i}x_{1i}) + \cdots\cdots + \beta_k \sum_{i=1}^{N} (x_{0i}x_{ki}) = \sum_{i=1}^{N} (y_i x_{0i}) \\[2ex]
\beta_0 \sum_{i=1}^{N} (x_{1i}x_{0i}) + \alpha_2 \sum_{i=1}^{N} (x_{1i})^2 + \cdots\cdots + \sum_{i=1}^{N} (x_{1i}x_{ki}) = \sum_{i=1}^{N} (y_i x_{1i}) \\[2ex]
\text{---} \\[1ex]
\beta_0 \sum_{i=1}^{N} (x_{ki}x_{0i}) + \alpha_2 \sum_{i=1}^{N} (x_{1i}x_{ki}^0) + \cdots\cdots + \sum_{i=1}^{N} (x_{ki})^2 = \sum_{i=1}^{N} (y_1^0 x_{Ni}^0)
\end{cases}
\tag{5.85}
$$

$$
X^T X B = X Y \tag{5.86}
$$

$$
B = (X^T X)^{-1} X Y \tag{5.87}
$$

$$
(X^T X)^{-1} =
\begin{bmatrix}
d_{00} & d_{01} & d_{02} & . & d_{0k} \\
d_{10} & d_{11} & d_{12} & . & d_{1k} \\
d_{20} & d_{21} & d_{22} & . & d_{2k} \\
. & . & . & . & . \\
d_{k0} & d_{k1} & d_{k3} & . & d_{kk}
\end{bmatrix}
\tag{5.88}
$$

$$
d_{jk} = \frac{\left(\sum_{i=1}^{N} x_{ki}x_{ji} \right)'}{\Delta}
\tag{5.89}
$$

It is not easy to compute the cofactors $\left(\sum_{i=1}^{N} x_{ki}x_{ji} \right)'$ and the determinant of the (X^{T}X) matrix multiplication. Therefore, the computation depends on the (X) matrix dimension and more specifically on the number of the independent variables of the process as well as on the number of experiments produced during the process of the statistical investigation. Frequently, the computation software of the problem cannot produce a solution for the coefficient matrix, even if we have carefully prepared data (controlled and verified). To overcome this situation, we must verify whether the inverse matrix of the (X) and (X^T) matrix multiplication presents a degenerated state. This undesirable situation appears when one or more correlation(s) exist(s) between the independent variables of the process. For this reason, when we have two factors with a strong correlation in the ensemble of independent variables, one of them will be excluded before developing the calculation algorithm to determine the correlation coefficients.

In order to obtain the value of the residual variance, we first define the matrix of the expected observations $\hat{Y} = XB$ and then we observe that the quadratic displacement between the measured and computed output values of the variables can be written as:

$$
[Y - \hat{Y}][Y - \hat{Y}]^T = \sum_{i=1}^{N} (y_i - \hat{y}_i)^2
$$

Now we introduce the matrix of the theoretical coefficients of the regression (coefficients of the relation (5.4)) here symbolized by Br. Therefore, the coefficient matrix B, defined above, is an estimation of the Br matrix and we can consequently write that the mean value of matrix B is matrix Br: $M(B) \rightarrow Br$ or $M[B - Br] \rightarrow 0$.

If we apply the concept of mean value to the matrix obtained from the multiplication of [B–Br] and [B–Br]^{T}, and using the definition for the variance and covariance of two variables, we obtain the result given by matrix (5.90):

$$M[(B - Br)(B - Br)^{T}] = \begin{bmatrix} \sigma^2_{\beta_0} & cov(\beta_0\beta_1) & cov(\beta_0\beta_2) & . & cov(\beta_0\beta_k) \\ cov(\beta_1\beta_0) & \sigma^2_{\beta_1} & cov(\beta_1\beta_2) & . & cov(\beta_1\beta_k) \\ cov(\beta_2\beta_0) & cov(\beta_2\beta_1) & \sigma^2_{\beta_2} & . & cov(\beta_2\beta_k) \\ . & . & . & & . \\ cov(\beta_k\beta_0) & cov(\beta_k\beta_1) & cov(\beta_k\beta_2) & . & \sigma^2_{\beta_k} \end{bmatrix}$$

(5.90)

It should be mentioned that the diagonal components of this matrix contain the theoretical variances of coefficients $\beta_j, j = 1, ..N$. Moreover, these variances are necessary to test the significance of the coefficients of the model. Indeed, when matching a model with an experimental study, matrix (5.90) is fundamental for testing the significance of the coefficients. Now, we have to consider the differences between the measured $y_i, i = 1, ..N$ and the expected mean values of the measurements introduced through the new vector column (Y_{ob}):

$$Y_{ob} = Y - M(Y) = \begin{bmatrix} y_1 - m(y_1) \\ y_2 - m(y_2) \\ . \\ y_N - m(y_N) \end{bmatrix}$$

(5.91)

Thus, the replacement of $B = (X^TX)^{-1}XY$(5.94) in the left-hand side of relation (5.90) results in: $M[(B - Br)(B - Br)^{T}] = M[[(XX^T)^{-1}X^TY_{ob}][(X^TX)^{-1}X^TY_{ob}]^{T}]$. Here, we can observe that (X^TX) is a diagonal symmetric matrix and, for that reason, we can write that $[(X^TX)^{-1}]^T = [(X^TX)^T]^{-1}$. Therefore, the relation $M[(B - Br)(B - Br)^{T}]$ can be written as $M[(B - Br)(B - Br)^{T}] = (X^TX)^{-1}M(Y_{ob}Y_{ob}^T)$. Because we generally have $\sigma^2_{y1} = \sigma^2_{y2} = ... = \sigma^2_{yN} = \sigma^2_y$ and due to the statistical independence of errors, we have $cov[(y_i - m(y_i))(y_l - m(y_l))]$ as zero for all $i \neq l$ and thus we can write the matrix $M(Y_{ob}Y_{ob}^T)$ as follows:

$$M(Y_{ob}Y_{ob}^T) = \begin{bmatrix} \sigma^2_{y1} & 0 & 0 & . & 0 \\ 0 & \sigma^2_{y2} & 0 & . & 0 \\ 0 & 0 & \sigma^2_{y3} & . & . \\ . & . & . & & . \\ 0 & 0 & 0 & . & \sigma^2_{yN} \end{bmatrix} = \begin{bmatrix} 1 & 0 & 0 & . & 0 \\ 0 & 1 & 0 & . & 0 \\ 0 & 0 & 1 & . & 0 \\ . & . & . & & . \\ 0 & 0 & 0 & . & 1 \end{bmatrix} \sigma^2_y$$

(5.92)

With this last observation, the calculation for $M[(B - Br)(B - Br)^{T}]$ results in:

$$M[(B - Br)(B - Br)^T] = (X^TX)^{-1}\sigma_y^2 \tag{5.93}$$

This result is very important because it shows how we compute the values of the elements of the matrix of mean errors $M[(B - Br)(B - Br)^T]$. These elements allow the calculation of the dispersions (variances) that characterize each β_j model coefficient of the process as shown in relation (5.94), which results from combining relations (5.93), (5.90) and (5.89):

$$\sigma_{\beta_j}^2 = d_{jj}\sigma_y^2 \; ; \quad cov(\beta_j\beta_k) = d_{jk}\sigma_y^2 \tag{5.94}$$

From a practical point of view, we should draw the readers' attention to the following significant and important specifics:

1. The $(X^TX)^{-1}$ matrix is the most important to identify the coefficients of the model and to estimate the mean values of errors associated to each β_j coefficient. This matrix is currently called the correlation matrix or error matrix.

2. This matrix does not have the state of a diagonal matrix and consequently, all the regression coefficients are in mutual correlation. So we cannot develop a different significance test for each of the coefficients. From this point of view it is not possible to use the t_j values given by relation (5.95) as the base of a procedure for the process factor arrangement:

$$t_j = \frac{|\beta_j|}{\sigma_y\sqrt{d_{jj}}} = \frac{|\beta_j|}{s_y\sqrt{d_{jj}}} \tag{5.95}$$

3. We can use the t_j values to start a heuristic procedure, which can be obtained from the regression expression of the non-significant coefficients. For this purpose, the following algorithm is used:
   a) the factor with the smallest t_j value is eliminated.
   b) if the residual variance decreases, then the exclusion is correct and thus, a new identification for the coefficients can be carried out. The opposite case shows that the excluded factor is important.
   c) new values for t_j will be obtained and a new elimination procedure can start.
   d) we close the procedure when is not possible to decrease the residual variance.
   e) the final remaining coefficients are the based estimations of the true coefficients.

Until now, no other procedures have been available for the enhancement of an initial proposed relationship between the regression variables.

5.4.5
Multiple Regression with Monomial Functions

In a multiple regression with monomial functions, the particularization of the relationship between the general process variables (5.3) gives the relation written below, where $f_j(x_j)$ is a continuous function:

$$y^{th} = y = f(x_1, x_2, .., \beta_0, \beta_1, ...\gamma) = \gamma f_1(x_1)f_2(x_2)f_3(x_3)....f_k(x_k)$$

This type of relationship between the dependent and all independent variables was first reported by Brandon [5.21]. In this form of function, we observe that the index (i) does not have a random position; thus, for $i = 1$, the function of the factor has a strong influence on the process, whereas, for $i = k$, the function of the factor has a slight influence on the process.

The algorithm that allows the identification of the functions and the γ constant can be described as follows:

1. An empirical regression line will be processed for the $y–x_1$ dependence with the statistical data from Table 5.11.
2. Thus, the dependence of $y_{x_1} = f_1(x_1)$ can now be appreciated and, using the classical least squares method, we can identify all the unknown coefficients.
3. A new set of values for the dependent variables of the process will be produced by dividing the old values by the corresponding $f_1(x_1)$ values, so that $y_1 = y/f_1(x_1)$. This new set of values of dependent variables are independent of factor x_1 and, as a consequence, we can write:
   $y_1^{th} = \gamma f_2(x_2)f_3(x_3)....f_k(x_k)$.
4. The first point of the algorithm can be repeated with respect to the $y_1–x_2$ interdependence. Consequently, we can write:
   $y_{x_2} = f_2(x_2)$;
5. We compute the coefficients of function $f_2(x_2)$ by the procedures recommended in item 2. and we build a new set of values for the dependent variables
   $y_2 = y_1/f_2(x_2) = y/[f_1(x_1)f_2(x_2)]$. These new values are independent with respect to x_1 and x_2;
6. The procedure continues with the identification of $f_3(x_3)$, ... $f_k(x_k)$ and we finally obtain the set of the last dependent variables as

$$y_k = \frac{y_{k-1}}{f_k(x_k)} = \frac{y}{f_1(x_1)f_2(x_2)...f_k(x_k)}.$$

It is easy to observe that vector y_k, gives its value to constant γ:

$$\hat{y}_k = \gamma = \frac{1}{N}\sum_{i=1}^{N} y_{ki}$$ because it is absolutely independent.

5.5
Experimental Design Methods

For all researchers, and especially for those working in experimental domains, a frequent requirement is summarized by the following phrase: *a maximum of information with a minimum of experiments*. This expression considers not only saving the researcher's time but also expensive reactants and energy. The use of experimental design or planning methods can guarantee not only to greatly reduce the number of experiments needed in an actual research but also to maintain the maximum information about the process. At the same time this technique gives the mathematical procedures of data processing for the complete characterization of the statistical model of a process [5.1, 5.13, 5.21–5.24].

The methodology of experimental design uses a terminology which is apparently different from the vocabulary frequently used in this chapter. Therefore, we call experimental conditions *factors* (or *factor* when we have only one); in fact, in Fig. 5.1, the experimental conditions are entirely included in the class of independent variables of the process. The word *level* (or *levels* when we have more than one), introduces here the values taken by the factors (factor). The term *response* is used to quantitatively characterize the observed output of the process when the levels of the factors are changed.

If we consider a process with k factors and if we suggest N_1 changes for the first factor, N_2 changes for the second factor, etc, then the total number of experiments will be $N_{ex} = N_1 N_2 .... N_k$. In fact, then, $N_1, N_2, ... N_k$ represent each factor level. The most frequent situation is to have $N_1 = N_2 = N_3 = ... = N_k = 2$ and in this case, we obtain the famous 2^k *method* for experimental planning. In fact, the method represents an optimal plan to describe the experiments using two levels for each process factor.

5.5.1
Experimental Design with Two Levels (2^k Plan)

The experimental research of a process with k factors and one response can be carried out considering all the combinations of the k factors with each factor at both levels. Thus, before starting the experimental research, we have a plan of the experiments which, for the mentioned conditions, is recognized as a *complete factorial experiment* (CFE) or 2^k *plan*. The levels of each of the various factors establish the frontiers of the process-investigated domain.

This abstract definition will be explained with the actual example of gaseous permeation through a zeolite/alumina composite membrane. Here, we must investigate the effect of the five following factors on the rate of permeation: the temperature (T) when the domain is between 200 and 400 °C, the trans-membrane pressure (Δp) when the domain is between 40 and 80 bar, the membrane porosity (ε) ranging from 0.08 to 0.18 m^3/m^3, the zeolite concentration within the porous structure (c_z) from 0.01 to 0.08 kg/kg and the molecular weight of the permeated gas (M) which is between 16 and 48 kg/kmol. With respect to the first factor (T), we can easily identify the value of the maximum level $z_1^{max} = 400\ °C$, the value of the minimum level $z_1^{min} = 200\ °C$, the value of the intermediate level $z_1^0 = 300\ °C$ and the factor (temperature) displacement which is considered as $\Delta z_1 = 100\ °C$. We can observe that

$$z_1^0 = \frac{z_1^{min} + z_1^{max}}{2} \quad \text{and} \quad \Delta z_1 = \frac{z_1^{max} - z_1^{min}}{2}$$

If we switch this observation to a general case we can write:

$$z_j^0 = \frac{z_j^{max} + z_j^{min}}{2} \quad , \quad \Delta z_j = \frac{z_j^{max} - z_j^{min}}{2} \tag{5.96}$$

Here z_j, $j = 1, k$ introduce the original values of the factors. The point with coordinates $(z_1^0, z_2^0, ...z_k^0)$ is recognized as the *centre of the experimental plan* or *fundamental level*. Δz_j introduces the *unity* or *variation interval* respect to the axis z_j, $j = 1, k$. At this point, we have the possibility to transform the dimensional coordinates $z_1, z_2, ...z_k$ to the dimensionless ones, which are introduced here by relation (5.97). We also call these relations *formulas*.

$$x_j = \frac{z_j - z_j^0}{\Delta z_j} \quad , \quad j = 1, 2, ..k \tag{5.97}$$

It is not difficult to observe that, by using this system of dimensionless coordinates for each factor, the upper level corresponds to +1, the lower level is –1 and the fundamental level of each factor is 0. Consequently, the values of the coordinates of the experimental plan centre will be zero. Indeed, the centre of the experiments and the origin of the system of coordinates have the same position. In our current example, we can consider that the membrane remains unchanged during the experiments, i.e. the membrane porosity (ε) and the zeolite concentration (c_z) are not included in the process factors.

Therefore, we have to analyse the variation of the rate of permeation according to the temperature (z_1), the trans-membrane pressure difference (z_2) and the gas molecular weight (z_3). Then, we have 3 factors each of which has two levels. Thus the number of experiments needed for the process investigation is $N = 2^3 = 8$. Table 5.13 gives the concrete plan of the experiments. The last column contains the output "y" values of the process (flow rates of permeation). Figure 5.8 shows a geometric interpretation for a 2^3 experimental plan where each cube corner defines an experiment with the specified dimensionless values of the factors. So as to process these statistical data with the procedures that use matrix calculations, we have to introduce here a fictive variable x_0, which has a permanent +1 value (see also Section 5.4.4).

Table 5.13 The matrix for a 2^3 experimental plan (example of gas permeation).

Natural values of factors				Dimensionless values of factors			Response values
Experiment number	z_1	z_2	z_3	x_1	x_2	x_3	Permeation flow rates $y * 10^6$ (kg/s)
1	200	40	16	−1	−1	−1	8
2	400	40	16	+1	−1	−1	11
3	200	80	16	−1	+1	−1	10
4	400	80	16	+1	+1	−1	18
5	200	40	44	−1	−1	+1	3
6	400	40	44	+1	−1	+1	5
7	200	80	44	−1	+1	+1	4
8	400	80	44	+1	+1	+1	7

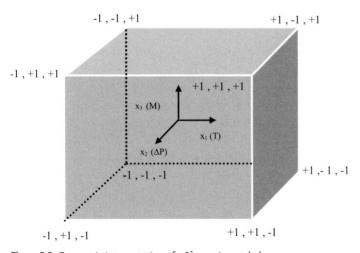

Figure 5.8 Geometric interpretation of a 2^3 experimental plan.

From a theoretical point of view, if we transform the matrix according to the 2^3 experimental plan, we obtain the state form shown in Table 5.14. This matrix has two important properties: the first is its orthogonality, the mathematical expression of which is:

$$\sum_{i=1}^{N} x_{li}x_{ju} = 0 \quad \forall \ l \neq j \ , \ l, u = 0, 1, ...k \tag{5.98}$$

The second is recognized as the normalization property, which shows that the sum of the dimensionless values of one factor is zero; besides, the sum of the square values of one factor is equal to the total number of experiments. Relations (5.99) and (5.100) give the mathematical expression of the norm property:

$$\sum_{i=1}^{N} x_{ji} = 0 \quad j \neq 0, \ j = 1, 2...k \tag{5.99}$$

$$\sum_{i=1}^{N} x_{ji}^2 = N \quad j = 0, 1, ...k \tag{5.100}$$

Table 5.14 Matrix for a 2^3 experimental plan with x_0 as fictive factor. Each line x_1,x_2,x_3 corresponds to one point of Fig. 5.8.

i	x_0	x_1	x_2	x_3	y
1	+1	−1	−1	−1	y_1
2	+1	+1	−1	−1	y_2
3	+1	−1	+1	−1	y_3
4	+1	+1	+1	−1	y_4
5	+1	−1	−1	+1	y_5
6	+1	+1	−1	+1	y_6
7	+1	−1	+1	+1	y_7
8	+1	+1	+1	+1	y_8

The orthogonality of the planning matrix, results in an easier computation of the matrix of regression coefficients. In this case, the matrix of the coefficients of the normal equation system (X^TX) has a diagonal state with the same value N for all diagonal elements. As a consequence of the mentioned properties, the elements of the inverse matrix $(X^TX)^{-1}$ have the values $d_{jj} = 1/N$, $d_{jk} = 0, j \neq k$.

In these conditions, we obtain the coefficients of the regression equation according to very simple relations as can be observed in the following matrix expression:

$$B = \begin{bmatrix} \beta_0 \\ \beta_1 \\ . \\ . \\ \beta_k \end{bmatrix} = (X^T X)^{-1} X^T Y = \begin{bmatrix} 1/N & 0 & 0 & . & 0 \\ 0 & 1/N & & . & 0 \\ 0 & 0 & 1/N & . & 0 \\ . & . & . & . & . \\ 0 & 0 & 0 & . & 1/N \end{bmatrix} \cdot \begin{bmatrix} \sum_{i=1}^{N} x_{0i} y_i \\ \sum_{i=1}^{N} x_{1i} y_i \\ . \\ . \\ \sum_{i=1}^{N} x_{ki} y_i \end{bmatrix}$$

$$= \begin{bmatrix} (\sum_{i=1}^{N} x_{0i} y_i)/N \\ (\sum_{i=1}^{N} x_{1i} y_i)/N \\ . \\ . \\ (\sum_{i=1}^{N} x_{0i} y_i)/N \end{bmatrix} \tag{5.101}$$

Each coefficient β_j of the regression relationship is given by the scalar multiplication and summation of the y column and the x_j column; a final multiplication by 1/N closes the β_j calculation ($\beta_j = \frac{1}{N} \sum_{i=1}^{N} x_{ji} y_i$, $j = 0, k$). Now, with the help of the experimental planning from Table 5.13, we can compute the multiple linear regression given by relation (5.102). Physically, this calculation corresponds to the assumption that the flow rate of permeation through a membrane depends linearly on the temperature, trans-membrane pressure and molecular weight of permeated gas.

$$y^{th} = y = f(x_1, x_2, x_3, \beta_0, \beta_1, \beta_2, \beta_3) = \beta_0 + \beta_1 x_1 + \beta_2 x_2 + \beta_3 x_2 \tag{5.102}$$

$$\begin{bmatrix} x_{1i} \\ -1 \\ +1 \\ -1 \\ +1 \\ -1 \\ +1 \\ -1 \\ +1 \end{bmatrix} * \begin{bmatrix} y_i \\ 9 \\ 11 \\ 10 \\ 18 \\ 3 \\ 5 \\ 4 \\ 7 \end{bmatrix} = \begin{bmatrix} x_{1i} y_i \\ -9 \\ +11 \\ -10 \\ +18 \\ -3 \\ +5 \\ -4 \\ +7 \end{bmatrix} \quad \sum_{i=1}^{8} x_{1i} y_i = 15 \quad \beta_1 = \left(\sum_{i=1}^{8} x_{1i} y_i\right)/N = 15/8 = 1.86$$

Table 5.15 contains the calculation results for all the coefficients of relation (5.102).

Table 5.15 The coefficients of relationship (5.102) according to the data from Table 5.13.

i	x_{0i}	x_{1i}	x_{2i}	x_{3i}	$x_{0i}y_i$	$x_{1i}y_i$	$x_{2i}y_i$	$x_{3i}y_i$	β_j
1	+1	−1	−1	−1	9	−9	−9	−9	
2	+1	+1	−1	−1	11	+11	−11	−11	$\beta_0 = 67/8 = 8.375$
3	+1	−1	+1	−1	10	−10	+10	−10	
4	+1	+1	+1	−1	18	+18	+18	−18	$\beta_1 = 15/8 = 1.875$
5	+1	−1	−1	+1	3	−3	−3	+3	
6	+1	+1	−1	+1	5	+5	−5	+5	$\beta_2 = 11/8 = 1.375$
7	+1	−1	+1	+1	4	−4	+4	+4	
8	+1	+1	+1	+1	7	+7	+7	+7	$\beta_3 = -29/8 = -3.625$
$\sum_{i=1}^{N=8}$	8	0	0	0	67	15	11	−29	

For a 2^3 plan, when we consider a more complete regression relationship in which the factors interact, we can write:

$$f(x_1, x_2, x_3, \beta_0, ..\beta_3, \beta_{12}, ..\beta_{23}, \beta_{123}) = \beta_0 + \beta_1 x_1 + \beta_2 x_2 + \beta_3 x_3 + \beta_{12} x_1 x_2$$
$$+ \beta_{13} x_1 x_3 + \beta_{23} x_2 x_3 + \beta_{123} x_1 x_2 x_3 \qquad (5.103)$$

Here, β_{12}, β_{13}, β_{23} correspond to the effect of double interactions (factor 1 with factor 2, etc) and β_{123} introduces the effect of triple interaction. Table 5.16 completes the values shown in Table 5.13 with the values needed to calculate the considered interactions whereas Table 5.17 shows the synthesized calculations of the interactions between the coefficients.

Table 5.16 The operation matrix for double and triple interaction effects.

i	x_{0i}	x_{1i}	x_{2i}	x_{3i}	$x_{1i}x_{2i}$	$x_{1i}x_{3i}$	$x_{2i}x_{3i}$	$x_{1i}x_{2i}x_{3i}$	y_i
1	+1	−1	−1	−1	+1	+1	+1	−1	9
2	+1	+1	−1	−1	−1	−1	+1	+1	11
3	+1	−1	+1	−1	−1	+1	−1	+1	10
4	+1	+1	+1	−1	+1	−1	−1	−1	18
5	+1	−1	−1	+1	+1	−1	−1	+1	3
6	+1	+1	−1	+1	−1	+1	−1	−1	5
7	+1	−1	+1	+1	−1	−1	+1	−1	4
8	+1	+1	+1	+1	+1	+1	+1	+1	7

Table 5.17 Calculation of the interaction coefficients for model (5.110).

i	$x_{1i}x_{2i}$	$x_{1i}x_{3i}$	$x_{2i}x_{3i}$	$x_{1i}x_{2i}x_{3i}$	y_i	$x_{1i}x_{2i}y_i$	$x_{1i}x_{3i}y_i$	$x_{2i}x_{3i}y_i$	$x_{1i}x_{2i}x_{3i}y_i$	$\beta_{12,etc}$
1	+1	+1	+1	−1	9	+9	+9	+9	−9	$\beta_{12} = 7/8 = 0.875$
2	−1	−1	+1	+1	11	−11	−11	+11	+11	
3	−1	+1	−1	+1	10	−10	+10	−10	+10	$\beta_{13} = -5/8 = -0.625$
4	+1	−1	−1	−1	18	+18	−18	−18	−18	
5	+1	−1	−1	+1	3	+3	−3	−3	+3	$\beta_{23} = -5/8 = -0.625$
6	−1	+1	−1	−1	5	−5	+5	−5	−5	
7	−1	−1	+1	−1	4	−4	−4	+4	−4	$\beta_{123} = -5/8 = -0.625$
8	+1	+1	+1	+1	7	+7	+7	+7	+7	
$\sum_{i=1}^{8}$	0	0	0	0	67	7	−5	−5	−5	

If one or more parallel trials are available for the data from Table 5.13, then for the pleasure of statistical calculation, we can compute new values for the given coefficients and consequently we can investigate their statistical behaviour. A real residual variance can then be established. Unfortunately, we do not have the repeated data for our problem of gaseous permeation through a porous membrane. It is known that the matrix $(X^TX)^{-1}$ has the values $d_{jj} = 1/N$, $d_{jk} = 0, j \neq k$ and that, consequently, the regression coefficients will not be correlated. In other words, they are independent of each other. Two important aspects are noticed from this observation: (i) we can test the significance of each coefficient in the regression relationship separately; (ii) the rejection of a non-significant coefficient from the regression relationship does not have any consequence on the values of the remaining coefficients.

Coefficients $\beta_j, \beta_{jl}, \beta_{jlm}$, $j \neq l$, $j \neq m$, j and l and m $= 1, 2, ...k$ obtained with the help of a CFE have the quality to be absolutely correct estimators of the theoretical coefficients as defined in relation (5.4). It is important to repeat that the value of each coefficient quantifies the participation of the corresponding factor to the response construction.

Because the diagonal elements of the correlation matrix $(X^TX)^{-1}$ have the same value, we can conclude (please see the mentioned relation) that they have been determined with the same precision. Indeed, we can write that all the square roots of the coefficient variances have the same value:

$$s_{\beta_j} = s_{\beta_{jl}} = s_{\beta_{jlm}} = \frac{s_{rp}}{\sqrt{N}} \tag{5.104}$$

Let us now go back to the problem of gaseous permeation and more precisely to the experimental part when we completed the data from Table 5.13 with the values from the permeation flow rate. These values are obtained from three experiments for the centre of plan 2^3:

$$y_1^0 = 10.5 \cdot 10^{-6} kg/(m^2 s); \quad y_2^0 = 11 \cdot 10^{-6} kg/(m^2 s); \quad y_3^0 = 10 \cdot 10^{-6} kg/(m^2 s).$$

We obtain all the square roots of the variances needed to test the significance of the coefficients with these data:

$$\bar{y}^0 = \sum_{i=1}^{3} y_i^0/3 = 10.5 \ ; \ s_{rp}^2 = \sum_{i=1}^{3} (y_i^0 - \bar{y}^0)^2/2 = 0.25 \ ; \ s_{rp} = 0.5 \ ; \ s_{\beta_i} = s_{\beta_{jl}}$$

$$= s_{\beta_{jlm}} = s_{rp}/\sqrt{N} = 0.5/\sqrt{8} = 0.177.$$

Table 5.18 contains the calculation concerning the significance of the regression coefficients from relation (5.110). However, respect to table 5.6, the rejection condition of the hypothesis has been changed so that we can compare the computed t value (t_j) with the t value corresponding to the accepted significance level $(t_{\alpha/2})$.

Table 5.18 The significance of the coefficients for the statistical model (5.103).

n	H_0	Student variable value: t_j	$t_{\alpha/2}$ for $\nu = 2$	t_j and $t_{\alpha/2}$	Verdict
1	$\beta_0 = 0$	$t_0 = |\beta_0|/s_{\beta_0} = 8.37/0.17 = 47.2$	4.3	$t_0 > t_{\alpha/2}$	rejected
2	$\beta_1 = 0$	$t_1 = |\beta_1|/s_{\beta_1} = 1.86/0.17 = 10.5$	4.3	$t_1 > t_{\alpha/2}$	rejected
3	$\beta_2 = 0$	$t_2 = |\beta_2|/s_{\beta_2} = 2.75/0.17 = 15.5$	4.3	$t_2 > t_{\alpha/2}$	rejected
4	$\beta_3 = 0$	$t_3 = |\beta_3|/s_{\beta_3} = 3.62/0.17 = 20.45$	4.3	$t_3 > t_{\alpha/2}$	rejected
5	$\beta_{12} = 0$	$t_{12} = |\beta_{12}|/s_{\beta_{12}} = 0.875/0.17 = 4.94$	4.3	$t_{12} > t_{\alpha/2}$	rejected
6	$\beta_{13} = 0$	$t_{13} = |\beta_{13}|/s_{\beta_{13}} = 0.625/0.17 = 3.5$	4.3	$t_{13} < t_{\alpha/2}$	accepted
7	$\beta_{23} = 0$	$t_{23} = |\beta_{23}|/s_{\beta_{23}} = 0.625/0.17 = 3.5$	4.3	$t_{23} < t_{\alpha/2}$	accepted
8	$\beta_{123} = 0$	$t_{123} = |\beta_{123}|/s_{\beta_{123}} = 0.625/0.17 = 3.5$	4.3	$t_{123} < t_{\alpha/2}$	accepted

The calculation from Table 5.18 shows that coefficients β_{13}, β_{23}, β_{123} have no importance for the model and can consequently be eliminated. From these final observations, the remaining model of gaseous permeation, can be represented in a dimensionless form by the relation (5.105). We must notice that, in these calculations, the values of the y column have been multiplied by 10^6.

$$\hat{y} = 8.37 + 1.85x_1 + 2.75x_2 - 3.632x_3 + 0.875x_1x_2 \tag{5.105}$$

At the end of the process of the statistical modelling, we have to test the significance of the model. Here is the case of the model for gaseous permeation through a porous membrane for which we compute:

- the value of the residual variance:

$$s_{rz}^2 = \left(\sum_{i=1}^{N} (y_i - \hat{y}_i)^2\right)/(N - n_\beta) = 7.14/3 = 1.78;$$

- the numerical value of the associated Fischer variable:
  $F = s_{rz}^2/s_{rp}^2 = 1.78/0.177 = 10;$
- the theoretical value of the associated Fischer variable corresponding to this concrete case:
  $\alpha = 0.05, v_1 = 3, v_2 = 2$ and $F_{3,2,0.05} = 19.16$.

Thanks to the assigned significance level, we can acknowledge the model to be adequate because we have $F < F_{3,2,0.05}$ $(10 < 19.6)$.

5.5.2
Two-level Experiment Plan with Fractionary Reply

Each actual experimental research has its specificity. From the first chapter up to the present paragraph, the process modelling has been requiring more and more statistical data. With an excess of statistical data we have a better residual and reproducibility in the calculation of variances and thus coefficients can be identified more precisely. Nevertheless, this excess is not absolutely necessary and it is known that reducing the volume of statistical data saves money. When we use a CFE in our research, we first assume that each process model regression relationship is a polynome in which the interactions of the factors are considered. For example, if the relationship of the variables of the model can be limited to the linear approximation then, to develop the model, it is not necessary to use an experimental investigation made of a complete CFE. We can indeed use only one part of a CFE for experimental investigation; this part of the CFE is recognized as a fractionary factorial experiment (FFE). Because an FFE must be orthogonal, we start from the next CFE below; from this start we make sure that the number of experiments in the regression relationship remains greater than the number of unknown coefficients. We consider that the purpose of a process including three factors is to obtain a linear approximation between the process variables because we assume that this process gives a good characterization of an interesting part of the response surface. Therefore, for this part of the response surface, we can write:

$$y^{th} = f(x_1, x_2, x_3, \beta_0, \beta_1, \beta_2, \beta_3) = \beta_0 + \beta_1 x_1 + \beta_2 x_2 + \beta_3 x_3 \tag{5.106}$$

To solve this problem where we have 3 unknowns, we can chose a type 2^2 CFE in which the $x_1 x_2$ column will be the plan for x_3. Table 5.19 gives CFE 2^2 whereas Table 5.20 shows the transformation of our problem into an FFE plan. Thus, from an initial number of $2^3 = 8$ experiments we will produce only 2^2 experiments; more generally we can say that, when we use a type 2^{k-1} FFE, we halve the initial minimum required number of experiments.

Table 5.19 The type 2^2 CFE matrix.

i	x_0	x_1	x_2	$x_1 x_2$	y
1	+1	+1	+1	+1	y_1
2	+1	+1	−1	−1	y_2
3	+1	−1	−1	+1	y_3
4	+1	−1	+1	−1	y_4

Table 5.20 The FFE plan from a type 2^2 CFE plan.

i	x_0	x_1	x_2	x_3	y
1	+1	+1	+1	+1	y_1
2	+1	+1	−1	−1	y_2
3	+1	−1	−1	+1	y_3
4	+1	−1	+1	−1	y_4

Using the experimental plan from Table 5.20 it is possible to estimate the constant terms and the three coefficients related to the linear terms from the regression relationship.

Practically, we cannot a priori postulate the nullity of the effects of the interaction. Indeed, we can accept the fact that some or all of the effects of the interaction are insignificant according to the linear effects but these are present. Then, from a practical point of view, when the coefficients corresponding to the effects of interaction are not zero and when we have the coefficients obtained by a 2^{3-1} plan, it is clear that these last coefficients include the participation of interactions on the major linear participations into the process response. The estimators of the general or theoretical coefficients are: $\beta_1^{th}, \beta_2^{th}, \beta_3^{th}, \beta_{12}^{th}, \beta_{13}^{th}, \beta_{23}^{th}$ and consequently, we can write:

$$\beta_1 \rightarrow \beta_1^{th} + \beta_{23}^{th} \quad \beta_2 \rightarrow \beta_2^{th} + \beta_{13}^{th} \quad \beta_3 \rightarrow \beta_3^{th} + \beta_{12}^{th} \tag{5.107}$$

In order to complete the FFE we can add a new column which contains the multiplication $x_1 x_3$ to Table 5.20. However, we observe that the elements of this multiplication and the elements of the x_2 column are the same; so we cannot complete the FFE. Thus we can also use the fact that, in Table 5.20, we have:

$$x_3 = x_1 x_2 \qquad (5.108)$$

If we multiply the relation above by x_3, we obtain $x_3{}^2 = x_1 x_2 x_3$ or $1 = x_1 x_2 x_3$, which is recognized as the contrast of the FFE plan. Now multiplying this contrast by x_1, x_2, x_3 yields the relations (5.109). These relations explain the relationships described in Eq. (5.107).

$$x_1 = x_1^2 x_2 x_3 = x_2 x_3 \qquad x_2 = x_1 x_3 \qquad x_3 = x_1 x_2 \qquad (5.109)$$

When we decide to work with an FFE plan and when we have more than three factors, a new problem appears because we then have more possibilities to build the plan. For an answer to the question that requires a choice of most favourable possibility, we use the resolution power of each one of the options. So we generate the first possibility for an FFE plan by choosing the production (generation) relation. We can then go on with the contrast relation through which we obtain all the actual relations that are similar to those given in (5.107).

This procedure will be repeated for all the possibilities of building an FFE plan. The decision will be made according to the researcher's interest as well as to the need to obtain as much information as possible about the investigated process.

We will complete this abstract discussion with the concrete case of a process with $k = 4$ factors taking CFE 2^3 as a basis for an FFE plan. To this end we have:

$$x_4 = x_1 x_2 x_3 \qquad (5.110)$$

or one out of the next three relations as a production relation:

$$x_4 = x_1 x_2 \qquad x_4 = x_1 x_3 \qquad x_4 = x_2 x_3 \qquad (5.111)$$

Table 5.21 gives the FFE matrix that is associated with the production relation (5.110). According to the procedure described above (showing the development of relations (5.109)), we produce the formal (5.112) system. It shows the correlation between the obtainable and theoretical coefficients of the regression relationships.

$$\begin{cases} x_1 = x_2 x_3 x_4 \rightarrow \beta_1 = \beta_1^{th} + \beta_{234}^{th} \\ x_2 = x_1 x_3 x_4 \rightarrow \beta_2 = \beta_2^{th} + \beta_{134}^{th} \\ x_3 = x_1 x_2 x_4 \rightarrow \beta_3 = \beta_3^{th} + \beta_{124}^{th} \\ x_4 = x_1 x_2 x_3 \rightarrow \beta_4 = \beta_4^{th} + \beta_{123}^{th} \\ x_1 x_2 = x_3 x_4 \rightarrow \beta_{12} = \beta_{12}^{th} + \beta_{34}^{th} \\ x_1 x_3 = x_2 x_4 \rightarrow \beta_{13} = \beta_{13}^{th} + \beta_{24}^{th} \\ x_1 x_4 = x_2 x_3 \rightarrow \beta_{14} = \beta_{14}^{th} + \beta_{23}^{th} \end{cases} \qquad (5.112)$$

Table 5.21 The FFE matrix from 2^3 plan and $x_4 = x_1 x_2 x_2$ as a production relation.

i	x_0	x_1	x_2	x_3	x_4	y
1	+1	+1	+1	+1	+1	y_1
2	+1	+1	+1	−1	−1	y_2
3	+1	−1	+1	−1	+1	y_3
4	+1	−1	+1	+1	−1	y_4
5	+1	−1	−1	+1	+1	y_5
6	+1	+1	−1	+1	−1	y_6
7	+1	+1	−1	−1	+1	y_7
8	+1	−1	−1	−1	−1	y_8

Considering the formal system (5.112), we observe that the triple interaction is indirectly considered here. It is doubtful that the actual results could confirm this class of interaction but if we can prove that they are present, then the plan from Table 5.21 can be suggested. Table 5.22 shows the FFE plan from the case when the first relation from the assembly (5.111) is the production relation. It is important to observe that all the binary interactions are indirectly considered in the formal system of the correlations of the obtainable and theoretical coefficients (5.113). Therefore, if the interest is to keep all the binary interactions of factors in the process model relationship, this FFE plan can be used successfully.

Table 5.22 The FFE matrix from a 2^3 plan and $x_4 = x_1 x_2$ as a production relation.

i	x_0	x_1	x_2	x_3	x_4	y
1	+1	+1	+1	+1	+1	y_1
2	+1	+1	+1	−1	+1	y_2
3	+1	−1	+1	−1	−1	y_3
4	+1	−1	+1	+1	−1	y_4
5	+1	−1	−1	+1	+1	y_5
6	+1	+1	−1	+1	−1	y_6
7	+1	+1	−1	−1	−1	y_7
8	+1	−1	−1	−1	+1	y_8

$$\begin{cases} x_1 = x_2x_4 \rightarrow \beta_1 = \beta_1^{th} + \beta_{24}^{th} \\ x_2 = x_1x_4 \rightarrow \beta_2 = \beta_2^{th} + \beta_{14}^{th} \\ x_3 = x_1x_2x_3x_4 \rightarrow \beta_3 = \beta_3^{th} + \beta_{1234}^{th} \\ x_4 = x_1x_2 \rightarrow \beta_4 = \beta_4^{th} + \beta_{12}^{th} \\ x_1x_3 = x_2x_3x_4 \rightarrow \beta_{13} = \beta_{13}^{th} + \beta_{234}^{th} \\ x_2x_3 = x_1x_3x_4 \rightarrow \beta_{23} = \beta_{23}^{th} + \beta_{134}^{th} \\ x_3x_4 = x_1x_2x_4 \rightarrow \beta_{34} = \beta_{34}^{th} + \beta_{124}^{th} \end{cases}$$

(5.113)

High-level FFEs such as, for example 1/4 or 1/8 from complete factorial experiments (CFEs) can be used for complex processes, especially if the effect on the response of some factors is the objective of the research. It is not difficult to decide that, if we have a problem with the k factors where the p linear effects compensate the effects of interaction, then the 2^{k-p} FFE can be used without any restriction.

The plan 2^{k-p} FFE keeps the advantages of the CFE 2^k plan, then:
- it is an orthogonal plan and consequently simple calculation for $\beta_0, \beta_1, \dots$ is used;
- all the regression coefficients keep their independence;
- each coefficient is computed as a result of all N experiments;
- the same minimal variance characterizes the determination of all regression coefficients.

We can then add a new "spherical" property to the properties of CFE 2^k and FFE 2^{k-p}. This new property can be used to characterize the quantity of planning information. To show the content of this property, by means of the independence of the regression relationship coefficients and according to the law governing the addition of variance for a linear regression, we can write:

$$s_{\bar{y}}^2 = s_{rz}^2 = s_{\beta_0}^2 + x_1^2 s_{\beta_1}^2 + x_2^2 s_{\beta_2}^2 + \dots + x_k^2 s_{\beta_k}^2$$

(5.114)

Because $s_{\beta_j}^2 = s_{rp}^2/N$, relation (5.114) becomes:

$$s_{\bar{y}}^2 = s_{rz}^2 = \frac{s_{rp}^2}{N}(1 + x_1^2 + x_2^2 + \dots + x_k^2) = \frac{s_{rp}^2}{N}(1 + \gamma^2) \qquad \gamma^2 = \sum_{j=0}^{N} x_j^2$$

(5.115)

Here, from a geometric viewpoint, γ is a sphere radius for the space of k dimension. When γ is significant, the residual variance $s_{\bar{y}}^2$ is also significant and, consequently, only a small quantity of information characterizes the process model.

5.5.3

Investigation of the Great Curvature Domain of the Response Surface: Sequential Experimental Planning

Figure 5.9 shows the response surface that gives the correlation between the dependent variable y (or η) and two independent factors (with values z_1 (or t) and z_2 (or c) respectively). The problem of this example concerns a chemical reaction

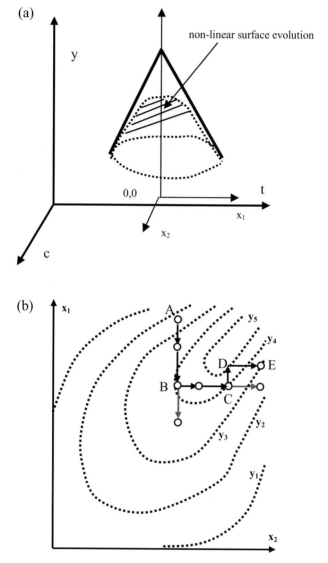

Figure 5.9 (a) Response surface for $k = 2$. (b) Sections of the response surface and of the gradual displacement towards the domain of the great surface curvature.

where the conversion (η) is a function of the concentration (c) and temperature (t) of the reactant. For more details, please see the data from Table 5.2. Considering Fig. 5.9(a), we can easily identify the two different domains: the first domain corresponds to the cases when y is linearly dependent on x_1 and x_2 (or near to a linear dependence); the second domain corresponds to the height of the curvature surface where the effects of the quadratic factors are significant.

We have the possibility, from a theoretical as well as from a practical point of view, to plan an experimental research so as to investigate this domain. Figure 5.9 (b) shows how we can gradually carry out these experiments.

We begin the experiments from an a priori starting point and according to the y variations. First, we keep the x_2 value unchanged and increase or decrease x_1. If y begins to decrease (point B from Fig. 5.9 (b)), we stop decreasing x_1 and go on increasing x_2 while keeping x_1 fixed. Then, we get to point C where we find out that x_2 must be maintained and x_1 changed.

It is clear that we can thus determine a way to the extreme point of the response surface curvature. At the same time, it is not difficult to observe that the ABCDE way is not a gradient. Despite its triviality, this method can be extended to more complex dependences (more than two variables) if we make amendments. It is important to note that each displacement required by this procedure is accomplished through an experiment; here the length of displacement is an apparently random variable since we cannot compute this value because we do not have any analytical or numerical expression of the response function. The response value is available at the end of the experiments.

The example shown above, introduces the necessity for a statistical investigation of the response surface near its great curvature domain. We can establish the proximity of the great curvature domain of the response surface by means of more complementary experiments in the centre of the experimental plan ($x_1 = 0, x_2 = 0,...x_k = 0$). In these conditions, we can compute $\bar{y}_0$, which, together with β_0 (computed by the expression recommended for a factorial experiment

$\beta_0 = \left(\sum_{i=1}^{N} x_{0i} y_i \right)/N = \left(\sum_{i=1}^{N} y_i \right)/N$), gives relative information about the curvature

of the surface response through relation (5.116).

$$\beta_0 - \bar{y}^0 \rightarrow \sum_{j=1}^{k} \beta_{jj}^{th} \tag{5.116}$$

It is well known that the domains of the great curvature of the response surface are characterized by non-linear variable relationships. The most frequently used state of these relationships corresponds to a two-degree polynomial. Thus, to express the response surface using a two-degree polynomial, we must have an experimental plan which considers one factor and a minimum of three different values. A complete factorial 3^k experiment requires a great number of experiments ($N = 3^k$; $k = 3$ $N = 27$; $k = 4$ $N = 81$). It is obvious that the reduction of the number of experiments is a major need here. We can consequently reduce the number of experiments if we accept the use of a composition plan (sequential plan) [5.25] for the experimental process. The core of a sequential plan is a CFE 2^k plan with k<5 or an FFE plan with k>5. If, by means of CFE or FFE plans, the regression analysis results in an inadequate regression relationship, we can carry out new experiments for the plan. To be classified as sequential plans, these supplementary experiments require:

1. The addition of a 2^k number of experiments (uniformly disposed on the system axes) to the 2^k CFE plan. The coordinates of these points will be $(\pm a, 0, 0....0), (0, \pm a, 0...0),$ ...$(0, 0, 0...0, \pm a)$ where a is the dimensionless distance from the plan centre to an additional point.
2. An increase in the number of experiments in the centre of the experimental plan (n_0).

For a process with k factors and one response, relation (5.117) can be used to estimate the number of experiments needed by a sequential plan:

$$N = 2^k + 2k + n_0 \quad \text{for} \ k < 5$$
$$N = 2^{k-1} + 2k + n_0 \quad \text{for} \ k > 5$$

(5.117)

The construction of a sequential plan with k = 2 is shown in Fig. 5.10. Points A B C D are the components of the 2^2 CFE and points A′ B′ C′ D′ are the components added to the basis plan. The notation n_0 in the centre of the plan shows that we must repeat the experiments. The recommended values of a dimensionless a, corresponds to the situation when the obtained composition plan keeps almost all the properties of the CFE plan.

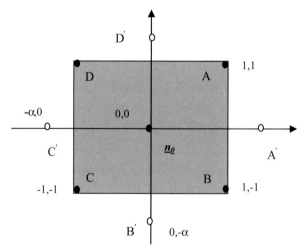

Figure 5.10 Composition of the plan based on a 2^2 CFE.

5.5.4
Second Order Orthogonal Plan

When we select the good value of the dimensionless α, then the corresponding sequential plan remains orthogonal like its CFE basic plan. At the same time, if we do not have any special request concerning a sequential plan, the number of experiments to determine fundamental factors can be drastically reduced to $n_0 = 1$. With $n_0 = 1$ and $k = 2$, we obtain the sequential plan shown in Table 5.23. However, with this general state this plan is not orthogonal because we have

$$\sum_{i=1}^{N} x_{0i} x_{ji}^2 \neq 0 \ , \quad \sum_{i=1}^{N} x_{ji}^2 x_{li} \neq 0 \tag{5.118}$$

Table 5.23 Sequential plan for a 2^2 CFE.

i	x_0	x_1	x_2	$x_1 x_2$	x_1^2	x_2^2	y
1	+1	+1	+1	+1	+1	+1	y_1
2	+1	+1	−1	−1	+1	+1	y_2
3	+1	−1	−1	+1	+1	+1	y_3
4	+1	−1	+1	−1	+1	+1	y_4
5	+1	$+\alpha$	0	0	α^2	α^2	y_5
6	+1	$-\alpha$	0	0	α^2	α^2	y_6
7	+1	0	$+\alpha$	0	0	0	y_7
8	+1	0	$-\alpha$	0	0	0	y_8
9	+1	0	0	0	0	0	y_9

In order to comply with the othogonality property, we have to transform the plan described in Table 5.23. For this purpose, we carry out the quadratic transformations of the data given in Table 5.23 by:

$$x_j' = x_j^2 - \frac{\sum_{i=1}^{N} x_{ji}^2}{N} = x_j^2 - \bar{x}_j^2 \tag{5.119}$$

With these transformations, we observe that:

$$\sum_{i=1}^{N} x_{0i} x_{ji}' = \sum_{i=1}^{N} x_{ji}^2 - N\bar{x}_j^2 = 0 \ , \quad \sum_{i=1}^{N} x_{ji}' x_{li}' \neq 0 \tag{5.120}$$

which is a fundamental approach to the orthogonal matrix of the planned experiments. Once the quadratic transformations have been carried out, we have to complete the orthogonal matrix. As far as we have a multiple equation system with α as unique unknown, we need to have a correlation matrix $(X^TX)^{-1}$ where all the non-diagonal elements are null. Table 5.24 has been obtained subsequent to the modified Halimov procedure [5.26]. This table gives the α values for the various factors and for a 2^{k-1} type basic CFE plan.

Table 5.24 Computed α values for a second order orthogonal plan.

	Number of independent factors				
	2	3	4	5	6
CFE basic plan	2^2	2^3	2^4	2^{5-1}	2^{6-1}
α	1	1.215	1.414	1.547	1.612

For $k = 2$, a second order orthogonal matrix plan is the state shown in Table 5.25. Due to the orthogonality of the matrix plan, the regression coefficients will be computed one after the other as follows:

$$\beta_j = \frac{\sum\limits_{i=1}^{N} x_{ji} y_i}{\sum\limits_{i=1}^{N} x_{ji}^2} \tag{5.121}$$

The relation (5.104) can be particularized to the general case of the second order orthogonal plan when we obtain the following relation for coefficients variances:

$$s_{\beta_j}^2 = s_{rp}^2 / \sum\limits_{i=1}^{N} x_{ji}^2 \tag{5.122}$$

So the regression coefficients have been calculated for an orthogonal composition matrix and as a consequence, for the quadratic effect, we obtain the next expressions:

$$\hat{y} = \beta_0' + \beta_1 x_1 + \beta_2 x_2 + \ldots + \beta_k x_k + \beta_{12} x_1 x_2 + \ldots +$$
$$\beta_{k-1k} x_{k-1} x_k + \beta_{11}(x_1^2 - \bar{x}_1^2) + \ldots + \beta_{kk}(x_k^2 - \bar{x}_k^2) \tag{5.123}$$

Therefore, the classic form of the regression relationship derives from calculating β_0 with relation (5.124):

$$\beta_0 = \beta_0' - \beta_{11}\bar{x}_1^2 - \beta_{22}\bar{x}_2^2 - \beta_{33}\bar{x}_3^2 - \ldots - \beta_{kk}\bar{x}_k^2 \tag{5.124}$$

The associated variance β_0 is thus taken into account from the addition law as follows:

$$s_{\beta_0}^2 = s_{\beta_0'}^2 = (\bar{x}_1^2) s_{\beta_{11}}^2 + (\bar{x}_2^2) s_{\beta_{22}}^2 + (\bar{x}_3^2) s_{\beta_{33}}^2 + \ldots + (\bar{x}_k^2) s_{\beta_{kk}}^2 \tag{5.125}$$

The use of the reproducibility variance allows the significance test of the coefficients of the final regression relationship:

$$\hat{y} = \beta_0 + \beta_1 x_1 + \beta_2 x_2 + .. + \beta_k x_k + \beta_{12} x_1 x_2 + ... + \beta_{k-1k} x_{k-1} x_k + \beta_{11} x_1^2 + ... \beta_{kk} x_k^2$$

$$(5.126)$$

Finally, we conclude this analysis with the estimation of the model confidence. For this purpose, we use a classical procedure which consists in calculating $F = s_{rz}^2/s_{rp}^2$, establishing $\upsilon_1 = N - n_\beta$, $\upsilon_2 = N - n_0$ and calculating the significance level (α) and $F_{\upsilon_1, \upsilon_2, \alpha}$ as well as comparing F and $F = s_{rz}^2/s_{rp}^2$ before making a decision.

It is important to emphasize that in the case of an orthogonal composition plan, as shown by relation (5.122), the various regression coefficients are not calculated with similar precisions.

Table 5.25 Orthogonal composition matrix from a CFE 2^2.

i	x_0	x_1	x_2	$x_1 x_2$	$x_1' = x_1^2 - \bar{x}_1^2$	$x_2' = x_2^2 - \bar{x}_2^2$	y
1	+1	+1	+1	+1	+1/3	+1/3	y_1
2	+1	+1	−1	−1	+1/3	+1/3	y_2
3	+1	−1	−1	+1	+1/3	+1/3	y_3
4	+1	−1	+1	−1	+1/3	+1/3	y_4
5	+1	+1	0	0	+1/3	−2/3	y_5
6	+1	−1	0	0	+1/3	−2/3	y_6
7	+1	0	+1	0	−2/3	+1/3	y_7
8	+1	0	−1	0	−2/3	+1/3	y_8
9	+1	0	0	0	−2/3	−2/3	y_9

5.5.4.1 Second Order Orthogonal Plan, Example of the Nitration of an Aromatic Hydrocarbon

The presentation of this example has two objectives: (i) to solve a problem where we use a second order orthogonal plan in a concrete case; (ii) to prove the power of statistical process modelling in the case of the non-continuous nitration of an aromatic hydrocarbon.

The initial step is the description of the process. Indeed, the nitration of the aromatic hydrocarbon occurs in a discontinuous reactor in a perfectly mixed state. The reaction takes place by contacting an aqueous phase containing nitric and sulfuric acids with an organic phase which initially contains the aromatic hydrocarbon. The aromatic hydrocarbon transformation degree (y) depends on the following factors of the process:

- the temperature of the reaction (t associated to z_1, $\prec t \succ = °C$);
- the time for reaction lasts (reaction time) (τ associated to z_2, $\prec \tau \succ = min$);
- the concentration of the sulfonitric mixture according to the total reaction mass (c_{sn} associated to z_3, $\prec c_{sn} \succ = \%$ g/g);
- the concentration of the nitric acid in the sulfonitric mixture (c_a associated to z_4, $\prec c_a \succ = \%$ g/g).

The fundamental level of the factors and their variation intervals have been established and are given in Table 5.26. We accept that the factors' domains cover the great curvature of the response surface. Consequently, a regression relationship with interaction effects is a priori acknowledged.

Table 5.26 Fundamental level and intervals of variation of the factors (example 5.5.4.1)

	z_1	z_2	z_3	z_4
z_j^0	50	40	60	40
Δz_j	25	20	20	15

To solve this problem we have to use a second order orthogonal plan based on a 2^4 CFE plan. According to Table 5.24, we can establish that, for α dimensionless values of factors, we can use the numerical value $\alpha = 1.414$. Table 5.27 contains all the data that are needed for the statistical calculation procedure of the coefficients, variances, confidence, etc., including the data of the dependent variables of the process (response data).

The transformation of the dimensional z_j into the dimensionless x_j has been made using relations (5.96) and (5.97). For the reproducibility variance, four complementary experiments are available in the centre of the plan. The degrees of transformation measured at the centre of the plan are: $y_1^0 = 61.8\%$, $y_2^0 = 59.3\%$, $y_3^0 = 58.7\%$, $y_4^0 = 69\%$.

Table 5.27 Composition matrix for a 2^4 CFE (Statistical data for the example 5.5.4.1).

n^0	x_0	x_1	x_2	x_3	x_4	x_1'	x_2'
1	+1	+1	+1	+1	+1	0.2	0.2
2	+1	−1	−1	+1	+1	0.2	0.2
3	+1	+1	−1	−1	+1	0.2	0.2
4	+1	−1	+1	−1	+1	0.2	0.2
5	+1	+1	−1	+1	−1	0.2	0.2
6	+1	−1	+1	+1	−1	0.2	0.2
7	+1	+1	+1	−1	−1	0.2	0.2
8	+1	−1	−1	−1	−1	0.2	0.2
9	+1	+1	−1	+1	+1	0.2	0.2
10	+1	−1	+1	+1	+1	0.2	0.2
11	+1	+1	+1	−1	+1	0.2	0.2
12	+1	−1	−1	−1	+1	0.2	0.2
13	+1	+1	+1	+1	−1	0.2	0.2
14	+1	−1	−	+1	−1	0.2	0.2
15	+1	+1	−1	−1	−1	0.2	0.2
16	+1	−1	+1	−1	−1	0.2	0.2
17	+1	0	0	0	0	−0.8	−0.8
18	+1	1.414	0	0	0	1.2	−0.8
19	+1	−1.4.14	0	0	0	1.2	−0.8
20	+1	0	1.414	0	0	−0.8	1.2
21	+1	0	−1.4.14	0	0	−0.8	1.2
22	+1	0	0	1.414	0	−0.8	−0.8
23	+1	0	0	−1.414	0	−0.8	−0.8
24	+1	0	0	0	1.414	−0.8	−0.8
25	+1	0	0	0	−1.414	−0.8	−0.8

Table 5.27 Continued.

I	X_3'	X_4'	X_1X_2	X_1X_3	X_1X_4	X_2X_3	X_2X_4	X_3X_4	y
1	0.2	0.2	+1	+1	+1	+1	+1	+1	86.9
2	0.2	0.2	+1	−1	−1	−1	−1	+1	40.0
3	0.2	0.2	−1	−1	+1	+1	−1	−1	66.0
4	0.2	0.2	−1	+1	−1	−1	+1	−1	34.4
5	0.2	0.2	−1	+1	−1	−1	+1	−1	76.6
6	0.2	0.2	−1	−1	+1	+1	−1	−1	55.7
7	0.2	0.2	+1	−1	−1	−1	−1	+1	91
8	0.2	0.2	+1	+1	+1	+1	+1	+1	43.6
9	0.2	0.2	−1	+1	+1	−1	−1	+1	74.1
10	0.2	0.2	−1	−1	−1	+1	+1	+1	52.0
11	0.2	0.2	+1	−1	−1	−1	+1	−1	74.5
12	0.2	0.2	+1	+1	+1	+1	−1	−1	29.6
13	0.2	0.2	+1	+1	−1	+1	−1	−1	94.8
14	0.2	0.2	+1	−1	+1	−1	+1	−1	49.6
15	0.2	0.2	−1	−1	−1	+1	+1	+1	68.6
16	0.2	0.2	−1	+1	+1	−1	−1	+1	51.8
17	−0.8	−0.8	0	0	0	0	0	0	61.8
18	−0.8	−0.8	0	0	0	0	0	0	95.4
19	−0.8	−0.8	0	0	0	0	0	0	41.7
20	−0.8	−0.8	0	0	0	0	0	0	79.0
21	−0.8	−0.8	0	0	0	0	0	0	42.4
22	1.2	−0.8	0	0	0	0	0	0	77.6
23	1.2	−0.8	0	0	0	0	0	0	58.0
24	−0.8	1.2	0	0	0	0	0	0	45.6
25	−0.8	1.2	0	0	0	0	0	0	52.3

The various steps of the statistical calculation are:
1. We begin with calculating the reproducibility:

$$\bar{y}^0 = \left(\sum_{i=1}^{4} y_i^0 \middle/ 4\right) = 60.95 \text{ and } s_{rp}^2 = \left(\sum_{i=1}^{4} (y_i^0 - \bar{y}^0)^2\right) \middle/ 3 = 5.95.$$

2. We use Eq. (5.121) to calculate the regression coefficients and Eq. (5.122) to determine the variances of the coefficients. The calculation is explained below and the results given in Table 5.28.

Table 5.28 Values of the coefficients and their variances (example 5.5.4.1).

	β_0'	β_1	β_2	β_3	β_4	β_{11}	β_{22}	
β_j	61.54	17.37	6.42	4.7	−4.37	4.5	1.3	
$s^2_{\beta j}$	0.245	0.245	0.245	0.245	0.245	0.746	0.746	
$s_{\beta j}$	0.545	0.545	0.545	0.545	0.545	0.864	0.864	

	β_{33}	β_{44}	β_{12}	β_{13}	β_{14}	β_{23}	β_{24}	β_{34}
β_j	4.09	−5.34	2.18	0.2	1.2	0.56	0.76	1.9
$s^2_{\beta j}$	0.746	0.746	0.372	0.373	0.372	0.372	0.372	0.372
$s_{\beta j}$	0.864	0.864	0.61	0.61	0.61	0.61	0.61	0.61

$$\beta_1 = \left(\sum_{i=1}^{25} x_{1i} y_i\right) \middle/ \left(\sum_{i=1}^{25} x_{1i}^2\right)$$

$$= \frac{86.9 - 40 + 66 - 34.4 + 76.6 - 55.7 + 91 - 47.6 + 74.1 - 52 + 74.5 - 29.6 + 94.8 - 49.6 + 68.6 - 51.8 + 1.414 * 95.4 - 1.141 * 41.7}{1 + 1 + 1 + 1 + 1 + 1 + 1 + 1 + 1 + 1 + 1 + 1 + 1 + 1 + 1 + 1 + 2 + 2}$$

$$= 17.37$$

$$\beta_{11} = \left(\sum_{i=1}^{25} x_{11}' y_i\right) \middle/ \left(\sum_{i=1}^{25} (x_{1i}')^2\right)$$

$$= \frac{0.2(86.9 + 40 + 66 + 34.4 + 76.6 + 55.7 + 91.0 + 47.6 + 74.1 + 52 + 74.5 + 29.6 + 91.1 + 49.6 + 68.6 + 51.8}{16 * (0.2)^2 + 7 * (0.8)^2 + 2 * (1.2)^2}$$

$$+ \frac{0.8 * 61.8 + 1.2 * 95.4 + 1.2 * 41.7 - 0.8 * 79 - 0.8 * 42.4 - 0.8 * 77.6 - 0.8 * 58.0 - 0.8 * 45.6 - 0.8 * 52.3}{16 * (0.2)^2 + 7 * (0.8)^2 + 2 * (1.2)^2}$$

$$= 4.5$$

$$\beta_{12} = \left(\sum_{1}^{25} (x_1 x_2)_i y_i \right) / \left(\sum_{1}^{25} (x_1 x_2)_i^2 \right)$$

$$= \frac{86.9 + 40 - 66 - 34.4 - 76.6 - 55.7 + 91 + 47.6 - 74.1 - 52 + 74.5 + 29.6 + 94.8 + 49.6 - 68.6 - 51.8}{16}$$

$$= 2.18$$

$$s_{\beta 1}^2 = s_{rp}^2 / \left(\sum_{i=1}^{25} x_{1i}^2 \right) = 5.95/20 = 0.245 \quad ,$$

$$s_{\beta 11}^2 = s_{rp}^2 / \left(\sum_{i=1}^{25} x_{1i}' \right)^2 = 5.95/(16 * 0.2^2 + 7 * 0.8^2 + 2 * 1.2^2) = 0.746 \ ,$$

$$s_{\beta 12}^2 = s_{rp}^2 / \left(\sum_{i=1}^{25} x_1 x_2 \right)_i^2 = 5.95 : /16 = 0.372 \quad \ldots\ldots$$

3. We verify the significance of each coefficient of the model with the Student test. In Table 5.29 the results of the tests are given. We can then observe that coefficients β_{22}, β_{14}, β_{23} and β_{34} are non-significant.

Table 5.29 Results of the Student test for the significance of the coefficients (example 5.5.4.1).

	t_0	t_1	t_2	t_3	t_4	t_{11}	t_{22}	t_{33}
$t_j = \left| \beta_j \right| / s_{\beta j}$		31.9	11.7	8.64	8.04	5.2	1.5	4.73
$t_{3,0.05}$	3.09	3.09	3.09	3.09	3.09	3.09	3.06	3.09
Conclusion	S	S	S	S	S	S	NS	S

	t_{44}	t_{12}	t_{13}	t_{14}	t_{23}	t_{24}	t_{34}
$t_j = \left| \beta_j \right| / s_{\beta j}$	6.22	3.57	3.18	1.97	0.91	1.25	3.8
$t_{3,0.05}$	3.09	3.09	3.09	3.09	3.09	3.09	3.09
Conclusion	S	S	S	NS	NS	NS	S

4. After elimination of the non-significant coefficients, we write the new model expression and then, we compute the residual variance:

$$\hat{y} = 61.54 + 17.37x_1 + 6.4x_2 + 4.7x_3 - 4.37x_4 + 2.18x_1x_2 + 1.9x_3x_4$$

$$= 4.5(x_1^2 - 0.8) + 4.9(x_2^2 - 0.8) - 5.34(x_4^2 - 0.8)$$

$$\hat{y} = 58.9 + 17.37x_1 + 6.4x_2 + 4.7x_3 - 4.37x_4 + 2.18x_1x_2 + 1.9x_3x_4 + 4.5x_1^2 + 4.09x_3^2$$

$$- 5.31x_4^2$$

$$s_{rz}^2 = \left(\sum_{i=1}^{25} (y_i - \hat{y}_i)^2 \right) / (N - n_\beta) = \left(\sum_{i=1}^{25} (y_i - \hat{y}_i)^2 \right) / (25 - 10) = 396.77/15$$

$$= 26.4$$

5. We check whether the obtained model is adequate or not:
   - $F = s_{rz}^2/s_{rp}^2 = 26.4/5.95 = 4.4$
   - $F_{v1,v2,\alpha} = F_{15,3,0.05} = 8.6$
   - Since we have $F_{v1,v2,\alpha} > F$, we admit that the established equation for the degree of transformation of the aromatic hydrocarbon is satisfactory.
6. Finally, we come back to the dimensional state of the factors. The result that can be used for the process optimization is:

$$\hat{y} = 64.87 - 21.68z_1 - 4.04z_2 - 34.31z_3 + 20.53z_4 + 0.00436z_1z_2 + 0.00633z_3z_4$$

$$+ 0.25z_1^2 + 0.2045z_3^2 - 0.354z_4^2$$

5.5.5
Second Order Complete Plan

Even though the second order orthogonal plan is not a rotatable plan (for instance see Eqs. (5.114) and (5.115)), the errors of the experimental responses (from the response surface) are smaller than those coming from the points computed by regression. It is possible to carry out a second order rotatable plan using the Box and Hunter [5.23, 5.27] observation which stipulates that the conditions to transform a sequential plan into a rotatable plan are concentrated in the dimensionless α value where $\alpha = 2^{k/4}$ for $k<5$ and $\alpha = 2^{(k-1)/4}$ for $k>5$ respectively. Simultaneously, the number of experiments at the centre of the experimental plan (n_0) must be increased in order to make it possible to stop the degeneration of the correlation matrix $(X^T X)^{-1}$. Table 5.30 contains the required values of dimensionless α and of n_0 for a second order rotatable plan.

Table 5.30 Values of dimensionless α and n_0 for a rotatable plan with k factors.

	Number of process factors							
	2	3	4	5	6	6	6	7
CFE basic plan	2^2	2^3	2^{4-1}	2^{5-1}	2^{5-1}	2^{6-1}	2^{6-1}	2^{7-1}
α	1.414	1.682	2.00	2.378	2.00	2.828	2.378	3.33
n_0	5	6	7	10	6	15	9	21

For $k = 2$ the values of a rotatable planning matrix of the second order are given in Table 5.31. This table derives from Table 5.23. It is important to observe that the complete second order-planning matrix is not orthogonal because we have $\sum_{i=1}^{N} x_{0i}x_{ji} \neq 0$ and $\sum_{i=1}^{N} x_{ji}x_{li} \neq 0$ (see relation (5.98) for the orthogonality property).

Table 5.31 Second order complete matrix from a 2^2 CFE.

i	x_0	x_1	x_2	$x_1 x_2$	x_1^2	x_2^2	y
1	+1	+1	+1	+1	+1	+1	y_1
2	+1	+1	−1	−1	+1	+1	y_2
3	+1	−1	−1	+1	+1	+1	y_3
4	+1	−1	+	−1	+1	+1	y_4
5	+1	+1.414	0	0	+2	0	y_5
6	+1	−1.414	0	0	+2	0	y_6
7	+1	0	+1.414	0	0	+2	y_7
8	+1	0	−1.414	0	0	+2	y_8
9	+1	0	0	0	0	0	y_9
10	+1	0	0	0	0	0	y_{10}
11	+1	0	0	0	0	0	y_{11}
12	+1	0	0	0	0	0	y_{12}

As a consequence, the β_{jj} coefficients will be linked with other coefficients and with the β_0 constant term. Moreover, to solve the problem of coefficients we must resolve the normal equation system by computing the inverse $(X^TX)^{-1}$ of the characteristic matrix (X^TX). As already noted in Section 5.4, the matrix of the coefficients and their associated variances are computed as follows:

$$B = (X^TX)^{-1}XY, \quad s_{\beta_i}^2 = d_{jj}s_{rp}^2 \tag{5.127}$$

For this case of second order complete plan, the specificity of the matrix of the coefficients results in an assembly of relations directly giving the regression values of the coefficients. In this example, where the complete second order plan is based on a 2^k CFE, these relations are written as follows:

$$\beta_0 = \frac{A}{N}\left[2\lambda_k^2(k+2)\sum_{i=1}^{N}x_{0i}y_i - 2\lambda_k C\sum_{j=1}^{K}\sum_{i=1}^{N}x_{ji}^2 y_i\right] \tag{5.128}$$

$$\beta_j = \frac{C}{N}\sum_{i=1}^{N}x_{ji}y_i \tag{5.129}$$

$$\beta_{jj} = \frac{A}{N}\left[C^2[(k+2)\lambda_k - k]\sum_{i=1}^{N}x_{ji}^2 y_i + C^2(1-\lambda_k)\sum_{j=1}^{k}\sum_{i=1}^{N}x_{ji}^2 y_i - 2\lambda_k C\sum_{i=1}^{N}x_{0i}y_i\right] \tag{5.130}$$

$$\beta_{lj} = \frac{C^2}{N\lambda_k}\sum_{i=1}^{N}x_{ji}x_{li}y_i \tag{5.131}$$

$$s_{\beta_0}^2 = \frac{2A\lambda_k^2(k+2)}{N}s_{rp}^2 \tag{5.132}$$

$$s_{\beta_j}^2 = \frac{A[(k+1)\lambda_k - (k-1)C^2]}{N}s_{rp}^2 \tag{5.133}$$

$$s_{\beta_i}^2 = \frac{C^2}{\lambda_k N}s_{rp}^2 \tag{5.134}$$

$$C = C_j = \frac{N}{\sum_{i=1}^{N}x_{ji}^2} \tag{5.135}$$

$$A = A_k = \frac{1}{2\lambda_k[(k+2)\lambda_k - k]} \tag{5.136}$$

$$\lambda_k = \frac{Nk\sum_{i=1}^{s}n_i\gamma_i^4}{(k+2)\left(\sum_{i=1}^{s}n_i\gamma_i^2\right)^2} \tag{5.137}$$

It should be mentioned that, in the calculation of parameter λ_k, s represents the number of spheres circumscribed to the experimental centre plan, γ_i is recognized as the radius of each i circumscribed sphere (see relation (5.115)) and n_i is the number of experimental points for the i sphere. It is evident that $\sum\limits_{i=1}^{s} n_i = N$, where N gives the total number of experiments in the plan. When we use a complete second order plan, it is not necessary to have parallel trials to calculate the reproducibility variance, because it is estimated through the experiments carried out at the centre of the experimental plan. The model adequacy also has to be examined with the next procedure:

1. We begin with calculating the sum of residual squares

$$S_{rp}^2 = \sum_{i=1}^{n_0} (y_i^0 - \overline{y}^0)^2 \text{ with the following degrees of freedom:}$$

$$v_1 = N - n_\beta = N - \frac{(k+1)(k+2)}{2};$$

2. We then compute the sum of the reproducibility squares

with the experimental centre plan: $S_{rp}^2 = \sum\limits_{i=1}^{n_0} (y_i^0 - \overline{y}^0)^2$,

where the degrees of freedom are: $v_2 = n_0 - 1$.

3. We define $S_{na}^2 = S_{rz}^2 - S_{rp}^2$ with $v_{na} = v_1 - v_2$ degrees of freedom as the sum of non-adequacy squares;

4. Finally, for a selected significance level, the computed Fischer variable value $F = (S_{na}^2/v_{na})/(S_{rp}^2/v_2)$ determines whether the model is adequate or not by comparison with the theoretical Fischer variable value $F_{v_{na},v_2,\alpha}$; when $F < F_{v_{na},v_2,\alpha}$ we agree to have an adequate model.

According to the testing of the significance of the model coefficients, we use the Student test where variances $s_{\beta_i}^2$ (relation (5.127)) are in fact S_{rp}^2/v_2. Due to the fact that the coefficients are linked, if one or more coefficients are eliminated, then a new determination can be carried out.

5.5.6
Use of Simplex Regular Plan for Experimental Research

The simplex regular plan can be introduced here with the following example: a scientist wants to experimentally obtain the displacement of a y variable towards an optimal value for a $y = f(x_1,x_2)$ dependence. When the analytical expression $y = f(x_1,x_2)$ is known, the problem becomes insignificant, and then experiments are not necessary. Figure 5.11A shows that this displacement follows the way of the greatest slope. In the actual case, when the function $f(x_1,x_2)$ is unknown, before starting the research, three questions require an answer: (i) How do we select the starting point? (ii) Which experimental and calculation procedure do we use to select the direction and position of a new point of the displacement? (iii) When do we stop the displacement?

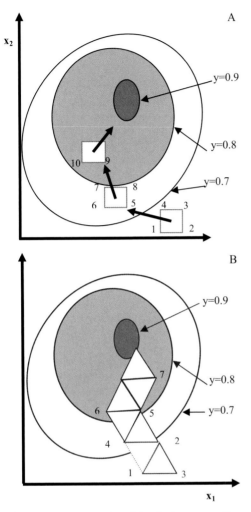

Figure 5.11 Representation of the displacement to the great curvature domain. A, according to the greatest slope method; B, according to the regular simplex method.

Question (ii) is certainly the most crucial. A possible answer to this question will be developed in the next section. The research has to begin with a small or local plan of experiments in order to describe the first movements from the starting point: when the first point of these previously planned experiments has been completed, the most non-favourable experiment will be rejected and it will be replaced by another experiment; thus we obtain, the displacement of the local group of experiments.

For a process with k factors, an abstract presentation of this procedure can be given as follows:

- We define a regular simplex plan as an assembly of k+1 equidistant points; for k = 1, the simplex is a segment; for k = 2, it is a triangle; for k = 3, we are faced with a regular tetrahedron, etc.
- Each simplex has a geometric centre placed at one point.
- When we replace the point rejected out of the group, in order to maintain a number of k+1 points, the next point will be the mirror image of the rejected point relative to the opposite face of the simplex.
- After the replacement of the rejected point, the simplex is rebuilt with a new geometric centre; only the experiments corresponding to the new point can be carried out to start the procedure (displacement and elimination) over again.

This procedure guarantees that, on the one hand, displacement towards the optimum point through the elimination of the less favourable points and, on the other hand, displacement through the maximum curvature of the response surface. For the example of a process with two factors, Fig. 5.9 B shows schematically the described procedure. The starting point of the regular simplex is triangle 123; point 3 is the less favourable response "y" and, consequently, it must be rejected; point 4 is the mirror image of point 3 according to the opposite face 12 of the simplex; thus, triangle 421 is the new simplex regular; here point 1 results in the less favourable y and as described above, we then choose point 5 which is the reflected image of point 4 (with respect to the opposite face of simplex 421).

If the dimensionless factors of an investigated process are distributed in a planning matrix (5.138) where x_j values are obtained using relation (5.139), then we can prove that the points of the matrix are organized as a regular simplex. Relation (5.140) corresponds to the distance from a point to its opposite face.

$$X = \begin{bmatrix} x_1 & x_2 & . & x_j & . & x_{k-1} & x_k \\ -x_1 & x_2 & . & x_j & . & x_{k-1} & x_k \\ 0 & -2x_2 & . & . & . & x_{k-1} & x_k \\ 0 & 0 & . & x_j & . & x_{k-1} & x_k \\ . & . & . & -jx_j & . & x_{k-1} & x_k \\ . & . & . & . & . & -(k-1)x_{k-1} & x_k \\ 0 & 0 & 0 & 0 & 0 & () & -kx_k \end{bmatrix} \quad (5.138)$$

$$x_j = \sqrt{\frac{1}{2j(j+1)}} \quad (5.139)$$

$$h_j = \frac{j+1}{\sqrt{2j(j+1)}} \quad (5.140)$$

For k factors, the number of experiments required by the simplex regular matrix is N = k+1. So, the class of saturated plans contains the simplex regular plan where the number of experiments and the number of the unknowns' coefficients are the same. For the process characterization in this example, we can only use the relationships of the linear regression. Concerning the simplex regular matrix

(5.138), we observe that it is an orthogonal matrix because we have $\sum\limits_{i=1}^{N} x_{ji}x_{li} = 0$, $\forall\, j \neq l$, $j, l = 1, 2, ..k$; $\sum\limits_{i=1}^{N} x_{ji} = 0$. However, in this case, we observe that the conditions $\sum\limits_{i=1}^{N} x_{ji}^2 = N$ are missing. Moreover, we can notice that:

$$\sum_{i=1}^{N} x_{ji}^2 = j\frac{1}{2j(j+1)} + j^2\frac{1}{2j(j+1)} = 0.5 \tag{5.141}$$

consequently, the correlation matrix of the regression coefficients can be written as follows:

$$(X^TX)^{-1} = \begin{bmatrix} 1/N & 0 & . & . & 0 \\ 0 & 2 & . & . & . \\ . & . & 2 & . & . \\ . & . & . & 2 & 0 \\ 0 & 0 & . & 0 & 2 \end{bmatrix} \tag{5.142}$$

then, the correlation matrix of the coefficients of the regression becomes:

$$\beta_0 = \left(\sum_{j=1}^{N} y_i\right)/N \ , \ \ \beta_j = 2\sum_{i=1}^{N} x_{ji}y_i \tag{5.143}$$

In the previous sections we have shown that the variances relative to the β_j coefficients for the orthogonal plans are: $s_{\beta_j}^2 = s_{rp}^2/\left(\sum\limits_{i=1}^{N} x_{ji}^2\right)$, and that, for a simple regular plan, these variances become $s_{\beta_j}^2 = s_{rp}^2/0.5 = 2s_{rp}^2$. This fact shows that the precision of a CFE plan is higher than the equivalent regular plan.

For practical use, the simplex regular plan must be drafted and computed before starting the experiment. For k process factors, this matrix plan contains k columns and k+1 lines; in the case of k = 6 the matrix (5.151) gives the following levels of the factors:

$$X = \begin{bmatrix} 0.5 & 0.289 & 0.204 & 0.158 & 0.129 & 0.109 \\ -0.5 & 0.289 & 0.204 & 0.158 & 0.129 & 0.109 \\ 0 & -0.578 & 0.204 & 0.158 & 0.129 & 0.109 \\ 0 & 0 & -0.612 & 0.158 & 0.129 & 0.109 \\ 0 & 0 & 0 & -0.632 & 0.129 & 0.109 \\ 0 & 0 & 0 & 0 & -0.645 & 0.109 \\ 0 & 0 & 0 & 0 & 0 & -0.654 \end{bmatrix} \tag{5.144}$$

The next observations will complete the understanding of this method when it is applied to the experimental scientific investigation of a real process:

1. When the experiments required by the initial simplex regular plan are completed then we eliminate the point that produces the most illogical or fool response values; by building the image of this point according to the opposite face of the simplex, we obtain the position of the new experimental point.

2. The position (coordinates) of the new experimental point can be determined as follows: (a) the jth coordinates of the new point $x_j^{(k+2)}$ are computed by relation (5.145), where $x_j^{(e)}$ is the jth coordinate of the rejected point and $x_j^{(c)}$ is the jth corresponding coordinate of the opposite face of the rejected point; (b) the jth coordinates of the centre of the opposite face of the excluded point are given by relation (5.146):

$$x_j^{(k+2)} = 2x_j^{(c)} - x_j^{(e)}$$ (5.145)

$$x_j^{(c)} = \left(\sum_{i=1,i\neq e}^{k+1} x_{ji} \right) / k$$ (5.146)

3. After each experiment a regression relationship can be obtained and analyzed using relation (5.143).
4. We can stop the experiments when the displacements of the factors do not result in a significant change in the process output.

To conclude this section, it is important to mention that the method of simplex regular plan is an open method. So, during its evolution, we can produce and add additional factors. This process can thus result in a transformation from a simplex regular plan with k columns and k+1 lines to a superior level with k+1 columns and k+2 lines. The concrete case described in the next section shows how we use this method and how we introduce a new factor into a previously established plan.

5.5.6.1 SRP Investigation of a Liquid–Solid Extraction in Batch

This example concerns a discontinuous (batch) liquid–solid extraction process. Here, the quantity of extracted species (y, $\prec y \succ = $ kgA/kg liq, A = type of species) depends on the following factors: the ratio of mixing phases (m_l/m_s-associated to z_1; $\prec m_l/m_s \succ = $ kg liq /kg solid), the contact time (τ associated to z_2; $\prec \tau \succ = $ min); the mixing rate ($w_a = \pi n d_a$-associated to z_3, $\prec w_a \succ = $ m/s, n-rotation speed, d_a – mixer diameter); the mean concentration of one species carrier, which is placed in the liquid phase (c_{sA}-associated to z_4, $\prec c_{sA} \succ = $ kg carrier/kg liq); the diameter of the solid particles (d-associated to z_5, $\prec d \succ = $ m). The temperature can be another important factor in the process, but initially we can consider that it is constant. Nevertheless, it will be considered as an additional factor in a second step of this analysis. The experiments are carried out with a solid containing 0.08 kg A/kg solid.

The fundamental levels of the factors and the variation intervals are shown in Table 5.32.

Table 5.32 Fundamental levels and variation intervals for the factors of the process.

	z_1	z_2	z_3	z_4	z_5
z_j^0	3	50	1.2	0.01	1.5×10^{-3}
Δz_j	1	20	0.6	0.004	0.5×10^{-3}

The objective of the problem is to obtain the values of the factors that correspond to a maximum concentration of the species (A) in the liquid phase.

To solve this problem we use the simplex regular method. For $k = 5$, the dimensionless matrix of experiments is obtained with relation (5.138). Thus, the matrix of the dimensionless factors is transformed into dimensional values with relations (5.96) and (5.97). Table 5.33 corresponds to this matrix, the last column of which contains the values of the process response. According to this table, the point placed in position 4 was found to be the least favourable for the process. However, before rejecting it, we have to build the coordinates of the new point by means of the image reflection of point number 4 (this point will be calculated to be number 7 from k+1+1). For this purpose, we use relations (5.145) and (5.146).

Table 5.33 Simplex regular plan with natural values of the factors (example 5.5.6.1).

i	z_1	z_2	z_3	z_4	$z_5 * 10^5$	y kgA/kg lq
1	3.5	55.7	1.32	0.0106	1.55	0.029
2	2.5	55.7	1.32	0.0106	1.55	0.042
3	3	39.4	1.32	0.0106	1.55	0.026
4	3	50	0.83	0.0106	1.55	0.023
5	3	50	1.2	0.0075	1.55	0.028
6	3	50	1.2	0.01	1.177	0.031

Now we show the results of these calculations, which began with the computation of the coordinate of the opposite face of the remaining simplex:

$$x_1^{(c)} = \left(\sum_{i=1, i \neq 4}^{6} x_{1i} \right) / 5 = (0.5 - 0.5 + 0 + 0 + 0)/5 = 0$$

$$x_2^{(c)} = \left(\sum_{i=1,i \neq 4}^{6} x_{2i} \right) / 5 = (0.289 + 0.289 - 0.528 + 0 + 0)/5 = 0$$

$$x_3^{(c)} = \left(\sum_{i=1,i \neq 4}^{6} x_{3i} \right) / 5 = (0.204 + 0.204 + 0.204 + 0 + 0)/5 = 0.612/5 = 0.122$$

$$x_4^{(c)} = \left(\sum_{i=1,i \neq 4}^{6} x_{4i} \right) / 5 = (0.158 + 0.158 + 0.158 - 0.632 + 0)/5 = -0.158/5$$

$$= -0.0317$$

$$x_5^{(c)} = \left(\sum_{i=1,i \neq 4}^{6} x_{5i} \right) / 5 = (0.129 + 0.129 + 0.129 + 0.129 - 0.645)/5 = 0.129/5$$

$$= -0.026$$

Now, the current dimensionless coordinate of the new point is obtained (see relation (5.145)) as follows:

$$x_1^{(k+2)} = x_1^{(7)} = 2x_1^{(c)} - x_1^{(e)} = 2x_1^{(c)} - x_1^{(4)} = 2*0 - 0 = 0$$

$$x_2^{(k+2)} = x_2^{(7)} = 2x_2^{(c)} - x_2^{(e)} = 2x_2^{(c)} - x_2^{(4)} = 2*0 - 0 = 0$$

$$x_3^{(k+2)} = x_3^{(7)} = 2x_3^{(c)} - x_3^{(e)} = 2x_3^{(c)} - x_3^{(4)} = 2*0.122 - (-0.612) = 0.8.56$$

$$x_4^{(k+2)} = x_4^{(7)} = 2x_4^{(c)} - x_4^{(e)} = 2x_4^{(c)} - x_4^{(4)} = -2*0.037 - 0.129 = -0.203$$

$$x_5^{(k+2)} = x_5^{(7)} = 2x_5^{(c)} - x_5^{(e)} = 2x_5^{(c)} - x_5^{(4)} = -2*0.026 - 0.109 = -0.164$$

Moreover, this new point is added to the remaining points and a new simplex (123567) will then be obtained. It is given in Table 5.34 where the factors are given in natural values. Relations (5.96) and (5.97) have been used to transform $x_1^{(k+2)} .... x_5^{(k+2)}$ into natural values.

Table 5.34 Simplex regular plan with values of natural factors (second step of example 5.5.6.1).

n^0	z_1	z_2	z_3	z_4	$z_5 * 10^5$	y kgA/kg lq
1	3.5	55.7	1.32	0.0106	1.55	0.029
2	2.5	55.7	1.32	0.0106	1.55	0.042
3	3	39.4	1.32	0.0106	1.55	0.026
5	3	50	1.2	0.0075	1.55	0.028
6	3	50	1.2	0.01	1.177	0.031
7	3	50	1.54	0.009	1.41	0.0325

In this table, we can observe that the 7th experiment has been produced and its corresponding y value has been given. We can notice that, in simplex 123567, point number 3 is the less favourable point for the process (in this case it is the point with the lowest yield). It should therefore be eliminated. Now we can proceed with the introduction of the temperature as a new process factor. In the previous experiments, the temperature was fixed at $z_6^0 = 45\ °C$. Initially, we consider that $z_6^0 = 45\ °C$ and we select the variation interval to be $\Delta z_6 = 15\ °C$. In this situation, if we apply Eq. (5.95), we have $x_6 = \dfrac{z_6 - 45}{15}$ and obviously $x_6^{(0)} = 0$. In order to develop the 6-dimensions simplex we use relation (5.140) and then we obtain $h_6 = (6+1)/\sqrt{2*6*(6+1)} = 0.764$. At this point, we can establish the values of the factors for the 8th experiment. For the first five factors the values are derived from the coordinates of the geometric centre of the simplex with 5 dimensions. These dimensionless values $x_1^{(8)}, x_2^{(8)}, ...x_5^{(8)}$ corroborate the procedure used for the calculation of the coordinates of a new point but, here, we consider that the coordinates of the rejected point are zero. The results of these computations are as follows:

$$x_1^{(c)} = \left(\sum_{i=1, i\neq 4}^{7} x_{1i} \right)/6 = (0.5 - 0.5 + 0 + 0 + 0 + 0)/6 = 0$$

$$x_2^{(c)} = \left(\sum_{i=1, i\neq 4}^{7} x_{2i} \right)/5 = (0.289 + 0.289 - 0.528 + 0 + 0 + 0)/6 = 0$$

$$x_3^{(c)} = \left(\sum_{i=1, i\neq 4}^{7} x_{3i} \right)/6 = (0.204 + 0.204 + 0.204 + 0 + 0 + 0.856)/6 = 1.468/6$$

$$= 0.245$$

$$x_4^{(c)} = \left(\sum_{i=1, i \neq 4}^{7} x_{4i} \right)/6 = (0.158 + 0.158 + 0.158 - 0.632 + 0 - 0.203)/6$$

$$= 0.158/5 = -0.060$$

$$x_5^{(c)} = \left(\sum_{i=1, i \neq 4}^{7} x_{5i} \right)/6 = (0.129 + 0.129 + 0.129 + 0.129 - 0.655 - 0.164)/6$$

$$= 0.109/5 = -0.048$$

$$x_1^{(k+2)} = x_1^{(8)} = 2x_1^{(c)} - x_1^{(e)} = 2x_1^{(c)} - x_1^{(4)} = 2*0 - 0 = 0$$

$$x_2^{(k+2)} = x_2^{(8)} = 2x_2^{(c)} - x_2^{(e)} = 2x_2^{(c)} - x_2^{(4)} = 2*0 - 0 = 0$$

$$x_3^{(k+2)} = x_3^{(8)} = 2x_3^{(c)} - x_3^{(e)} = 2x_3^{(c)} - x_3^{(4)} = 2*0.245 - 0 = 0.49$$

$$x_4^{(k+2)} = x_4^{(8)} = 2x_4^{(c)} - x_4^{(e)} = 2x_4^{(c)} - x_4^{(4)} = 2*(-0.06) - 0 = -0.12$$

$$x_5^{(k+2)} = x_5^{(8)} = 2x_5^{(c)} - x_5^{(e)} = 2x_5^{(c)} - x_5^{(4)} = -2*0.048 - 0.109 = -0.096$$

For $z_6^{(8)}$ we obtain $z_6^{(8)} + z_6 x_6^{(8)} = z_6^{(8)} + z_6(x_6^{(0)} + h_6) = 45 + 15(0 + 0.764) = 52.2\,°C$. The 8th experiment together with the 123567 points gives the simplex 1235678, which is written with the values of the dimensional factors given in Table 5.35.

Table 5.35 Simplex matrix plan after the introduction of a new factor (example 5.5.6.1).

i	z_1	z_2	z_3	z_4	$z_5 * 10^5$	z_6	y kgA/kg lq
1	3.5	55.7	1.32	0.0106	1.55	45	0.029
2	2.5	55.7	1.32	0.0106	1.55	45	0.042
3	3	39.4	1.32	0.0106	1.55	45	0.026
5	3	50	1.2	0.0075	1.55	45	0.028
6	3	50	1.2	0.01	1.177	45	0.031
7	3	50	1.54	0.009	1.41	45	0.0325
8	3	50	1.49	0.0095	1.45	52.2	*waited !*

After carrying out the concrete experiment required by the 8th simplex point, the process analysis continues according to the exemplified procedure, which will stop when y cannot be increased anymore.

5.5.7
On-line Process Analysis by the EVOP Method

On-line investigation methods for statistical analysis are used when the performances of a continuous process carried out in a pilot unit or in an apparatus, have to be improved. The Evolutionary Operation Process (EVOP) method [5.7, 5.27, 5.28, 5.31] is the most famous method for on-line process analysis. The name of this method comes from its analogy with biological evolution. This analogy is based on the observation of the natural selection process in which a small variation in independent life factors is responsible for genetic mutations and thus for the evolution of species.

The objective of the EVOP method is to obtain changes in the factors of the process so as to get a more favourable state of the process outputs by means of on-line process investigation. This research is made up of small changes and programmed step-by-step. Due to the small changes in the factors, it is possible to have situations in which the effects on the output process variables can be difficult to detect because they are covered by the random effects (see the Fig. 5.1). To compensate for this difficulty in the EVOP method, the process analysis is carried out from one stage (phase) to another under a condition that imposes more iterative cycles for each phase. For each cycle, a variable number of experiments with unchanged values for the factors is important for controlling the propagation of errors. At each phase, all the experiments produced correspond to an a priori selected CFE or FFE plan. After completing the experiments required by one stage, we process their statistical data and make the necessary decision concerning the position and starting conditions for the next phase of the process analysis.

The number of cycles for each stage must be thoroughly selected because then the interest is to observe the small changes occurring simultaneously with the permanent random fluctuations in the process output. The data from a cycle are transferred to the next cycle to complete the new phase by calculation of the mean values and variances. It is well known that the errors in the mean value of n independent observations are $\sqrt{n}$ smaller than the error of an isolated measure. Therefore, this fact sustains the transfer of data from one cycle to the next one.

Figure 5.12 gives a graphic introduction to the EVOP method for the example of a process with two factors; it is important to notice that, in a real case, the displacements of the factors have to be smaller than those suggested in Fig. 5.12. Despite its apparent freedom, the EVOP method imposes some strict rules; some of them are described below:

1. For each phase, the number of cycles is not imposed by a rule or by a mathematical relation.
2. At the beginning of the second cycle, the calculation of the total effects of the analysis is obligatory; the total effects in

the example given by Fig. 5.12 are calculated by the following relation:

$$ET^a = \frac{1}{5}\left(\bar{y}_2^a + \bar{y}_3^a + \bar{y}_4^a + \bar{y}_4^a - 4\bar{y}_1^a\right) \tag{5.147}$$

Here, each $\bar{y}_i^a$ value is the mean value for all the cycles carried out with respect to the actual phase of the analysis.

3. At the beginning of the second cycle of each phase, each mean value $\bar{y}_i^a$ will be completed by its confidence intervals.
4. The cycles are stopped when the intervals of confidence for all $\bar{y}_i^a$, remain unchanged.
5. The analysis of the chain of the ET^a values completed with the split up of the total effect can help in selecting the next phase.

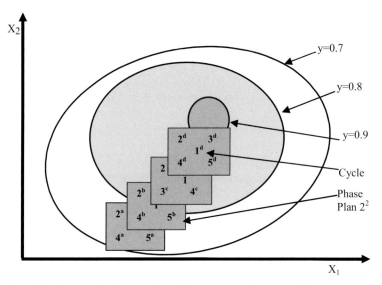

Figure 5.12 EVOP method particularized for a process with two variables.

This research method can be better illustrated by a concrete example. The investigated process example described in the next section is an organic synthesis, which takes place in a perfectly mixed reactor.

5.5.7.1 EVOP Analysis of an Organic Synthesis

We consider the case of a discontinuous organic synthesis, which occurs in a liquid medium undergoing intensive agitation; the temperature is controlled by an external heating device. The process efficiency is characterized by the conversion defined here as the ratio between the quantity of the useful species obtained and the theoretical quantity of the same species. This last value is fixed by the thermodynamics and the reaction conditions. When using the EVOP method, we mean to observe the effects of the temperature and of the reaction time on the conversion. We can consider that all the other factors of the process, such as the mixing intensity, the concentrations of the reactants and catalyst, etc. remain constant, which is required by the technological considerations of the process.

We assume that the standard temperature and reaction time are fixed to 85 °C and 180 min. but small changes (± 5 °C and 10 min) have been observed to affect the process efficiency. However, these variations do not affect the process drastically. Moreover, to begin the analysis we can observe a similitude between this concrete case and the example shown in Fig. 5.12. Indeed, the working plan is a CFE 2^2 which is noted as $1^a2^a3^a4^a5^a$ in Fig. 5.12. The superscript [a] indicates that we are in the first phase of the EVOP procedure. The dimensionless coordinates for each point of the CFE2^2 plan are: $1^a(0,0)$, $2^a(-1,1)$, $3^a(1,1)$, $4^a(1,-1)$, $5^a(-1,-1)$. We can identify the first coordinate of 1^a to 5^a point of the CFE 2^2 plan which is $x_1 = (t - 85)/5$ and the second point coordinate is $x_2 = (\tau - 180)/10$. Table 5.36 contains the results for the first four cycles of the first phase of the particular EVOP method.

Table 5.36 Reaction conversion for four cycles of the first phase of the EVOP method.

E C	1^a	2^a	3^a	4^a	5^a
1	59.6	65.1	65.3	62.0	62.1
2	62.1	61.3	67.6	65.5	65.8
3	63.5	61.7	62.6	67.9	62.8
4	63.7	60.5	67.2	63.2	62.8

If we consider the coordinates of the points of the CFE plan, we observe that points 3^a and 4^a are the maximum values of x_1, whereas points 3^a and 5^a have the maximum values for x_2. Consequently, the effects of the factors and of their interactions will be written as follows:

$$EA^a = \frac{1}{2}(\bar{y}_3 + \bar{y}_4 - \bar{y}_2 - \bar{y}_5) \tag{5.148}$$

$$EB^a = \frac{1}{2}(\bar{y}_3 + \bar{y}_5 - \bar{y}_2 - \bar{y}_4) \tag{5.149}$$

$$EAB^a = \frac{1}{2}(\bar{y}_2 + \bar{y}_3 - \bar{y}_4 - \bar{y}_5) \tag{5.150}$$

We frequently use the concepts of mean values and variances in the application of the EVOP method. Before showing the concrete computations of this actual application, we need to recall here the expression for the confidence interval of a mean value: $\mu = \bar{x} \pm t_\alpha s/\sqrt{n}$ where s is the variance, $\bar{x}$ is the mean value of the selection, n gives the selection dimension and t_α is the value of the Student random variable with a significance level equal to α and with $v = n - 1$ degrees of freedom. Table 5.37 shows the EVOP evolution from one cycle to another respect to the data given in Table 5.36. The computations from Table 5.37 show that:

- The succession of cycles produces an important reduction in the mean deviation values and, at the same time, the confidence intervals tend to reach a final stable state.
- The effect of each factor and of its interactions on the process response (the conversion in our case) begins to be observable after running a suitable number of cycles.
- At the end of the fourth cycle, an increase in the conversion caused by the increase in temperature occurs; this observation is sustained by the positive values of the confidence interval for the mean effect of the temperature: $\mu_A = EA^a(+/-)ts/(n_c)^{0.5} = (3.6, 7.9)$.
- The positive total effect recorded after the fourth cycle, cannot sustain a further increase in the reaction time because the confidence interval for the mean effect of this factor contains a negative and a positive value: $\mu_B = EB^a(+/-)ts/(n_c)^{0.5} = (-1.0, 3.0)$; moreover we can observe that the interaction of both studied factors (temperature and reaction time) has a negative effect: $\mu_{AB} = EAB^a(+/-)ts/(n_c)^{0.5} = (-4, 0)$.
- with the situation given by the data from Table 5.37, we have two possibilities for the evolution of the research: (i) we can start with a new phase where the temperature will be increased; (ii) or we can increase the number of cycles in the actual phase so as to obtain more confidence with respect to the positive effect of the temperature.

Table 5.37 Calculation sheet for the analysis of an EVOP process (example 5.5.7.1). o.d – old deviations, n.d – new deviations.

Calculation elements Cycle = 1, n_c = 1		Experiment conversion mean value					Mean deviations (m.d)
		1ᵃ	2ᵃ	3ᵃ	4ᵃ	5ᵃ	Sum of o.d: S_a = xx
1	Sum of old cycles S	–	–	–	–	–	Precedent m.d: s_a = xx
2	Mean value of the previous cycles M	–	–	–	–	–	Sum of n.d: S_n = xx

Table 5.37 Continued.

Calculation elements Cycle = 1, n_c = 1	Experiment conversion mean value					Mean deviations (m.d)
3 New results N	59.6	64.1	65.8	62.0	62.1	Mean value of n.d: $S_n = xx$
4 Differences (2) – (3) D	–	–	–	–	–	$S = \sqrt{\sum_{i=1}^{n} D_i^2}$, $s = S/\sqrt{n-1}$
5 New sum (1) + (3) SN	59.6	64.1	65.3	62.0	62.1	
6 New mean value $(SN)/n_c$						

Calculations of the mean effects	The confidence intervals
$EA^a = \dfrac{1}{2}(\bar{y}_3 + \bar{y}_4 - \bar{y}_2 - \bar{y}_5) = 0.55$	$\mu_A = EA^a(+/-)ts/(n_c)^{0.5} = (xx, xx)$
$EB^a = \dfrac{1}{2}(\bar{y}_3 + \bar{y}_5 - \bar{y}_2 - \bar{y}_4) = 0.65$	$\mu_B = EB^a(+/-)ts/(n_c)^{0.5} = (xx, xx)$
$EAB^a = \dfrac{1}{2}(\bar{y}_2 + \bar{y}_3 - \bar{y}_4 - \bar{y}_5) = 2.65$	$\mu_{AB} = EAB^a(+/-)ts/(n_c)^{0.5} = (xx, xx)$
$ET^a = \dfrac{1}{5}(\bar{y}_2^a + \bar{y}_3^a + \bar{y}_4^a + \bar{y}_4^a - 4\bar{y}_1^a) = 3.02$	$t = t_{5n_c-1,\alpha} =$

Calculation elements Cycle = 2 , n_c = 2	Experiment conversion mean value					Mean deviations (m.d)
	1^a	2^a	3^a	4^a	5^a	Sum of a.d: $S_a = xx$
1 Sum of the previous cycles S	59.6	64.1	65.8	62.0	62.1	Precedent m.d: $s_o = xx$

2   Mean value of
    previous cycles
    M
	59.6	64.1	65.8	62.0	62.1	Sum of n.d: $S_n = 6.72$

3 New results N	62.1	61.3	67.6	65.5	65.8	Mean value of n.d: $S_n = 3.36$
4 Differences (2) – (3) D	–2.5	2.8	–2.3	–3.5	–3.7	$S = \sqrt{\sum_{i=1}^{n} D_i^2}$, $s = S/\sqrt{n-1}$
5 New sum (1)+(3) SN	121.7	125.4	132.9	127.5	127.9	
6 New mean value $(SN)/n_c$	60.8	62.7	66.4	63.7	63.9	

Table 5.37 Continued.

Calculations of the mean effects	The confidence intervals
$EA^a = \dfrac{1}{2}(\bar{y}_3 + \bar{y}_4 - \bar{y}_2 - \bar{y}_5) = 1.75$	$\mu_A = EA^a(+/-)ts/(n_c)^{0.5} = (-2.17, 5.67)$
$EB^a = \dfrac{1}{2}(\bar{y}_3 + \bar{y}_5 - \bar{y}_2 - \bar{y}_4) = 1.95$	$\mu_B = EB^a(+/-)ts/(n_c)^{0.5} = (-1.97, 5.89)$
$EAB^a = \dfrac{1}{2}(\bar{y}_2 + \bar{y}_3 - \bar{y}_4 - \bar{y}_5) = 0.75$	$\mu_{AB} = EAB^a(+/-)ts/(n_c)^{0.5} = (-3.17, 4.67)$
$ET^a = \dfrac{1}{5}(\bar{y}_2^a + \bar{y}_3^a + \bar{y}_4^a + \bar{y}_4^a - 4\bar{y}_1^a) = 2.70$	$t = t_{5n_c - 1,\alpha} = t_{9,0.05} = 3.69$

Calculation elements	Experiment conversion mean value					Mean deviations (m.d)
Cycle = 3 , n_c = 3						
	1^a	2^a	3^a	4^a	5^a	Sum of o.d: S_o = 9.31
1 Sum of the previous cycles S	121.7	125.4	132.9	127.5	127.9	Precedent m.d: s_o = 3.36
2 Mean value of previous cycles M	60.8	62.7	66.4	63.7	63.9	Sum of n.d: S_n = 6.44
3 New results N	63.5	61.7	62.6	67.9	62.8	Mean value of n.d: S_n = 3.1
4 Differences (2) – (3) D	–2.7	1	3.8	–4.2	1.1	$S = \sqrt{\sum_{i=1}^{n} D_i^2}$, $s = S/\sqrt{n-1}$
5 New sum (1)+(3) SN	185.2	187.1	195.5	195.4	190.7	
6 New mean value $(SN)/n_c$	61.7	62.4	65.2	65.1	63.6	

Calculations of the mean effects	The confidence intervals
$EA^a = \dfrac{1}{2}(\bar{y}_3 + \bar{y}_4 - \bar{y}_2 - \bar{y}_5) = 2.15$	$\mu_A = EA^a(+/-)ts/(n_c)^{0.5} = (-0.69, 4.99)$
$EB^a = \dfrac{1}{2}(\bar{y}_3 + \bar{y}_5 - \bar{y}_2 - \bar{y}_4) = 0.65$	$\mu_B = EB^a(+/-)ts/(n_c)^{0.5} = (-2.19, 3.49)$
$EAB^a = \dfrac{1}{2}(\bar{y}_2 + \bar{y}_3 - \bar{y}_4 - \bar{y}_5) = -0.55$	$\mu_{AB} = EAB^a(+/-)ts/(n_c)^{0.5} = (-3.39, 2.29)$
$ET^a = \dfrac{1}{5}(\bar{y}_2^a + \bar{y}_3^a + \bar{y}_4^a + \bar{y}_4^a - 4\bar{y}_1^a) = 1.90$	$t = t_{5n_c - 1,\alpha} = t_{14,0.05} = 3.32$

Table 5.37 Continued.

Calculation elements Cycle = 4 , n_c = 4		1ᵃ	2ᵃ	3ᵃ	4ᵃ	5ᵃ	Mean deviations (m.d)
			Experiment conversion mean value				
1	Sum of the previous cycles S	185.2	187.1	195.5	195.4	190.7	Sum of o.d: S_o = 12.0 Precedent m.d: s_o = 3.1
2	Mean value of previous cycles M	61.7	62.4	65.2	65.1	63.6	Sum of n.d: S_n = 3.982
3	New results N	63.7	60.5	67.2	63.2	62.8	Mean value of n.d: s_n = 2.82
4	Differences (2) – (3) D	–2.0	1.9	–2.0	1.9	0.8	$S = \sqrt{\sum_{i=1}^{n} D_i^2}$, $s = S/\sqrt{n-1}$
5	New sum (1) + (3) SN	248.7	247.6	262.7	258.6	253.5	
6	New mean value (SN)/n_c	62.2	61.9	65.2	65.7	63.4	

Calculations of the mean effects	The confidence intervals
$EA^a = \frac{1}{2}(\bar{y}_3 + \bar{y}_4 - \bar{y}_2 - \bar{y}_5) = 5.6$	$\mu_A = EA^a(+/-)ts/(n_c)^{0.5} = (3.6, 7.9)$
$EB^a = \frac{1}{2}(\bar{y}_3 + \bar{y}_5 - \bar{y}_2 - \bar{y}_4) = 1.0$	$\mu_B = EB^a(+/-)ts/(n_c)^{0.5} = (-1.0, 3.0)$
$EAB^a = \frac{1}{2}(\bar{y}_2 + \bar{y}_3 - \bar{y}_4 - \bar{y}_5) = -2$	$\mu_{AB} = EAB^a(+/-)ts/(n_c)^{0.5} = (-4, 0)$
$ET^a = \frac{1}{5}(\bar{y}_2^a + \bar{y}_3^a + \bar{y}_4^a + \bar{y}_4^a - 4\bar{y}_1^a) = 7.4$	$t = t_{5n_c-1,\alpha} = t_{19,0.05} = 3.17$

5.5.7.2 Some Supplementary Observations

The example presented above successfully illustrates how we develop and use the EVOP method for a discontinuous process. When we have a continuous process, it is suggested to transform it artificially into a discontinuous process. For this purpose, we must take into consideration all the factors of the process representing flow rates according to a fixed period of time. With these transformations we can control the effect of the random factors that influence the continuous process. If, for example, we consider the case of a continuous reactor, then, the conversion can be obtained from the analysis of 5 to 6 samples (each selected at a fixed period of time), when the corresponding input and output quantities are related to the reactor. Additionally, the other reactor factors are not different from those of the discontinuous process. The case of the continuous reactor can easily be extended to all separation apparatuses or pilot units working continuously.

In all experimental process investigations, where the final decision is the result of the hypotheses based on a comparison of the variances, we must know whether the observed variances are related to the process or to the experimental analysis procedure. Indeed, it is quite important to determine, when an experimental research is being carried out, whether we have to use a method or an instrument of analysis that produces an artificially high variance on the measured parameters.

Before the era of modern computers, the EVOP process investigation was used successfully to improve the efficiency of many chemical engineering processes. Now its use is receding due to the competition from process mathematical modelling and simulation. However, biochemical and life processes are two large domains where the use of the EVOP investigation can still bring spectacular results.

5.6
Analysis of Variances and Interaction of Factors

The objective of the statistical analysis of variances is to separate the effects produced by the dependent variables in the factors of the process. At the same time, this separation is associated with a procedure of hypotheses testing what allows to reject the factors (or groups of factors) which do not significantly influence the process. The basic mathematical principle of the analysis of variances consists in obtaining statistical data according to an accepted criterion. This criterion is complemented with the use of specific procedures that show the particular influence or effects of the grouping criterion on dependent variables.

Besides, after identifying the effects, it is necessary to compare variances of the process produced by the variation of the factors and the variances of the process produced by the random factors [5.5, 5.8, 5.29–5.31].

The number of criteria that determines the grouping of the data is strictly dependent on the number of the factors of the process accepted for the investigation.

These abstract concepts will be illustrated in the next section with the example of a catalytic chemical reaction in which we consider that different type of catalysts are available to perform the reaction and where the conversion for a fixed contact time is the dependent variable of the process. If we consider that all the other factors of the process stay unchanged, then, we can take into account a single variable factor of the process: the type of catalyst. The basis of the mono factor variance analysis concerns the collected data containing the maximum number of conversion measurements respect to each type of catalyst. Now, if the temperature is also considered as an independent variable (factor), for each fixed temperature, the collected data must show the conversion values for each catalyst. Now, we can arrange the data in order to start the analysis of the variances of two factors. Obviously, this example can be generalized to the case of k factors (analysis of the variances with k-factors). If the residual variance increases from one experiment to the other, the effect of each factor is not summative, then we can claim that, in this case, we have an effect of the interaction factors.

For a process with more than two factors, we can consider the interactions of different factors theoretically. However, in real cases only two and a maximum of three factors interactions are accepted. All the examples selected in what follows consider the same major problem: how do we reject the non-significant factors out of the large range of factors of the process.

5.6.1
Analysis of the Variances for a Monofactor Process

The analysis of the variances of a monofactor process can be used for the indirect testing of both mean values obtained when the process factors take m discrete values. Table 5.38 introduces the preparation of the data for the analysis of the variances of a monofactor process. We can note that each value of the factor must produce m measurements of the process response.

The data arrangement shown in Table 5.38 can hint that the observable differences from one value to the other, from one column to the other are caused by the factor changes and by the problems of reproducibility.

Table 5.38 Experimental data arrangement for starting the analysis of the variances of a monofactor.

Factor value	$x = a_1$	$x = a_2$	$x = a_j$	......	$x = a_m$
Trial					
1	v_{11}	v_{21}	v_{j1}	......	v_{m1}
2	v_{12}	v_{22}	v_{j2}	......	v_{m2}
3	v_{13}	v_{23}	v_{j3}	......	v_{m3}
......	......	......	......	......	......
i	v_{1i}	v_{2i}	v_{ji}	......	v_{mi}
......	......	......	......	......	......
n	v_{1n}	v_{2n}	v_{jn}	......	v_{mn}
Total	v_1	v_2	v_j	......	v_m

In the table, the differences between the columns result from the change in the values of the factors and the differences between the lines give the reproducibility problems of the experiments. The total variance (s^2) associated to the table data, here given by relation (5.151), must be divided according to its components: the variances of inter-lines (or reproducibility variances) and variances of inter-columns (or variances caused by the factor).

$$s^2 = \frac{\left[\sum_{j=1}^{m}\sum_{i=1}^{n}(v_{ji} - v^=)^2\right]}{mn - 1} = \frac{\left[\sum_{j=1}^{m}\sum_{i=1}^{n}v_{ji}^2 - \frac{1}{mn}\left(\sum_{j=1}^{m}\sum_{i=1}^{n}v_{ij}\right)^2\right]}{mn - 1} \tag{5.151}$$

The result of this division is given in Table 5.39 where the starting data to complete the table have been obtained using the sums S_1, S_2 and S_3:
• the sum of all the squares of all observations (S_1):

$$S_1 = \sum_{j=1}^{m}\sum_{i=1}^{n}v_{ji}^2 \tag{5.152}$$

• the sum of the squares of the total of each column divided by the number of observations (S_2):

$$S_2 = \frac{\sum_{j=1}^{m}v_j^2}{n} \tag{5.153}$$

• the sum of the squares of the all added experimental observations divided by the total number of observations (S_3):

$$S_3 = \frac{\left(\sum_{j=1}^{m}\sum_{i=1}^{n}v_{ij}\right)^2}{mn} = \frac{\left(\sum_{j=1}^{m}v_j\right)^2}{mn} \tag{5.154}$$

Table 5.39 Analysis of the variances for a monofactor.

Variance origin	Sums of the differences	Degrees of freedom	Variances	Computed value of the Fischer variable	Theoretical value of the Fischer variable
Between the columns	$S_2 - S_3$	$m - 1$	$s_1^2 = \dfrac{S_2 - S_3}{m - 1}$	$F = \dfrac{s_1^2}{s_2^2}$	$F_{m-1, m(n-1), \alpha}$
Between the lines	$S_1 - S_2$	$m(n - 1)$	$s_2^2 = \dfrac{S_1 - S_2}{m(n - 1)}$		
Total	$S_1 - S_3$	$mn - 1$			

Table 5.39 also contains the indications and calculations required to verify the zero hypothesis. This hypothesis considers the equality of the variance containing the effect of the factor on the process response (s_1^2) with the variance that shows the experimental reproducibility (s_2^2).

According to the aspects of the statistical hypothesis about the equality of two variances (see also Section 5.3) we accept the zero hypothesis if the computation shows that $F < F_{m-1,m(n-1),\alpha}$. If we refuse the zero hypothesis, then we accept that the considered factor of the process has an important influence on the response.

The numerical application described above, concerns the catalytic oxidation of SO_2 where six different catalysts are tested. The main purpose is to select the most active catalysts out of the six given in this table. All the other parameters that characterize the reaction have been maintained constant during the experiments and eight measurements have been produced for each type of catalyst. Table 5.40 presents the SO_2 transformation degrees obtained. Before reaching a conclusion about these results, we have to verify whether the different transformation degrees obtained with the six catalysts are significant or not.

Table 5.40 SO_2 transformation degree for six different catalysts.
Integral reactor $l/d = 50$, $l = 1$ m, $c_{SO2} = 8\%v/v$, $c_{o2} = 10\%v/v$,
N_2 inert gas, $d_p = 0.003$ m, $w_f = 0.1$ m/s.

Catalyst m	1	2	3	4	5	6
Trial number n						
1	25.1	22.8	25.5	24.5	25.5	24.7
2	27.0	23.8	27.9	25.2	28.7	27.1
3	29.6	27.1	28.8	27.7	26.2	26.0
4	26.6	22.7	26.9	26.9	25.7	26.2
5	25.2	22.8	25.4	27.1	27.2	25.7
6	28.3	27.4	30.0	30.6	27.9	29.2
7	24.7	22.2	29.6	26.4	25.6	28.0
8	25.1	25.1	23.5	26.6	28.5	24.4
Total	211.6	193.9	217.6	215.0	215.3	211.2
Mean value	26.5	24.1	27.2	26.9	26.9	26.4

To begin the analysis, we consider the zero hypothesis (in which the degrees of transformation reached with the different catalysts are similar) and to verify it, we make the computations required in Table 5.39. Then we have: $S_1 = 33\,511.11$, $S_2 = 33\,368.53$, $S_3 = 33\,322.20$, $S_2 - S_3 = 46.33$, $S_1 - S_2 = 142.58$, $S_1 - S_3 = 188.91$,

$m - 1 = 5$, $m(n - 1) = 42$, $mn = 47$, $s_1^2 = 46.33/5 = 9.27$, $s_2^2 = 144.58/42 = 3.16$, $F = 9.27/3.16 = 2.93$.

The theoretical value of the Fischer random variable corresponding to the confidence level $\alpha = 0.05$ is 2.44 (it is a solution of the equation

$$1 - \alpha = \int_0^{F_{m-1,m(n-1),\alpha}} f_{m-1,m(n-1)}(F)dF).$$ Now we can observe that $F = 2.93 \succ F_{5,35,0.05}$

$= 2.44$ and consequently we can reject the zero hypothesis, which suggests the equality of the reproducibility variance and of the variance due to the change in catalyst. In other words, we can claim that each catalyst tested has a different influence on the SO_2 transformation degree.

5.6.2
Analysis of the Variances for Two Factors Processes

When we investigate the effect of two factors on a process response, then the collected data will be as shown in Table 5.41. Here the differences between the observed values along one line present the effect of the change of x_1 from α_1 to α_m, whereas the differences between the observed values along one column are the result of the change of x_2 from β_1 to β_n. Each value of the table represents an observation that corresponds to a grouping of factors. Here, we can have one or more measurements of the process response, but frequently only one measurement is used.

Table 5.41 Arrangement of the experimental data to start the analysis of two-factor variances.

Values for the first factor	$x_1 = \alpha_1$	$x_1 = \alpha_2$	$x_1 = \alpha_j$	......	$x_1 = \alpha_m$	total
Values for the second factor						
$x_2 = \beta_1$	v_{11}	v_{21}	v_{j1}	......	v_{m1}	v_{l1}
$x_2 = \beta_2$	v_{12}	v_{22}	v_{j2}	......	v_{m2}	v_{l2}
$x_2 = \beta_3$	v_{13}	v_{23}	v_{j3}	......	v_{m3}	v_{l3}
	......	....	......	......	......	
$x_2 = \beta_i$	v_{1i}	v_{2i}	v_{ji}	......	v_{mi}	v_{li}
	....	....	......	......	......	
$x_2 = \beta_n$	v_{1n}	v_{2n}	v_{jn}	......	v_{mn}	v_{lm}
Total	v_{c1}	v_{c2}	v_{cj}	......	v_{cm}	

In this case, conversely to the residual variance, we can propose two zero hypotheses: the first is H_{10}: "the variance of the response values determined by the change of factor x_1 has the same value as the residual variance"; the second one is H_{20}: "the variance of the response values (when x_2 factor changes) is similar to the residual variance". With these hypotheses we indirectly start the validation of two others assumptions: (i) the equality of the mean values of the lines (related to H_{10}), (ii) the equality of the mean values of the columns (related to $H_{20}°$).

The splitting of the total variance into parts associated to Table 5.41 follows a procedure similar to that for the analysis of the variances of a monofactor process, as previously explained. In this case, we introduce the sums of the squares S_1, S_2, S_3, S_4, S_r that are defined using Eqs. (5.155)–(5.159). Then, we compute the variances of the data of the lines (s_1^2), the variances of the data of the columns (s_2^2) and the residual variance of all data (s_{rz}^2). Then, the sums for the computation of the analysis of the variances of two factors processes are:

- the sum of all squares for all experimental data:

$$S_1 = \sum_{j=1}^{m} \sum_{i=1}^{n} v_{ji}^2 \tag{5.155}$$

- the sum of the squares of all the added columns divided by the number of observations from a column:

$$S_2 = \frac{\sum_{j=1}^{m} v_{cj}^2}{n} \tag{5.156}$$

- the sum of the squares of all the added lines divided by the number of observations from a line:

$$S_3 = \frac{\sum_{i=1}^{n} v_{li}^2}{m} \tag{5.157}$$

- the sum of the squares of all added experimental observations divided by the number of total observations:

$$S_4 = \frac{\left(\sum_{j=1}^{m} \sum_{i=1}^{n} v_{ij}\right)^2}{mn} = \frac{\left(\sum_{j=1}^{m} v_{cj}\right)^2}{mn} \tag{5.158}$$

- the sum of the residual squares:

$$S_r = S_1 + S_4 - S_2 - S_3 \tag{5.159}$$

It is not difficult to observe, when we compare this example with the analysis of variances of a monofactor processes, that sum S_3 is the only one to be completely new. The other sums, such as S_1 and S_2, remain unchanged or are named differently (here, S_4 is similar to the S_3 of the analysis of variances for a monofactor process). The corresponding number of degrees of freedom is attached to S_2, S_3

and S_{rz}. They are respectively $m - 1$ for S_2, $n - 1$ for S_3 and $(m - 1)(n - 1)$ for S_{rz}. These degrees of freedom will be associated to the Fischer random variable while the proposed hypotheses are being tested. Using the same principle as used for S_2, S_3 and S_{rz}, we can establish that $mn - 1$ corresponds to the number of degrees of freedom for sum S_4. With these observations, we can completely synthesize the analysis of variances for two factors processes, as shown in Table 5.42. The hypotheses $H_{10} : \sigma_1^2 = \sigma_{rz}^2 \Leftrightarrow s_1^2 = s_{rz}^2$ and $H_{20} : \sigma_2^2 = \sigma_{rz}^2 \Leftrightarrow s_2^2 = s_{rz}^2$ will be accepted when $F_1 < F_{(m-1),(n-1)(m-1),\alpha}^{(1)}$ and $F_2 < F_{(n-1),(n-1)(m-1),\alpha}^{(2)}$. It is possible to have situations where we accept one hypothesis and reject the second one. In this last case, we have to accept that both considered factors play an important role in the process response.

The analysis of the catalytic oxidation of SO_2 developed previously in this chapter, can be completed as follows: (i) the experiments with catalysts number 2 and number 6 are eliminated; (ii) new experiments are introduced in order to consider the temperature as a process factor. All the other factors of the catalytic process keep the values from Table 5.40. In Table 5.43 we present a new set of experimental results in order to obtain more knowledge of the effect of the type of catalyst and the temperature on the degree of oxidation. The correspondence between the different types of catalysts reported in Tables 5.43 and 5.40 are respectively: $1 \rightarrow 1$, $2 \rightarrow 3$, $3 \rightarrow 4$, $4 \rightarrow 5$. As has been explained above, the inlet gas composition, the gas flow rate and the length of the catalytic bed remain unchanged for all experiments, the last limitation is imposed in order to obtain the smallest errors in the measurements for the process response [5.32].

Table 5.42 Synthesis of the analysis of the variances of two factors.

Origin of the variance	Differences of sums	Number of degrees of freedom	Variances		Computed value of the Fischer variable	Theoretical value of the Fischer variable	Decision
Between the columns	$S_2 - S_4$	$m - 1$	$s_1^2 = \dfrac{S_4 - S_2}{m - 1}$		$F_1 = s_1^2/s_r^2$	$F_{(m-1),(n-1)(m-1),\alpha}^{(1)}$	$F_1 < F^{(1)}$ accept H_{10}
Between the lines	$S_3 - S_4$	$n - 1$	$s_2^2 = \dfrac{S_3 - S_4}{n - 1}$		$F_2 = s_2^2/s_r^2$	$F_{(n-1),(n-1)(m-1),\alpha}^{(2)}$	$F_2 < F^{(2)}$ accept H_{20}
Residual	$S_{rz} = S_1 + S_4 - S_2 - S_3$	$(m-1)(n-1)$	$s_{rz}^2 = \dfrac{S_{rz}}{(m-1)(n-1)}$		$H_{10} : \sigma_1^2 = \sigma_{rz}^2 \Leftrightarrow s_1^2 = s_{rz}^2$ $H_{20} : \sigma_2^2 = \sigma_{rz}^2 \Leftrightarrow s_2^2 = s_{rz}^2$		
Total	$S_1 - S_4$	$mn - 1$					

Table 5.43 Comparison of SO_2 oxidation degree with different catalysts and at various temperatures.

Catalyst type	$x_1 = 1$	$x_2 = 2$	$x_3 = 3$	$x_4 = 4$	Total of each line
Temperature of reaction					
$x_2 = 440\,°C$	25	28	22	24	$v_{l1} = 99$
$x_2 = 450\,°C$	27	29	23	23	$v_{l2} = 102$
$x_2 = 460\,°C$	30	32	26	29	$v_{l3} = 117$
Total of each column	$v_{c1} = 82$	$v_{c2} = 89$	$v_{c3} = 71$	$v_{c4} = 76$	

With the experimental data from Table 5.43, we intend to show whether both the type of catalyst and the temperature have an important influence on the oxidation degree of sulfur dioxide. We begin with calculating the sums from Table 5.42. Then, we have:

$S_1 = (25^2+27^2+....+23^2+29^2) = 8538$, $S_2 = (82^2+89^2+71^2+76^2) = 8487.3$, $S_3 = (99^2+102^2+117^2) = 8473.5$, $S_4 = (25+27+...+23+29)^2 = 8427.0$, $S_r = S_1 + S_4 - S_2 - S_3 = 8538 + 8427.0 - 8487.3 - 8473.5 = 4.2$, $S_2 - S_4 = 60.3$ with $m - 1 = 3$ degrees of freedom, $S_3 - S_4 = 46.5$ with $n - 1 = 2$ degrees of freedom, $S_{rz} = 4.2$ with $(m - 1)(n - 1) = 6$ degrees of freedom, $s^2_1 = 60.3/3 = 20.1$, $s^2_2 = 46.5/2 = 23.3$, $s^2_{rz} = 4.2/6 = 0.7$, $F_1 = s^2_1/s^2_{r} = 20.1/0.7 = 28.8$, $F_2 = s^2_2/s^2_r = 23.3/0.7 = 33.3$, $F^{(1)}_{m-1,(m-1)(n-1),a} = F^{(1)}_{3,6,0.05} = 4.786$, $F^{(2)}_{n-1,(m-1)(n-1),a} = F^{(1)}_{2,6,0.05} = 5.14$.

The results of the computations are given in Table 5.44, which is a particularization of the general Table 5.42. The last three columns of Table 5.44, give the testing calculations for H_{10} and H_{20} showing that these hypotheses are rejected. We can thus observe that there are important differences between the residual variance and the variance due to the change in the type of catalyst and temperature. In other words, both factors are important factors in this process. It should be mentioned that the analysis of variances does not give a quantitative response detailing the exact type of catalyst or/and the temperature to be used for the best yield.

When the investigated process shows a small residual variance we can consider that the variance results from the action of small random factors. At the same time, this small variance is a good indication of an excellent reproducibility of the experimental measurements. Conversely, a great residual variance can show that the measurements are characterized by poor reproducibility. However, this situation can also result from one or more unexpected or unconsidered factors; this situation can be encountered when the interactions between the factors (parameters) have been neglected. In these cases, the variance of the interactions represents an important part of the overall residual variance.

Table 5.44 Synthesis of the analysis of variances for two factors – Example 5.6.2.

Origin of the variance	Differences of sums	Degrees of freedom	Variances	Computed value of the Fischer variable	Theoretical value of the Fischer variable	Decision
Between the columns	$S_2 - S_4 = 60.3$	$m - 1 = 3$	$s_1^2 = \dfrac{S_4 - S_2}{m-1} = 20.1$	$F_1 = s_1^2/s_r^2$ $= 28.7$	$F^{(1)}_{(m-1),(n-1)(m-1),\alpha}$ $= 4.76$	$F_1 > F^{(1)}$ Reject H_{10}
Between the lines	$S_3 - S_4 = 46.2$	$n - 1 = 2$	$s_2^2 = \dfrac{S_3 - S_4}{n-1} = 23.3$	$F_2 = s_2^2/s_r^2$ $= 33.3$	$F^{(2)}_{(n-1),(n-1)(m-1),\alpha}$ $= 5.14$	$F_2 > F^{(2)}$ Reject H_{20}
Residual	$S_r =$ $S_1 + S_4 - S_2 -$ $S_3 = 4.2$	$(m-1)$ $(n-1) = 6$	$s_{rz}^2 =$ $\dfrac{S_{rz}}{(m-1)(n-1)} = 0.7$	$H_{10} : \sigma_1^2 = \sigma_{rz}^2 \Leftrightarrow s_1^2 = s_{rz}^2$ $H_{20} : \sigma_2^2 = \sigma_{rz}^2 \Leftrightarrow s_2^2 = s_{rz}^2$		
Total	$S_1 - S_4 = 110$	$mn - 1 = 11$				

5.6.3
Interactions Between the Factors of a Process

To illustrate the interaction of factors in a concrete process, we will consider the example of a process with two factors which are called A and B. The experimental investigation of the considered process is made using a CFE 2^2 plan. Both parameters, A and B will present the levels A_1 and A_2, B_1 and B_2, respectively, and, consequently, the process response has four values which are a_1, a_2, b_1, b_2, (the subscripts 1 and 2 indicate the higher and lower level of the factor). With these four values, we can develop the analysis of variances for two factors. First, we have to divide the residual variance into two parts: the first shows that the differences between the measured values of the responses are due to the experimental problems of the reproducibility; and the second indicates the action of the interaction of the factors on the responses of the process. For this separation we need a great number of measurements for each grouping of factors. So, for point A_1B_1, where the values of the dimensionless factors are $x_1 = -1$ and $x_2 = -1$, we obtain more values of the process response; moreover, we have the same problem for the other following points: A_1B_2 ($x_1 = -1$, $x_2 = 1$), A_2B_1 ($x_1 = 1$, $x_2 = -1$), A_2B_2 ($x_1 = 1$, $x_2 = 1$). The solution to this problem will result in the possibility to compute the variance caused by the reproducibility. In other words, we will be able to appreciate the effect of small random factors on the process response.

The data concerning this example are shown in Table 5.45. For the development of the analysis of variances, we use sums S_1, S_2, S_3, S_4 which have already been introduced with the analysis with two factors. The supplementary sum S_5 (Eq.

(5.160)), which is the total sum of the squares sum of the repeated values for each experimental point, is also considered here:

$$S_5 = \frac{\sum\limits_{i=1}^{n}\sum\limits_{j=1}^{m}\left(\sum\limits_{k=1}^{p} v_{ij}^{(k)}\right)^2}{mn-1} \tag{5.160}$$

Table 5.45 Data for the analysis of variances of two factors with interaction effects.

Values of factor A	$x_1 = \alpha_1$	$x_1 = \alpha_2$	......	$x_1 = \alpha_j$	......	$x_1 = \alpha_m$	Total
Values of factor B							
$x_2 = \beta_1$	$v_{11}^{(1)}$ $v_{11}^{(p)}$	$v_{12}^{(1)}$ $v_{12}^{(p)}$	....	$v_{1j}^{(1)}$ $v_{1j}^{(p)}$	....	$v_{1m}^{(1)}$ $v_{1m}^{(p)}$	v_{l1}
$x_2 = \beta_2$	$v_{21}^{(1)}$ $v_{21}^{(p)}$	$v_{22}^{(1)}$ $v_{22}^{(p)}$	....	$v_{2j}^{(1)}$ $v_{2j}^{(p)}$	....	$v_{2m}^{(1)}$ $v_{2m}^{(p)}$	v_{l2}
......	......	......	......	......	......	......	......
$x_2 = \beta_i$	$v_{i1}^{(1)}$ $v_{i1}^{(p)}$	$v_{i2}^{(1)}$ $v_{i2}^{(p)}$	....	$v_{ij}^{(1)}$ $v_{ij}^{(p)}$	......	$v_{im}^{(1)}$ $v_{im}^{(p)}$	v_{li}
......	......	......	......	......	......	......	......
$x_2 = \beta_n$	$v_{n1}^{(1)}$ $v_{n1}^{(p)}$	$v_{n2}^{(1)}$ $v_{n2}^{(p)}$	....	$v_{nj}^{(1)}$ $v_{nj}^{(p)}$	....	$v_{nm}^{(1)}$ $v_{nm}^{(p)}$	v_{lm}
Total	v_{c1}	v_{c2}	......	v_{cj}	......	v_{cm}	

Table 5.46 contains a summary to analyze these variances. Here the basic problem is the testing of the following statistical hypotheses: $H_{10} : \sigma_1^2 \equiv \sigma_A^2 = \sigma_{rz}^2$, $\Leftrightarrow s_1^2 = s_{rz}^2$, $H_{20} : \sigma_2^2 \equiv \sigma_B^2 = \sigma_{rz}^2 \Leftrightarrow s_2^2 = s_{rz}^2$, $H_{120} : \sigma_{12}^2 \equiv \sigma_{AB}^2 = \sigma_{rz}^2 \Leftrightarrow s_{12}^2 = s_{rz}^2$ these hypotheses can be described as follows:

- the variance of the data produced by changes in factor A and the variance of the residual data are similar, then, all data represent the same population (H_{10});
- the variance of the data produced by changes in factor B and the variance of the residual data are similar, then, factor B is not significant for the evolution of the process (H_{20});
- the variance of the data produced by the interactions between factors A and B and the residual data variance are similar, then, the interaction factor has no effect on the process output (H_{120}).

Table 5.46 Summary of the analysis of variances for two factors with interaction effects.

Origin of variance	Differences of sums	Number of freedom degrees	Variances		Computed value of the Fischer variable	Theoretical value of the Fischer variable	Decision
Between the columns	$S_2 - S_4$	$m-1$	$s_1^2 = \dfrac{S_2 - S_4}{m-1}$		$F_1 = s_1^2/s_r^2$	$F_{(m-1),(n-1)(m-1),\alpha}^{(1)}$	$F_1 < F^{(1)}$ Accept H_{10}
Between the lines	$S_3 - S_4$	$n-1$	$s_2^2 = \dfrac{S_3 - S_4}{n-1}$		$F_2 = s_2^2/s_r^2$	$F_{(n-1),(n-1)(m-1),\alpha}^{(2)}$	$F_2 < F^{(2)}$ Accept H_{20}
Interaction AB	$S_{12} = S_5 + S_4 - S_2 - S_3$	$(m-1).$ $(n-1)$	$s_{12}^2 = \dfrac{S_{12}}{(m-1)(n-1)}$		$F_{12} = s_{12}^2/s_r^2$	$F_{(m-1)(n-1),mp(n-1),\alpha}^{(12)}$	$F_{12} < F^{(12)}$ Accept H_{120}
Residual	$S_{rz} = S_1 - S_5$	$mp(n-1)$	$s_{rz}^2 = \dfrac{S_{rz}}{mp(n-1)}$				

We can observe in Table 5.43 that the maximum yield is obtained with catalyst number two ($x_2 = 2$), the response obtained with this catalyst can be analyzed deeply with respect to other process parameters such as the input reactor gas flow rate and the temperature. Two different values or levels of these parameters will be considered whereas other parameters or factors will remain constant (Table 5.40). Table 5.47 gives the experimental data after the arrangement required by Table 5.45 together with the partial and total mean values of SO_2 oxidation degree.

Table 5.47 Data for the analysis of variances for two factors with interaction effects. Example of SO_2 oxidation factors: temperature (T) and flow rate (G).

Flow rate	$G_1 = 0.1 m^3/(m^3 cat\ s)$		$G_2 = 0.14 m^3/(m^3 cat\ s)$		Total	Mean
Temperature						
$T_1 = 450$	21.2		22.65			
	21.5	63.75	22.55	68.4	132.15	22.02
	21.05	21.27	23.20	22.8		
$T_2 = 470$	21.65		22.3			
	21.95	65.90	22.2	67.1	133.0	22.15
	22.30	21.76	22.7	22.36		
Total	129.65		135.20		265.15	
Mean	21.60		22.58			

The interest here is to verify whether the temperature, the flow rate and their interactions produce changes in the SO_2 oxidation degree. For the case when factors interact, it is interesting to determine what the favourable direction for factors variation is.

The problem is firstly investigated by making the necessary calculations to analyze the effect of the factors and the interactions with the following procedure:

- we identify: $m = 2$, $n = 2$, $p = 3$;
- we compute: $S_1 = (21.2^2 + 21.5^2 + ... + 22.2^2 + 22.7^2) = 5862.27$, $S_2 = (129.6^2 + 135.2^2)/6 = 5864.12$, $S_3 = (132.15^2 + 133^2)/6 = 5858$, $S_4 = (265.15^2)/12 = 5858.7$, $S_5 = (63.75^2 + 65.9^2 + 68.4^2 + 67.1^2)/3 = 5887.61$, $S_2 - S_4 = 5.41$, $S_3 - S_4 = 0.05$, $S_{12} = S_{AB} = S_5 + S_4 - S_3 - S_2 = 23.44$, $S_1 - S_5 = 3.56$, $s^2_1 = 5.41/(2-1) = 5.42$, $s^2_2 = 0.05/(2-1) = 0.05$, $s^2_{12} = 23.41/(1*1) = 23.41$, $s^2_{rz} = 3.56/(2*3*1) = 0.59$, $F_1 = 5.41/0.59 = 9.18$, $F_2 = 0.05/0.59 = 0.08$, $F_{12} = 23.41/0.59 = 39.9$, $F_{1,6,0.05} = 5.99$.
- we compute all the data for Table 5.48 where we verify hypotheses H_{10}, H_{20}, H_{120};
- we identify that the change in flow rate and the interaction temperature–flow rate are important for the sulphur dioxide oxidation degree.

Table 5.48 Numerical example introduced in Table 5.47.

Variances and degrees of freedom	Hypotheses	Computed value of the Fischer variable	Theoretical value of the Fischer variable	Decision
$s^2_1 = 5.41$, $v_1 = 1$	$H_{10} : \sigma^2_1 \equiv \sigma^2_A = \sigma^2_{rz}$ $\Leftrightarrow s^2_1 = s^2_{rz}$	$F_1 = s^2_1/s^2_{rz} = 9.18$	$F_{1,6,0.05} = 5.99$	Refuse
$s^2_2 = 0.05$, $v_2 = 1$	$H_{20} : \sigma^2_2 \equiv \sigma^2_B = \sigma^2_{rz}$ $\Leftrightarrow s^2_2 = s^2_{rz}$	$F_2 = s^2_2/s^2_{rz} = 0.084$	$F_{1,6,0.05} = 5.99$	Accept
$s^2_{12} = 23.44$, $v_{12} = 1$ $s^2_{rz} = 0.59$, $v_1 = 6$	$H_{120} : \sigma^2_{12} \equiv \sigma^2_{AB} = \sigma^2_{rz}$ $\Leftrightarrow s^2_{12} = s^2_{rz}$	$F_{12} = s^2_{12}/s^2_{rz} = 39.9$	$F_{1,6,0.05} = 5.99$	Refuse

The analysis of the effects of the interaction contains the calculation of the confidence interval with respect to the increase in SO_2 conversion when – for temperature T_1 – the flow rate varies between G_1 and G_2 and when – for flow rate G_1 – the temperature varies between T_1 and T_2. If d_i is the mean value of the increase in the SO_2 conversion degree for case i ($i = 1 \rightarrow T_1$ = constant and the flow rate changes between and G_2), then the confidence interval for this mean value will be:

$$I_i = \left\langle d_i - - t_{p+p,\alpha} s_{rz}/\sqrt{2p} \; , \; d_i - + t_{p+p,\alpha} s_{rz}/\sqrt{2p} \right\rangle \tag{5.161}$$

where $t_{p+p,\alpha}$ is the student random variable value with 2p degrees of freedom and $1 - \alpha$ of confidence level.

In our case $t_{2p,\alpha} = t_{6,0.05} = 4.317$; $s_{rz} = (0.59)^{1/2} = 0.77$; the calculation of the intervals of confidence from Table 5.49 shows that we do not have a complete argument to suggest the variation of factors. This conclusion is sustained by the fact that we have negative and positive values for each confidence interval.

Table 5.49 The confidence intervals for the increase of the SO_2 oxidation degree.

Flow rate	$G_1 = 0.1 \ m^3/(m^3 cat \ s)$	$G_1 = 0.1 \ m^3/(m^3 cat \ s)$	$\overline{d}_i$	I_i
Temperature	mean value	mean value		
$T_1 = 450\,°C$	21.27	22.8	1.53	(−0.57, 3.63)
$T_2 = 470\,°C$	27.76	22.35	0.6	(−1.5, 2.7)

5.6.3.1 Interaction Analysis for a CFE 2^n Plan

When we use a CFE 2^2 plan to determine the interaction effects, we introduce associated variances that can be easily used to produce answers to the aspects concerning the interaction between the factors of the process [5.33, 5.34].

It is known that the analysis of variances shows which factors and interactions must be kept and which must be rejected. At the same time, the analysis of the significance for the coefficients of the statistical model of the process gives the same results: rejection of the non-significant factors and interactions from the model and consequently from the experimental process analysis. Here, apparently, we have two competitive statistical methods for the same problem. In fact, the use of the analysis of the variances before starting the regression analysis, guarantees an excellent basis to select the relationship between the variables of the process. Otherwise, a previous analysis of the dispersion (variances) drives the regression analysis to the cases when its development is made with non-saturated plans. After these necessary explanations, we can start the problem of detecting the interactions of the factors for a concrete process by showing the terminology used. For the example of the process with factors A, B and C, this terminology is given in Table 5.50. Here the values of the dependent variable of the process (process responses) are symbolically particularized according to the higher states (levels) of the factors.

Table 5.50 Terminology used for the interaction analysis using a CFE 2^3 plan.

C levels	C_1				C_2			
B levels	B_1		B_2		B_1		B_2	
A levels	A_1	A_2	A_1	A_2	A_1	A_2	A_1	A_2
Response values	(1)	a	b	ab	c	ac	bc	abc

In order to obtain the effect on the response values of factor A when it varies from level A_1 to A_2, we must extract the results obtained with A_1 from the results obtained with A_2. According to Table 5.50, we can write the following relations:

$$EA = (a - (1)) + (ab - b) + (ac - c) + (abc - bc)$$

$$EA = abc + ab + ac + a - bc - b - c - (1) \tag{5.162}$$

It is easy to observe that we subtract all results from the sum of responses that contain symbol "a". By the same procedure, we can write the effect of factors B and C. It results in:

$$EB = abc + ab + bc + b - ac - a - c - (1) \tag{5.163}$$

$$EC = abc + ac + bc + c - ab - a - b - (1) \tag{5.164}$$

The interaction effect AB is obtained by subtracting the effect of A at the level B_1 from the effect of factor A at level B_2. This is written mathematically as follows:

$$EAB = [(abc - bc) + (ab - b)] - [(ac - c) + (a - (1))]$$

$$EAB = abc + ab + c + (1) - ac - bc - a - b \tag{5.165}$$

The remaining interaction effects AC and BC are written using the same definition. Then, we obtain the following relations:

$$EAC = abc + ac + b + (1) - ab - bc - a - c \tag{5.166}$$

$$EBC = abc + bc + a + (1) - ab - ac - b - c \tag{5.167}$$

If we consider a formal vector E which includes all the effects on the process response, then we can build relation (5.168) which includes all the relations from (5.162) to (5.167):

$$[E] = \begin{bmatrix} EA \\ EB \\ EC \\ EAB \\ EAC \\ EBC \\ EABC \end{bmatrix} = \begin{bmatrix} -(1) + a - b + ab - c + ac - bc + abc \\ -(1) - a + b + ab - c - ac + bc + abc \\ -(1) - a - b - ab + c + ac + bc + abc \\ +(1) - a - b + ab + c - ac - bc + abc \\ +(1) - a + b - ab - c + ac - bc + abc \\ +(1) + a - b - ab - c - ac + bc + abc \\ -(1) + a + b - ab + c - ac - bc + abc \end{bmatrix} \qquad (5.168)$$

In our example, we can keep the order of the values given in Table 5.50. However, if we change this order, then the expressions for relations (5.162)–(5.167) must agree with this change. Relation (5.168) can easily be written using the 2^3 matrix plan. Nevertheless, here, we have to consider the first point of the plan with negative coordinates. Table 5.51 shows the variation inside the factorial cube which is at the origin of relation (5.168). It is observable that the multiplication of the response column with columns A, B,..., ABC gives the corresponding partial effects EA, EB,.....,EABC.

Table 5.51 Use of the CFE 2^3 for the development of relation (5.168).

I	A	B	C	AB	AC	BC	ABC	y_i
1	−1	−1	−1	+1	+1	+1	−1	$y_1 = (1)$
2	+1	−1	−1	−1	−1	+1	+1	$y_2 = a$
3	−1	+1	−1	−1	+1	−1	+1	$y_3 = b$
4	+1	+1	1	+1	−1	−1	−1	$y_4 = ab$
5	−1	−1	+1	+1	−1	−1	+1	$y_5 = c$
6	+1	−1	+1	−1	+1	−1	−1	$y_6 = ac$
7	−1	+1	+1	−1	−1	+1	−1	$y_7 = bc$
8	+1	+1	+1	+1	+1	+1	+1	$y_8 = abc$

starting from marked point

As has been developed above, the analysis of variances imposes the calculation of the variances due to the changes and interactions between the factors. In addition, we also have to verify the next seven hypotheses where $\sigma_{rz}^2 = s_{rz}^2$ is assumed to be $\sigma_{ABC}^2 = s_{ABC}^2$:

$$H_A : \sigma_A^2 = \sigma_{rz}^2 \Leftrightarrow s_A^2 = s_{rz}^2, H_B : \sigma_B^2 = \sigma_{rz}^2 \Leftrightarrow s_B^2 = s_{rz}^2, H_C : \sigma_C^2 = \sigma_{rz}^2 \Leftrightarrow s_C^2 = s_{rz}^2$$

$$H_{AB} : \sigma_{AB}^2 = \sigma_{rz}^2 \Leftrightarrow s_{AB}^2 = s_{rz}^2, H_{AC} : \sigma_{AC}^2 = \sigma_{rz}^2 \Leftrightarrow s_{AC}^2 = s_{rz}^2, H_{BC} : \sigma_{BC}^2 = \sigma_{rz}^2$$

$$\Leftrightarrow s_{BC}^2 = s_{rz}^2$$

The acceptance of a hypothesis from those mentioned above corresponds to accepting the fact that the factor or interaction linked to the hypothesis is not important in the investigated process. In this example, the sums of the squares used for the production of the analysis of variances, is made with a CFE 2^3 plan (Table 5.52), they are expressed using the partial effects as follows: $S_A = (EA)^2/8$, $S_B = (EB)^2/8$, $S_C = (EC)^2/8$, $S_{AB} = (EAB)^2/8$, $S_{AC} = (EAC)^2/8$, $S_{BC} = (EBC)^2/8$, $S_{ABC} = (EABC)^2/8$.

Table 5.52 Synthesis of the analysis of variances for a CFE 2^3 plan.

Origin of the variance	Sums of the squares	Degrees of freedom	Variances	Computed value of the Fischer variable	Theoretical value of the Fischer variable	Decision
Change of factor A	S_A	1	$s^2_A = S_A/1$	$F_A = s^2_A/s^2_{rz}$	$F_{1,1,\alpha}$	$F_A < F_{1,1,\alpha}$ Accept H_A
Change of factor B	S_B	1	$s^2_B = S_B/1$	$F_B = s^2_B/s^2_{rz}$	$F_{1,1,\alpha}$	$F_B < F_{1,1,\alpha}$ Accept H_B
Change of factor C	S_c	1	$s^2_c = S_c/1$	$F_C = s^2_B/s^2_{rz}$	$F_{1,1,\alpha}$	$F_C < F_{1,1,\alpha}$ Accept H_c
Interaction A B	S_{AB}	1	$s^2_{AB} = S_{AB}/1$	$F_{AB} = s^2_{AB}/s^2_{rz}$	$F_{1,1,\alpha}$	$F_{AB} < F_{1,1,\alpha}$ Accept H_{AB}
Interaction A C	S_{AC}	1	$s^2_{AC} = S_{AC}/1$	$F_{AC} = s^2_{AC}/s^2_{rz}$	$F_{1,1,\alpha}$	$F_{AC} < F_{1,1,\alpha}$ Accept H_{AC}
Interaction B C	S_{BC}	1	$s^2_{BC} = {}_{BC}/1$	$F_{BC} = s^2_{BC}/s^2_{rz}$	$F_{1,1,\alpha}$	$F_{BC} < F_{1,1,\alpha}$ Accept H_{BC}
Interaction A BC (residual)	S_{ABC}	1	$s^2_{ABC} = S_{ABC}/1$ $s^2_{ABC} = S_{ABC}/1 = s^2_{rz}$			
Total	$S_T = S_1 - S_2$ 7		xxxx	xxxx		

The analysis of variances using a CFE 2^n plan in which, for each experimental point, we produce only one measurement, frequently presents an important residual variance. This result is a consequence of the fact that each point is the result of a particular combination of interaction effects. If, for each experimental point of the plan, we produce more experiments, then we have the normal possibility to compute a real residual variance (5.169). In this situation, the sum is successfully used as shown in Table 5.52 for the residual variance computation.

$$S_{rz} = \frac{\sum\limits_{i=1}^{2^n} \sum\limits_{k=1}^{r-1} d_{ik}^2}{r} \qquad (5.169)$$

In relation (5.169), d_{ik} represents the difference between two values from the total values produced at point "i" ($k = 1, r$).

In the following example the application of this computation procedure is developed. The analysis of variances is carried out for the air oxidation of an aromatic hydrocarbon. In this process, where air is bubbled in the reaction vessel, we obtain two products: a desired compound and a secondary undesired compound. Here, it is important to know how the transformation degree of the hydrocarbon evolves towards the by-product when different process parameters (factors) are varied as follows:

- the catalyst concentration (A) varies from $A_1 = 0.1\%$ g/g to $A_2 = 0.4\%$ g/g
- the bubbling time (B) for air flow (0.01 m^3/m$^3_{liquid}$ s) varies from $B_1 = 60$ min to $B_2 = 70$ min
- the reaction temperature(C) varies from $C_1 = 50\,°C$ to $C_2 = 60\,°C$

Table 5.53 gives the experimental results of the hydrocarbon conversion in a by-product. With the data below, we can characterize the particular effect of each parameter on the process output (hydrocarbon oxidation degree in an undesired compound) and the conclusion expected here is to suggest a proposal for the enhancement of the efficiency of the process.

Table 5.53 Analysis of the variances made for a 2^2 plan for an aromatic hydrocarbon oxidation in an undesired by-product.

$C_1 = 50\,°C$				$C_2 = 60\,°C$			
$B_1 = 60$ min		$B_2 = 70$ min		$B_1 = 60$ min		$B_2 = 70$ min	
$A_1 = 0.1\%$	$A_2 = 0.4\%$	$A_1 = 0.1\%$	$A_2 = 0.4\%$	$A_1 = 0.1\%$	$A_2 = 0.4\%$	$A_1 = 0.1\%$	$A_2 = 0.4\%$
12.6	13.5	13.4	14.9	13.2	17.7	15.9	19.2
13.1	12.0	12.4	13.4	15.7	18.2	16.4	18.7
S 25.7	S 25.5	S 25.8	S 28.3	S 28.9	S 35.9	S 32.3	S 37.9
(1)	a	b	ab	c	ac	bc	abc

The necessary computations for this example are organized as follows:

1. We compute the values of the particular effects with relations (5.168). The results are: EA = 14.9, EB = 8.3, EAB = 1.3, EC = 29.7, EAC = 10.3, EBC = 2.5, EABC = –4.5;

2. The associated sums of squares have the values:
   $S_A = 14.9^2/(2*8) = 3.937$, $S_B = 8.3^2/(2*8) = 4.305$,
   $S_C = 29.7^2/(2*8) = 55.13$, $S_{AB} = 1.3^2/(2*8) = 0.105$,
   $S_{AC} = 10.3^2/(2*8) = 6.630$, $S_{BC} = 2.5^2/(2*8) = 0.39$,
   $S_{ABC} = 4.5^2/(2*8) = 1.050$, $S_1 = (13.5^2 + 12.0^2 + ... + 19.2^2 + 18.7^2) = 3696.82$, $S_2 = ((13.5 + 12.0 + ... + 19.2 + 18.7)^2)/16 = 3609$, $S_T = S_1 - S_2 = 87.82$;

3. The sum of the residual squares has been computed according to relation (5.169): $S_{rz} = (0.5^2 + 1.5^2 + 1^2 + 0.5^2 + 2.5^2 + 0.5^2 + 0.5^2 + 0.5^2)/2 = 6.35$.

4. All the values of the sums S_A, S_B,....S_{ABC} are one (1) for the associated number of degrees of freedom. So variances s_A, s_B,...s_{ABC} have the same values as the corresponding sums; the sum of residual squares associates value $v = 8$ to the number of the degrees of freedom. This fact gives value $s^2_{rz} = S_r/8 = 0.797$ for the residual variances.

5. The computed values for the associated Fischer variable (see also Table 5.52) for the variances of the factors and their interactions present the next values: $F_A = 17.4$, $F_B = 5.4$, $F_A = 17.4$, $F_C = 69.2$, $F_{AB} = 0.13$, $F_{AC} = 8.3$, $F_{BC} = 0.49$, $F_{ABC} = 1.3$.

6. The theoretical value of the Fischer random variable associated to this actual case is $F_{1,8,0.05} = 5.32$; By comparing this value with the computed values of the Fischer variable given here, we can decide that factors A, B, C as well as interaction AC determine the hydrocarbon oxidation degree in the undesired product.

7. Because we observe that factor B has an independent influence on the output of the process and considering the data from Table 5.54, we can assert that, in order to obtain small values of the conversion to by-product, it is not recommended to increase the value of B. We can compute the change in the degree of hydrocarbon oxidation in the undesired product when factors A and C increase. Indeed, this computation can result in a recommendation concerning the increase in A and C. The next mean values of the output variable are thus obtained in the points where we have only A and C, namely: A_1C_1, A_1C_2, A_2C_1, A_2C_2: $m_{A1C1} = (12.6 + 13.1 + 13.4 + 12.4)/4 = 12.875$, $m_{A1C2} = 15.30$, $m_{A2C1} = 13.45$, $m_{A2C2} = 18.45$. The changes in the oxidation degree associated to these mean values are: $d_1 = m_{A2C1} - m_{A1C1} = 0.575$ and $d_2 = m_{A2C2} - m_{A2C1} = 3.15$. Then, the mean value of the oxidation degree change is $d = (d_1 + d_2)/2 = 1.86$. This value is included within confidence interval $I = (0.4, 3.32)$ according to relation (5.161). Then, if we increase A or C or A and C, we will increase the conversion of the aromatic hydrocarbon in the undesired by-product.

5.6.3.2 **Interaction Analysis Using a High Level Factorial Plan**

Sometimes we may encounter situations requiring the analysis of the effects of the factors on the output variables of a process by working with more than two levels for one or more factors. The analysis of variances for this type of process is associated with a difficult methodology of data processing and interpretation. However, the method can be simplified if, at the starting point, we split the primary experimental data table into different tables in which each factor presents only two levels. We then analyze each table according to the methods presented in the previous paragraphs. The splitting up procedure is explained with a concrete example.

A small perfectly mixed discontinuous reactor is used at laboratory scale to conduct the Friedel–Crafts reaction $Ar–H + RCl \xrightarrow{AlCl_3} Ar–R + HCl$. Three factors and two or more levels of each parameter have been used in an experimental plan in order to separate and compare their influence on the aromatic hydrocarbon conversion. The following factors and levels have been used:

- reaction time (A) which has two levels: $A_1 = 10$ h, $A_2 = 7$ h;
- the particular time when the catalyst is introduced into the reactor or "timing" (B) with three levels: $B_1 = 2$ h, $B_2 = 3$ h, $B_3 = 4$ h;
- the mixing intensity (C) given here by the rotation speed of the mixer driver, which has been modified according to the following rotation levels: $C_1 = 10$ rot/min, $C_2 = 15$ rot/min, $C_3 = 20$ rot/min, $C_4 = 25$ rot/min.

The measurements of the hydrocarbon transformation are given in Table 5.54. Before using these measurements, we need to obtain data showing the interactions and the combination of factors producing the best process efficiency. Before beginning the analysis, we will divide the initial data into different fractional tables, each one with two factors. The data translation is very simply done by subtracting a constant number (such as 60 for example) from each value of the table. Then, the new table of data (Table 5.55) will be split up using the following algorithm:

1. The first variable factor (factor C) is taken from Table 5.54 by summing its values (all different levels) into only one which will give the new value of the process for the two other factors.
2. The obtained table will be noted with the interaction name of the non-rejected factors; so if C is rejected, the name of the partial table will be AB;
3. We repeat steps 1 and 2 for factor B, then we obtain the partial table AC. For factor A, table BC is produced.

Table 5.54 Friedel–Crafts reaction efficiency in an experimental plan with 3 factors and 4 levels.

	A₁			A₂		
	B_1	B_2	B_3	B_1	B_2	B_3
C_1	74.3	68.7	65.1	67.7	68.2	70.5
C_2	73.6	65.9	65.7	66.5	69.3	71.0
C_3	72.3	65.5	66.9	65.6	69.8	71.3
C_4	70.4	65.3	67.8	65.3	71.0	71.1

Table 5.55 Translation of data from Table 5.54.

	A₁			A₂		
	B_1	B_2	B_3	B_1	B_2	B_3
C_1	14.2	8.7	5.1	7.7	8.2	12.5
C_2	13.6	5.9	5.7	6.5	9.3	11.0
C_3	12.3	5.5	6.9	5.6	9.8	11.3
C_4	10.4	5.3	7.8	5.3	11.0	11.1

The computation for the division of Table 5.55 is:
- Elimination of factor C: partial table AB. For each point of the partial table AB (2*3 points), we compute the value of the response. Then, we have:
  A_1B_1 = 14.3 + 13.6 + 12.3 + 10.4 = 50.6, A_1B_2 = 8.7 + 5.9 + 5.5 + 5.3 = 25.4, A_1B_3 = 5.1 + 5.7 + 6.9 + 7.8 = 25.5, etc.;
- Elimination of factor B: partial table AC. In this case, with the same procedure used for partial table AB, we obtain:
  A_1C_1 = 14.3 + 8.7 + 5.1 = 20.1, A_2C_1 = 7.7 + 8.2 + 10.5 = 26.4, A_1C_2 = 13.6 + 5.9 + 5.7 = 25.2, etc.
- Elimination of factor A: partial table BC: as explained above,
  B_1C_1 = 14.3 + 7.7 = 22.0, B_1C_2 = 13.6 + 6.5 = 20.1, B_1C_3 = 12.3 + 5.6 = 17.9, etc.

The results of these calculation are summarized in Table 5.56, which is composed of three different partial tables: AB, AC, BC. This new set of data will be used for the final analysis of variances. For each partial table, the analysis of the variances of two factors will be carried out. Additionally, the values of the sums of the squares needed by the procedure of analyzing the variances (see Table 5.42) will be computed. As far as each value in the partial tables is the result of an addition of many original data, all the sums of the squares for each of these tables, will be divided by the number of data used to produce the values. For example, partial table AB results from the elimination of factor C, because the C factor has four levels then all the sums of the squares associated to this table will be divided by four (number of factor levels). The addition that characterizes the interaction is obtained by the difference between the sum of the total squares and the sum of the squares containing the main effects.

Table 5.56 Division of Table 5.55 into three tables.

Two factors, table AB

	B_1	B_2	B_3	total
A_1	50.6	25.4	25.5	101.5
A_2	25.1	38.3	43.9	107.3
total	75.7	63.7	69.4	208.8

Two factors, table AC

	C_1	C_2	C_3	C_4	total
A_1	28.1	25.2	24.7	23.5	101.5
A_2	26.4	26.8	26.7	27.4	107.3
total	54.5	52.0	51.4	50.9	207.8

Two factors, table BC

	C_1	C_2	C_3	C_4	total
B_1	22.0	20.1	17.9	15.7	75.7
B_2	16.9	15.2	15.3	16.3	63.7
B_3	15.6	16.7	18.2	18.9	69.4
total	54.5	52.0	51.4	50.9	206.8

Now we can compute sums S_1, S_2, S_3, S_4 (see Table 5.42), which specifically concern partial table AB from Table 5.56. So we have: $S_{1(AB)} = (50.6^2 + 25.4^2 + ... + 43.9^2)/(4*1) = 1971.6$, $S_{2(AB)} = (101.5^2 + 107.3^2)/(4*3) = 1817.96$, $S_{3(AB)} = (75.7^2 + 63.7^2 + 69.7^2)/(4*2) = 1826.07$, $S_{4(AB)} = (50.6 + 25.4 + ... + 43.9)/(4*6) = 1816.47$.

Then we can calculate the following sums (see Table 5.42): $S_A = S_{2(AB)} - S_{4(AB)} = 1.49$, $S_B = S_{3(AB)} - S_{4(AB)} = 9.58$, $S_{T(AB)} = S_{1(AB)} - S_{4(AB)} = 155.20$ and so $S_{AB} = S_{T(AB)} - (S_A + S_B) = 144.13$.

By a similar procedure, we obtain the sums of squares S_1, S_2, S_3, S_4 when factor B (three levels) has been eliminated. These are: $S_{1(AC)} = (28.1^2 + 26.4^2 + ... + 23.3^2 + 27.4^2)/(3*1) = 1821.94$, $S_{2(AC)} = S_{2(AB)} = (101.5^2 + 107.3^2)/(4*3) = 1817.96$, $S_{3(AC)} = (54.5^2 + 52.0^2 + 51.4^2 + 50.9^2)/(3*2) = 1817.83$, $S_{4(AC)} = S_{4(AB)} = (50.6 + 25.4 + .. + 43.9)/(4*6) = 1816.47$, $S_C = S_{2(AC)} - S_{4(AC)} = 1.36$, $S_{T(AC)} = S_{1(AC)} - S_{4(AC)} = 5.47$. For the sum of squares that characterizes interaction AC, we have: $S_{AC} = S_{T(Ac)} - (S_A + S_C) = 2.62$. For the third partial table, the computations of these sums give: $S_{1(BC)} = (22.0^2 + 16.9^2 + ... + 16.3^2 + 18.9^2)/(2*1) = 1841.82$, $S_{2(BC)} = (75.7^2 + 63.7^2 + 69.4^2)/(4*2) = 1826.07$, $S_{3(BC)} = S_{3(AC)} = (54.5^2 + 52.0^2 + 51.4^2 + 50.9^2)/(3*2) = 1817.83$, $S_{4(BC)} = S_{4(AC)} = S_{4(AB)} = 1816.47$. Whereas, for the sum of the squares that characterizes the BC interaction, we have: $S_{BC} = S_{T(BC)} - (S_B + S_C) = 14.41$.

For this application, the residual sum of squares is obtained by eliminating the sums of squares for A, B, C, AB, AC, BC from the total sum of the squares $S_T = S_1 - S_4$ where S_1 and S_4 are computed with the data from the original table (Table 5.55). Therefore, we obtain $S_1 = 14.3^2 + 8.7^2 + ... 11.3^2 + 11.1^2 = 2000.6$, $S_4 = (14.3 + 8.7 + .. + 11.2 + 11.1)^2/24 = 1816.47$, $S_T = 184.13$. Consequently, the residual sum of squares and their associated degrees of freedom will be: $S_{rz} = S_T - (S_A + S_B + S_C + S_{AB} + S_{AC} + S_{BC}) = 10.90$, $\nu = (2 - 1)(3 - 1)(4 - 1) = 6$.

Now we have all the necessary sums for the development of the analysis of variances. However, we first have to verify the following hypotheses:

- there is no significant difference between the variance due to the action of factor A and the residual variance

$$H_A : \sigma_A^2 = \sigma_{rz}^2 \Leftrightarrow s_A^2 = s_{rz}^2$$

- there is no significant difference between the variance due to the action of factor B and the residual variance

$$H_B : \sigma_B^2 = \sigma_{rz}^2 \Leftrightarrow s_B^2 = s_{rz}^2$$

- there is no significant difference between the variance due to the action of factor C and the residual variance

$$H_C : \sigma_C^2 = \sigma_{rz}^2 \Leftrightarrow s_C^2 = s_{rz}^2$$

- the interaction between factors A and B cannot lead to a new different statistical population

$$H_{AB} : \sigma_{AB}^2 = \sigma_{rz}^2 \Leftrightarrow s_{AB}^2 = s_{rz}^2$$

- the interaction between factors A and C cannot lead to a new different statistical population

$$H_{AC} : \sigma^2_{AC} = \sigma^2_{rz} \Leftrightarrow s^2_{AC} = s^2_{rz}$$

- the interaction of the factors B and C cannot lead to a new different statistical population

$$H_{BC} : \sigma^2_{BC} = \sigma^2_{rz} \Leftrightarrow s^2_{BC} = s^2_{rz}$$

Table 5.57 contains the synthesis of the analysis of variances for the problem of the Friedel–Crafts reaction. It is easy to observe that hypotheses H_A, H_B, H_C, H_{AC} and H_{BC} have been accepted. So, with respect to the specified state of the factors, the efficiency of the Friedel–Crafts reaction depends only on interaction AB (reaction time and timing (B)).

Table 5.57 Analysis of variances, example 5.6.3 (dependence of Friedel–Crafts reaction efficiency on temperature, reaction time and particular time of introduction of the catalyst).

Origin of variance	Sums for variance	Degrees of freedom	Variances	Computed value of the Fischer variable	Theoretical value of the Fischer variable	Decision
Change of factor A	$S_A = 1.49$	1	$s^2_A = S_A/1$ $= 1.49$	$F_A = s^2_A/s^2_{rz}$ $= 1.49$	$F_{1,6,\alpha} = 5.99$	$F_A < F_{1,6,\alpha}$ Accept H_A
Change of factor B	$S_B = 9.58$	2	$s^2_B = S_B/2$ $= 4.49$	$F_B = s^2_B/s^2_{rz}$ $= 2.63$	$F_{2,6,\alpha} = 5.14$	$F_B < F_{2,6,\alpha}$ Accept H_B
Change of factor C	$S_C = 1.36$	3	$s^2_C = S_C/3$ $= 0.45$	$F_C = s^2_B/s^2_{rz}$ $= 0.25$	$F_{3,6,\alpha} = 4.76$	$F_C < F_{3,6,\alpha}$ Accep H_c
Interaction A B	$S_{AB} = 144.13$	2	$s^2_{AB} = S_{AB}/2$ $= 72.06$	$F_{AB} = s^2_{AB}/s^2_{rz}$ $= 39.56$	$F_{2,6,\alpha} = 4.76$	$F_{AB} > F_{2,6,\alpha}$ Refuse H_{AB}
Interaction A C	$S_{AC} = 2.62$	3	$s^2_{AC} = S_{AC}/3$ $= 0.87$	$F_{AC} = s^2_{AC}/s^2_{rz}$ $= 0.48$	$F_{3,6,\alpha} = 4.76$	$F_{AC} < F_{3,6,\alpha}$ Accept H_{AC}
Interaction B C	$S_{BC} = 14.41$	6	$s^2_{BC} = S_{BC}/6$ $= 2.41$	$F_{BC} = s^2_{BC}/s^2_{rz}$ $= 1.32$	$F_{6,6,\alpha} = 4.28$	$F_{BC} < F_{6,6,\alpha}$ Accept H_{BC}
Residual	$S_r = 10.90$	6	$s^2_{rz} = S_{rz}/6$ $= 1.82$			

The interval of confidence of the variation of the reaction efficiency can be calculated using Table 5.56. Then, considering the AB interaction, we can compute the mean efficiency of the reaction for positions A_1B_1, A_1B_2, A_1B_3; we consequently have $m_{A_1B_1} = 50.5/4 = 12.65$, $m_{A_1B_2} = 25.4/4 = 6.35$, $m_{A_1B_3} 25.5/4 = 6.36$ and therefore the variations associated to the efficiency of the reaction are: $d_{12} = m_{A_1B_1} - m_{A_1B_2} = 6.3$, $d_{13} = m_{A_1B_1} - m_{A_1B_3} = 6.29$. The mean value of the variation of the reaction efficiency will be $d_B = (d_{12} + d_{13})/2 = 6.295$ and we compute the confidence interval for this mean value. The calculation of the theoretical value of the Student variable for $\alpha = 0.05$ and $\upsilon = 6$(this is the number of degrees of freedom associated to the residual variance) is $1 - \alpha = \int_o^t f_9(s)ds$ and $t = t_{6,0.05} = 2.47$ and therefore now, using relation (5.161), we can compute the confidence interval for this variation of reaction efficiency. The result is $I_B = (6.295 - 2.47*(1.82)^{0.5}/(2*3), 6.295 + 2.47*(1.82)^{0.5}/(2*3)) = (4.95, 7.65)$. In other words, we can say that, for reaction time $A_1 = 10$ h, if we change the timing of introduction of the catalyst from $B_1 = 2$ h to $B_2 = 4$ h, then we obtain a variation of the reaction efficiency between 4.95 and 7.65%. The case when factor A has level A_2 and factor B changes between B_1 and B_3 can be approached by the same procedure. The final conclusion of this analysis shows that the level A_1 for factor A, the level B_1 for factor B and any level for factor C are enough to ensure the conditions for the most favourable reaction efficiency.

5.6.3.3 Analysis of the Effects of Systematic Influences

The external systematic influence is common in experimental research when the quality of the raw materials and of the chemicals undergo minor changes and/or when the first data were obtained in one experimental unit and the remaining measurements were carried out in a similar but not identical apparatus.

In these situations, we cannot start the analysis of data without separating the effect of the external systematic influence from the unprocessed new data. In other words, we must separate the variations due to the actions of some factors with systematic influences from the original data. For this purpose, the methods of Latin squares and of effects of unification of factors have been developed in the plan of experiments.

In the method of Latin squares, the experimental plan, given by the matrix of experiments, is a square table in which the first line contains the different levels of the first factor of the process whereas the levels for the second factor are given in the first column. The rest of the table contains capital letters from the Latin alphabet, which represent the order in which the experiments are carried out (example: for pressure level P_1, four experiments for the temperature levels T_1, T_2, T_3, T_4 occur in the following sequence: A, B, C, A where A has been established as the first experiment, B as the second experiment, etc). The suffixes of these Latin capital letters introduce the different levels of the factors. Table 5.58 presents the schema of a plan of Latin squares. We can complete the description of this plan showing that the values of the process response can be written in each letter box once the experiment has been carried out. Indeed, we utilize three indexes for the theoretical utterance of a numerical value of the process response (v). For example, for $v_{ij}^{(A)}$, the i index shows that the level of the P factor is P_i, the j index gives level T_j for factor T and the final superscript A shows that the progression of the experiments must be A. Considering Table 5.58, it is important to observe that the experiments complete each box placed in an intersection between a line and a column with only a single value.

Table 5.58 Data for the Latin squares method for a process with three factors.

First factor T (temperature)	T_1	T_3	T_3	T_4
Second factor P (pressure)				
P_1	A	B	C	D
P_2	B	C	D	A
P_3	C	D	A	B
P_4	D	A	B	C

The correct use of the Latin squares method imposes a completely random order of execution of the experiments. As far as the experiment required in the box table is randomly chosen and as a single value of the process response is introduced into the box, we guarantee the random spreading of the effect produced by the factor which presents a systematic influence.

Once the levels of the factors have been selected, we can begin to write the plan introducing the order of the experiments by using: (i) the random changes between lines or between columns; (ii) the line variations using a random number generator; (iii) the extraction from a black box. Using one of these procedures to select the order of the experiments allows one to respect the conditions imposed by the random spreading of the effect produced by the factors.

The variance analysis for a plan with Latin squares is not different from the general case previously discussed in Section 5.6.3. Therefore we must compute the following sums:

S_1 – sum of the squares of all individual observations;

S_2 – sum of the squares of the sums of the columns divided by the number of observations in a column;

S_3 – sum of the squares of the sums of the lines divided by the number of observations in a line;

S_4 – sum of the squares of the sums of observations with the same Latin letter divided by the number of the observations having the same letter;

S_5 – the square of the sum of all observations divided by the number of observations.

Indeed, these sums allow the calculation of the variances due to each of the factors introducing the columns, the lines and the letter in the plan. We can introduce the statistical hypotheses about the effect of the factors on the process response using the variances of the factors with respect to the residual variance. This residual variance is computed by $s_{rz}^2 = (S_1 + S_5 - S_2 - S_3)/[(n-1)(n-1)]$ where n is the box number in a line (or a column). We can then verify the following hypotheses:

- The effect on the process response of the factor which changes the columns of the plan is not important. Mathematically we can write:

$$H_C : \sigma_C^2 = \sigma_{rz}^2 \Leftrightarrow s_C^2 = s_{rz}^2;$$

- The effect on the process response of the factor which changes the lines in a plan is not important. Therefore, we can write:

$$H_L : \sigma_L^2 = \sigma_{rz}^2 \Leftrightarrow s_L^2 = s_{rz}^2;$$

- The factor which changes the letter in the plan does not have a considerable influence on the process response. Then, according to the updated cases, we can write:

$$H_A : \sigma_A^2 = \sigma_{rz}^2 \Leftrightarrow s_A^2 = s_{rz}^2.$$

Now every reader knows that to check a hypothesis in which we compare two variances, we have to use the Fischer test. Here the computed value of a Fischer random variable is compared with its theoretical value particularized by the concrete degrees of freedom (v_1, v_2) and the confidence level $1 - \alpha$. Table 5.59 presents the synthesis of the analysis of variances for this case of the Latin squares method.

The following example will illustrate this method. The reaction considered is the chlorination of an organic liquid in a small laboratory scale reactor which works under agitation and at a constant chlorine pressure; the temperature of the reactor is controlled by a liquid circulating in a double-shell. The analysis of the reactor product shows the presence of some undesired components. The concentrations of the desired product and by-products are determined by the temperature, the chlorination degree (more precisely the reaction time) and by the catalyst concentration. Various procedures can be used for the addition of the catalyst: the whole catalyst is poured in one go, by fractions diluted with reactants, etc; the objective is to obtain a catalyst concentration between 0.1 and 0.3% g/g. The addition of the catalyst can be considered as an example of systematic influence and then its effect on the concentration of the by-products can be analyzed by the Latin squares method. Five levels are selected for the temperature and for the chlorination degree, which are considered as factors which do not have a systematic influence. The catalyst addition procedure and its concentration with respect to the reaction mixture can be introduced as a process factor with systematic influence by a group of five letters: A, B, C, D, E. Table 5.60 gives the factorial program obtained after the experiments have been extracted from a black box. This table contains all the measured concentrations of undesired products after each experiment.

Table 5.59 Analysis of the variances for the case of Latin squares method.

Origin of the variance	Sums of the squares	Degrees of freedom	Variances	Computed value of the Fischer variable	Theoretical value of the Fischer variable	Decision
Effect of a factor that changes the columns	$S_2 - S_5 = S_C$	$n - 1$	$s^2_C = S_C/(n-1)$	$F_C = s^2_C/s^2_{rz}$	$F_{n-1,(n-1)(n-2),\alpha}$	$F_C < F_{n-1,(n-1)(n-2),\alpha}$ H_C accepted
Effect of a factor that changes the lines	$S_3 - S_5 = S_L$	$n - 1$	$s^2_L = S_L/(n-1)$	$F_L = s^2_L/s^2_{rz}$	$F_{n-1,(n-1)(n-2),\alpha}$	$F_L < F_{n-1,(n-1)(n-2),\alpha}$ H_L accepted
Effect of a factor that changes the letter	$S_4 - S_5 = S_A$	$n - 1$	$s^2_A = S_A/(n-1)$	$F_A = s^2_A/s^2_{rz}$	$F_{n-1,(n-1)(n-2),\alpha}$	$F_A < F_{n-1,(n-1)(n-2),\alpha}$ H_c accepted
Residual	$S_r = S_1 + S_5 - (S_2 + S_4)$	$(n-1)*(n-2)$	$s^2_{rz} = S_r/[(n-1)(n-2)]$		Without the power enabling one to identify interaction effects	
Total	$S_1 - S_5$	$n^2 - 1$				

Table 5.60 Factorial plan for the Latin squares method – case of chlorination of an organic liquid.

Temperature	90	70	50	60	80	Total L	Total letter
Chlorination degree							
40	B r 39.5	A 28.9	D 11.6	C 13.9	E 22.6	116.5	A = 89.4
35	E 32.2	C 25.5	B r 10.0	D 12.5	A 14.6	94.8	B = 129.7
45	D 57.6	E 41.0	C 12.1	A 14.7	B r 25.9	151.3	C = 136.8
30	A 19.6	D 21.2	E 10.3	B 9.8	C 12.1	72.9	D = 138.3
50	C 73.2	B r 44.5	A 11.7	E r 19.7	D 35.4	184.5	E = 125.8
Total C	222.0	161.1	55.7	70.6	110.6	620	

The data in the table above will first be used to determine whether the addition procedure as well as the major factors of the process influence the dependent process variable (concentration of the undesired components in the reaction product). With the purpose being to obtain the real residual variance from the experiments, Table 5.60 (the boxes of which contain an r), have been repeated. These results are shown in Table 5.61.

Table 5.61 Results of repeated experiments, example 5.6.3.

Position (T,D)	(90,40)	(70,50)	(50,35)	(60,50)	(80,45)
Old value	39.5	44.5	10.0	19.7	25.9
New value	36.9	42.4	11.3	20.5	24.1

As previously explained, the first step to solve this application is the computation of the sums required by Table 5.59. Then we obtain:

S_1 – the sum of the squares of the individual observations:
$$S_1 = 39.5^2 + 28.9^2 + \ldots + 19.7^2 + 35.4^2 = 21717.42;$$
S_2 – the sum of the squares of the sums of the columns divided by the number of observations of the column:
$$S_2 = (222^2 + 161^2 + 55.7^2 + 70.6^2 + 110.6^2)/5 = 19104.84;$$
S_3 – the sum of the squares of the sums of the lines divided by the number of observations of the line:
$$S_3 = (116.5^2 + 94.8^2 + 151.3^2 + 72.9^2 + 184.5^2)/5 = 16961.13;$$
S_4 – the sum of the squares of sums with the same letter divided by the number of observations which have the same letter:
$$S_4 = (89.4^2 + 129.7^2 + 156.8^2 + 138.3^2 + 125.8^2)/5$$
$$= 15696.28$$
S_5 – the square of the sum of all observations divided by the total number of observations:
$$S_5 = (39.5 + 28.9 + \ldots + 19.7 + 35.4)^2/25 = 620.0^2/25$$
$$= 15376.$$

The results of the analysis using the data of Table 5.59 are given in Table 5.62. Considering the decision column, we conclude that two zero hypotheses have been refused and one has been accepted.

Table 5.62 Analysis of the variances for the Latin squares method, example 5.6.3.

Origin of the variance	Sums of squares	Degrees of freedom	Variances	Computed value of the Fischer variable	Theoretical value of Fischer variable	Decision
Effect of factor that change the columns	$S_2 - S_5 = S_C$ $S_C = 3728.4$	$n-1$ $n-1=4$	$s^2_C = S_C/(n-1)$ $s^2_C = 933.82$	$F_C = s^2_C/s^2_{rz}$ $F_C = 15.82$	$F_{n-1,(n-1)(n-2),\alpha}$ $F_{4,12,0.05} = 3.26$	$F_C >$ $F_{n-1,(n-1)(n-2),\alpha}$ H_C rejected
Effect of a factor that changes the lines	$S_3 - S_5 = S_L$ $S_L = 1585.3$	$n-1$ $n-1=4$	$s^2_L = S_L/(n-1)$ $s^2_L = 396.28$	$F_L = s^2_L/s^2_{rz}$ $F_L = 6.71$	$F_{n-1,(n-1)(n-2),\alpha}$ $F_{4,12,0.05} = 3.26$	$F_L >$ $F_{n-1,(n-1)(n-2),\alpha}$ H_L rejected
Effect of a factor that changes the letter	$S_4 - S_5 = S_A$ $S_A = 320.24$	$n-1$ $n-1=4$	$s^2_A = S_A/(n-1)$ $s^2_A = 80.06$	$F_A = s^2_A/s^2_{rz}$ $F_A = 1.35$	$F_{n-1,(n-1)(n-2),\alpha}$ $F_{4,12,0.05} = 3.26$	$F_A <$ $F_{n-1,(n-1)(n-2),\alpha}$ H_c accepted
Residual	$S_{rz} = S_1 + S_5 -$ $(S_2 + S_4)$ $S_r = 707.20$	$(n-1)*(n-2)$ $= 12$	$s^2_{rz} = S_{rz}/[(n-1)(n-2)] = 58.91$		It is not possible to identify the effects of double interactions.	
Total	$S_1 - S_5$	$n^2 - 1 = 24$				

It is important to note that the effect of the factor that changes the letter in the Latin squares table is negligible. Then, for the investigated chlorination reaction both the concentration of the catalyst (between 0.1 and 0.3% g/g) and its process of addition do not have any effect on the concentration of the by-products. Nevertheless, this conclusion cannot be definitive because we can find from Table 5.62 that we have a high residual variance. In this case, we can suggest that the interaction effects are certainly included in the residual variance.

The real residual variance frequently named "reproducibility variance" can be determined by repeating all the experiments but this can turn out to be quite expensive. The Latin squares method offers the advantage of accepting the repetition of a small number of experiments with the condition to use a totally random procedure for the selection of the experiments. With the data from Table 5.61 and using the relation $s^2_{rz} = (\sum_{i=1}^{n_C} \sum_{j=1}^{n_L-1} d^2_{ij})/[(n_C(n_L - 1))]$, where d_{ij} are the differences between the observed values for all the n_c columns and n_L lines (where the new experiments can be found), we obtain: $s^2_{rz} = (2.6^2 + 1.3^2 + 1.8^2 + 2.1^2 + 0.8^2)/(5*1)$ =1.56. Five degrees of freedom characterize this new computed variance.

Now, it is clear that the residual variance from Table 5.62 contains one or more interaction effects. Moreover, for this application or, more precisely, for the data given for the particularization of the Latin squares method, a partial response has been obtained. Consequently, a new research plan must be suggested in order to answer our problem.

The method of the effects of the unification of factors considers that, for a fixed plan of experiments, we can produce different groups where each contains experiments presenting the same systematic influence [5.8, 5.13, 5.23, 5.35, 5.36]. To introduce this method, we can consider the case of a process with three factors analyzed with a CFE 2^3 plan of experiments. In our example, we will take into account the systematic influence of a new factor D. To begin this analysis, we will use the initial plan with eight experiments with the condition to separate these experiments into two blocks or groups:

- the first block is bound with the first level of the factor of systematic influence and the second block corresponds to the next level of the factor of systematic influence;
- we accept both blocks to be related by a triple interaction variance (s^2_{ABC}).

For this case of separation into two groups or blocks, it is important to determine the experiments from the 2^3 plan which are contained in block D_1 and those contained in D_2.

Table 5.63 shows the detailed separation of the experiments into groups. Each experiment corresponding to a different block is identified by a current name and by a code. The experiments with the sign + in the ABC column correspond to the block D_1, the remaining experiments to block D_2.

Table 5.63 The division of a CFE 2^3 plan into two blocks.

i	A	B	C	AB	AC	BC	ABC	y_i	
1	−1	−1	−1	+1	+1	+1	−1	$y_1 = (1)$	Block D_1
2	+1	−1	−1	−1	−1	+1	+1	$y_2 = a$	Experiments: 2, 3, 5, 8 Codified names: a, b, c, abc
3	−1	+1	−1	−1	+1	−1	+1	$y_3 = b$	
4	+1	+1	−1	+1	−1	−1	−1	$y_4 = ab$	Block D_2
5	−1	−1	+1	+1	−1	−1	+1	$y_5 = c$	Experiments: 1, 4, 6, 7
6	+1	−1	+1	−1	+1	−1	−1	$y_6 = ac$	Codified names: (1), ab, ac, bc
7	−1	+1	+1	−1	−1	+1	−1	$y_7 = bc$	
8	+1	+1	+1	+1	+1	+1	+1	$y_8 = abc$	

When we have the possibility to obtain the real residual variances (2–3 experiments repeated in the D_1 and D_2 blocks), we can suggest to validate the following hypothesis: $H_{ABC} : \sigma^2_{ABC} = \sigma^2_{rz} \Leftrightarrow s^2_{ABC} = s^2_{rz}$ and if it is rejected, we can conclude an important or crucial effect on the process response of the factor which shows a systematic influence.

The justification for our consideration showing that the action of a factor with systematic influence is concentrated in the relation which binds the blocks (frequently named contrast) is sustained by the following observations:

- if we accept that block D_1 increases the process response, then, with respect to the D_2 block, the results will be:
  $(a + d)$, $(b + d)$,
  $(c + d)$, $(abc + d)$;
- with Eq. (5.168) we obtain
  $EA = -(1) + (a + d) - (b + d) + ab - (c + d) + ac -$
  $bc + (abc + d) = -(1) + a - b + ab - c + ac - bc + abc$. Similar
  expressions are thus obtained for the effects EB, EC, EAB, EAC,
  EBC; these effects are not affected by the increase of the response
  in block D_1.
- for the contrast we obtain $EABC = -(1) + (a + d) + (b + d) - ab +$
  $(c + d) - ac - bc + (abc + d) = -(1) + a + b - ab + c - ac - bc + abc +$
  $4d$; this result shows a displacement with 4d; so the variance due
  to this interaction is the only variance obtained when we utilize a
  two block division for a CFE 2^3 plan.

A division into four blocks made from two unification relations, is also possible with a CFE 2^3 plan where the systematic influence of one or more factors is considered. If interactions AB and AC give the unification relations, then, by using the block division procedure used above (Table 5.63), the following blocks will be obtained:

Block 1 or block + +: experiments 1 and 8 with code names (1) and abc;
Block 2 or block – –: experiments 2 and 7 with code names a and bc;
Block 3 or block – +: experiments 3 and 6 with code names b and ac;
Block 4 or block + –: experiments 4 and 5 with code names ab and c.

In this division example, if interactions AB and AC influence the process response, we can conclude that the displacement of the process response contains the effect of a systematic influence.

The examples where a CFE 2^3 plan has been divided into two or four blocks are not explicit enough to develop the idea that the relations of the unification of blocks are selected randomly. In the next example, a CFE 2^4 plan is developed with the purpose being to show the procedures to select the unification relations of inter-blocks. In this plan, the actions showing a systematic influence will be divided into two blocks or into four blocks with, respectively, eight experiments or four experiments per block. We start this new analysis by building the CFE 2^4 plan. Table 5.64 contains this CFE 2^4 plan and also gives the division of the two blocks when we use the ABCD interaction as a unification relation.

Table 5.64 The separation of a CFE 2^4 into two blocks.

i	A	B	C	D	AB	AC	AD	BC	BD	CD	ABC	ABD	ACD	BCD	ABCD	y_i
1	−1	−1	−1	−1	+1	+1	+1	+1	+1	+1	−1	−1	−1	−1	+1	$y_1 = (1)$
2	+1	−1	−1	−1	−1	−1	−1	+1	+1	+1	+1	+1	+1	−1	−1	$y_2 = a$
3	−1	+1	−1	−1	−1	+1	+1	−1	−1	+1	+1	+1	−1	+1	−1	$y_3 = b$
4	+1	+1	−1	−1	+1	−1	−1	−1	−1	+1	−1	−1	+1	+1	+1	$y_4 = ab$
5	−1	−1	+1	−1	+1	−1	+1	−1	+1	−1	+1	−1	+1	+1	−1	$y_5 = c$
6	+1	−1	+1	−1	−1	+1	−1	−1	+1	−1	−1	+1	−1	+1	+1	$y_6 = ac$
7	−1	+1	+1	−1	−1	−1	−1	+1	−1	−1	−1	+1	+1	−1	+1	$y_7 = bc$
8	+1	+1	+1	−1	+1	+1	+1	+1	−1	−1	+1	−1	−1	−1	−1	$y_8 = abc$
9	−1	−1	−1	+1	+1	+1	−1	+1	−1	−1	−	+1	+1	+1	−1	$y_9 = d$
10	+1	−1	−1	+1	−1	−1	+1	+1	−1	−1	+1	−1	−1	+1	+1	$y_{10} = ad$
11	−1	+1	−1	+1	−1	+1	−1	−1	+1	−1	+1	−1	+1	−1	+1	$y_{11} = bd$
12	+1	+1	−1	+1	+1	−1	+1	−1	+1	−1	−1	+1	−1	−1	−1	$y_{12} = abd$
13	−1	−1	+1	+1	+1	−1	−1	−1	−1	+1	+1	+1	−1	−1	+1	$y_{13} = cd$
14	+1	−1	+1	+1	−1	+1	+1	−1	−1	+1	−1	−1	+1	−1	−1	$y_{14} = acd$
15	−1	+1	+1	+1	−1	−1	−1	+1	+1	+1	−1	−1	−1	+1	−1	$y_{15} = bcd$
16	+1	+1	+1	+1	+1	+1	+1	+1	+1	+1	+1	+1	+1	+1	+1	$y_{16} = abcd$

Block E_1 or Block E_+:	Block E_2 or Block E_-:	ABCD
Experiences: 1, 4, 6, 7, 10, 11, 13, 16	Experiences: 2, 3, 5, 8, 9, 12, 14, 15	unification
Codes: (1), ab, ac, bc, ad, bd, cd, abcd	Codes: a, b, c, abc, d, abd, acd, bcd	

If we now suppose that the aim is to divide the CFE plan 2^4 into four blocks, we can select one of the following unification relations: (i) ABCD coupled with one from the three order interactions (ABC, ACD, BCD, etc.); (ii) ABCD coupled with one from the two-order interactions (AB, AC, AD, etc.); (iii) two interactions of three-order, etc. To establish which coupling is the most favourable, it is necessary to know what type of information disappears in each case. For this purpose we show here some of the multiplications of the ABCD interaction relations with their possible coupling interaction relations where $A^2 = B^2 = C^2 = D^2 = 1$.

$$\text{I. } ABCD * BCD = A \quad \text{II. } ABCD * ABC = D \quad \text{III. } ABCD * ACD = B$$
$$\text{IV. } ABCD * AB = CD \quad \text{V. } ABCD * AC = BD \quad \text{VI. } ABCD * AD = BC.... \tag{5.170}$$

From this result we can then conclude that: (i) for a *four–three* coupling in the data processing, the information about the effect of the direct factor (A, B, C, D) action on the process response disappears; it is obvious that, for actual cases, it is difficult to accept this situation; (ii) when a *four–two* coupling occurs, the information that shows the effect of one interaction of order two disappears, however, this situation can sometimes be accepted in actual cases; (iii) it is not difficult to show that for a *three–three* coupling we obtain the case of the *four–two* coupling.

The division of the CFE 2^4 plan into four blocks by means of the *four–two* couple is useful to identify the weakest order two interactions that can be used with the order four interactions as unification relations. At the same time, we can also analyze the *three–three* couple obtained with the most non-important order three interactions. In fact, it is easy to accept that, for an investigated process, the effects on the process response of the order three interactions are non-important for most actual situations. Indeed, when for a CFE 2^4 plan, the ABC and BCD interactions are the weakest, these interactions can be selected as relations for the unification of the inter-blocks. Then, we can rapidly produce the division into four blocks: $E_1 = E_{--}, E_2 = E_{-+}, E_3 = E_{+-}, E_4 = E_{++}$. Table 5.65 shows the blocks and the corresponding experiments with their usual numbers and codes.

Table 5.65 The blocks repartition of a CFE 2^4 plan using the contrasts ABC, BCD.

ABC	–1	+1	+1	–1	+1	–1	–1	+1	–1	+1	+1	–1	+1	–1	–1	+1
BCD	–1	–1	+1	+1	+1	+1	–1	–1	+1	+1	–1	–1	–1	–1	+1	+1
y_i	(1)	a	b	ab	c	ac	bc	abc	d	ad	bd	abd	cd	acd	bcd	abcd

Block E1 = E–		Block E2 = E–+		Block E3 = E+–		Block E1 = E+	
(1)	number 1	ab	number 4	a	number 2	b	number 2
bc	number 7	ac	number 6	abc	number 8	c	number 5
abd	number 12	d	number 9	bd	number 11	ad	number 10
acd	number 14	bcd	number 15	cd	number 13	abcd	number 16

For the cases of 2^5 and 2^6 CFE plans, the division into blocks must respect the principles previously shown for a 2^4 plan. Considering a 2^5 plan, the recommended contrast couplings are of the *three–three–four* type. If the coupling chain is ABC–ADE–BCDE, then the main block F_1 (analogue to E_1 for the case of a CFE 2^4) will contain experiments (1)–bc–de–abd–acd–abc–ace–bcde.

We establish the repartition of the experiments for the remaining blocks by multiplying the F_1 chain by a, b and c; for these products we have $a^2 = b^2 = c^2 = d^2 = 1$. So, the F_2 will contain the next chain of experiments: a–abc–ade–bd–cd–bc–ce–abde. The following application presents an actual case for a CFE 2^4 plan where the separation has been obtained according to the contrasts ABC and BCD.

Numerical application. This application concerns the conversion of one reactant by an esterification reaction occurring in a discontinuous and stirred reactor. It is a function of the temperature (factor A), the alcohol–acid molar ratio (factor B), the reaction time (factor C) and the catalyst concentration (factor D). A CFE 2^4 plan is used to investigate the different effects of the factors. The levels of the factors have been established in order to obtain a good reactant conversion. These levels are: temperatures: $A_1 = 110\,°C$, $A_2 = 130\,°C$; alcohol–acid molar ratio: $B_1 = 1.2$, $B_2 = 1.5$; reaction time: $C_1 = 3$ h, $C_2 = 4$ h; catalyst concentration: $D_1 = 1\%$ g/g, $D_2 = 2\%$ g/g.

Three different qualities of alcohol have been used: recycled, distilled or rectified. It is easy to observe that the quality of the alcohol introduces a systematic influence towards factor B in the esterification reaction. Indeed, the development of the experimental research is made with a 2^4 plan with four blocks. The ABC and BCD have the contrasts considered for the blocks division. For the experiments grouped in block E_1, the first type of alcohol has been used. Distilled alcohol is the reactant used in the experiments of the second block (E_2) and the rectified alcohol for the experiments of the last two blocks (E_3, E_4). Table 5.66 presents the initial data where the division of the blocks is not visible.

Table 5.66 The conversion for an esterification reaction in a CFE 2^4 plan.

| Esterification reaction 4 blocks | D_1 | | | | D_2 | | | |
| | C_1 | | C_2 | | C_1 | | C_2 | |
	B_1	B_2	B_1	B_2	B_1	B_2	B_1	B_2
A_1	(1)	b	c	bc	d	bd	cd	bcd
	28	31	26	32	33	33	36	38
A_2	a	ab	ac	abc	ad	abd	acd	abcd
	20	24	20	30	24	24	37	31

Table 5.67 shows the conversions characterizing each block and the corresponding columns of the sums, these data are necessary to compute the variance due to the division into blocks. Indeed, these sums will be used for the computation of the square sums showing the differences in the reaction conversion produced by the alcohol quality (S_M).

Table 5.67 Presentation of the blocks in the example of an esterification reaction.

Block E_1		Block E_2		Block E_3		Block E_4	
Experiment code	Conversion	Experiment code	Conversion	Experiment code	Conversion	Experiment code	Conversion
(1)	28	a	20	b	31	d	33
abd	24	bd	33	ad	24	ab	24
acd	37	cd	36	abcd	31	ac	20
bc	32	abc	30	c	26	bcd	38
Total	121	Total	119	Total	112	Total	115

The computation for the analysis of the variances is carried out following the procedure described in Section 5.6.3.1. When we begin to complete Table 5.52 as recommended by this procedure, we can observe that we must add the effect of the D factor as well as its interactions. Nevertheless, in this table, we cannot add the unification of the interactions accepted by the data provided by the division into blocks. In addition to the data described here, we have to realize the following computations in order to complete Table 5.52:

- the sum of the squares of the sums of the conversion obtained for each block divided by the number of blocks:

$$S_1 = (121^2 + 119^2 + 112^2 + 117^2)/4 = 13642.75;$$

- the sum of the squares of each conversion divided by the total number of the experiments:

$$S_2 = (28^2 + 31^2 + 26^2 + ... + 24^2 + 37^2 + 31^2)/16 = 13630.56;$$

- the sum of the squares showing the differences due to the alcohol quality:

$$S_M = S_1 - S_2 = 12.19;$$

- the sum of squares due to the unification of the interactions by using the following algorithm:
  (a) we compute EABC and EBCD using the general procedure particularized to the data in Table 5.64:
  EABC = $-(1) + a + b - ab + c - ac - bc + abc - d + ad + bd -$ abd + cd - acd - bcd + abcd = $-28 + 20 + 31 - 24 + 26 - 20 -$ $32 + 30 - 33 + 24 + 33 - 24 + 36 - 30 - 38 + 37 = 0;$

EBCD = – (1) – a + b + ab + c + ac – bc – abc + d + ad – bd –
abd – cd – acd + bcd + abcd = –28 – 20 + 31 + 24 + 26 + 20 –
32 –30 + 33 + 24 – 33 – 24 – 36 – 37 + 38 + 31 = –13;

(b) we calculate S_{ABC} and S_{BCD} using the values of the effects
EABC and EBCD:
$S_{ABC} = (EABC)^2/16 = 0$, $S_{BCD} = (EBCD)^2/16 = 13^2/16 = 10.65$;

(c) we finish the algorithm by computing the squares sum of
S_{ABC} and S_{BCD}: $S_{INT} = S_{ABC} + S_{BCD} = 10.65$.

At this point, we have to verify the correctness of the selection of the unification relations. When $S_M \cong S_{INT}$ we can conclude that our selection for the unification relations is good; in this case, we can also note that the calculations have been made without errors. Otherwise, if computation errors have not been detected, we have to observe that the selected interactions for the unification of blocks are strong and then they cannot be used as unification interactions. In this case, we have to carry out a new experimental research with a new plan. However, part of the experiments realized in the previous plan can be recuperated. Table 5.68 contains the synthesis of the analysis of the variances for the current example of an esterification reaction. We observe that, for the evolution of the factors, the molar ratio of reactants (B) prevails, whereas all other interactions, except interaction AC (temperature–reaction time), do not have an important influence on the process response (on the reaction conversion). This statement is sustained by all zero hypotheses accepted and reported in Table 5.68. It should be mentioned that the alcohol quality does not have a systematic influence on the esterification reaction efficiency. Indeed, the reaction can be carried out with the cheapest alcohol. As a conclusion, the analysis of the variances has shown that conversion enhancement can be obtained by increasing the temperature, reaction time and, catalyst concentration, independently or simultaneously.

Table 5.68 Synthesis of the variance analysis for CFE 2^4, example of an esterification reaction.

Origin of the variance	Sums of the differences	Degrees of freedom	Variances	Computed value of the Fischer variable	Theoretical value of the Fischer variable	Decision
Temperature variation A	$S_A = 138.06$	1	$s^2_A = 138.06$	$F_A = s^2_A/s^2_{rz}$ $= 20.38$	$F_{1,6,\alpha} = 5.99$	$F_A > F_{1,6,\alpha}$ H_A refused
Molar ratio variation B	$S_B = 22.56$	1	$s^2_B = 22.05$	$F_B = s^2_B/s^2_{rz}$ $= 2.63$	$F_{1,6,\alpha} = 5.99$	$F_B < F_{1,6,\alpha}$ H_B accepted
Reaction time variation C	$S_C = 115.56$	1	$s^2_C = 115.56$	$F_c = s^2_B/s^2_{rz}$ $= 17.36$	$F_{1,6,\alpha} = 5.99$	$F_C > F_{1,6,\alpha}$ H_C refused
Catalyst conc. variation D	$S_D = 76.56$		$s^2_D = 6.56$	$F_D = s^2_D/s^2_{rz}$ $= 11.29$	$F_{1,6,\alpha} = 5.99$	$F_D > F_{1,6,\alpha}$ H_C refused
Alcohol type M	$S_M = 12.19$	3	$s^2_M = 4.06$	$F_M = s^2_M/s^2_{rz}$ $= 0.59$	$F_{3,6,\alpha} = 4.76$	$F_M < F_{1,6,\alpha}$ H_M accepted
Interaction AC	$S_{AC} = 52.56$	1	$s^2_{AC} = 52.56$	$F_{AC} = s^2_{AB}/s^2_{rz}$ $= 7.62$	$F_{1,6,\alpha} = 5.99$	$F_{AC} > F_{1,6,\alpha}$ H_{AB} refused
Interaction AB	$S_{AB} = 10.56$	1	$s^2_{AB} = 10.56$	$F_{AB} = s^2_{AB}/s^2_{rz}$ $= 1.55$	$F_{1,6,\alpha} = 5.99$	$F_{AB} < F_{1,6,\alpha}$ H_{AB} accepted
Interaction BC	$S_{BC} = 1.56$	1	$s^2_{BC} = 1.56$	$F_{BC} = s^2_{AB}/s^2_{rz}$ $= 0.23$	$F_{1,6,\alpha} = 5.99$	$F_{BC} < F_{1,6,\alpha}$ H_{BC} accepted
Interaction BD	$S_{BD} = 18.06$	1	$s^2_{BD} = 18.06$	$F_{BD} = s^2_{AB}/s^2_{rz}$ $= 2.66$	$F_{1,6,\alpha} = 5.99$	$F_{BD} < F_{1,6,\alpha}$ H_{BD} accepted
Interaction CD	$S_{CD} = 10.56$	1	$s^2_{CD} = 10.56$	$F_{CD} = s^2_{AB}/s^2_{rz}$ $= 1.55$	$F_{1,6,\alpha} = 5.99$	$F_{CD} < F_{1,6,\alpha}$ H_{CD} accepted
Interaction ABD	$S_{ABD} = 0.56$	1	$s^2_{ABD} = 0.56$	$F_{ABD} = s^2_{AB}/s^2_{rz}$ $= 0.07$	$F_{1,6,\alpha} = 5.99$	$F_{ABD} < F_{1,6,\alpha}$ H_{ABD} accepted
Interaction ACD	$S_{ACD} = 6.00$	1	$s^2_{ACD} = 6.00$	$F_{ACD} = s^2_{AC}/s^2_{rz}$ $= 0.88$	$F_{1,6,\alpha} = 5.99$	$F_{ACD} < F_{1,6,\alpha}$ H_{ACD} accepted
Interaction ABCD	$S_{ABCD} = 0.06$	1	$s^2_{ABCD} = 0.06$	$F_{ABCD} = s^2_{BC}/s^2_{rz}$ $= 0.01$	$F_{1,6,\alpha} = 5.99$	$F_{ACBD} < F_{1,6,\alpha}$ H_C accepted
Residual	$S_r = 47.36$	6	$s^2_{rz} = 6.77$	All interactions without AC		

5.7
Use of Neural Net Computing Statistical Modelling

At the beginning of this chapter, we introduced statistical models based on the general principle of the Taylor function decomposition, which can be recognized as non-parametric kinetic model. Indeed, this approximation is acceptable because the parameters of the statistical models do not generally have a direct contact with the reality of a physical process. Consequently, statistical models must be included in the general class of connectionist models (models which directly connect the dependent and independent process variables based only on their numerical values). In this section we will discuss the necessary methodologies to obtain the same type of model but using artificial neural networks (ANN). This type of connectionist model has been inspired by the structure and function of animals' natural neural networks.

Neural nets are computing programs that behave externally as multi-input multi-output computing blocks. Although artificial neural networks were initially devised for parallel processing, they are being used on sequential machines (von Neumann) as well.

They have been used successfully in several diverse engineering fields [5.37–5.39], such as process control engineering [5.40, 5.41] and non-parametric statistics [5.42–5.44]. A neural network is readily programmed for kinetic prediction where many strongly interacting factors do affect the process rate or when data are either incomplete, not defined or even lacking. With reference to the "black box" used by classical statistics to describe the action of internal parameters of processes and their interaction on the process exit, the ANN methodology is strongly different because it explains the mechanism working inside the black box.

5.7.1
Short Review of Artificial Neural Networks

As mentioned in the introduction, ANNs are models inspired by the structure and the functions of the biological neurons, since they can also recognize patterns, disordered structure data and can learn from observation.

A network is composed of units or simple named nodes, which represent the neuron bodies. These units are interconnected by links that act like the axons and dendrites of their biological counterparts. A particular type of interconnected neural net is shown in Fig. 5.12. In this case, it has one input layer of three units (leftmost circles), a central or hidden layer (five circles) and one output (exit) layer (rightmost) unit. This structure is designed for each particular application, so the number of the artificial neurons in each layer and the number of the central layers is not a priori fixed.

The system behaves like synaptic connections where each value of a connection is multiplied by a connecting weight and then the obtained value is transferred to another unit, where all the connecting inputs are added. If the total sum exceeds a certain threshold value (also called offset or bias), the neuron begins to fire [5.45, 5.46].

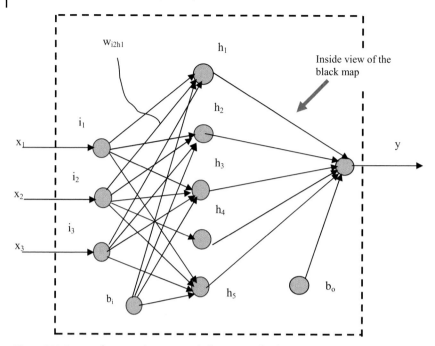

Figure 5.13 Layers of units and connection links in an artificial neuronal network. I_1–i_3: input neurons, h_1–h_5: hidden neurons, b_i, b_o: exit and output bias neurons, w_{i2h1}: weight of transmission, i_2–h_1, x_1–x_3: input process variables and y: output process variable.

The changes brought about in the pattern of neurons constitute the basis for learning.

In biological neurons, learning is carried out by changing the synaptic resistance associated to a change in the activation pattern of neurons.

Neural networks are able to learn because they can change the connection weights between two units which are in direct contact. After learning, the knowledge is somehow stored in the weights.

However, artificial neurons are much simpler than natural ones, the analogy serves to highlight an important feature of ANNs: the ability to learn through training. Just as the brain learns to infer from observations, an ANN learns the key features of a process through repeated training with data and, like in natural learning, its performance improves as it gains experience with a process.

The Latin expression "repetitio est mater studiorum", can be used here to describe the learning process with an ANN. The ANN repeats on and on, gradually adjusting the output to the imposed data output.

5.7.2
Structure and Threshold Functions for Neural Networks

The information flow between two biological neurons is affected by a variable synaptic resistance. In artificial systems, each connecting link has an associated weight. If two units are linked by a connection, the activation value of the emitting units is multiplied by the connecting weight before reaching the receiving unit.

The input value for an arbitrary unit, j, is then the sum of all activations coming from the units of the preceding layer, multiplied by the respective weights, w_{kj}, plus the bias value θ_{ij}. Thus, the total input to unit j, will be written as (5.171) where n represents the number of the neurons preceding neuron j and o_k shows the output.

$$I_j = \sum_{k=1}^{n} w_{kj}o_k + \theta_{ij} \tag{5.171}$$

Even though, most networks use the same type of input, their output generation may differ. In general, the output is computed by means of a transfer function, also called activation function. Concerning the behaviour of the transfer function, a gradual approach is required [5.47]. Therefore, a continuous threshold function is selected, chiefly because its continuity and derivability at all points are required features for the current optimization of the algorithms of learning. This type of function is well suited to the learning procedure that will be described later. A typical continuous threshold function is the following exponential sigmoid:

$$o_j = \frac{1}{1 + e^{-\beta I_j}} \tag{5.172}$$

where o_j is the activation value of neuron j, I_j is the total input to neuron j (as calculated by relation 5.171) and β is a constant which frequently takes a unitary value. The use of β, allows some modifications of the width of the region of the sigmoid, a feature which is useful in setting the learning ability of the net. Table 5.69 shows some other sorts of threshold functions that can be successfully used for developing an application.

Table 5.69 Common threshold functions used in ANN modeling.

Type	Function expression	Symbol signification
Linear	$o_j = \begin{cases} a + bI_j , & 0 \leq I_j \prec 1 \\ 1 , & I_j \geq 1 \end{cases}$	I_j = total input to neuron j
Saturating linear	$o_j = \begin{cases} I_j , & 0 \prec I_j \prec 1 \\ 1 , & I_j \geq 1 \\ 0 , & I_j \leq 0 \end{cases}$	o_j = output from neuron j
Sigmoid classic	$o_j = \dfrac{1}{1 + e^{-I_j}}$	I^T_j – transpose of I_j
Hyperbolic tangent	$o_j = \tanh(I_j)$	σ_j standard deviation of I_j
Radial basis	$o_j = \exp[-I^T_j / (2\sigma^2_j)]$	a, b – numerical constants

The function of the neural net depends not only on the information and processing mode of each isolated unit, but also on its overall topology. The topology considered in Fig. 5.12 must to be considered only as a didactic example. In the case of questions about the necessity of the hidden layer, we can easily give an answer: a hidden layer allows one to increase the network memory and provides some flexibility in the learning process. With the very simple topology considered in Fig. 5.13, the net is able to map linear and nonlinear relationships between inputs and outputs. The number of units in the input layer is determined by the variables that affect the response (x_1, x_2, x_3 in Fig. 5.13). The number of units in the hidden layer will be established during the learning process from a compromise between predicting errors and the number of iterations needed to attain them. In addition to the above units, two bias units are used (in Fig. 5.13, one for the hidden layer and one for the output unit). Their inputs are zero and their outputs or activation values are equal to one. Their use provides the threshold values to the hidden layer and to the output unit.

It is not difficult to observe that the application of an ANN to a problem involves four steps:

1. selection of the network topology (i.e. the layout of the neurons and their inter-connections),
2. specification of the transformation operator for each neuron from the topology,
3. initial assignment of weights w_{kj}, which are updated as the network learns,
4. initial learning, called training, which involves choosing the data and the training method.

As neural network theory has been developed, the empiricism associated with the choices at each step, has given ways to heuristic rules and guidelines [5.48, 5.49]. Nevertheless, experience still plays an important part in designing a network. The network depicted in Fig. 5.13 is the most commonly used and is called the feed-forward network because all signals flow forward.

Even though a number of techniques have been developed for the development of networks, they still remain iterative trial and error procedures. The heuristic approach described here can be used to reduce the trial and error selection process.

A hidden layer, with its appropriate units is capable of mapping any input presentation [5.50] and is thus necessary to restrict the topology to one layer only. So as to determine the optimum hidden units, the learning rate (v_l) and the momentum term (m_τ) will be assigned arbitrarily but with constant values and the gain term will be fixed at a value of one. With all the parameters fixed, various net topologies exhibit the same trends relative to each other, "vis-à-vis" the overall absolute error as a function of the number of iterations in the training mode [5.51]. Thus, it was found possible to determine the optimum net architecture within 50–100 iterations and without using the whole graph which describes the variation of the absolute error as a function of the number of iterations for each topology.

The number of iterations will be used as the criterion whenever on-line predictions are to be made, such as for chemical process control where computation time is important. The selection of v_l and m_τ and the gain term is essentially a trial and error procedure. Contrary to the usual approach, each of these parameters has not been fixed at a constant value for the entire training period. These were initially assigned with arbitrary values ($v_l = 0.8$, $m_\tau = 0.8$, gain $= 1$ for example, although these values are not a priori imposed). Then, they were updated while the parameters "jolted" the overall absolute error out of the local minima, which is typically encountered in the mechanism of the descendent gradient. Once the net parameters and the net architecture have been fixed, the minimum number of training data sets required for adequate mapping will be determined by trial and error procedures. The net is then ready to learn the data presented using the back-propagation algorithm.

5.7.3
Back-propagation Algorithm

As described below, the required behaviour is taught to the neural net by back propagation. This procedure is carried out by exposing the network to sets consisting of one input vector and its corresponding output vector. By an iterated procedure of trial and error, the convergence to determine the weight values that minimizes a prescribed error value is then achieved.

Back propagation is a kind of rapid descendent method of optimization. However, some authors prefer other optimization algorithms rather than back propagation, for example the Levenberg–Marquardt method [5.52]. The back-propagation algorithm with the delta rule is called a supervised learning method, because weights are adapted to minimize the error between the desired outputs and those calculated by the network. The error is calculated, for convenience, from the following expression in terms of squared deviations:

$$\Phi(w_{kj}^p) = \frac{1}{2} \sum_{p=1}^{r} \sum_{j=1}^{n} (\varsigma_j^p - o_j^p)^2 \tag{5.173}$$

where ς_j^p is the desired value output unit j for the sample pair p, o_j^p is the observed value for the same unit j and sample pair p, and p is the sum index for the total number of pairs r.

The adjustment on weights w_{kj} is done using the sensitivity of the error with respect to that weight, as:

$$\Delta w_{kj} = -\gamma \frac{\partial \Phi}{\partial w_{kj}} \tag{5.174}$$

The expression for the weight change is obtained from Eqs. (5.171) and (5.172) replacing them in the relation (5.174):

$$\Delta w_{hj} = \alpha \gamma (\varsigma_j - o_j) o_j (1 - o_j) o_h \tag{5.175}$$

where the indexes h and j refer to the nodes of the hidden and of the output layer respectively. Equation (5.175) allows one to modify the weights between the hidden and the output layers. On the other hand, the group of equations for the change of the weights between the hidden and output layer is obtained also, as:

$$\Delta w_{in\,k} = \alpha\gamma o_{in} o_k(1 - o_k) \sum_j \delta_j w_{hj} \qquad (5.176)$$

where the subscripts "in, h, j" now refer to the input layer, the hidden layer and the output layer, respectively. As for the output layer, δ_j can be expressed as:

$$\delta_j = \alpha(\varsigma_j - o_j)o_j(1 - o_j) \qquad (5.177)$$

In this case α and γ represent, respectively, the rate factor in output and the scaling factor of the net. These relations are also related to the sigmoid threshold function.

In order to modify the weights between the input and hidden layer it is necessary to know the weights between the hidden units and the output units. Therefore, during back propagation, first we change the connection weights between the output and hidden layer, and then we change the remaining weights conversely to the direction of information flow during the normal operation of the network: from hidden layer to input layer backwards.

While training is performed, the weights are initialized with values between – 0.5 and 0.5 [5.48, 5.53] using a random procedure. The input–output experimental pairs are successively shown to the net and the weights are changed simultaneously. When all pairs have been shown to the network, the error is computed. If it is larger than the value allowed, all the pairs are shown to the network again in order to induce more changes in the weights. This cycle is repeated until convergence is achieved.

5.7.4
Application of ANNs in Chemical Engineering

The ANN techniques can be successfully applied in the field of chemical engineering. The examples presented here give a brief overview of the capacity of ANNs to solve some chemical engineering problems. Readers interested in investigating this topic further can refer to Bulsari's book [5.41] or to other authors referenced in the bibliography [5.39, 5.54]. In addition, to complete the information presented here, an important number of Internet sites can be used as well as some of the current chemical engineering scientific publications, which have given important attention to this subject.

Three major ways can be identified for the use of an ANN in chemical engineering:

1. as a substitute for the complicated models of transport phenomena or stochastic based models;

2. as data support, especially for the equilibrium and kinetic data needed by models based on transport phenomena. This kind of model is recognized as a hybrid neural –regression model;
3. as a model for control and process operation.

In all the above-mentioned cases, once the learning processes have been completed, ANNs have an assistant function which gives one or many answers to an argument of the complex modeled process (parameters, factors or independent variables). When we use an ANN as a substitute for models for stochastic or complicated transport phenomena, the learning process must be as shown in Fig. 5.14 which shows the coupling of an ANN and a complicated mathematical process. The mathematical model gives the input and output vectors for the ANN, which, in normal cases, are represented by the measured data. When the learning process has been completed, the process mathematical model (PMM) and the optimizing algorithm (OA) are decoupled and the ANN is ready to produce the simulation results for the process. This procedure is also used to produce the ANN simulators needed for the control of the processes or their usual automatic operation.

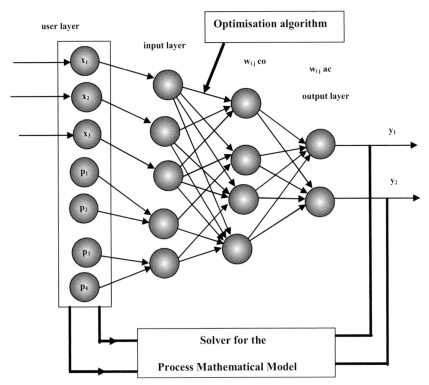

Figure 5.14 The architecture of assembly ANN-PMM-OA for the learning step (ANN – substitute for the complicated PMM).

In Fig. 5.14, it is shown that a previous formal user layer is necessary before using the input layer [5.55, 5.56]. Nevertheless, it is not necessary to have the same number of units in the input layer and in the user layer when each unit introduces a parameter or a process variable for the model.

The design of the input layer, which is a parameter for the ANN topology, will be coupled to the general problem of the topology of the ANN design.

The hybrid neural–regression model, shown in Fig. 5.15, uses one or more ANN(s) as generator for some of the numerical values needed by the base model process. As shown in Eq. (5.172) and in the equations reported in Table 5.69, the obtained answer of the net is uncertain, particularly for the inputs near zero. For those situations, the ANN will be assisted by one or more regression equations. However, why should a regression equation be used instead of neural net computing alone? In fact, the neural system is capable of giving a precise guess in the case of a kinetic yield at given P, T and one or more y_i^τ or y_i^x variables (see Fig. 5.15) and time or position needed by the complex modeled process. However, this response is not reliable when the value to be found is the time derivative of a neural net-generated curve because in this curve there exist points for which there is a maximum of the rate function.

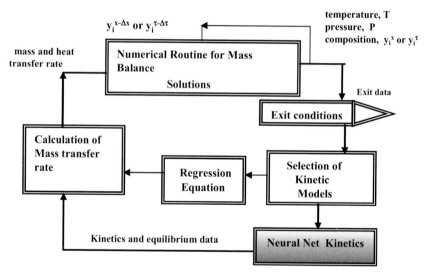

Figure 5.15 Principle of mass transfer model integration with a neural net (hybrid neural-regression model).

Neural networks can be used when traditional computing techniques can also be applied, but they can perform some calculations that would otherwise be very difficult to realize with current computing techniques. In particular, they can design a model from training data. A neural network can also be adapted to perform many different analogue functions such as pattern recognition, image processing, and trend analysis. These tasks are difficult to perform with conventional digital program computers. An important function of the neural network is its ability to discover the trends in a collection of data. Indeed, trend analysis is very important in dynamic chemical engineering problems, in process control, as well as in chemical formulation, data mining, and decision support. Neural networks are also particularly useful as data sensor analysis and processing of industrial units or integrated chemical plants as well as in commercial activity in industrial chemistry. If the researcher has a good algorithm capable of completely describing the problem, then traditional calculation techniques can, in most cases, give the best solution, but if no algorithm or other digital solution exists to address a complex problem with many variables, then a neural network that learns from examples may provide a more effective solution to the problem.

References

5.1 W. L. Gore, *Statistical Methods for Chemical Experimentation*, Interscience Publishers, New York, 1952.

5.2 A C. Bonnet, L. N. Franklin, *Statistical Analysis in Chemistry and the Chemical Industry*, John-Wiley, New York, 1954.

5.3 V. Youden, *Statistical Methods for Chemistry*, John Wiley, New York, 1955.

5.4 O. L. Davies, *Statistical Methods in Research and Production*, Oliver and Body, London, 1957.

5.5 V. V. Nalimov, *The Application of Mathematical Statistics by Chemical Analysis*, Pergamon Press, Oxford, 1963.

5.6 K. C. Peng, *The Design of Scientific Experiments*, Addison-Wesley, New York, 1967.

5.7 G. E. P. Box, N. Drapper, *Evolutionary Operation*, John Wiley, New York, 1969.

5.8 A. Gluck, *Mathematical Methods for Chemical Industry*, Technical Book, Bucharest, 1971.

5.9 R. Calcutt, R. Body, *Statistics for Analytical Chemists*, Chapman and Hall, London, 1983.

5.10 I. N. Miller, J. Miller, *Statistics for Analytical Chemistry*, Prentice-Hall, Hemel Hempstead, 1993.

5.11 W. P. Gardiner, *Statistical Analysis Methods for Chemists: A Software-Based Approach*, Royal Society of Chemistry, Cambridge, 1997.

5.12 C. W. Robert, J. A. Melvin (Eds.), *Handbook of Chemistry and Physics*, CRC Press Inc, Boca Raton, FL, 1983.

5.13 V. V. Kafarov, *Cybernetic Methods for Technologic Chemistry*, Mir, 1969.

5.14 W. Cornell, *Experiments with Mixture: Designs, Models and the Analysis of Mixture Data*, John Wiley, New York, 1990.

5.15 T. Dobre, O. Floarea, *Momentum Transfer*, Matrix-Rom, Bucharest, 1997.

5.16 M. Tyron, *Theory of Experimental Errors*, Technical Book, Bucharest, 1974.

5.17 R. Chaqui, *Truth, Possibility and Probability*, Elsevier, Amsterdam, 2001.

5.18 N. R. Drapper, H. Smith, *Applied Regression Analysis*, John Wiley, New York, 1981.

5.19 O. Iordache, *Mathematical Methods for Chemical Engineering-Statistics Methods*; Polytechnic Institute of Bucharest, Bucharest, 1982.

5.20 J. N. Kappar, *Mathematical Modeling*, John Wiley, New York, 1988.

5.21 W. C. Hamilton, *Statistics in Physical Science*, The Ronald Press Co, New York, 1964.

5.22 R. Mihail, *Introduction to Experimental Planning Strategy*, Technical Book, Bucharest, 1983.

5.23 R. Carlson, *Design and Optimization in Organic Synthesis*, Elsevier, Amsterdam, 1997.

5.24 P. G. Maier, R. F. Zund (Eds.), *Statistical Method in Analytical Chemistry*, Wiley-Interscience, New York, 2000.

5.25 G. E. P. Box, K. B. Wilson, *J. R. Soc., Ser. B*, **1951**, *13*(1), 1–15.

5.26 V. Halimov, N. Cernova, *Mathematical Statistics for Chemical Analysis*, Nauka, Moscow, 1965.

5.27 G. E. P. Box, J. S. Hunter, *Ann. Math. Stat.*, **1957**, *28*, 1,195.

5.28 G. W. Lowe, *Trans. Inst. Chem. Eng.*, **1964**, *42* (9), 1334–1342.

5.29 W. E. Deming, *Statistical Adjustment of Data*, Dover, New York, 1964.

5.30 P. R. Bevington, *Data Reduction and Errors Analysis for Physical Science*, McGraw-Hill, New York, 1969.

5.31 C. K. Bayne, T. B. Rubin, *Practical Experimental Design and Optimization Methods for Chemists*, VCH, Deerfield Beach Florida, 1986.

5.32 O. Muntean, G. Bozga, *Chemical Reactors*, Technical Book, Bucharest, 2001.

5.33 M. Giovanni, H. Christoph, W. W. Bernd, R. Darskus, *Anal. Chem.*, **1997**, *69* (4), 601–606.

5.34 J. Tellinghuisen, *J. Phys. Chem. A*, **2001**, *105*, 3917–3924.

5.35 S. Gosh, C. R. Rao (Eds.), *Handbook of Statistics: Design and Analysis of Experiments*, Elsevier, Amsterdam, 1996.

5.36 J. Krauth, *Experimental Design*, Elsevier, Amsterdam, 2000.

5.37 P. G. Lisboa (Ed.), *Neural Network, Current Applications*, Chapman and Hall, London, 1992.

5.38 R. Maus, J. Keyes (Eds.), *Handbook of Expert System in Manufacturing: Neural Nets for Custom Formulation*, McGraw-Hill, New York, 1991.

5.39 L. Fausett, *Fundamentals of Neural Network*, Prentice-Hall, New York, 1994.

5.40 N. V. Bath, I. J. McAvoy, *Comput. Chem. Eng.*, **1990**, *14* (4/5), 573–584.

5.41 A. B. Bulsari (Ed.), *Neural Networks for Chemical Engineering*, Elsevier, Amsterdam, 1995.

5.42 H. S. Stern, *Technometrics*, **1996**, *38*, 205–220.

5.43 F. Blayo, M. Verleysen, *Artificial Neuronal Networks*, PUF, Paris, 1996.

5.44 R. Hecht-Nielsen, *Neurocomputing*, Addison-Wesley, Amsterdam, 1990.

5.45 K. Gurney, *An Introduction to Neural Network*, UCLA Press, Los Angeles, 1997.

5.46 G. Montague, J. Morris, *Trends. Biotechnol*, **1994**, *6* (12), 312–325.

5.47 (a) D. van Camp, *Come istruire una rete artificiale di neuroni*, Le Scienze n° 291, pp. 134–136, Le Scienze spa, Milano, 1992; (b) T. Masters, *Practical Neural Network in C_{++}*, Academic Press, London, 1993.

5.48 N. V. Bath, T. J. McAvoy, *Comput. Chem. Eng.*, **1992**, *16* (2/3), 271–281.

5.49 M. A. Sartori, J. A. Panos, *IEEE Trans. Neural Networks*, **1991**, *2*, 4–16.

5.50 R. Sharma, D. Singhal, R. Ghosh, A. Dwiedi, *Comput. Chem. Eng.*, **1999**, *23* (5/6), 385–391.

5.51 A. B. Bulsari, S. Palasaori, *Neural Comput. Appl.*, **1993**, *1*,160–165.

5.52 M. L. Mavrovounitios, S. Chang, *Comput. Chem. Eng.*, **1992**, *1* (6), 283–291.

5.53 J. Delleirs (Ed.), *Neural Networks in QSAR and Drug Design*, Academic Press, New York, 1996.

5.54 P. R. Patnaik, *Biotechnol. Adv.*, **1999**, *17*, 477–489.

5.55 S. Zouling, L. K. Marjatta, S. Palosaori, *Chem. Eng. J.*, **2001**, *81*, 101–108.

6
Similitude, Dimensional Analysis and Modelling

Although many practical engineering problems involving momentum, heat and mass transport can be modelled and solved using the equations and procedures described in the preceding chapters, an important number of them can be solved only by relating a mathematical model to experimentally obtained data.

In fact, it is probably fair to say that very few problems involving real momentum, heat, and mass flow can be solved by mathematical analysis alone. The solution to many practical problems is achieved using a combination of theoretical analysis and experimental data. Thus engineers working on chemical and biochemical engineering problems should be familiar with the experimental approach to these problems. They have to interpret and make use of the data obtained from others and have to be able to plan and execute the strictly necessary experiments in their own laboratories. In this chapter, we show some techniques and ideas which are important in the planning and execution of chemical and biochemical experimental research. The basic considerations of dimensional analysis and similitude theory are also used in order to help the engineer to understand and correlate the data that have been obtained by other researchers.

One of the goals of the experimental research is to analyze the systems in order to make them as widely applicable as possible. To achieve this, the concept of similitude is often used. For example, the measurements taken on one system (for example in a laboratory unit) could be used to describe the behaviour of other similar systems (e.g. industrial units). The laboratory systems are usually thought of as *models* and are used to study the phenomenon of interest under carefully controlled conditions, Empirical formulations can be developed, or specific predictions of one or more characteristics of some other similar systems can be made from the study of these models. The establishment of systematic and well-defined relationships between the laboratory model and the "other" systems is necessary to succeed with this approach. The correlation of experimental data based on dimensional analysis and similitude produces models, which have the same qualities as the transfer based, stochastic or statistical models described in the previous chapters. However, dimensional analysis and similitude do not have a theoretical basis, as is the case for the models studied previously.

Chemical Engineering. Tanase G. Dobre and José G. Sanchez Marcano
Copyright © 2007 WILEY-VCH Verlag GmbH & Co. KGaA, Weinheim
ISBN: 978-3-527-30607-7

6.1
Dimensional Analysis in Chemical Engineering

In order to explain dimensional analysis in chemical engineering, we present a typical problem of chemical engineering that requires an experimental approach. Consider the steady flow of an incompressible Newtonian fluid through a long, smooth-walled, horizontal and circular pipe which is heated from the outside.

In this system two important characteristics are of interest to an engineer designing the pipeline:
1. the pressure drop per unit length along the pipe as a result of friction,
2. the heat transfer coefficient that shows the kinetics of heat transfer from the pipe wall to the bulk fluid.

The first step in planning an experiment to study this problem would take into consideration the choice of factors, or variables that affect the pressure drop ($\Delta p/l$) and the heat transfer coefficient (α). As a first approach, we can consider the effects of temperature and pressure separately. In fact, the temperature variation has no direct effect on the pressure drop but has an effect on the fluid's physical properties.

We can formulate that the pressure drop is a function of the pipe diameter, d, the fluid density, ρ, the fluid viscosity, η, and the mean velocity at which the fluid is flowing in the pipe (w). Thus, we can express this relationship as:

$$\Delta p/l = f(d, \rho, \eta, w) \tag{6.1}$$

The heat transfer coefficient is considered as a function of the parameters previously described and of the two thermal properties of the liquid: the heat capacity, c_p, and the thermal conductivity, λ:

$$\alpha = f(d, \rho, \eta, c_p, \lambda, w) \tag{6.2}$$

To carry out the experiments in a meaningful and systematic way, it will be necessary, first, to consider one of the parameters as a variable while keeping the others constant and then to measure the corresponding pressure drop. The same type of experiment is carried out for the measurement of the heat transfer coefficient. Contrary to the mass transport pressure drop, which could be measured directly, the heat transfer coefficient is obtained indirectly by measuring the temperature of the wall and of the fluid at the entrance and exit of the pipe. The determination of the functional relationship between $\Delta p/l$, α and the various parameters that influence the process is illustrated in Fig. 6.1.

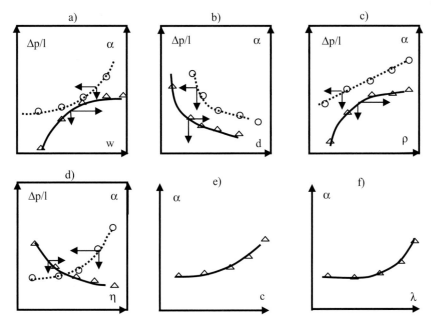

Figure 6.1 Illustrative plots showing the dependence of $\Delta p/l$ and α on the state of different process factors. (a) d, ρ, η, c_p, λ constant, (b) w, ρ, η, c_p, λ constant, (c) w, d, η, c_p, λ constant , (d) w, d, ρ, c_p, λ constant, (e) w, d, ρ, η, λ constant, (f) w, d, η, c_p, ρ constant.

Some of the results shown in this figure have to be obtained from experiments that are very difficult to carry out. For example, to obtain the data illustrated in Fig. 6.1(c) we must vary the liquid density while keeping the viscosity constant. For the data needed in Fig. 6.1(e), the thermal conductivity has to be varied while the density, the thermal capacity and viscosity are kept constant. These curves are actually almost impossible to obtain experimentally because the majority of the studied parameters are dependent on each other. This problem could be solved using a much simpler approach with the dimensionless variables that are described below. In fact, we can combine the different parameters described in Eqs. (6.1) and (6.2) in non-dimensional combinations of variables (called dimensionless groups, products' criteria)

$$\frac{\Delta p}{\rho w^2}\frac{d}{l} = \phi\left(\frac{wd\rho}{\eta}\right) \tag{6.3}$$

and

$$\frac{\alpha d}{\lambda} = \Phi\left(\frac{wd\rho}{\eta}, \frac{c_p\eta}{\lambda}\right) \tag{6.4}$$

Thus, instead of working with five parameters for the estimation of $\Delta p/l$, we have only two. In the case of α, which depends on seven parameters, this has been reduced to three dimensionless variables.

In the first case, the experimental work will simply consist of variation of the dimensionless product $wd\rho/\eta$ and determination of the corresponding value of $\Delta p/(\rho w^2)(d/l)$. The results of the experiments can then be represented by a single universal curve, as illustrated in Fig. 6.2(a). Varying the dimensionless product $wd\rho/\eta$ and determining, for the dimensionless group $c_p\eta/\lambda$, the corresponding value of $\alpha d/\lambda$, makes it possible to obtain the results shown in Fig. 6.2(b) for the pipe heat transfer. From these results we can conclude that carrying out the experimental work will be much simpler, easier, and cheaper. The basis of these simplifications lies in consideration of the involved variables' dimensions. It is known that the physical quantities can be given in terms of basic dimensions such as mass, M, length, L, time, T, temperature, θ, quantity of substance N and light intensity, Λ. The derivation systems of basic dimensions also coexist between them, F L T θ N Λ are the most common.

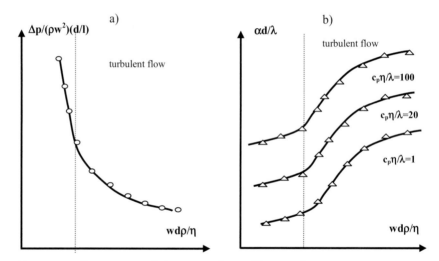

Figure 6.2 An illustrative example for pressure drop and heat transfer coefficient evaluation using dimensionless groups: (a) dimensionless pressure drop, (b) dimensionless heat transfer coefficient.

For example Newton's second law, $F = ma$, can be written as:

$$[F] = [m][a] = M\,L\,T^{-2} \tag{6.5}$$

Here, the brackets are used to indicate an operation using the basic dimension of the variables. It is not difficult to obtain the dimension formulae for the variables presented in the previously discussed examples; these are:

$[\Delta p/L] = ML^{-2}\,T^{-2}$, $[w] = LT^{-1}$, $[d] = L$, $[\rho] = ML^{-3}$, $[\eta] = ML^{-1}T^{-1}$, $[c_p] = L^2T^{-2}\theta^{-1}$, $[\lambda] = MLT^{-3}\theta^{-1}$, $[\alpha] = MT^{-3}\theta^{-1}$.

A rapid check of the groups' dimension, which appears in relationships (6.3) and (6.4), shows that they actually are dimensionless criteria:

$$\left[\frac{\Delta p}{\rho w^2}\frac{d}{l}\right] = \frac{ML^{-2}T^{-2}L}{ML^{-3}(LT^{-1})^2} = M^0L^0T^0 \ , \quad \left[\frac{wd\rho}{\eta}\right] = \frac{(LT^{-1})L(ML^{-3})}{ML^{-1}T^{-1}} = M^0L^0T^0$$

and

$$\left[\frac{\alpha d}{\lambda}\right] = \frac{(MT^{-3}\theta^{-1})L}{MLT^{-3}\theta^{-1}} = M^0L^0T^0\theta^0 \ , \quad \left[\frac{c_p\eta}{\lambda}\right] = \frac{(L^2T^{-2}\theta^{-1})(ML^{-1}T^{-1})}{MLT^{-3}\theta^{-1}} = M^0L^0T^0\theta^0$$

With this methodology, not only has the number of variables been reduced, but also the new groups are dimensionless combinations of variables, which means that the results presented in Fig. 6.2 will be independent of the system of units used. This type of analysis is called dimensional analysis. The basis for its application to a wide variety of problems is found in the Buckingham Pi Theorem described in the next section. Dimensional analysis is also used for other applications such as:
- establishing the dimensional formula for the derived physical variables,
- verifying the dimensional homogeneity of the physical relationships and equations used for the characterization of a process,
- verifying whether the units of measurement used for process variables are correct.

6.2
Vaschy–Buckingham Pi Theorem

When researchers want to use dimensional analysis of a process, the first and fundamental question they have to answer concerns the number of dimensionless groups that are required to replace the original list of process variables. The answer to this question is given by the basic theorem of dimensional analysis, which is stated as follows:

"If a process is characterized by an equation involving m physical variables, then this equation can be reduced to a relationship between m − n independent dimensionless groups, where n represents the number of basic dimensions used to describe the variable".

The dimensionless groups are frequently called "pi terms" due to the symbol used by Buckingham [6.1] to define the fact that the dimensionless group is a product. Their first modern presentation was given by Vaschy [6.2], even though several early investigators, including Rayleigh, contributed to the development of the pi theorem. In spite of the simplicity of the pi Theorem, its improvement is not simple. This will not, however, be presented here, because the detailed mathematical improvement is beyond the scope of this chapter. Many books give a more detailed treatment of the pi theorem and dimensional analysis [6.3–6.15].

The pi theorem is based on the idea of the dimensional homogeneity of the process equations or on the relationships that characterize one particular process. From this point of view, all the coefficients of statistical models that have already been discussed in Chapter 5 have a physical dimension, because the dependent and the independent process variables have a physical dimension. Essentially, we assume that any physically meaningful equation, which characterizes one process and which involves m variables, such as $y_1 = f(x_1, x_2....x_m)$ presents, for each term contained on the right-hand side, the same dimension as for the left-hand side. This equation could be transformed into a set of dimensionless products (pi terms):

$$\Pi_1 = \phi(\Pi_2, \Pi_3...\Pi_{m-n}) \tag{6.6}$$

The required number of pi terms is lower than the number of original n variables, where n is determined by the minimum number of basic dimensions required to describe the original list of variables. For common momentum and mass transfer, the basic dimensions are usually represented by M, L, and T. For heat transfer processes, four basic dimensions – M, L, T, θ – have to be used. Moreover, in a few rare cases, the variables could be described by a combination of basic dimensions such as, for any flow processes, M/T^2 and L. The use of the pi theorem may appear to be mysterious and complicated, although there are systematic and relatively simple procedures to develop the pi theorem for a given problem.

6.2.1
Determination of Pi Groups

Several methods can be used to form the dimensionless pi terms in a dimensional analysis. The most important are those applying a systematic determination of the pi terms, but they can be used only when the terms are dimensionless and independent. These methods, which will be described in detail later, are called "method of base non-complete group" [6.16] or "method of repeating variables" [6.17]. The determination of pi groups must be considered as the beginning of modelling for a process using dimensional analysis. We can consider that a model is completely established if a general characteristic process function, obtained after the application of this method, can be particularized by experimental data.

One of the simplest analyses consists in dividing the method into a series of distinct steps that can be followed for any given problem. The description given below is very similar to the methodology generally applied in the production of a mathematical model process as previously presented in Chapter 2.

Step 1: List all the variables that are involved in the problem (process)

This step is one of the most difficult and is, of course, extremely important because all pertinent variables have to be included in the analysis. The term variable includes any physical quantity, dimensional and apparently non-dimensional constant that plays a role in the phenomenon under investigation. The determination of the variables must take into account practical knowledge of the problem as well as the physical laws governing the phenomenon. Variables typically include the parameters that are necessary not only to describe the geometry of the system (such as the diameter of the pipe in the example below), but also to define the fluid properties (such as the density, viscosity, thermal capacity, thermal conductivity of the fluid, the diffusion coefficient for one species in the working fluid, etc.) as well as to indicate the external effects that influence the system (such as the driving pressure drop in the further discussed cases).

These general types of variables are intended to be as broad as possible in order to be helpful in identification. However, in some cases, the variables may not easily fit into one of these categories. This is why each problem has to be carefully analyzed.

Two conditions are very important during this analysis. First, generally, the researchers wish to have a minimum number of variables in order to minimize the experimental work. Secondly, these variables have to be independent. For example, for a problem of flow in a pipe, the geometric dimensions such as the pipe diameter and the section flow, could both be considered as variables. However, only the pipe diameter will be considered in the list of variables because the section flow already contains the basic geometric dimension.

Step 2: Establishment of the dimensional formula for each variable from the selected list

For a typical chemical engineering problem, the dimensions considered are generally M, L, T and θ. The dimensions F, L, T, θ can also be used but, in this case, especially for heat transfer problems and for coupled heat and mass transfer processes, complicated dimensional formulae are derived. To establish a dimensional formula for a variable, it is necessary to have a relationship containing this variable. This relationship can be independent of the process to which the dimensional analysis is applied. The use of tables containing dimensional formulae for physical variables can also be effective.

Step 3: Determination of the required number of pi terms

This step can be accomplished by means of the pi theorem which indicates that the number of pi terms is equal to m – n, where m (determined in step 1) is the number of selected variables and n (determined in step 2) is the number of basic dimensions required to describe these variables. The reference dimensions usually correspond to the basic dimensions and can be determined by a careful inspection of the variables' dimensions obtained in step 2. As previously noted, the basic dimensions rarely appear combined, which results in a lower number of reference dimensions than the number of basic ones.

Step 4: Selection of a non-complete group containing the same number of variables and basic dimensions

Here we select some variables from the original list in order to combine them with the remaining variables to form the pi term. The variables contained in the non-complete group do not change during the process of pi term production. All the required reference (basic) dimensions must be included within the non-complete group of repeating variables. Each repeating variable must be dimensionally independent of the others (a similar consideration is taken into account when the dimensions of one repeating variable cannot be reproduced by any combination of the exponent product of the remaining repeating variables). In fact, we can conclude that the repeating variables cannot be combined with other repeating variables to form dimensionless criteria.

For any given problem, we are usually interested in determining how one particular variable influences (and is influenced by) other variables. A one-dimensional analysis accepts only one dependent variable. It is recommended not to choose the dependent variable as one of the repeating variables, since the repeating variable will generally appear in more than one pi group term and then the variable separation cannot be carried out easily.

Step 5: Development of the pi terms one at a time by multiplying a non-repeating variable by a non-complete group which has the repeating variable necessary to obtain the arbitrary different exponents

Essentially, each pi term will be of the form $x_i x_1^\alpha x_2^\beta x_3^\gamma$ where x_i is a non-repeating variable and x_1, x_2, x_3 represent the repeating variables of the non-complete group. The exponents α, β, γ are determined in order to give a dimensionless combination. The case presented here corresponds to a process where variables are introduced with three basic dimensions (M, L, T). For heat transfer and the coupling of heat and mass transfer processes, the form used for a pi term is $x_i x_1^\alpha x_2^\beta x_3^\gamma x_4^\delta$. The values of the exponents α, β, γ are determined in this step by generating a system of linear algebraic equations containing these exponents. The basis for the development of the system is represented by the condition of the dimensionless pi group.

Step 6: Checking all the resulting pi terms to make sure they are dimensionless

In order to prove that the pi terms are correctly formulated, their dimensionless condition should be confirmed by replacing the variables in the dimensional formula by the pi terms. This step can be carried out by writing the variables in terms of M, L, T, θ. If the dimensional analysis has been produced using F, L, T, θ as basic dimensions, then check the formula to make sure that the pi terms are dimensionless.

Step 7: Establishment of the final form as a relationship among the pi terms

The most frequently used form of the final dimensional analysis is written as Eq. (6.6) where Π_1 will contain the dependent variables in the numerator. It should be emphasized that, if you have started out with a good list of variables (and the other steps of the analysis have been completed correctly), the relationship in terms of the pi groups can be used as a basis to describe the investigated problem. All we need to do is work with the pi groups and not with the individual variables. However, it should be clearly noted that the functional relationship between the pi groups has to be determined experimentally. The result is a relationship criterion able to show the main behaviour of the analyzed system or process. The chemical engineering research methodologies can also result in obtaining a theoretical relationship criterion using various theoretical bases [6.18–6.20]

To illustrate the steps described above, we will consider the problem already introduced at the beginning of this chapter, which was concerned with the pressure drop and heat transfer of an incompressible Newtonian fluid flowing in a pipe.

The first problem is the classical example used to show the scientific force of the dimensional analysis – and especially of the pi theorem. Remember that we are interested in the pressure drop per unit length ($\Delta p/l$) along the pipe. According to the experimenter's knowledge of the problem and to step 1, we must list all the pertinent variables that are involved; in this problem, it was assumed that:

$$\Delta p/l = f(d, \rho, \eta, w)$$

where d is the pipe diameter, ρ and η are the fluid density and viscosity, and w is the mean fluid velocity.

In step 2, we express all the variables in terms of basic dimensions. Using M, L, T as basic dimensions, it follows that:

$$[\Delta p/l] = \frac{[F/S]}{[l]} = \frac{MLT^{-2}L^{-2}}{L} = ML^{-2}T^{-2}$$

$$[d] = L$$

$$[\rho] = ML^{-3}$$

$$[\eta] = ML^{-1}T^{-1}$$

$$[w] = LT^{-1}$$

We could also use F, L, and T as basic dimensions. Now, we can apply the pi theorem to determine the required number of pi terms (step 3). An inspection of the variable dimensions obtained in step 2 reveals that the three basic dimensions are all required to describe the variables. Since there are five (m = 5) variables (do not forget to count the dependent variable, $\Delta p/l$) and three required reference dimensions (n = 3), then, according to the pi theorem, two pi groups (5 – 3) will be required.

We need to select three out of the four variables (d, ρ, η, w) in the list of the incomplete group with repeating variables (step 4) to be used to form the pi terms. Remember that we do not want to use the dependent variable as one of the repeating variables. Generally, we will try to select the dimensionally simplest repeating variables. For example, if one of the variables has a length dimension, we can choose it as one of the repeating variables. We can note that this incomplete group has to contain all the basic dimensions established by step 2. For this step, we use d, ρ and w as repeating variables in the incomplete group.

We are now ready to form the two pi groups and to identify the exponents associated with the repeating variables from the incomplete group (step 5). Typically, we will start with the dependent variable and combine it with the repeating variables to form the first pi term:

$$\Pi_1 = (\Delta p/l)d^{\alpha}\rho^{\beta}w^{\gamma} \tag{6.7}$$

This combination has to be dimensionless and, in the particular example, only M, L and T are presented:

$$[\Pi_1] = [(\Delta p/l)d^{\alpha}\rho^{\beta}w^{\gamma}] \tag{6.8}$$

The dimensional relationship (6.8) is developed into Eq. (6.9). Then, exponents α, β, γ must be determined so that the resulting exponent of each of the basic dimensions M, L and T, is zero (it gives a dimensionless combination). Thus, we can also write the relationship (6.10):

$$M^0L^0T^0 = ML^{-2}T^{-2}(L)^{\alpha}(ML^{-3})^{\beta}(LT^{-1})^{\gamma} = M^{(1+\beta)}L^{(-2+\alpha-3\beta+\gamma)}T^{(-2-\gamma)} \tag{6.9}$$

$$\begin{cases} 1+\beta = 0 \\ -2+\alpha - 3\beta + \gamma = 0 \\ -2-\gamma = 0 \end{cases} \tag{6.10}$$

Solution of the equation system (6.10) gives the desired values for α, β, γ. It is easy to observe that the following solution is obtained: $\alpha = 1$, $\beta = -1$, $\gamma = -2$. Therefore, the pi group is:

$$\Pi_1 = \frac{\Delta p}{\rho w^2}\frac{d}{l}$$

This procedure is now repeated for the remaining non-repeating variables. In this example, there is only one additional variable (η):

$$\Pi_2 = \eta d^{\alpha}\rho^{\beta}w^{\gamma} \tag{6.11}$$

By analogy with Eqs. (6.8) and (6.9), we can write Eqs. (6.12) and (6.13) which allow one to build a system of linear algebraic equations (6.14). This system gives the values of α, β, γ associated with Π_2.

$$[\Pi_2] = [\eta d^\alpha \rho^\beta w^\lambda] \tag{6.12}$$

$$M^0 L^0 T^0 = ML^{-1}T^{-1}(L)^\alpha (ML^{-3})^\beta (LT^{-1})^\gamma = M^{(1+\beta)} L^{(-1+\alpha-3\beta+\gamma)} T^{(-1-\gamma)} \tag{6.13}$$

$$\begin{cases} 1 + \beta = 0 \\ -1 + \alpha - 3\beta + \gamma = 0 \\ -1 - \gamma = 0 \end{cases} \tag{6.14}$$

Solving Eq. (6.14), it follows that $\alpha = -1$, $\beta = -1$, $\gamma = -1$ and:

$$\Pi_2 = \frac{\eta}{wd\rho} \tag{6.15}$$

At this point, we can check the dimensionless condition of the pi groups (step 6). However, before checking, we have to write the dimensional formulae for the variables contained in the selected list using the basic dimensions F, L, T. To obtain this transformation in the dimensional formulae used in step 2, the relationship $F = MLT^{-2}$ is used to replace the mass (M). The result obtained is:

$$[\Delta p/l] = FL^{-3} \ , \ [d] = L \ , \ [\rho] = FL^{-4}T^2 \ , \ [\eta] = FL^{-2}T \ , \ [w] = LT^{-1}$$

Now we can check whether the obtained pi groups are dimensionless:

$$[\Pi_1] = \left[\frac{\Delta p}{1} \frac{d}{\rho w^2}\right] = \frac{(FL^{-3})(L)}{(FL^{-4}T^2)(LT^{-1})^2} = F^0 L^0 T^0$$

$$[\Pi_2] = \left[\frac{\eta}{wd\rho}\right] = \frac{FL^{-2}T}{(LT^{-1})(L)(FL^{-4}T^2)} = F^0 L^0 T^0$$

or alternatively,

$$[\Pi_1] = \left[\frac{\Delta p}{1} \frac{d}{\rho w^2}\right] = \frac{ML^{-2}T^{-2}(L)}{(ML^{-3})(LT^{-1})^2} = M^0 L^0 T^0$$

$$[\Pi_2] = \left[\frac{\eta}{wd\rho}\right] = \frac{ML^{-1}T^{-1}}{(LT^{-1})(L)(ML^{-3})} = M^0 L^0 T^0$$

Finally (step 7), we can express the result of dimensional analysis as:

$$\frac{\Delta p}{\rho w^2} \frac{d}{1} = \phi\left(\frac{\eta}{\rho wd}\right) \tag{6.16}$$

This result indicates that this problem can be studied in terms of these two pi terms, rather than in terms of the original five variables. Nevertheless, the dimensional analysis will not provide the form of the function ϕ. This can be obtained from a suitable set of experiments. The power form for ϕ has been successfully used in chemical engineering literature. Thus, Eq. (6.16) can be particularized into Eq. (6.17) and, after the introduction of the Reynolds number, Eqs. (6.18) and (6.19) are obtained. Eq. (6.19) is the famous Fanning expression for the fluid pressure drop in the pipe. We can also derive the friction factor, λ_f, from Eqs. (6.18) and (6.19):

$$\frac{\Delta p}{\rho w^2} \frac{d}{l} = c \left(\frac{\eta}{\rho w d} \right)^P \tag{6.17}$$

$$\Delta p = c \, Re^{-P} \frac{l}{d} \frac{w^2}{2} \rho \tag{6.18}$$

$$\Delta p = \lambda_f \frac{l}{d} \frac{w^2}{2} \rho \tag{6.19}$$

The second problem, introduced at the beginning of this chapter and discussed here, is meant to show how – with the presented 7-step algorithm – we can obtain a simple dimensionless relationship between the various process variables affecting the heat transfer between the wall and the fluid.

Step 1 is rapidly resolved, based on the discussion of these problems at the beginning of this chapter (Fig. 6.1). The list of variables considers that:

$$\alpha = F(d, \rho, \eta, c_p, \lambda, w)$$

where the definition of each variable has been presented above.

Step 2 requires expressing all variables in terms of the basic dimensions. Using M, L, T, and θ as basic dimensions, the process variables show the dimensional formulae:

$$[\alpha] = MT^{-3}\theta^{-1}$$

$$[d] = L$$

$$[\rho] = ML^{-3}$$

$$[\eta] = ML^{-1}T^{-1}$$

$$\left[c_p \right] = L^2 T^{-2} \theta^{-1}$$

$$[\lambda] = MLT^{-3}\theta^{-1}$$

$$[w] = LT^{-1}$$

A similar result is obtained if we use F, L, T, θ as basic dimensions. As previously described, the M dimension is replaced by F. In the case of $[\rho]$, the basic dimensions of M, L, T, and θ are replaced by F, L, T, and θ. From $F = MLT^{-2}$ we obtain

$M = FL^{-1}T^2$ which is used in the $[\rho]$ formula to finally obtain: $[\rho] = FL^{-1}T^2L^{-3} = FL^{-4}T^2$.

Step 3 begins with determining the number of basic dimensions (M, L, T, θ). In this case n = 4 and with m = 7 (the number of variables considered in the first step) we conclude that the number of pi groups, n_π, is: $n_\pi = m - n = 7 - 4 = 3$.

In order to start step 4, we need to choose an incomplete group composed of n variables; the variables of this incomplete group will be coupled one by one with the remaining variables. Remember that we do not want to use the dependent variable as one of the repeating variables. At the same time, the incomplete group of repeating variables has to include all basic dimensions. We have chosen an incomplete group which includes d, ρ, η and λ because it has a very high number of variables with simple dimensional formulae. We are now ready to form three pi terms (step 5). To do so, we have to begin with the dependent variable and combine it with the repeating variables. Therefore, the first pi term is:

$$\Pi_1 = \alpha d^\beta \rho^\gamma \eta^\delta \lambda^\varepsilon \tag{6.20}$$

Since this combination has to be dimensionless, we can write:

$$[\Pi_1] = [\alpha d^\beta \rho^\gamma \eta^\delta \lambda^\varepsilon] \tag{6.21}$$

or:

$$M^0L^0T^0\theta^0 = MT^{-3}\theta^{-1}(L)^\beta(ML^{-3})^\gamma(ML^{-1}T^{-1})^\delta(MLT^{-3}\theta^{-1})^\varepsilon \tag{6.22}$$

Respectively:

$$M^0L^0T^0\theta^0 = M^{(1+\gamma+\delta+\varepsilon)}L^{(\beta-3\gamma-\delta+\varepsilon)}T^{(-3\beta-\delta-3\varepsilon)}\theta^{(-\varepsilon-1)} \tag{6.23}$$

Now we can identify the exponents β, γ, δ and ε of the basic dimensions using the equality between the exponents of the basic dimensions on the left-hand side and the exponents on the right-hand side (Eq. (6.23)). Then we obtain the next system of linear equations:

$$\begin{cases} 1 + \gamma + \delta + \varepsilon = 0 \\ \beta - 3\gamma - \delta + \varepsilon = 0 \\ -3\beta - \delta - 3\varepsilon = 0 \\ 1 - \varepsilon = 0 \end{cases} \tag{6.24}$$

The solution of this system of algebraic equations gives the desired values for β, γ, δ and ε. It is simple to obtain $\varepsilon = -1$, $\beta = 1$, $\gamma = 0$, $\delta = 0$ and therefore to write:

$$\Pi_1 = Nu = \frac{\alpha d}{\lambda} \tag{6.25}$$

which is the classical Nusselt dimensionless number (Nu), currently used in heat transfer processes. Step 5 must be repeated in order to obtain the dimensionless groups Π_2 and Π_3. We still have some variables able to be coupled with the incomplete groups, which are the flow rate (w) and the liquid thermal capacity (c_p).

If the selected variable is the flow rate, we can write the dimensionless expression:

$$\Pi_2 = w d^\beta \rho^\gamma \eta^\delta \lambda^\varepsilon \tag{6.26}$$

In this case, Eqs. (6.21)–(6.24) used for complete identification of this group, show the following particularizations:

$$[\Pi_2] = [w d^\beta \rho^\gamma \eta^\delta \lambda^\varepsilon] \tag{6.27}$$

$$M^0 L^0 T^0 \theta^0 = LT^{-1}(L)^\beta (ML^{-3})^\gamma (ML^{-1}T^{-1})^\delta (MLT^{-3}\theta^{-1})^\varepsilon \tag{6.28}$$

$$M^0 L^0 T^0 \theta^0 = M^{(\gamma+\delta+\varepsilon)} L^{(1+\beta-3\gamma-\delta+\varepsilon)} T^{(-1-\delta-3\varepsilon)} \theta^{(-\varepsilon)} \tag{6.29}$$

$$\begin{cases} \gamma + \delta + \varepsilon = 0 \\ 1 + \beta - 3\gamma - \delta + \varepsilon = 0 \\ -1 - \delta - 3\varepsilon = 0 \\ \varepsilon = 0 \end{cases} \tag{6.30}$$

The solution of this system of algebraic equations gives the new values for β, γ, δ, and ε adapted to the Π_2 group. It is then simple to obtain $\varepsilon = -0$, $\beta = 1$, $\gamma = 1$, $\delta = -1$ and therefore:

$$\Pi_2 = Re = \frac{w d \rho}{\eta} \tag{6.31}$$

If we carry out step 5 again, we obtain the group formed by coupling the liquid thermal capacity with the incomplete group of repeating variables. After the usual procedure Π_3 is written as:

$$\Pi_2 = c_p d^\beta \rho^\gamma \eta^\delta \lambda^\varepsilon \tag{6.32}$$

The new values needed for β, γ, δ and ε, will be obtained by applying the algorithm to the Π_3 group in Eq. (6.32). The next relationships show the following particularization:

$$[\Pi_3] = \left[c_p d^\beta \rho^\gamma \eta^\delta \lambda^\varepsilon \right] \tag{6.33}$$

$$M^0 L^0 T^0 \theta^0 = L^2 T^{-2}\theta^{-1}(L)^\beta (ML^{-3})^\gamma (ML^{-1}T^{-1})^\delta (MLT^{-3}\theta^{-1})^\varepsilon \tag{6.34}$$

$$M^0L^0T^0\theta^0 = M^{(\gamma+\delta+\varepsilon)}L^{(2+\beta-3\gamma-\delta+\varepsilon)}T^{(-2-\delta-3\varepsilon)}\theta^{(-1-\varepsilon)} \tag{6.35}$$

$$\begin{cases} \gamma + \delta + \varepsilon = 0 \\ 2 + \beta - 3\gamma - \delta + \varepsilon = 0 \\ -2 - \delta - 3\varepsilon = 0 \\ -1 - \varepsilon = 0 \end{cases} \tag{6.36}$$

It is then simple to obtain the new values of β, γ, δ and ε adapted to the Π_3 group. These are:

$\varepsilon = -1$, $\delta = 1$, $\gamma = 0$ and $\beta = 0$ and therefore:

$$\Pi_3 = Pr = \frac{c_p \eta}{\lambda} \tag{6.37}$$

This dimensionless group is recognized as the Prandtl number, which is currently used in heat transfer processes. This number is very important when the boundary layer theory is applied because it shows the relationship between the corresponding thickness of the heat transfer boundary layer and the hydrodynamic boundary layer [6.12].

The next step consists in making sure that the pi groups obtained are dimensionless (step 6). As explained above, the dimensional formulae for the variables contained in this selected list will be produced in the case of basic dimensions F, L, T and θ. Therefore, in the dimensional equations used in step 2, mass M will be replaced by force F using the relationship $F = MLT^{-2}$:

$$[\alpha] = FL^{-1}T^{-1}\theta^{-1} ,\ [d] = L ,\ [\rho] = FL^{-4}T^2 ,\ [\eta] = FL^{-2}T ,\ [w] = LT^{-1} ,$$
$$\left[c_p\right] = L^2T^{-2}\theta^{-1} ,\ [\lambda] = FT^{-1}\theta^{-1}$$

Now, let us check whether the obtained pi groups (Nu, Re, Pr) are dimensionless:

$$[\Pi_1] = [Nu] = \left[\frac{\alpha d}{\lambda}\right] = \frac{(FL^{-1}T^{-1}\theta^{-1})L}{FT^{-1}\theta^{-1}} = F^0L^0T^0\theta^0$$

$$[\Pi_2] = [Re] = \left[\frac{wd\rho}{\eta}\right] = \frac{(LT^{-1})(L)(FL^{-4}T^2)}{FL^{-2}T} = F^0L^0T^0\theta^0$$

$$[\Pi_3] = [Pr] = \left[\frac{c_p\eta}{\lambda}\right] = \frac{(L^2T^{-2}\theta^{-1})(FL^{-2}T)}{FT^{-1}\theta^{-1}} = F^0L^0T^0\theta^0$$

or alternatively,

$$[\Pi_1] = [Nu] = \left[\frac{\alpha d}{\lambda}\right] = \frac{(MT^{-3}\theta^{-1})L}{MLT^{-3}\theta^{-1}} = M^0L^0T^0\theta^0$$

$$[\Pi_2] = [Re] = \left[\frac{wd\rho}{\eta}\right] = \frac{(LT^{-1})(L)(ML^{-3})}{ML^{-1}T^{-1}} = M^0L^0T^0\theta^0$$

$$[\Pi_3] = [Pr] = \left[\frac{c_p\eta}{\lambda}\right] = \frac{(L^2T^{-2}\theta^{-1})(ML^{-1}T^{-1})}{MLT^{-3}\theta^{-1}} = M^0L^0T^0\theta^0$$

Finally (step 7), we can express the result of dimensional analysis as:

$$\frac{\alpha d}{\lambda} = \Phi\left(\frac{wd\rho}{\eta}, \frac{c_p\eta}{\lambda}\right) \tag{6.38}$$

Equation (6.38), which contains function Φ, has already been proved theoretically [6.12] and experimentally [6.18]. The famous relationship (6.39), which is applicable when pipe flow is fully developed, is currently used to characterize the heat transfer kinetics in other similar examples:

$$Nu = 0.023\, Re^{0.8}\, Pr^{0.33} \tag{6.39}$$

To summarize, the methodology to be followed in performing a dimensional analysis using the method of incomplete groups of repeating variables, consists in following this series of steps:
- Step 1: List all variables that are involved in the investigated phenomenon. This step needs a very good knowledge of these variables.
- Step 2: Each variable has to be described by its dimensional formula.
- Step 3: Establish the required number of pi groups.
- Step 4: Select the incomplete group of repeating variables. The number of repeating variables and basic dimensions involved in the problem are identical.
- Step 5: Form the pi term by multiplying one of the non-repeating variables by the incomplete group where the repeating variables have arbitrary powers and identify the actual pi expression. Repeat this step for all non-repeating variables.
- Step 6: Check all the resulting pi terms to make sure they are dimensionless.
- Step 7: Express the final form as a relationship among pi terms and add supplementary commentaries if necessary.

6.3
Chemical Engineering Problems Particularized by Dimensional Analysis

The two cases analyzed above give a model used to produce a particularization of the dimensional analysis to a chemical engineering problem. It has been observed that dimensional analysis is a good tool to rapidly elaborate a dimensionless frame for a system on which the experiments could be carried out by measuring the suggested variables. A further advantage lies in the "scale invariance" of dimensionless groups, thus enabling the only reliable scaling-up of the analyzed phenomena.

In the description of the various steps, it has been established that there are only two real problems in dealing with dimensional analysis. The first problem is the listing of all the relevant parameters that describe the process. Because chemical engineering processes are influenced by a high number of parameters, it is not easy to establish a good list of variables. The second problem is the determination of the process characteristics and of the real operational numbers, particularly in the case of large-scale factors. From the viewpoint of dimensional analysis, the descriptive chemical engineering model based on graphic representations is frequently effective in obtaining the correct interpretation of a process. We shall develop this problem in the following examples.

6.3.1
Dimensional Analysis for Mass Transfer by Natural Convection in Finite Space

We introduce this problem with two particular examples. The first is the etching of a metal placket immersed in a large specifically formulated liquid, with no gas production. The second is the drying of a recently built wall. In both cases, we have a non-observable flow and a particularization of the dimensional analysis is required.

These two examples do not appear to have any similarities. Nevertheless, after a deep analysis, we can conclude that both cases consist of a natural convection process produced by a concentration gradient.

This is presented schematically in Fig. 6.3, which also shows that the kinetics of these processes is described by the transport rate of A from the wall to the adjacent media. Using Fig. 6.3, we can establish that two elementary processes are presented in this system. The first is the flow induced by the concentration gradient and the second is the mass transfer sustained by the processes on the surface (a chemical reaction in the case of the metal placket immersed in a specifically formulated liquid and the transport through the porosity in the case of the drying wall). The case presented here corresponds to the situation when, in respect of the bulk density, the fluid density begins to decrease near the wall. This generates the displacement of the media and the specific ascension force, which is equivalent to the density difference. This phenomenon depends on the concentration difference in fluid A $\Delta c_A = (c_{Ap} - c_{A\infty})$. From Fig. 6.3 we can write a list of process variables:

$$k = f(H, \rho, \eta, D_A, g\beta_d \Delta c_A) \tag{6.40}$$

where g represents the gravitational acceleration, β_d is the density coefficient of the density–concentration dependence and Δc_A is the gradient for the natural convection. It is then easy to observe that the product $\beta_d \Delta c_A$ is dimensionless.

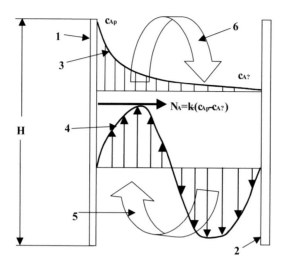

Figure 6.3 Mass transfer mechanism of natural convection between a placket and an adjacent medium. 1: placket or drying wall, 2: limit of adjacent medium, 3: concentration of A, 4: fluid velocity, 5 and 6: fluid global displacement. System properties: Geometric properties: H: height of placket. Fluid properties: C_A: concentration of A, density (ρ), viscosity (η), diffusion coefficient of A (D_A). Displacement properties: specific ascension force ($g\beta_c \Delta c_A$). Interaction properties: mass transfer coefficient (k_c).

We can now complete the first step of the dimensional analysis. The dimensions of the variables, using the MLT system for basic dimensions, are:

$$[k] = LT^{-1}$$

$$[H] = L$$

$$[\rho] = ML^{-3}$$

$$[\eta] = ML^{-1}T^{-1}$$

$$[D_A] = L^2T^{-1}$$

$$[g\beta_c \Delta c_A] = LT^{-2}$$

We can observe that all the basic dimensions (also specific to moment and mass transfer) are required to define the six variables, taking into consideration that, according to the Buckingham pi theorem, three pi terms will be needed (six variables minus three basic dimensions, $m - n = 6 - 3$).

The next step is the selection of three repeating variables such as H, ρ, and η to form the incomplete group of repeating variables. A quick inspection of these reveals that they are dimensionally independent, since each of them contains a basic dimension not included in the others. Starting with the dependent variable k_c, the first pi term can be formed combining k_c with the repeating variables so that:

$$\Pi_1 = k_c H^\alpha \rho^\beta \eta^\gamma \tag{6.41}$$

in terms of dimensions we have:

$$[\Pi_1] = [k_c H^\alpha \rho^\beta \eta^\gamma] \tag{6.42}$$

$$M^0 L^0 T^0 = LT^{-1}(L)^\alpha (ML^{-3})^\beta (ML^{-1}T^{-1})^\gamma \tag{6.43}$$

or

$$M^0 L^0 T^0 = M^{(\beta+\gamma)} L^{(1+\alpha-3\beta-\gamma)} T^{(-1-\gamma)} \tag{6.44}$$

the dimensionless condition of Π_1 implies:

$$\begin{cases} \beta + \gamma = 0 \\ 1 + \alpha - 3\beta - \gamma = 0 \\ -1 - \gamma = 0 \end{cases} \tag{6.45}$$

and, therefore $\alpha = 1$, $\beta = 1$ and $\gamma = -1$, the pi term then becomes:

$$\Pi_1 = \frac{k_c H \rho}{\eta} \tag{6.46}$$

The procedure is then repeated with the second non-repeating variable, D_A:

$$\Pi_2 = D_A H^\alpha \rho^\beta \eta^\gamma \tag{6.47}$$

It follows that

$$[\Pi_2] = [D_A H^\alpha \rho^\beta \eta^\gamma] \tag{6.48}$$

$$M^0 L^0 T^0 = L^2 T^{-1}(L)^\alpha (ML^{-3})^\beta (ML^{-1}T^{-1})^\gamma \tag{6.49}$$

$$M^0 L^0 T^0 = M^{(\beta+\gamma)} L^{(2+\alpha-3\beta-\gamma)} T^{(-1-\gamma)} \tag{6.50}$$

and

$$\begin{cases} \beta + \gamma = 0 \\ 2 + \alpha - 3\beta - \gamma = 0 \\ -1 - \gamma = 0 \end{cases} \tag{6.51}$$

The solution of this system is: $\alpha = 0$, $\beta = 1$ and $\gamma = -1$, and therefore:

$$\Pi_2 = Sc = \frac{D_A \rho}{\eta} \tag{6.52}$$

where symbol Sc introduces the Schmidt criterion which is frequently used in mass transfer problems. In the theory of boundary layers, the Schmidt criterion gives the relationship between the diffusion and hydrodynamic boundary layers. Figure 6.3 can be completed considering the additional thickness of the boundary layers formed at the placket wall and adjacent medium. The remaining non-repeating variable is $g\beta_c \Delta c_A$, where the third pi term is:

$$\Pi_3 = g\beta_c \Delta c_A H^\alpha \rho^\beta \eta^\gamma \tag{6.53}$$

and

$$[\Pi_3] = [g\beta_c \Delta c_A H^\alpha \rho^\beta \eta^\gamma] \tag{6.54}$$

$$M^0 L^0 T^0 = LT^{-2}(L)^\alpha (ML^{-3})^\beta (ML^{-1}T^{-1})^\gamma \tag{6.55}$$

$$M^0 L^0 T^0 = M^{(\beta+\gamma)} L^{(1+\alpha-3\beta-\gamma)} T^{(-2-\gamma)} \tag{6.56}$$

and, therefore,

$$\begin{cases} \beta + \gamma = 0 \\ 1 + \alpha - 3\beta - \gamma = 0 \\ -2 - \gamma = 0 \end{cases} \tag{6.57}$$

Solving this system, we obtain $\alpha = 3$, $\beta = 2$ and $\gamma = -2$ and we can write:

$$\Pi_3 = Gr_d = \frac{g\beta_c \Delta c_A H^2 \rho^2}{\eta^2} \tag{6.58}$$

Here Gr_d is the diffusion Grassoff number. It represents the natural convection displacement based on the concentration difference.

We have obtained the three required pi terms, which have to be checked in order to make sure that they are dimensionless. To do so, we use F, L and T, which will also verify the correctness of the original dimensions used for the variables. As explained earlier, we first have to replace M by F in the dimensional variable formula. Then the result is:

$[k] = LT^{-1}, [H] = L, [\rho] = FL^{-4}T^2, [\eta] = FL^{-2}T, [D_A] = L^2T^{-1}, [g\beta_c\Delta c_A] = LT^{-2}$

The dimensionless verification gives:

$$[\Pi_1] = \left[\frac{k_c H\rho}{\eta}\right] = \frac{(LT^{-1})(L)(FL^{-4}T^2)}{FL^{-2}T} = F^0L^0T^0$$

$$[\Pi_2] = \left[\frac{D_A\rho}{\eta}\right] = \frac{(L^2T^{-1})(FL^{-4}T^2)}{FL^{-2}T} = F^0L^0T^0$$

$$[\Pi_3] = \left[\frac{g\beta_c\Delta c_A H^3\rho^2}{\eta^2}\right] = \frac{(LT^{-2})(L)^3(FL^{-4}T^2)^2}{(FL^{-2}T)^2} = F^0L^0T^0$$

If this analysis results in a bad agreement with the dimensionless condition, we have to go back to the original list of variables and check the dimensional formula of each variable as well as the algebra used to obtain the exponents α, β and γ.

Before finishing the application, we show that each pi group obtained can be replaced by a combination between this pi number and others. So, if we divide Π_1 by Π_3, we obtain:

$$\Pi_4 = Sh = \frac{\Pi_1}{\Pi_2} = \frac{k_c H\rho}{\eta}\frac{\eta}{D_A\rho} = \frac{k_c H}{D_A} \qquad (6.59)$$

where Sh represents the Sherwood number which encrypts the mass transfer kinetics of the investigated process. Finally, we can represent the results of the dimensional analysis particularization in the form of:

$$Sh = f(Gr_d, Sc) \qquad (6.60)$$

However, at this stage of the analysis, the form and nature of the function f are unknown. To continue, we will have to perform a set of experiments or we can use one theoretical method able to show this function.

6.3.2
Dimensional Analysis Applied to Mixing Liquids

Mixing various components in a liquid medium is a chemical engineering operation with large industrial applications. Some examples of these applications are: paint production, resin and pigment mixing, gas–liquid transfer or reaction by bubbling in liquid, solid dissolution and solid crystallization in mixed liquid media, homogenous and heterogeneous chemical reactions involving liquid agitated media, aerobic and anaerobic biochemical reactions with molecular transformations in the liquid phase. These examples show the importance of the optimization of mixing liquids for the chemical industry.

An important number of factors having a key influence on this unit operation [6.19, 6.20], together with the examples described above, show how difficult it is to formulate a complete and unitary mixing theory responding to the various technical questions such as mixing time, distribution of the residence time, power consumption, heat and mass transfer kinetics in mixed media, scaling-up of a laboratory mixing plant etc.

In order to simplify the problem, we will apply dimensional analysis to liquid mixing in a particular case. The studied example will take into account the interactions showing:

- dependence of the power consumption with respect to process factors,
- dependence of the mixing time with respect to process factors,
- dependence of the mass transfer kinetics with respect to process factors in the case of dissolving suspended solids,
- dependence of the heat transfer kinetics with respect to process factors in the case of a wall heated by an agitated liquid.

The first necessary condition [6.21] to be taken into account in all particularization cases is the use of general mixing parameters (factors related to the geometry of agitation, the properties of liquid media, the type of agitators and rotation speed) as well as the use of the specific factors of the studied application. For example, in the case of suspended solid dissolution, we can consider the mass transfer coefficient for dissolving suspended solids, the mean dimension of the suspended solid particles, and the diffusion coefficient of the dissolved species in the liquid.

In this chapter, we present two particularizations: the first concerns the dependence of the power consumption on the considered influencing factors; the second shows the relationship between the mixing time and its affecting factors.

In order to establish the list of variables, we use the explicative Fig. 6.4 in both cases. It especially shows the geometry of agitation, allowing the introduction of geometric, material and dynamic factors.

In the first example, we considered that the power consumed by an agitator N, depends on the agitator diameter d, on the geometric position of the agitator in the liquid tank – expressed by the coordinates H, D, h, as well as on the rotation speed of the agitator n, and on the liquid physical properties (density ρ, viscosity η, and superficial tension σ). The interest here consists in formulating a relationship between the power consumption and the different affecting factors.

Considering Fig. 6.4, we can write (step 1 of the application procedure of the dimensional analysis) the following list of variables:

$$N = f(d, D, H, h, b, n, \rho, \eta, \sigma) \tag{6.61}$$

Now we can write (step 2) all variables in terms of basic dimensions. Using M, L and T it follows that:

$$[N] = ML^2T^{-3}, \ [d] = L, \ [D] = L, \ [H] = L, \ [h] = L, \ [b] = L, \ [n] = T^{-1}, \ [\rho] = ML^{-3},$$
$$[\eta] = ML^{-1}T^{-1}, [\sigma] = MT^{-2}$$

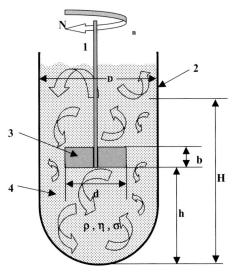

Figure 6.4 Schematic representation of mixing in liquid media.
1: Axis of paddle agitator, 2: tank of mixing system, 3: paddle of mixing system,
4: mixed liquid medium. System properties: geometric: agitator diameter (d) tank
diameter (D), liquid height (H), paddle width (b), bottom paddle position (h);
fluid: density (ρ), viscosity (η), superficial tension (σ); displacement: rotation
speed (n); interaction: power consumption (N).

By applying the pi theorem (step 3), we obtain that the number of pi groups
required is 7 because m = 10 (process variables) and n = 3 (basic dimensions). The
repeating variables of the incomplete group have been selected according to d, ρ
and n and to the considerations of step 4. We can now form all the pi groups one
at a time. Typically, we begin with the coupling of the dependent variable (power
consumption, N) with the incomplete group. The formulation of the first pi term
is:

$$\Pi_1 = Nd^\alpha \rho^\beta n^\gamma \tag{6.62}$$

By applying the dimensional formulation to this relationship we have:

$$M^0 L^0 T^0 = ML^2 T^{-3} (L)^\alpha (ML^{-3})^\beta (T^{-1})^\gamma \tag{6.63}$$

and

$$M^0 L^0 T^0 = M^{(1+\beta)} L^{(2+\alpha-3\beta)} T^{(-3-\gamma)} \tag{6.64}$$

and, consequently, the system of equations obtained with the exponents is:

$$\begin{cases} 1 + \beta = 0 \\ 2 + \alpha - 3\beta = 0 \\ -3 - \gamma = 0 \end{cases} \tag{6.65}$$

The solution of this system gives the desired values for α, β and γ. It follows that $\alpha = -5$, $\beta = -1$ and $\gamma = -3$ and therefore:

$$\Pi_1 = K_N = \frac{N}{d^5 n^3 \rho} \tag{6.66}$$

By repeating this calculation for the first independent variable, which has not been used as repeating variables in the incomplete group (the diameter of the vessel D), we have:

$$\Pi_2 = Dd^\alpha \rho^\beta n^\gamma \tag{6.67}$$

$$M^0 L^0 T^0 = L(L)^\alpha (ML^{-3})^\beta (T^{-1})^\gamma \tag{6.68}$$

$$M^0 L^0 T^0 = M^0 L^{(1+\alpha-3\beta)} T^{(-\gamma)} \tag{6.69}$$

$$\begin{cases} \beta = 0 \\ 1 + \alpha - 3\beta = 0 \\ \gamma = 0 \end{cases} \tag{6.70}$$

With these values for α, β and γ ($\alpha = -1$, $\beta = 0$ and $\gamma = 0$) the second pi group is:

$$\Pi_2 = \frac{D}{d} \tag{6.71}$$

For the other geometric factors, we obtain the next dimensionless relationships:

$$\Pi_3 = \frac{H}{d}, \ \Pi_4 = \frac{h}{d}, \ \Pi_5 = \frac{b}{d}$$

The remaining two pi groups are now identified. For the non-repeating variable η, the dimensional analysis calculation shows that:

$$\Pi_6 = \eta d^\alpha \rho^\beta n^\gamma \tag{6.72}$$

$$M^0 L^0 T^0 = ML^{-1} T^{-1} (L)^\alpha (ML^{-3})^\beta (T^{-1})^\gamma \tag{6.73}$$

$$M^0 L^0 T^0 = M^{(1+\beta)} L^{(-1+\alpha-3\beta)} T^{(-1-\gamma)} \tag{6.74}$$

$$\begin{cases} 1 + \beta = 0 \\ -1 + \alpha - 3\beta = 0 \\ -1 - \gamma = 0 \end{cases} \tag{6.75}$$

The solution to this system is $\alpha = -2$, $\beta = -1$ and $\gamma = -1$ and the sixth pi group can be written as:

$$\Pi_6 = Re = \frac{nd^2\rho}{\eta} \tag{6.76}$$

This criterion is recognized as the Reynolds number for mixing in a fluid.

The last non-repeating independent variable included in the list of variables gives the next formulation for the seventh pi group and generates all the calculation procedures for the identification of α, β and γ:

$$\Pi_7 = \sigma d^\alpha \rho^\beta n^\gamma \tag{6.77}$$

$$M^0 L^0 T^0 = MT^{-2}(L)^\alpha (ML^{-3})^\beta (T^{-1})^\gamma \tag{6.78}$$

$$M^0 L^0 T^0 = M^{(1+\beta)} L^{(\alpha-3\beta)} T^{(-2-\gamma)} \tag{6.79}$$

$$\begin{cases} 1 + \beta = 0 \\ \alpha - 3\beta = 0 \\ -2 - \gamma = 0 \end{cases} \tag{6.80}$$

The group identified by the introduction of α, β and γ values ($\alpha = 3$, $\beta = -1$ and $\gamma = -2$) into Eq. (6.77) is called the Weber number for mixing in a fluid. We can observe that, in this case, as in the previous one for the Re number, the original pi groups are transformed by the inversion of the terms of their algebraic fraction:

$$\Pi_7 = We = \frac{n^2 \rho d^3}{\sigma} \tag{6.81}$$

As in the previous examples, the next step (step 7) of the dimensional analysis procedure (which is not presented here) allows one to confirm that the obtained criteria are dimensionless. Now, finally, we can state the result of the dimensional analysis as:

$$K_N = f\left(\frac{D}{d}, \frac{H}{d}, \frac{h}{d}, \frac{b}{d}, Re, We\right) \tag{6.82}$$

The transformation of this relationship into the frequently used relationship for the theoretical power consumption for mixing in a fluid (Eq. (6.83)) is easily obtained. The We group relationship with K_N and the geometry dependence of the mixing constants a and b are needed for this transformation:

$$N = ad^{5-b}n^{3-b}\rho^{1-b}\eta^b \tag{6.83}$$

When the mixing time τ_M represents the dependent variable of the mixing in the fluid, all the independent variables used for the power consumption remain as variables affecting the mixing time. We also have to introduce a specific independent variable which is the A diffusion coefficient (D_A). The totality of the variables for this case will be:

$$\tau_M = f(d, D, H, h, b, n, \rho, \eta, \sigma, D_A) \tag{6.84}$$

Because the dimensional formula of D_A does not introduce a new basic dimension, we can establish that, in this case, the number of pi groups is 8 (eleven physical variables and three basic dimensions). If we use the incomplete group of the repeating variables, as in the case of the dependence of the power consumption factors, then we have to replace the K_N group by a group introduced by the new dependent variable (τ_M) and complete the established seven with a new group which includes the D_A factor. In this case, the formulation of the first pi group is given by Eq. (6.85):

$$\Pi_1 = \tau_M d^\alpha \rho^\beta n^\gamma \tag{6.85}$$

Applying the dimensional analysis procedure, we identify $\alpha = 0$, $\beta = 0$ and $\gamma = -1$ thus:

$$\Pi_1 = \tau_M n \tag{6.86}$$

As far as the Π_2–Π_7 groups are the same as those identified in the case of the power-factor dependence, we can identify the eighth pi group:

$$\Pi_8 = D_A d^\alpha \rho^\beta n^\gamma \tag{6.87}$$

Exponents α, β and γ have been identified by the following relationships:

$$M^0 L^0 T^0 = L^2 T^{-1} (L)^\alpha (ML^{-3})^\beta (T^{-1})^\gamma \tag{6.88}$$

$$M^0 L^0 T^0 = M^\beta L^{(2+\alpha-3\beta)} T^{(-1-\gamma)} \tag{6.89}$$

$$\begin{cases} \beta = 0 \\ 2 + \alpha - 3\beta = 0 \\ -1 - \gamma = 0 \end{cases} \tag{6.90}$$

from Eq. (6.90) we get that $\alpha = -2$, $\beta = 0$ and $\gamma = -1$ and that the Π_8 expression could be written as:

$$\Pi_8 = \frac{D_A}{nd^2} \tag{6.91}$$

This criterion is recognized as the Fourier number for mixing time in liquid media. Finally, for this case, we can express the result of dimensional analysis as:

$$\tau_M n = f\left(\frac{D}{d}, \frac{H}{d}, \frac{h}{d}, \frac{b}{d}, \frac{nd^2\rho}{\eta}, \frac{n^2\rho d^3}{\sigma}, \frac{D_A}{nd^2}\right) \tag{6.92}$$

This result indicates that this problem can generally be studied in terms of eight pi terms, or – for a fixed geometry – in terms of four pi terms, instead of the original eleven variables we started with. It also shows the complexity of this currently used chemical engineering operation.

6.4
Supplementary Comments about Dimensional Analysis

Despite the fact that other methods can be used to identify pi groups [6.22], we think that the method of the incomplete group of repeating variables explained in the preceding section, provides a systematic procedure for performing a dimensional analysis that can be easy enough for beginners. Pi terms can also be formed by inspection, as will be briefly discussed in the next sections. Regardless of the basis of dimensional analysis application for a concrete case, certain aspects of this important tool must seem a little baffling and mysterious to beginners and sometimes to experienced researchers as well.

In this section, we will show some of the guidelines required for a logical good start in a particular dimensional analysis. First, we need to have a good knowledge of the case being studied; this condition is one of the most important for successful application of this method. Some methodology guidelines will also be presented to establish a mathematical model (see, for example, the case of the conditions of univocity for the mathematical model of a particular process.)

6.4.1
Selection of Variables

One of the most important and difficult steps when applying dimensional analysis to any given problem, is the selection of the variables that are involved (see for example the introduction in the natural convection application presented in the preceding section). No simple procedure allows the variables to be easily identified. Generally, one must rely on a good understanding of the phenomena involved and of their governing physical laws. If extraneous variables are included, too many pi terms appear in the final solution, and it may then be difficult, and time and money consuming, to eliminate them experimentally. However, when important variables are omitted, an incorrect result will be produced.

These two aspects (introduction of extraneous variables and omission of important variables) show that enough time and attention has to be given when the variables are determined. Most chemical engineering problems involve certain simplifying assumptions that have an influence on the variables to be considered. Usually, a suitable balance between simplicity and accuracy is a required goal. The accuracy of the solution to be chosen depends on the objective of the study. For example, if we are only concerned with the general trends of the process, some variables that are thought to have a minor influence could be neglected for simplicity.

For all the engineering branches that use dimensional analysis as a methodology, the pertinent variables of one process can be classified into four groups:
- the variables describing the geometry of the system when the process occurs,
- the variables showing the properties of the materials involved in the evolution of the process being analyzed,
- the variables showing the internal dynamics of the process,
- the variables imposed by the external effects and having an important influence on the process dynamics.

6.4.1.1 Variables Imposed by the Geometry of the System

The geometric characteristics can usually be described by a series of lengths and angles. The application related to the mixing in a liquid medium (described above) shows the importance of geometry variables in a dimensional analysis problem. As in the above-mentioned case, the geometry of the system plays an important role in the majority of chemical engineering problems. Thus, a sufficient number of geometric variables must be included to describe the system. These variables can usually be identified quickly.

6.4.1.2 Variables Imposed by the Properties of the Materials

Fluid flow, heating and composition, which change by reaction or by transfer at one interface, represent the specificity of the chemical engineering processes. The response of a system to the applied effects that generate the mentioned cases depends on the nature of the materials involved in the process. All the properties of the materials such as density, viscosity, thermal capacity, conductivity, species diffusivity or others relating the external effects to the process response must be included as variables. The identification of these variables is not always an easy task. A typical case concerns the variation of the properties of the materials, in a nonlinear dependence with the operation variables. For example, when studying the flow of complex non-Newtonian fluids such as melted polymers in an externally heated conduct, their non-classical properties and their state regarding the effect of temperature make it difficult to select the properties of the materials.

6.4.1.3 Dynamic Internal Effects

Variables, such as the heat or mass transfer coefficients from or to the interface or the flow friction coefficient for a given geometry, represent variables that can be included in this group. They have a dynamic effect on the process state and generally represent the dependent variables of the process.

6.4.1.4 Dynamic External Effects

This definition is used to identify any visible variables that produce or tend to produce a change in the process. Pressure, velocity, gravity and external heating are some of the most frequently used variables from this group.

Since we wish to keep the number of variables to a minimum, it is important to have a selected list which contains only independent variables. For example, in the case of a flow problem, if we introduce the equivalent flow diameter (d_e), we do not have to introduce the flow area (A) nor the wetted perimeter (P) into the list of variables, because both variables have already been taken into consideration by the equivalent flow diameter ($d_e = 4A/P$). Generally, if we have a problem in which the variables are:

$$f(y, x_1, x_2, x_3, x_4, ...x_n) \qquad (6.93)$$

and it is known that an additional relationship exists among some of the variables, for example:

$$x_3 = f(x_4, ...x_n) \qquad (6.94)$$

then x_3 is not required and can be omitted. Conversely, if it is known that the variables $x_4, x_5,...x_n$ can only be taken into account through the relationship expressed by the functional dependence (6.94), then the variables $x_4, x_5,...x_n$ can be replaced by the single variable x_3, thus reducing the number of variables.

In addition to these supplementary comments about dimensional analysis, we can also discuss the following points, which are necessary to establish the list of variables. To do so, indeed, we have to:

1. Define the problem clearly using a descriptive model and auxiliary graphic presentation. Establish the main variable of interest (which is the dependent variable of the process).
2. Consider the basic laws that govern the phenomenon or accept an empirical theory describing the essential aspects of the investigated process as an open procedure for identifying independent variables.
3. Start the identification of the variables process by grouping them into the four groups of variables presented above (geometry, material properties, internal dynamic effects and external dynamic effects).
4. Verify whether other variables not included in the four groups of variables are important and must be considered and ensure that the dimensional constant, which can be introduced in the list of variables, has been accepted.
5. Make sure that all variables are independent and, to this end, the relationships among the subsets of the variables must be carefully observed.

6.5
Uniqueness of Pi Terms

A review of the method of an incomplete group of repeating variables used for identifying pi terms reveals that the specific pi terms obtained depend on the somewhat arbitrary selection of this incomplete group. For example, in the problem of studying the heat transfer from a wall to a fluid flowing in the pipe, we have selected d, ρ, η, and λ as repeating variables. This has led to the formulation of the problem in terms of pi terms:

$$\frac{\alpha d}{\lambda} = F\left(\frac{wd\rho}{\eta}, \frac{c_p \eta}{\lambda}\right) \tag{6.95}$$

What will the result be if we select d, ρ, η, and c_p as repeating variables? A quick check will reveal that the pi term involving the heat transfer coefficient (α) becomes:

$$\Pi_1 = \frac{\alpha d}{\eta c_p}$$

and the next pi terms remain the same. Thus, we can express the second result as:

$$\frac{\alpha d}{\eta c_p} = F_1\left(\frac{wd\rho}{\eta}, \frac{c_p \eta}{\lambda}\right) \tag{6.96}$$

Both results are correct, and will lead to the same final equation for α. Note, however, that the functions F and F_1 in Eqs. (6.95) and (6.96) will be different because the dependent pi terms are different for both relationships. From this example, we can conclude that there is no unique set of pi terms arising from a dimensional analysis. Nevertheless, the required number of pi terms has been fixed, and once a correct set has been determined, other possible sets can be developed by a combination of the products of the powers of the original set. This is a classical algebra problem, which shows that, if we have n independent variables (the pi terms obtained by the incomplete group method *are* independent variables), then each of these can be modified by a combination of the others and the resulting new set has n independent variables. For example, if we have a problem involving three pi terms: $\Pi_1 = F(\Pi_2, \Pi_3)$, we could form a new set from this initial one by combining the pi terms in order to form the new pi term Π_2', and to give $\Pi_2' = \Pi_2^\alpha \Pi_3^\beta$, where α and β are arbitrary exponents. For $\alpha = -1$ and $\beta = 0$ we obtain the inversion of the Π_2 expression group. Then the relationship between the dimensionless groups could be expressed as:

$$\Pi_1 = F_1(\Pi_2', \Pi_3) \quad \text{or} \quad \Pi_1 = F_2(\Pi_2, \Pi_2')$$

It must be emphasized, however, that the required number of pi terms cannot be reduced by this manipulation; only their form is altered. Thanks to this technique, we can see that the pi terms in Eq. (6.96) can be obtained from those presented in Eq. (6.95); then, if we multiply Π_1 from Eq. (6.95) by Π_3^{-1} we have:

$$\left(\frac{\alpha d}{\lambda}\right)\left(\frac{c_p \eta}{\lambda}\right)^{-1} = \frac{\alpha d}{\eta c_p}$$

which is the Π_1 of Eq. (6.96).

One may ask: which form is the best for the pi groups? Usually, we recommend keeping pi terms as simple as possible. In addition, it is easier to use the pi terms that could be improved by the experimental methodology used. The final choice remains arbitrary and generally depends on the researcher's background and experience.

6.6
Identification of Pi Groups Using the Inspection Method

The previously presented method of the incomplete group of repeating variables, provides a systematic procedure which, when properly executed, provides a correct, complete and unique set of pi terms. In other words, this method offers an excellent algorithm for the calculus. In this case, only the list of variables has to be determined by the researcher. Since the only restrictions for the pi terms are to be (a) correct in number, (b) dimensionless, and (c) independent, it is possible to produce other identifying procedures. One of them is the production of pi terms by inspection, without resorting to a more formal methodology.

To illustrate this approach, we will consider a new example: the case of a simple tubular membrane reactor for which we wish to show the dependence between the conversion (η_r) of the main reactant and other variables which influence the process.

The membrane reactor shown in Fig. 6.5 consists of a tubular shell containing a tubular porous membrane. It defines two compartments, the inner and the outer (shell) compartments. The reactants are fed into the inner compartment where the reaction takes place. We can observe that when the reactants flow along the reactor, one or more of the reaction participants can diffuse through the porous membrane to the outer side. In this case, we assume that only one participant presents a radial diffusion. This process affects the local concentration state and the reaction rate that determine the state of the main reactant conversion. The rate of reaction of the wall diffusing species is influenced by the transfer resistance of the boundary layer ($1/k_c$) and by the wall thickness resistance (δ/D_p).

As geometric variables, we can consider the diameter (d) and the length of the tubular reactor (l). The apparent constant rate of the chemical reaction (k_r) and the diffusion coefficient (D_m) of the species diffusing through the wall could be chosen as the internal dynamic variables of the process. The variables showing the properties of the materials (density and viscosity) as well as the variables characterizing the flow and the velocity (w) for example, can be considered, but these

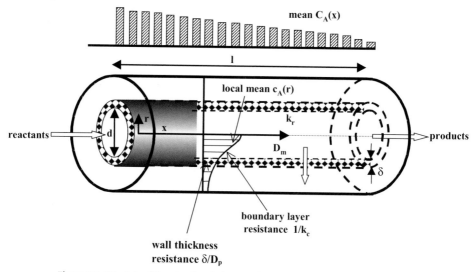

mean $C_A(x)$

l

local mean $c_A(r)$

k_r

reactants

d

r

x

D_m

products

δ

boundary layer
resistance $1/k_c$

wall thickness
resistance δ/D_p

Figure 6.5 Principle of the simple tubular membrane reactor.

are already introduced by the mass transfer coefficient (k_c). With this descriptive introduction, we can appreciate that, in this case, the variables are:

$$\eta_r = f(d, l, k_r, k_c, D_m, D_p/\delta) \tag{6.97}$$

Using M, L and T as basic dimensions, the following dimensional formulae of the variables are obtained:

$$[\eta_r] = M^0 L^0 T^0$$

$$[d] = L$$

$$[l] = L$$

$$[k_r] = T^{-1}$$

$$[k_c] = LT^{-1}$$

$$[D_m] = L^2 T^{-1}$$

$$\left[D_p/\delta \right] = LT^{-1}$$

In this dimensional analysis problem, five pi terms are needed because we have seven variables and two reference dimensions. The first pi term is represented by the conversion of the main reactant because this variable is dimensionless. The construction of the second pi group begins with variable d. This has a length dimension (L) and to form a dimensionless group, it must be multiplied by a variable the dimension of which is L^{-1}:

$$\Pi_2 = \frac{d}{l}$$

Inspecting the remaining variable, we observe that $[k_r d^2] = L^2 T^{-1}$. Then, by multiplying $k_r d^2$ by $1/D_m$, we obtain the third pi group:

$$\Pi_3 = \frac{D_m}{k_r d^2}$$

The formulation of the fourth pi group (Π_4) takes into account the observation that k_c and D_p/δ present the same dimensional formula and that their ratio is therefore dimensionless:

$$\Pi_4 = \frac{k_c \delta}{D_m}$$

The last pi group (Π_5) can be obtained by multiplying $k_c d$ (which has $L^2 T^{-1}$ dimension) by $1/D_m$, which also results in a dimensionless formula:

$$\Pi_5 = \frac{k_c d}{D_m}$$

The last three pi groups are well known in chemical engineering (Π_3 is recognized as the Fourier reaction number (Fo_r), Π_4 is the famous Biot diffusion number (Bi_d) and Π_5 is the Sherwood number (Sh)).

Relationship (6.98) shows the last result of this particularized case of dimensional analysis.

$$\eta_r = f(d/l, Fo_r, Bi_d, Sh) \tag{6.98}$$

It is important to note that when pi terms are formulated by inspection, we have to be certain that they are all independent. In this case or in any other general case, no pi group could result from the combination of two or more formulated pi groups. The inspection procedure of forming pi groups is essentially equivalent to the incomplete group method but it is less structured.

6.7
Common Dimensionless Groups and Their Relationships

Approximately three hundred dimensionless groups [6.23] are used to describe the most important problems that characterize chemical engineering processes. Out of these, only a limited number is frequently used and can be classified according to the flow involved in the investigated process, the transport and interface transfer of one property (species, enthalpy, pressure) and the interactions of the transport mechanisms of the properties. In order to be considered in this analysis, the dimensionless groups have to present the following general characteristics:

- Each dimensionless group provides a physical interpretation, which can be helpful in assessing its influence in a particular application.
- The dimensionless groups that characterize a particular application are correlated with others by the dimensional analysis relationship.
- When the process involves a transfer through an interface, some of the relationships between the dimensionless groups can be considered as relationships between the kinetic transfer and the interface properties.

6.7.1
Physical Significance of Dimensionless Groups

The physical interpretation of each dimensionless group is not an easy task. Because each dimensionless group can present various physical interpretations, the study of each particular dimensionless pi term has to be carefully carried out.

To illustrate this, we will discuss the example shown in Fig. 6.6, which presents one deformable fluid particle moving along a streamline. We can describe this system taking into account inertia, resistive (viscous) force and weight force. The magnitude of the inertia force along the streamline can be written as:

$$F_i = ma_s = m\frac{dw_s}{d\tau} = m\frac{dw_s}{ds}\frac{ds}{d\tau} = mw_s\frac{dw_s}{ds} \tag{6.99}$$

where ds is measured along the streamline and m is the particle mass. Based on the fact that a streamline is representative of a flow geometry when a mean flow rate, w, and a characteristic length are known, we can produce the dimensionless transformation for w_s and dw_s/ds. The dimensionless velocity and streamline position are respectively $w_{as} = w_s/w$ and $s_a = s/l$. Then Eq. (6.99) becomes:

$$F_i = m\frac{w^2}{l}w_{as}\frac{dw_{as}}{ds_a} \tag{6.100}$$

The weight force is described by $F_g = mg$, then the ratio between the inertia and the gravitational force is:

$$\frac{F_i}{F_g} = \frac{w^2}{gl}w_{as}\frac{dw_{as}}{ds_a} \tag{6.101}$$

The ratio between forces F_i/F_g is proportional to w^2/gl and its square root $(w/\sqrt{gl})$ is recognized as the Froude number. Its physical interpretation is the index of the relative importance of the inertial forces acting on the fluid particles with respect to the weight of these particles.

Now we consider the resistive force characterizing the movement of the particle along the streamline expressed as the product between tensor τ_{ss} and its normal surface A ($A = m/\rho.s_d$, where s_d is the apparent height of the deformed particle)

$$F_{rs} = \tau_{ss}A = \eta \frac{dw_s}{ds} \frac{m}{\rho s_d} \tag{6.102}$$

Using the dimensionless velocity and streamline position, and completing these values with the dimensionless height of the deformable particle $s_{da} = s_d/l$, Eq. (6.102) can be written as:

$$F_{rs} = \frac{\eta}{\rho l^2} \frac{m}{s_{da}} \frac{dw_{as}}{ds_a} \tag{6.103}$$

Then, the ratio between the inertia and the resistive forces is:

$$\frac{F_i}{F_{rs}} = \frac{wl\rho}{\eta} \frac{w_{as}}{s_{da}} \frac{dw_{as}}{ds_a} = Re \frac{w_{as}}{s_{da}} \frac{dw_{as}}{ds_a} \tag{6.104}$$

Here we can identify the Reynolds number (Re), which is a measure or an index of the relative importance of the inertial and resistive (viscous) forces acting on the fluid. If we write the general expression for the s direction rate of one property when the transport is molecular and convective, we have:

$$\vec{J}_{tAs} = -D_{\Gamma A} \frac{d\vec{\Gamma}_A}{ds} + \vec{w}_s \Gamma_A$$

we can obtain another physical interpretation for the Reynolds number, after particularizing the momentum transfer and replacing the corresponding terms ($D_{\Gamma A} = \nu = \eta/\rho$, $\Gamma_A = \rho w_s$, $\vec{J}_{tAs} = \vec{\tau}_{tsy}$). This particularization gives:

$$\vec{\tau}_{tsy} = -\eta \frac{d\vec{w}_s}{dy} + \vec{w}_s(\rho w_s) \tag{6.105}$$

Using the dimensionless velocity, we can write Eq. (6.106), which presents the ratio between the right-hand side terms of Eq. (6.105).

$$\frac{w_s(\rho w_s)}{\eta \dfrac{dw_s}{dy}} = cw_{as} \frac{\rho wl}{\eta} = cw_{as} Re \tag{6.106}$$

Equation (6.106) shows that the Reynolds number expresses the relationship between the momentum quantity supplied by the convection and the momentum quantity supplied by the molecular movement. At the same time, because the convective mechanism can be associated with the presence of the turbulence, we can consider the following ratio:

$$Re = \frac{\text{Momentum quantity transferred by turbulent mechanism}}{\text{Momentum quantity transferred by molecular mechanism}}$$

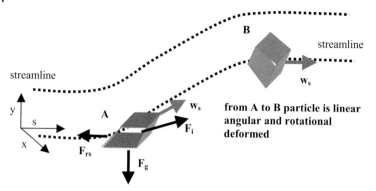

Figure 6.6 Particle moving along a streamline.

The case of the Reynolds number discussed above shows that the physical interpretation of one dimensionless group is not unique. Generally, the interpretation of dimensionless groups used in the flow area in terms of different energies involved in the process, can be obtained starting with the Bernoulli flow equation. The relationship existing between the terms of this equation introduces one dimensionless group.

6.6.2
The Dimensionless Relationship as Kinetic Interface Property Transfer Relationship

We begin this section by analyzing the case of free convection in an infinite medium. The example chosen is shown in Fig. 6.7. A two-dimensional surface with constant temperature t_p transfers heat to the adjacent infinite media. As a result of the temperature difference between the surface and the media, a natural convection flow is induced. A dimensional analysis applied to this problem shows that:

$$Nu = f(Gr_t)$$

where the Nusselt number (Nu) and the Grashof number for thermal convection (Gr_t) are given by:

$$Nu = \frac{\alpha d}{\lambda} \; , \quad Gr_t = \frac{g\beta_t \Delta t H^3 \rho^2}{\eta^2} \qquad (6.107)$$

The goal of this analysis is to obtain a relationship describing the kinetics of the heat transfer from the heated two-dimensional plate to the adjacent medium. This relationship is one dimensionless pi group. Moreover, we can use this example as a guide for the introduction of the relationships existing between dimensionless groups such as the relationships for the property transfer kinetics. To write the mathematical model for the problem of infinite medium natural heat convection, we use the particularization of the property transport equations. The corresponding equations were previously established in Chapter 3.

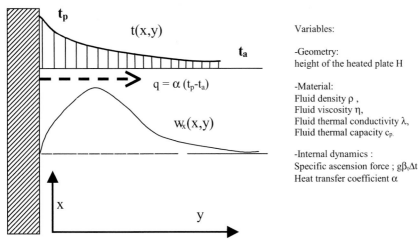

Figure 6.7 Heat transfer by natural convection from a plate to an infinite medium.

The momentum and energy transfer equation for the presented case may be written as:

$$w_x \frac{\partial w_x}{\partial x} + w_y \frac{\partial w_x}{\partial y} = \upsilon \frac{\partial^2 w_x}{\partial y^2} + \beta_t g(t - t_\infty) \tag{6.108}$$

$$w_x \frac{\partial t}{\partial x} + w_y \frac{\partial t}{\partial y} = a \frac{\partial^2 t}{\partial y^2} \tag{6.109}$$

$$\frac{\partial w_x}{\partial x} + \frac{\partial w_y}{\partial y} = 0 \tag{6.110}$$

where $\upsilon = \eta/\rho$ is the kinematic viscosity and $a = \lambda/\rho\, c_p$ is the thermal diffusivity of the medium.

The boundary conditions attached to the problem are:

$$y = 0,\ 0 < x < H,\ w_x = 0,\ w_y = 0,\ t = t_p \tag{6.111}$$

$$y = \infty,\ 0 < x < H,\ w_x = 0,\ t = t_\infty \tag{6.112}$$

$$x = 0,\ y = 0,\ w_x = w_y = 0,\ t = t_p \tag{6.113}$$

We can now introduce the following dimensionless notation:

$$Gr_x = \frac{g\beta_t(t_p - t_\infty)x^3}{\upsilon^2} \quad , \quad \xi = \frac{y}{x}\left(\frac{Gr_x}{4}\right)^{1/4} \tag{6.114}$$

where Gr_x is the local Grashof number and ξ is a combination of the cartesian coordinate. Therefore, if we use the method of the stream function ψ for the transformation of the original model, we can write:

$$\psi = 4\upsilon \left(\frac{Gr_x}{4}\right)^{1/4} \varphi(\xi) \tag{6.115}$$

and then:

$$w_x = \frac{d\psi}{d\xi}\frac{d\xi}{dy} = \left(\frac{g\beta_t(t_p - t_\infty)}{4\upsilon^2}\right) 4\upsilon x^{1/2}\varphi'(\xi) \tag{6.116}$$

$$w_y = -\frac{d\psi}{d\xi}\frac{d\xi}{dx} = \upsilon x^{-1/4} \left(\frac{g\beta_t(t_p - t_\infty)}{4\upsilon^2}\right)^{1/4} [\xi\varphi'(\xi) - 3\varphi(\xi)] \tag{6.117}$$

With these conditions, the equations of the original model can be written as:

$$\varphi''(\xi) + 3\varphi(\xi)\varphi'(\xi) - 2[\varphi'(\xi)]^2 + \theta(\xi) = 0 \tag{6.118}$$

$$\theta''(\xi) + Pr\,\varphi(\xi)\theta'(\xi) = 0 \tag{6.119}$$

where $\theta(\xi) = \dfrac{t(\xi) - t_\infty}{t_p - t_\infty}$ is the dimensionless temperature and Pr represents the Prandtl number.

The boundary conditions are now the following:

$$\xi = 0 \quad \varphi = \varphi' = 0 \; ; \; \theta = 1 \tag{6.120}$$

$$\xi = \infty \quad \varphi' = 0 \; ; \; \theta = 0 \tag{6.121}$$

With the approximation of $\varphi''(0)$ and $\theta'(0)$, the model represented by assembling Eqs. (6.118)–(6.121) can be readily solved by an adequate numeric method. The Prandtl number is, in this case, a parameter of numerical integration.

The heat transfer kinetics is represented by the heat flux produced and transferred by the plate.

$$q = \lambda \left(\frac{dt}{dy}\right)_{y=0} = \alpha_x(t_p - t_\infty) \tag{6.122}$$

The preceding expression can also be written as:

$$Nu_x = \frac{\alpha_x x}{\lambda} = \frac{\lambda\left(\dfrac{dt}{dy}\right)_{y=0} x}{\lambda(t_p - t_\infty)} = \frac{x\left(\dfrac{dt}{dy}\right)_{y=0}}{(t_p - t_\infty)} \tag{6.123}$$

When we introduce the combined variable ξ and the dimensionless temperature $\theta(\xi)$ into Eq. (6.123) we have:

$$\text{Nu}_x = \frac{x\left[\frac{d\theta(\xi)}{d\xi}\frac{d\xi}{dy}\frac{dt}{d\theta}\right]_{\xi=0}}{(t_p - t_\infty)} = -\theta'(0)\left(\frac{\text{Gr}_x}{4}\right)^{1/4} \tag{6.124}$$

and

$$\text{Nu} = \frac{1}{H}\int_0^H \text{Nu}_x dx = -0.4040\theta(0)\text{Gr}^{1/4} \tag{6.125}$$

The values used to calculate the temperature gradient and the velocity gradient near the heated vertical plate are given in Table 6.1.

Table 6.1 Some values of the temperature and velocity gradients at the surface in the case of natural convection heat transfer.

$\text{Pr} = c_p\eta/\lambda$	0.01	0.793	1	2	10	100	1000
$-\theta'(0)$	0.0812	0.5080	0.5671	0.7165	1.1694	2.191	3.966
$\varphi''(0)$	0.9862	0.6741	0.6421	0.5713	0.4192	0.2517	0.1450

When the heated medium is air ($\text{Pr} \approx 0.793$) Eq. (6.125) takes the value of 0.508 for $-\theta'(0)$, then we have:

$$\text{Nu} = 0.205\text{Gr}_t^{1/4} \tag{6.126}$$

In the Nusselt and Grashof numbers, the height of the heated plate is the characteristic length.

The example presented here allows these important conclusions:

- when we solve the transport property equation, we obtain a dimensionless relationship which characterizes the kinetics of the transfer for the property near the interface,
- the form of the obtained relationship can be simplified to a power type dependence.

Now we will consider the case of a transferable property for the contact between two phases. The transfer kinetics is characterized by the two transfer coefficients of the property given by Eq. (3.15). If we analyze the transport process with reference only to one phase, then we can write:

$$k_\Gamma = \frac{-D_\Gamma\left[\frac{d\Gamma}{dx}\right]_{x=x_i}}{(\Gamma_\infty - \Gamma_{i_i})} \tag{6.127}$$

where the index of the phase definition has been omitted and index i indicates the interface position. If the preceding relationship is multiplied by the ratio between the characteristic length and the diffusion coefficient of the property (l/D_γ), the relationship becomes:

$$\frac{k_\Gamma l}{D_\Gamma} = \frac{\left[\dfrac{d\Gamma}{d(x/l)}\right]_{x=x_i}}{(\Gamma_i - \Gamma_{\infty_i})} \tag{6.128}$$

This equation can also be written as:

$$Nu_\Gamma = \left(\frac{d\Gamma_a}{dx_a}\right)_{x_a=x_{ai}} \tag{6.129}$$

where Γ_a is the concentration of the dimensionless property and x_a represents the dimensionless transport coordinate. Nu_Γ is used here as a generalized Nusselt number and gives the transport kinetics for any kind of property (heat, species, etc.). The dimensionless groups' relationship that is able to explain the property gradient near the interface, in terms of other dimensionless groups characterizing the process, can be obtained from Eq. (6.129) if we consider that the interface is a plane given by the equation $x = x_i$. For this separated phase (for example, the left side of the interface), the flow is considered as two-dimensional with a normal and parallel direction with respect to the interface. We consider that the steady state flow and the participating natural convection are not excluded. The considered flow is similar to that shown in Fig. 6.6. The continuity of x and y could be written using the Navier–Stokes equations:

$$\frac{\partial w_x}{\partial x} + \frac{\partial w_y}{\partial y} = 0 \tag{6.130}$$

$$w_x \frac{\partial w_x}{\partial x} + w_y \frac{\partial w_x}{\partial y} = -\frac{1}{\rho}\frac{\partial p}{\partial x} + \upsilon\left(\frac{\partial^2 w_x}{\partial x^2} + \frac{\partial^2 w_x}{\partial y^2}\right) \tag{6.131}$$

$$w_x \frac{\partial w_y}{\partial x} + w_y \frac{\partial w_y}{\partial y} = g_y \beta_\Gamma \Delta\Gamma + \frac{1}{\rho}\frac{\partial p}{\partial y} + \upsilon\left(\frac{\partial^2 w_y}{\partial x^2} + \frac{\partial^2 w_y}{\partial y^2}\right) \tag{6.132}$$

The flow equations are completed with the corresponding boundary conditions, which, for example, show:
- a maximum velocity at the interface coupled with a constant pressure:

$$x = x_i \ , \ \ -y_v \leq y \leq y_v \ ; \ w_x = 0 \ , \ \ \frac{dw_y}{dx} = 0 \ , \ p = p_i \tag{6.133}$$

- the absence of the velocity component for planes $y = y_v$ and $y = -y_v$ (normal planes at the interface) coupled with the linear pressure state:

$$0 \leq x \leq x_i \ , \ y = -y_v \ , \ y = y_v \ ; \ w_x = w_y = 0 \ , \ p = p_i + \rho g(x_i - x) \tag{6.134}$$

- a particular velocity state for plane x = 0:

$$x = 0, \quad -y_v \leq y \leq y_v \; ; \; w_y = f(y), \; w_x = 0 \tag{6.135}$$

The convective-diffusion equation characterizes the transport of the property for the fluid placed on the left of the interface. Here, the property participates in a reaction process, which is described by simple kinetics. Then, the particularization of the convective property transport equation becomes:

$$w_x \frac{\partial \Gamma}{\partial x} + w_y \frac{\partial \Gamma}{\partial y} = D_\Gamma \left(\frac{\partial^2 \Gamma}{\partial x^2} + \frac{\partial^2 \Gamma}{\partial y^2} \right) - k_{rr} \Gamma \tag{6.136}$$

The first boundary condition for the convective-diffusion equation shows that, at the interface, the property flux is written using the transfer property coefficient:

$$x = x_i, \quad -y_v \leq y \leq y_v \; ; \; k_\Gamma (\Gamma_i - \Gamma_0) = D_\Gamma \frac{d\Gamma}{dx} \tag{6.137}$$

If, for the second boundary condition, we consider a constant concentration (Γ_0) of the property at plane x = 0, we can write:

$$x = 0, \quad -y_v \leq y \leq y_v \; ; \; \Gamma = \Gamma_0 \tag{6.138}$$

The third boundary condition considers that planes ys = s–y_v and ys = sy_v are impermeable to the transferred property:

$$0 \leq x \leq x_i, \quad y = -y_v, \; y = y_v \; ; \; \frac{d\Gamma}{dy} = 0 \tag{6.139}$$

The equations described above could be written in a dimensionless form taking into account different dimensionless parameters. They include a geometrical dimension such as the dimensionless coordinates $x_a = x/l$ and $y_a = y/l$; the dimensionless velocity, the pressure and property concentration:

$$w_x^a = \frac{w_x}{w}, \quad w_y^a = \frac{w_y}{w}, \quad p_a = \frac{p}{\Delta p}, \quad \Gamma_a = \frac{\Gamma - \Gamma_0}{\Gamma_i - \Gamma_0} \tag{6.140}$$

where w is a stable, real or computed velocity, characteristic of the system and Δp is the differential pressure ($p_0 - p_i$). With these dimensionless definitions, the basic model equations become:

$$\frac{\partial w_x^a}{\partial x_a} + \frac{\partial w_y^a}{\partial y_a} = 0 \tag{6.141}$$

$$\frac{w^2}{1} w_x^a \frac{\partial w_x^a}{\partial x_a} + \frac{w^2}{1} w_y^a \frac{\partial w_x^a}{\partial y} = \frac{\Delta p}{\rho l} \frac{\partial p_a}{\partial x_a} + \frac{\upsilon w}{l^2} \left(\frac{\partial^2 w_x^a}{\partial x_a^2} + \frac{\partial^2 w_x^a}{\partial y_a^2} \right) \tag{6.142}$$

$$\frac{w^2}{1} w_x^a \frac{\partial w_y^a}{\partial x_a} + \frac{w^2}{1} w_y^a \frac{\partial w_y^a}{\partial y_a} = g_y \beta_\Gamma \Delta \Gamma + \frac{\Delta p}{\rho l} \frac{\partial p_a}{\partial y} + \frac{\upsilon w}{l^2} \left(\frac{\partial^2 w_y^a}{\partial x_a^2} + \frac{\partial^2 w_y^a}{\partial y_a^2} \right) \tag{6.143}$$

$$\frac{w(\Gamma_i - \Gamma_0)}{1} w_x^a \frac{\partial \Gamma_a}{\partial x_a} + \frac{w(\Gamma_i - \Gamma_0)}{1} w_y^a \frac{\partial \Gamma_a}{\partial y_a} = \frac{D_\Gamma(\Gamma_i - \Gamma_0)}{1^2} \left(\frac{\partial^2 \Gamma_a}{\partial x_a^2} + \frac{\partial^2 \Gamma_a}{\partial y_a^2} \right)$$
$$- k_{rr}[\Gamma_0 + \Gamma_a(\Gamma_i - \Gamma_0)] \tag{6.144}$$

If we multiply Eqs. (6.142) and (6.143) by $\dfrac{1^2}{\upsilon w}$ and Eq. (6.144) by $\dfrac{1^2}{D_\Gamma(\Gamma_i - \Gamma_0)}$, we obtain the new forms of this set of equations. These are dimensionless and consequently include coefficients, which are dimensionless groups or combinations of the dimensionless groups. Assembling Eqs. (6.145) and (6.154) shows the initial model in its dimensionless form:

$$\frac{\partial w_x^a}{\partial x_a} + \frac{\partial w_y^a}{\partial y_a} = 0 \tag{6.145}$$

$$Re\left(w_x^a \frac{\partial w_x^a}{\partial x_a} + w_y^a \frac{\partial w_x^a}{\partial y} \right) = Eu.\ Re.\frac{\partial p_a}{\partial x_a} + \left(\frac{\partial^2 w_x^a}{\partial x_a^2} + \frac{\partial^2 w_x^a}{\partial y_a^2} \right) \tag{6.146}$$

$$Re\left(w_x^a \frac{\partial w_y^a}{\partial x_a} + w_y^a \frac{\partial w_y^a}{\partial y_a} \right) = Gr_\Gamma\ Re^{-1} + Eu.\ Re.\frac{\partial p_a}{\partial y} + \left(\frac{\partial^2 w_y^a}{\partial x_a^2} + \frac{\partial^2 w_y^a}{\partial y_a^2} \right) \tag{6.147}$$

$$Re.Pr_\Gamma\left(w_x^a \frac{\partial \Gamma_a}{\partial x_a} + w_y^a \frac{\partial \Gamma_a}{\partial y_a} \right) = \left(\frac{\partial^2 \Gamma_a}{\partial x_a^2} + \frac{\partial^2 \Gamma_a}{\partial y_a^2} \right) - Fo_{rr}\Gamma_a \tag{6.148}$$

$$x_a = x_i^a\ ,\ -y_v^a \le y_a \le y_v^a\ ;\ w_x^a = 0\ ,\ \frac{dw_y^a}{dx_a} = 0\ ,\ p = p_i^a \tag{6.149}$$

$$0 \le x_a \le x_i^a\ ,\ y_a = -y_v^a\ ,\ y_a = y_v^a\ ;\ w_x^a = w_y^a = 0\ ,\ p_a = p_i^a + \rho g(x_i^a - x_a).l/\Delta p \tag{6.150}$$

$$x_a = 0\ ,\ -y_v^a \le y_a \le y_v^a\ ;\ w_y^a = f(y_a)\ ,\ w_x^a = 0 \tag{6.151}$$

$$x_a = x_i^a\ ,\ -y_v^a \le y_a \le y_v^a\ ;\ Nu_\Gamma = \frac{d\Gamma_a}{dx_a} \tag{6.152}$$

$$x_a = 0\ ,\ -y_v^a \le y_a \le y_v^a\ ;\ \Gamma_a = 0 \tag{6.153}$$

$$0 \le x_a \le x_i^a\ ,\ y_a = -y_v^a,\ y_a = y_v^a\ ;\ \frac{d\Gamma_a}{dy_a} = 0 \tag{6.154}$$

The formal solution for this complete dimensionless model can be obtained when the flow equations can be resolved separately. Then we obtain:

$$w_x^a = f(Eu,\ Re,\ Gr_\Gamma,\ x_a,\ y_a) \tag{6.155}$$

$$w_y^a = g(Eu, Re, Gr_\Gamma, x_a, y_a) \tag{6.156}$$

where f and g define any particular function.

The solution for the concentration state of the transferable property can be written as:

$$\Gamma_a = h(Re, Pr_\Gamma, Fo_{r\Gamma}, w_x^a, w_y^a, x_a, y_a) \tag{6.157}$$

The substitution of Eqs. (6.155) and (6.156) into Eq. (6.157) gives a new form to the concentration state of the transferable property:

$$\Gamma_a = F(Eu, Gr_\Gamma, Re, Pr_\Gamma, Fo_{r\Gamma}, x_a, y_a) \tag{6.158}$$

Using this last relationship, we can now appreciate the value of the concentration dimensionless gradient of the transferable property near the interface. Then, the result is:

$$\left(\frac{d\Gamma_a}{dx_a}\right)_{x_a = x_a^i} = G(Eu, Gr_\Gamma, Re, \underset{\Gamma}{Pr}, Fo_{r\Gamma}, i_x, i_y) \tag{6.159}$$

where G is the F function derivative and i_x and i_y are the geometric simplex (ratio between the interface coordinates and the characteristic geometrical length).

The combination of Eqs. (6.130) and (6.159) gives a relationship between the general dimensionless groups characterizing the interface kinetic transfer of one property:

$$Nu_\Gamma = G(Eu, Gr_\Gamma, Re, Pr_\Gamma, Fo_{r\Gamma}, i_x, i_y) \tag{6.160}$$

We can conclude that the transfer intensity is determined by pressure (introduced by the Euler number (Eu)) as well as by natural convection (expressed by the non-particularized Grashof number (Gr$_\Gamma$)), by controlled convection (given by the Reynolds number (Re)), by chemical reaction (expressed by the reaction Fourier number (Fo$_{r\Gamma}$)), by the transport properties of the medium (assigned by the Prandtl number (Pr$_\Gamma$)) and finally by the geometry of the system (shown by the geometrical simplex i_x, i_y). Moreover, some of these actions are over represented because they cannot be used together. For example, the pressure action produces a controlled flow, which is characterized by the Reynolds number. However, in Eq. (6.160), the Euler number is not an independent parameter and can consequently be eliminated. Another example shows that, in the case of an important convective action (turbulent flow), the effect of the natural convection can be neglected. The same consideration shows that, in the cases of pure natural convection flow, the Reynolds number is not important. Finally, in the case of gas transfer at moderate pressures and temperatures, the generalized Prandtl number presents a constant value and, consequently, its influence in the kinetic relationship is not important.

Table 6.2 sums up these considerations in the case of transfer with no chemical reaction. We observe, for example, that the geometry of the system, the Reynolds number and the generalized Prandtl number, determine the intensity of the property transfer in a liquid medium with a turbulent flow.

Table 6.2 Relationships of the kinetic transfer dimensionless groups.

Fluid	Type of flow	Particularization of Eq. (6.160)	Particularization of heat transfer	Particularization of mass transfer
Liquid	natural convection	$Nu_\Gamma = G(Gr_\Gamma, Pr_\Gamma, i_x, i_y)$	$Nu = G(Gr_t, Pr, i_x, i_y)$	$Sh = G(Gr_d, Sc, i_x, i_y)$
	forced convection	$Nu_\Gamma = G(Re, Pr_\Gamma, i_x, i_y)$	$Nu = G(Gr_t, Pr, i_x, i_y)$	$Sh = G(Re, Sc, i_x, i_y)$
Gas	natural convection	$Nu_\Gamma = G(Gr_\Gamma, i_x, i_y)$	$Nu = G(Gr_t, i_x, i_y)$	$Sh = G(Gr_d, i_x, i_y)$
	forced convection	$Nu_\Gamma = G(Re, i_x, i_y)$	$Nu = G(Re, i_x, i_y)$	$Sh = G(Re, i_x, i_y)$

It is important to note that the classification presented above is not unique. Indeed, each particular case has its G function. For example, when a chemical reaction occurs, the generalized Fourier number ($Fo_{r\Gamma}$) and its particularization for heat and mass transfer can be introduced as a G function argument.

6.6.3
Physical Interpretation of the Nu, Pr, Sh and Sc Numbers

This section will present one of the possible physical interpretations of these important dimensionless numbers. First, to show the meaning of Nusselt number, we consider the heat transfer flux in the x direction in the case of a pure molecular mechanism compared with the heat transfer characterizing the process when convection is important. The corresponding fluxes are then written as:

$$q_m = -\lambda \left(\frac{dt}{dx} \right)_{x=x_i} \tag{6.161}$$

$$q_c = \alpha(t_i - t_\infty) \tag{6.162}$$

where α, λ and t have been defined above (for instance, see Fig. 6.7). Index i indicates the position of the interface. By analogy with the model already used to determine the significance of the Reynolds number, we calculate the ratio between both heat fluxes which is represented by the following relationship:

$$\frac{q_c}{q_m} = \frac{\alpha l}{\lambda} \frac{(t_\infty - t_i)}{\left(\dfrac{dt}{d(x/l)}\right)_{x=x_i}} = Nu\left(\frac{dt_a}{dx_a}\right)_{x_a = x_i^a} \tag{6.163}$$

The result shows that the physical significance of the Nusselt number is:

$$Nu = \frac{\text{quantity of heat transferred by the convective mechanism}}{\text{quantity of heat transferred by the molecular mechanism}}$$

As for the Prandtl number, we consider the heat transfer flux which can be written with the use of the fluid enthalpy (Eq. (6.164)) and the molecular momentum flux given by Eq. (6.165):

$$q_m = -\frac{\lambda}{\rho c_p}\left(\frac{d(\rho c_p t)}{dx}\right)_{x=x_i} \tag{6.164}$$

$$\tau_{myx} = -\frac{\eta}{\rho}\left(\frac{d(\rho w_y)}{dx}\right)_{x=x_i} \tag{6.165}$$

The ratio between both fluxes shows that the Prandtl number is an index giving the relative quantity of the momentum transported by the molecular mechanism and of the heat transported by the same mechanism at the interface:

$$\frac{\tau_{mxy}}{q_m} = \frac{c_p \eta}{\lambda}\left(\frac{d(\rho w_y)}{d(\rho c_p t)}\right)_{x=x_i} = Pr\left(\frac{d(\rho w_y)}{d(\rho c_p t)}\right)_{x=x_i} \tag{6.166}$$

As for the significance of the Sherwood number, the following mass transfer fluxes for species A are used:
- the flux of component A transported to the interface by pure diffusion (molecular mechanism):

$$N_{Am} = D_A\left(\frac{d(\rho\omega_A)}{dx}\right)_{x=x_i} \tag{6.167}$$

- the flux of component A transported to the interface by natural and provoked/induced/forced convection;

$$N_{Ac} = k_c\left((\rho\omega_a)_i - (\rho\omega_A)_\infty\right) \tag{6.168}$$

The ratio between both fluxes shows that the Sherwood number can be considered as an index of the relative participation of the convective and molecular mechanisms to the transport process.

The next relationship gives the mathematical form of this physical interpretation:

$$\frac{N_{Ac}}{N_{Am}} = \frac{k_c\left((\rho\omega_A)_i - (\rho\omega_A)_\infty\right)}{D_A\left(\dfrac{d(\rho\omega_A)}{dx}\right)_{x=x_i}} = \frac{k_c l}{D_A}\frac{\left((\rho\omega_A)_i - (\rho\omega_A)_\infty\right)}{\left(\dfrac{d(\rho\omega_A)}{d(x/l)}\right)_{x=x_i}} = Sh\frac{1}{\left(\dfrac{d\omega_A^a}{dx_a}\right)_{x_a = x_i^a}} \tag{6.169}$$

$$Sh = \frac{\text{Quantity of the species A transported to the interface by convective mechanism}}{\text{Quantity of the species A transported to the interface by molecular mechanism}}$$

The Schmidt number is the mass transfer analogue of the Prandtl number. Indeed, by analogy, we can note that the Schmidt number is a measure characterizing the ratio between the quantity of the momentum transported to the interface by the molecular mechanism and the quantity of species A transported to the interface by the same mechanism. Equation (6.171) shows this statement:

$$\frac{\tau_{mxy}}{N_{Am}} = \frac{\eta}{\rho D}\left(\frac{d(\rho w_y)}{d(\rho \omega_A)}\right)_{x=x_i} = Sc\left(\frac{d(w_y)}{d(\omega_A)}\right)_{x=x_i} \tag{6.170}$$

6.6.4
Dimensionless Groups for Interactive Processes

When a process is produced under the divergent or convergent action of two different forces, the ratio between them represents a dimensionless number. The heat and mass transfer enhanced by the supplementary action of a pulsating field (vibration of apparatus, pulsation of one (or two) phase flow(s), ultrasound action etc.) has been experimented and applied in some cases [6.25–6.27]. Then, the new dimensionless number $I_\omega = \dfrac{g}{\omega^2 A}$ has to be added to the list of dimensionless groups presented above in this chapter.

As an example of an interactive process, we can mention the rapid drying of a porous material when heat and mass transport occur simultaneously. This case corresponds to very intensive drying such as high frequency or conductive drying. In this case, a rapid transfer of humidity from the liquid to the vapour state, associated with local change in pressure, induces a rapid vapour flow in the porous structure. To establish the equations of water transport, we assume that the gradient is established from the material matrix to the outside. It results in additional moisture and heat transfer induced by the hydrodynamic (filtration) motion of liquid and vapours. The total pressure gradient within the material appears as the result of evaporation and of the resistance of the porous skeleton during vapour motion. The air from the adjacent medium flows by molecular and slip diffusion in the capillarity of the system.

In the case of a high-rate heat- and mass-transfer process, heat and mass flows are not described by the classical Onsager equations (as, for instance, in Eq. (3.12)):

$$\vec{j}_k = \sum_i L_{ki}\vec{X}i \tag{6.171}$$

but by the following generalized equation:

$$\vec{j}_k = \sum_i L_{ki}\vec{X}i + L_k^{(r)}\frac{\partial \vec{j}_k}{\partial \tau} \tag{6.172}$$

for example, the Fourier heat-conduction equation $\vec{q}_m = -\lambda \vec{\nabla} t$ will be replaced by:

$$\vec{j}_q = \vec{q} = \lambda \vec{\nabla} t - \tau_{rq} \frac{\partial \vec{q}}{\partial \tau} \tag{6.173}$$

Equation (6.173) is valid only for one-dimensional problems. For a multidimensional study, it can be used as an approximation where the relaxation period of the thermal stress τ_{rq} is defined as one experimental constant. A similar relationship is used for moisture diffusion. The term $L_k^{(r)} \frac{\partial \vec{j}_k}{\partial \tau}$ corresponds to the finite propagation velocity of a certain substance. The stress relaxation period τ_{rk} of substance k (mass, heat, etc.) is defined by:

$$\tau_{rk} = \frac{D_k}{v_k^2} \tag{6.174}$$

where v_k is the finite propagation velocity and D_k the diffusivity of substance k. From Eq. (6.172), we can now describe the flux of property with the local Γ_k concentration as:

$$\vec{j}_{tk} = \sum_{i=1}^{n} \left(D_{\Gamma kl} \vec{n}_i \vec{\nabla} \Gamma_{kl} \right) + \vec{w} \vec{\nabla} \vec{\Gamma}_k - \tau_{rk} \frac{\partial \vec{j}_k}{\partial \tau} \tag{6.175}$$

When we particularize this relationship for the mass transport of the humidity into a porous medium ($\vec{w} = 0$, because there is no microscopic displacement), we can observe the superposition of the thermo-diffusion and of the diffusive filtration (where p is the humidity flowing by filtration) over the pure diffusion process:

$$\vec{J}_{tu} = D_m \vec{\nabla} u + D_m \delta \vec{\nabla} t + D_m \delta_p \vec{\nabla} p \tag{6.176}$$

Based on Eq. (6.176), we obtain the next particularization for the general conservation relationship (see, for instance, Eq. (3.6)), which was established in Chapter 3:

$$\tau_{rk} \frac{\partial^2 \Gamma_k}{\partial \tau^2} + \left(\frac{\partial \Gamma_k}{\partial \tau} + \vec{w} \vec{\nabla} \vec{\Gamma}_k \right) = \text{div} \left(\sum_{i=1}^{n} L_{kl} \vec{n}_i \vec{\nabla} \Gamma_{kl} \right) + j_{vk} \tag{6.177}$$

It is not difficult to particularize this relationship for the three simultaneous processes occurring in the porous medium. The result is the Luikov [6.28] complete relationships:

$$\tau_{ru} \frac{\partial^2 u}{\partial \tau^2} + \frac{\partial u}{\partial \tau} = K_{11} \nabla^2 u + K_{12} \nabla^2 t + K_{13} \nabla^2 p \tag{6.178}$$

$$\tau_{rq} \frac{\partial^2 t}{\partial \tau^2} + \frac{\partial t}{\partial \tau} = K_{21} \nabla^2 u + K_{22} \nabla^2 t + K_{23} \nabla^2 p \tag{6.179}$$

$$\tau_{rp} \frac{\partial^2 p}{\partial \tau^2} + \frac{\partial p}{\partial \tau} = K_{31} \nabla^2 u + K_{32} \nabla^2 t + K_{33} \nabla^2 p \tag{6.180}$$

where the coefficients K_{ij} (i, j = 1,2,3) correspond to:

$$K_{11} = D_m , \; K_{12} = D_m\delta = (D_{m1}^t + D_{m2}^t), \; K_{13} = D_m\delta_p = k_p/\rho_0 \qquad (6.181)$$

$$K_{21} = D_m\frac{r\varepsilon}{c}, \; K_{22} = a, \; K_{32} = D_m\frac{\varepsilon r\delta_p}{c} \qquad (6.182)$$

$$K_{31} = -D_m\frac{\varepsilon}{c_{ph}}, \; K_{32} = -D_m\frac{\varepsilon\delta}{c_{ph}}, \; K_{33} = D_p - D_m\frac{\varepsilon\delta_p}{c_{ph}} \qquad (6.183)$$

where k_p is the filtration moisture-coefficient defined by the equation $\vec{J}_p = -k_p\vec{\nabla}p$; δ_p is the dimensionless filtration moisture flow, $\delta_p = k_p/D_m\rho_0$; a_p is the convective filtration diffusion coefficient, $a_p = \dfrac{k_p}{c_{ph}\rho_0}$; c_{ph} is the coefficient of humid air capacity in a porous material defined by the relationship $d(u_1 + u_2) = c_{ph}dp$; u_1 is the material moisture in the vapour state, u_2 is the moisture in the liquid state; ε is the dimensionless fraction defined by

$$\varepsilon = \frac{D_{m1}}{D_{m1} + D_{m2}} = \frac{D_{m1}}{D_m};$$ a is the thermal diffusivity, and D_{m1}^t and D_{m2}^t are respec-

tively the thermo-diffusion coefficients of vapour and liquid humidity.

Taking into account the above description, the mass transfer similarity numbers, which characterize this process, can be formulated. The following similarity numbers can then be formulated from the differential moisture transfer equations ((6.178) – (6.183)):

1. The homochronism of the transfer numbers of the field potential referred to as Fourier numbers:

$$Fo_q = \frac{a\tau}{l^2}, \; Fo_m = \frac{D_m\tau}{l^2}, \; Fo_p = \frac{a_p\tau}{l^2} \qquad (6.184)$$

These dimensionless groups are related by the criteria Lu and Lu_p (drying Luikov dimensionless groups).

2. The mass transfer relaxation Fourier number:

$$Fo_{rm} = \frac{D_m\tau_{rm}}{l^2} \qquad (6.185)$$

It is then important to specify that this number is formed by known magnitudes. The relaxation period of mass stress is about 10^4 times the thermal stress relaxation. The Fourier mass transfer number is, therefore, many times greater than the Fourier heat transfer relaxation number: $Fo_{rq} = a\tau_{rq}/l^2$.

3. The diffusion moisture-transfer number, with respect to the heat diffusion or the moisture- flow diffusion number (drying Luikov number [6.23, 6.28]):

$$Lu = \frac{D_m}{a}, \; Lu_p = \frac{D_p}{a} \tag{6.186}$$

The Lu number is the ratio between the mass diffusion coefficient and the heat diffusion coefficient. It can be interpreted as the ratio between the propagation velocity of the
iso-concentration surface and the isothermal surface. In
other words, it characterizes the inertia of the temperature
field inertia, with respect to the moisture content field (the
heat and moisture transfers inertia number). The Lu_p diffusive filtration number is the ratio between the diffusive filtration field potential (internal pressure field potential) and the
temperature field propagation.

For some moist materials, the Lu number increases with
the moisture content following a slow linear dependence.
From Eq. (6.184), we can appreciate that, for Lu>1, the propagation velocity of the mass transfer potential is greater than
the propagation velocity of the temperature field potential.
The value of the diffusive filtration number of moisture Lu_p
is normally much higher than one. The total internal pressure relaxation of the vapour–gas mixture in a capillary porous body is 2–3 orders of magnitude higher than the relaxation of the temperature field. The relationships between Fourier numbers may be expressed in terms of Lu and Lu_p:

$$Fo_m = Fo_q Lu, \; Fo_p = Fo_q Lu_p \tag{6.187}$$

4. The Kossovich (Ko) and Posnov (Pn and Pn_p) numbers
   [(6.23), (6.28)] defined by Eqs. (6.188)–(6.190) are obtained
   from the drying model (Eqs. (6.178)–(6.183) completed with
   specific initial and boundary conditions). Moreover, they are
   converted into a dimensionless form by applying the pi theorem:

$$Ko = \frac{r\Delta u}{c_q \Delta t} \tag{6.188}$$

$$Pn = \frac{\delta \Delta t}{\Delta u} = \frac{D_m^t \rho_0 \Delta t}{\rho_0 D_m \Delta u} \tag{6.189}$$

$$Pn_p = \frac{\delta_p \Delta p}{\Delta u} = \frac{k_p \Delta p}{\rho_0 D_m \Delta u} \tag{6.190}$$

The Ko number shows the relationship between the heat
consumed by the liquid evaporation ($r\Delta u$) and the heating of
the moist body ($c_q \Delta t$). The Pn number is an index of the

ratio between the quantity of the humidity transported by the thermal-diffusive mechanism and the pure diffusive mechanism. A similar consideration can be advanced with respect to the Pn_p number.

5. Using the property of the dimensionless groups which show that a mathematical combination of such groups gives a dimensionless group, we introduce the Feodorov number (Fe) [6.23, 6.28] which is as a dimensionless number describing the drying process of a porous material:

$$Fe = \varepsilon KoPn = \frac{\varepsilon \delta r}{c_q} \tag{6.191}$$

This number is independent of the heat and mass transfer potentials because it is defined by coefficients ε, δ, r and c_q (the last two are, respectively, the vaporization latent heat of moisture and the specific heat capacity of the moist body).

6. The Rebinder (Rb) number is formulated using the methodology described above in item 5. This dimensionless number is given as the ratio between the dimensionless temperature coefficient of drying and the Kossovich number:

$$Rb = \frac{B}{Ko} = \frac{b\dfrac{\Delta u}{\Delta t}}{\dfrac{r\Delta u}{c_q \Delta t}} = \frac{c_q b}{r} \tag{6.192}$$

As in the case of the Fedorov number, the Rebinder number is independent of the choice of the heat and mass transfer potentials. This number is part of the fundamental heat balance of the drying process. Unlike the Pn number, the temperature-drying coefficient describes the changes occurring in the integral mean temperature ($\bar{t}$) and in the mean moisture content ($\bar{u}$). In other words, it relates the kinetic properties of integral heat with moisture transfer properties, whereas, the Pn number is concerned with local changes in u and t.

7. The Biot numbers of heat and mass transfer could be obtained from the boundary conditions of a third kind:
   - the heat transfer Biot number:

$$Bi_q = \frac{\dfrac{\alpha_{ex}}{\rho_{ex} c_{pex}}l}{a} \cong \frac{\alpha_{ex}l}{\lambda_m} \tag{6.193}$$

   - the mass transfer Biot number:

$$Bi_m = \frac{k_{cex}l}{D_m} \tag{6.194}$$

Considering these Biot numbers, we can observe that they are similar to the Nusselt and Sherwood numbers. The only difference between these dimensionless numbers is the transfer coefficient property characterizing the Biot numbers' transfer kinetics for the external phase (α_{ex}: heat transfer coefficient for the external phase, k_{cex}: mass transfer coefficient for the external phase). We can conclude that the Biot number is an index of the transfer resistances of the contacting phases.

When the boundary conditions of the third kind cannot be established for heat ($q_{int}(\tau)$) and mass flux ($N_{Aint}(\tau)$) flows, then Bi_q and Bi_m are substituted by two Kirpichev (Ki) [6.23, 6.28] numbers:

$$Ki_q = \frac{q_{int}(\tau)l}{\lambda_o\Delta t}, \; Ki_m = \frac{N_{Aint}(\tau)l}{D_m\Delta u} \tag{6.195}$$

If fluxes $q_{int}(\tau)$ and $N_{Aint}(\tau)$ are defined by the Newton laws, these numbers (Bi and Ki) are related by simple equations:

$$Ki_q = Bi_q\frac{t_c - t_{int}}{\Delta t}, \; Ki_m = Bi_m\frac{u_c - t_{int}}{\Delta u} \tag{6.196}$$

where indexes c and int indicate the central and interface position of the moist body. Quantities Δt, Δu and Δp, appearing in the heat and mass transfer similarity numbers, are chosen taking the conditions of the problem into account.

The problem here discussed can be considered as an example, which can be generalized when the different elementary processes interact. At the same time, it shows the large potentialities of the chemical engineering methodologies in defining and using dimensionless groups for process characterization. All newly introduced dimensionless groups can also be obtained through an adequate dimensional analysis using the pi theorem procedure described earlier in this chapter.

6.6.5
Common Dimensionless Groups in Chemical Engineering

It is not easy to produce a basic list of the dimensionless groups frequently utilized in chemical engineering problems. This is due to the very large number of dimensionless groups that characterize the totality of chemical engineering processes. Table 6.3 gives a list of variables, which are commonly encountered in this type of analysis. Obviously, the list is not exhaustive but indicates a broad range of variables typically found in chemical engineering problems. Moreover, we can combine these variables with some of the common dimensionless groups given in Table 6.4.

Table 6.3 Some common variables typically used in chemical engineering.

Symbol	Definition
l	characteristic length
β_d	concentration convection coefficient
ρ	density
D	diffusion coefficient
ΔH_r	enthalpy of reaction
g	gravity acceleration
	heat transfer coefficient
k_c	mass transfer coefficient
ω	oscillation frequency
$p, \Delta p$	pressure or pressure difference
k_r	reaction kinetics constant
c	speed of sound
σ	surface tension
c_p or c_v	thermal capacity
λ	thermal conductivity
β_t	thermal convection coefficient
w	velocity
η	viscosity
R	universal constant of gases

Table 6.4 Some dimensionless groups typically used in chemical engineering.

Name and symbol	Definition formula	Physical interpretation	Type of application
Reynolds number Re	$\dfrac{wl\rho}{\eta}$	$\dfrac{\text{momentum quantity transfered by turbulent mechanism}}{\text{momentum quantity transferd by molecular mechanism}}$	all types of momentum, heat and mass transfer with forced convection
Froude number Fr	$\dfrac{w}{\sqrt{gl}}$	$\dfrac{\text{inertia force}}{\text{gravitational force}}$	flow with a free surface, pipe and packed bed two phase flow
Euler number Eu	$\dfrac{\Delta p}{\rho w^2}$	$\dfrac{\text{energy involved by the surface forces}}{\text{energy involved by the inertia forces}}$	problems in which pressure or pressure differences are of interest (jets from nozzles, injectors etc.)

Name and symbol	Definition formula	Physical interpretation	Type of application
Mach number Ma	$\dfrac{w}{c}$	$\dfrac{\text{inertia force}}{\text{compressibility force}}$	flow in which the compressibility of the fluid is important
Strouhal number St	$\dfrac{\omega l}{w}$	$\dfrac{\text{local inertia force}}{\text{global inertia force}}$	rotational steady and unsteady flow
Weber number We	$\dfrac{\rho w^2 l}{\sigma}$	$\dfrac{\text{energy involved by the inertia forces}}{\text{energy involved by the surface tension forces}}$	problems in which surface tension is important (bubbles, drops and particles)
Archimedes number Ar	$\dfrac{gl^3 \rho \Delta \rho}{\eta^2}$	$\dfrac{\text{archimedian force}}{\text{viscous force}}$	flow of the particles, drops and bubbles in liquid and gaseous media, fluidized and spurted bed
Grashof thermal number Gr_t	$\dfrac{gl^3 \beta_t \Delta t \rho^2}{\eta^2}$	$\dfrac{\text{thermal convection force}}{\text{viscous force}}$	flow and heat transfer by natural thermal convection
Grashof diffusion number Gr_t	$\dfrac{gl^3 \beta_d \Delta C \rho^2}{\eta^2}$	$\dfrac{\text{concentration convection forces}}{\text{viscous forces}}$	flow and mass transfer by natural convection
Nusselt number Nu	$\dfrac{\alpha l}{\lambda}$	$\dfrac{\text{quantity of the heat transfered by convection}}{\text{quantity of heat transfered by the molecular mechanism}}$	all heat transfer problems
Prandtl number Pr	$\dfrac{c_p \eta}{\lambda}$	$\dfrac{\text{momentum quantity transfered by molecular mechanism}}{\text{heat quantity transfered by molecular mechanism}}$	all heat transfer problems in forced convection
Biot number Bi	$\dfrac{\delta/\lambda}{1/\alpha}$	$\dfrac{\text{conductive resistance media}}{\text{convective resistance media}}$	all heat transfer problems with interface flux condition
Fourier reaction number Fo_r	$\dfrac{k_r l^2}{D^2}$	$\dfrac{\text{quantity of species consumed by reaction}}{\text{quantity of species transported by molecular mechanism}}$	mass transfer problems with chemical reaction
Schmidt number Sc	$\dfrac{\eta}{\rho D}$	$\dfrac{\text{momentum quantity transfered by molecular mechanism}}{\text{species quantity transfered by molecular mechanism}}$	all mass transfer problems with forced convection
Sherwood number Sh	$\dfrac{k_c l}{D}$	$\dfrac{\text{species quantity transported by convective mechanism}}{\text{species quantity transported by molecular mechanism}}$	all mass transfer problems

Some additional details or commentaries about these important dimensionless groups are discussed in the next sections.

Reynolds Number (Re)

This dimensionless number is undoubtedly the most famous parameter in chemical engineering and fluid mechanics. It was named after Osborne Reynolds (1842–1912), a British engineer, who, with his famous experiment called the "Reynolds experiment" (1892) showed for the first time that this combination of variables could be used as a criterion to characterize laminar and turbulent flows. The Reynolds number is the measure of the ratio between the inertia and the viscous forces of a fluid element. The flow occurring in different systems such as monophase, two-phase and three-phase flows in a packed bed, two-phase and three-phase flows in trays can also be characterized by different Reynolds numbers. In addition, in chemical engineering, all the kinetic relationships, where forced convection is present (see for instance Table 6.2), can be described according to the Reynolds number. For example, the general kinetic relationship $Nu_\Gamma = G(\, Re, Pr_\Gamma, i_x, i_y)$ will show a particular form of the function G for a laminar flow and another particular form of the same function when turbulent flow occurs. The small values of the Reynolds number indicate that the viscous forces are dominant in the system and that we can consequently eliminate the participation of the forced convective mechanism in the flux property equation. It results in simplified forms of the equations related to flow field and to property field. In some cases, this type of simplification allows an analytical solution. Another example can be shown in the case of flow occurring over an immersed body (packed bed, sedimentation, etc.). In this example, for very large Reynolds numbers, inertial effects predominate over viscous effects and it may be possible to neglect the effect of viscosity and consider the problem as if it involved a 'non-viscous' fluid. In this case the Navier–Stokes equations can easily be reduced to the Euler equations for flow.

Froude Number (Fr)

The Froude number is named after William Froude (1810–1879), a British civil engineer, mathematician, and naval expert who pioneered the use of towing tanks of ship design. In some scientific papers the Froude number is defined as the square of the mathematical equation considered here, (for instance, see Table 6.8). This dimensionless number shows the importance of the gravitational force in some chemical engineering processes. This is typically the case for natural convection and surface flows. Figure 6.8 shows two examples of chemical engineering processes, which are used to separate or put into contact two different phase fluids. In case A, where a two-phase flow occurs in a contacting tray device, the Froude number is used to characterize the hydrodynamics and stability of the flow, whereas, in case B, where we have a co-current two-phase pipe flow, it defines the different states of the flow.

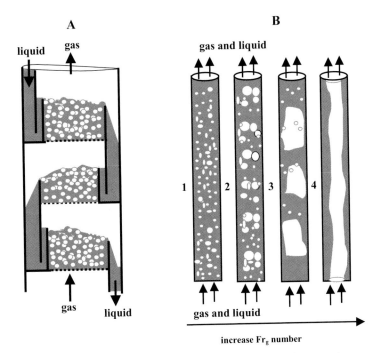

Figure 6.8 Two specific chemical engineering processes where the Fr number is applied. A: device with trays for contacting two phases; B: two-phase pipe flow in co-current configuration. 1: bubble flow; 2: aggregated flow; 3: plug flow; 4: annular flow.

Euler Number (Eu)

The Froude number described above is frequently used for the description of radial and axial flows in liquid media when the pressure difference along a mixing device is important. When cavitation problems are present, the dimensionless group $(p_r - p_v)/\rho w^2$ – called the Euler number – is commonly used. Here p_v is the liquid vapour saturation pressure and p_r is a reference pressure. This number is named after the Swiss mathematician Leonhard Euler (1707–1783) who performed the pioneering work showing the relationship between pressure and flow (basic static fluid equations and ideal fluid flow equations, which are recognized as Euler equations).

Mach Number (Ma)

The Austrian physicist E. Mach (1838–1916) is the recognized founder of this dimensionless group. This number is not very useful for most of the chemical engineering flow problems because it considers that the flowing fluid density is not affected by the field flow. In chemical engineering processes the Mach number takes values lower than 0.3. This means that this type of process is placed on the boundary between flows without and with compressibility effects. However, in some cases where the sound field has been introduced as the active factor to enhance a specific unit operation (for example absorption) characterization using this number could be useful. This number is the more commonly used parameter in the fields of gas dynamics and aerodynamics.

Strouhal Number (St)

This number is used to characterize the stationary and unsteady oscillatory flow when the oscillatory field frequency presents a significant value. This type of flow can be generated for example, when a fluid is transported by piston pumps. In this case, the frequency flow parameters could be described by a combination of Strouhal number and Reynolds number:

$$Ff = St.\,Re = \frac{\rho \omega l^2}{\eta}$$

A second example can be generated when an intensive flow over a body produces closed field lines (called a vortex) at variable distances. This effect was observed by Strouhal (1850–1912) when some flow of air over wires produced a song. The Strouhal's 'singing wires' give the measure of the frequency that characterizes the vortex flow. This type of flow has been used to produce the so-called grid turbulence, which has various applications in the forced cooling of electronic devices [6.29].

Weber Number (We)

The Weber number is used when the surface tension forces acting on a fluid element are important. The Weber number for this special flow case is introduced by applying the pi theorem particularized to mixing in a liquid medium. In this case, it characterizes the ratio between the surface forces along the paddle that retain the flowing fluid element and the inertial forces that displace the flowing fluid element. This dimensionless number may be useful for the characterization of thin film flow and for the formation and breaking of droplets and bubbles. However, not all the problems involving a flow with an interface will require the inclusion of the surface tension.

When surface tension differences appear or are produced between some points or some small regions of an interface, the flow produced is called the Marangoni flow or flow with Marangoni effect. The Marangoni number, used to characterize the flow shown on Fig. 6.9, is a combination of the Reynolds number, the Weber number and the Schmidt number:

$$Mn = Re^a We^b Sc^c$$

if we make the assumption that the Mn number is dimensionless we obtain

$$a = 2,\, b = -1,\, c = 1$$

Then the Marangoni number could be written as:

$$Mn = \frac{\Delta\sigma}{\rho D}\frac{l}{}$$

where $\Delta\sigma$ will be $\partial\sigma/\partial t * \Delta t$ when the Marangoni flow is caused by a temperature gradient or where $\Delta\sigma$ will be $\partial\sigma/\partial c * \Delta c$ when it is produced by the concentration gradient.

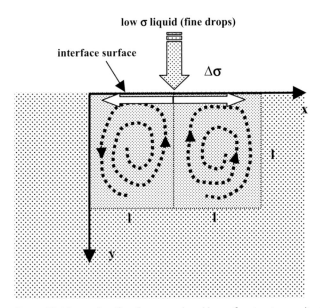

Figure 6.9 Flow and surface forces produced by surface tension gradients ($\Delta\sigma$).

Grashof Numbers (Gr_t, Gr_d)
These two numbers (diffusion Grashof number and thermal Grashof number) are used to characterize the natural convection produced by a thermal or concentration gradient. At moderate temperature or concentration gradient values, the natural convection flow keeps the properties of a laminar flow. However, higher gradient values of temperature and/or concentration can be caused by a turbulent natural convection flow. All the flows associated with heat and mass transfer processes, which occur without an important external action, are characterized with the Grashof number. Grashof numbers are also used in meteorology. Grashof numbers and the Froude number are the most used dimensionless groups that include gravitational acceleration as a physical parameter.

Nusselt Number (Nu)
This number is the main dimensionless group for heat transfer problems. With the partial heat transfer coefficient as physical parameter, it characterizes the kinetics of interface heat transfer. Unfortunately we cannot generally appreciate

the intensity of the heat transfer from the Nu number values because, in some cases, the intensity of the heat transfer process is not directly related to the Nu value. For example, the Nusselt number computed for liquid boiling, which is a very intensive heat transfer operation, can be lower than the Nusselt number computed for heat transfer from a heating device to adjacent media. This is caused by the differences between the characteristic lengths. Table 6.2 shows that the Nusselt number is coupled with other dimensionless groups by various relationships. Different ways are used in chemical engineering to particularize these relationships: the analytical and numerical solution of the equations of heat and mass transfer at steady state, particularized for a system of simple geometry; the particularizations of the boundary layer theory for a heat transfer case; the particularization of the transfer analogies for an actual heat transfer case and finally the experimental data correlation.

Prandtl Number (Pr)

The significance of the Prandtl number has been given earlier in this chapter. Another meaning is given by the boundary layer heat transfer theory and it shows that we can consider the Prandtl number as a relationship between the heat boundary layer and the hydrodynamics boundary layer associated in a concrete case.

We should also note that the boundary layer is the region where the solid interacts mechanically and thermally with the surrounding flow. A practical spin-off of Prandtl's recognition of the boundary layer is the understanding of the mechanisms of skin friction and heat transfer. This number is named after Ludwig Prandtl (1875–1953). Indeed, his discovery of the boundary layer is regarded as one of the most important breakthroughs of all time in fluid mechanics and has earned Prandtl the title of "Father of Modern Fluid Mechanics". The heat and mass transfer analogies frequently used in chemical engineering are based on the Prandtl theory (Prandtl and Prandtl–Taylor boundary layer analogies). For heat transfer in gaseous media at moderate pressures, the Prandtl number can be neglected since, in this case, its values are between 0.7 and 1.

Schmidt Number (Sc)

As explained earlier, with respect to the heat and mass transfer analogies, the Schmidt number is the Prandtl number analogue. Both dimensionless numbers can be appreciated as dimensionless material properties (they only contain transport media properties). For gases, the Sc number is unity, for normal liquids it is 600–1800. The refined metals and salts can have a Sc number over 10 000.

Sherwood Number (Sh)

Initially called the diffusion Nusselt number, this number characterizes the mass transfer kinetics when expressed in dimensionless terms.

All the statements given before for the Nusselt number have the same significance for the Sherwood number if we change the words "couple heat transfer" to "couple mass transfer". We have to specify that, as far as the Sherwood number is concerned, we can use the mass transfer coefficient apparent values. This is the case for tray columns when the mass transfer coefficients are reported for the geometric active tray area, not for the real mass transfer tray area. This is very useful in problems when a precise value of the real mass transfer area cannot be precisely established.

6.7
Particularization of the Relationship of Dimensionless Groups Using Experimental Data

After establishment of the general dimensional analysis relationship for a concrete case, we have to formulate an adequate theory or an experimental investigation that will show the specific relationships between dimensionless groups. At the same time, the obtained relationships have to be justified as usable. The beginning of this chapter shows that dimensional analysis is an aid in the efficient interpretation of experimental data. As previously shown, a dimensional analysis cannot provide a complete answer to any specific relationship among the groups which are unknown. The general methods used to produce groups of relationships have been mentioned earlier; they are:

1. the analytical solution of the equations of all transfer properties for a particular example,
2. the numerical solution of the equations of all transfer properties supplemented with a correlation and regression analysis with respect to the relationships between the dimensionless groups,
3. the particularization of the boundary layer theory if, for the studied case, a stable boundary layer can be defined,
4. the particularization of the transfer analogies and their experimental validation,
5. the development of a consistent experimental research programme supplemented with a correlation and regression analysis with respect to the relationships between the dimensionless groups.

In the last case, when the determination of the relationships between the dimensionless groups is based on suitable experimental data, all the methods presented here can be used successfully. The degree of difficulty involved in this process depends on the number of pi terms and the nature of the experiments (for example experiments that require a change in the geometric simplex cannot be accepted because they require new experimental plants).

For a chemical engineering problem, the concrete activity of the correlation of experimental data shows the following two particularities:

1. No more than three, exceptionally four, pi non-geometric groups characterize the majority of analyzed problems.

2. The power form of the dimensionless pi groups is the one most used from among the possible relationships.

The simplest problem concerns only one pi term. The complexity of the analysis increases rapidly with increasing number of pi terms because then the choice of the experimental plan related to the proposed relationship for the dimensionless pi groups cannot be solved by an automatic procedure.

6.7.1
One Dimensionless Group Problem

When a one-dimensional analysis problem shows that the difference $m - n$ ($m =$ number of process variables, $n =$ number of basic dimensions associated with the process variables) is unitary, then only one pi term is required to describe the process. The functional relationship that can be used for one pi term is:

$$\Pi_1 = C$$

where C is a constant. The value of the constant, however, has to be determined experimentally. Normally, only one experiment is needed for the identification of C.

The case of Stokes settling velocity is considered as an illustrative example. If we assume that the stationary settling velocity, w_0, of a small particle flowing into a liquid or gaseous medium is a function of its diameter, d, specific weight, $g\Delta\rho$, and the viscosity of the gaseous or liquid medium, it follows that:

$$w_0 = f(d, g\Delta\rho, \eta)$$

and the dimensions of the variables are:

$$[w_0] = LT^{-1} \; [d] = L \; [g\Delta\rho] = [g(\rho_p - \rho)] = ML^{-1}T^{-2} \; [\eta] = ML^{-1}T^{-1}.$$

We observe that four variables and three basic dimensions (M, L, T) are required to describe the variables. For this problem, one pi term (group) can be produced according to the pi theorem. This pi group can easily be expressed as:

$$\Pi_1 = \frac{w_0\eta}{g\Delta\rho d^2}$$

Since there is only one pi group, it follows that:

$$\frac{w_0\eta}{g\Delta\rho d^2} = C$$

or

$$w_0 = Cgd^2 \frac{\Delta\rho}{\eta}$$

Thus, for a given particle and fluid, the gravitational settling velocity varies directly with d^2, $\Delta\rho$, and with $1/\eta$. However, we cannot predict the value of the settling velocity since the constant C is unknown. In this case, we have to carry out experiments to measure the particle velocity and diameter, the density difference and the fluid viscosity. We can run a single experimental test but we will certainly have to repeat it several times in order to obtain a reliable value for C. It should be emphasized that once the value of C is determined, it is not necessary to run similar tests using different spherical particles and fluids because C is a universal constant. Indeed, the settling velocity of the small particles is a function only of the diameter and the specific weight of the particles and the fluid viscosity.

An approximate solution to this problem can also be obtained with the particularization of the Hadamard–Rybczynski problem [6.30, 6.31] from which it is found that C = 1/18 so that:

$$w_0 = \frac{1}{18}gd^2 \frac{(\rho_p - \rho)}{\eta} \tag{6.197}$$

This relationship is commonly called the "Stokes settling velocity" and is applicable for $Re = w_0 d\rho/\eta < 1$ and when particle interactions are not present during the settling process.

6.7.2
Data Correlation for Problems with Two Dimensionless Groups

The chemical engineering processes which can be characterized by two dimensionless groups are important, especially for heat and mass transfer with gaseous media, as shown in Table 6.4. If the phenomenon can be described with two pi terms we have:

$$\Pi_1 = \Phi(\Pi_2)$$

the form of the Φ function can be identified by varying Π_2 experimentally and measuring the corresponding Π_1 value. The results can be conveniently presented in graphical form as in Fig. 6.10. Here the uniqueness of the relationship between Π_1 and Π_2, is shown.

Nevertheless, since it is an empirical relationship, we can only conclude that it is valid over the range of Π_2 dealt with by the experiments. It would be unwise to extrapolate beyond this range since, as illustrated with the dashed lines in the figure, the nature of the phenomenon can dramatically change if the range of Π_2 is extended.

For the valid range, we clearly obtain a curve with a break (B), then for each part, we will produce a particularization of the function Φ from the general Π_1 and Π_2 relationship.

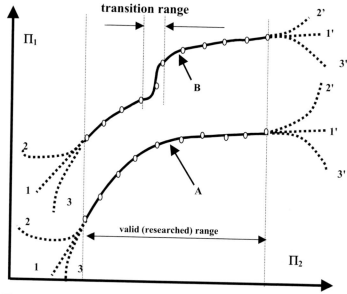

Figure 6.10 Π_1 versus Π_2 and illustration of the effect of extrapolation data over the valid range (A: continuous curve, B: curve with a break; 1,1', 2, 2', 3; 3' possible extrapolations).

The break in the curve can be associated with a fundamental change in the process mechanism (such as, for example: a change from laminar to turbulent flow, a change from moderate natural convection to turbulent natural convection, etc.)

If we assume that the function Φ from the general Π_1 and Π_2 relationship is a power expression, then the relationship $\Pi_1 = a\Pi_2^b$ will be obtained. If we apply the logarithm to this relationship, we can identify a and b using a normalized linear system Eq. (5.15).

To illustrate this methodology, we show the case of pressure drop per unit length for one-phase flow in a packed bed. In laboratories, the pressure drop is measured over a 0.1 m length of packed bed using an apparatus as shown in Fig. 6.11. The fluid used is water at 20 °C (ρ = 1000 kg/m³, η = 10⁻³ kg/ms). While the tests are carried out the velocity is varied and the corresponding pressure drop is measured. Table 6.5 shows the results of these tests.

Table 6.5 Measurements of packed bed pressure drop for the experimental device from Fig. 6.11.

Fictive water velocity (m/s)	0.01	0.02	0.03	0.04	0.05	0.06	0.07	0.08	0.09	0.1
Pressure drop for 1m bed height (N/m²)	10930	21350	32760	43736	142727	198217	261759	332893	417838	497497

water from a constant level tank

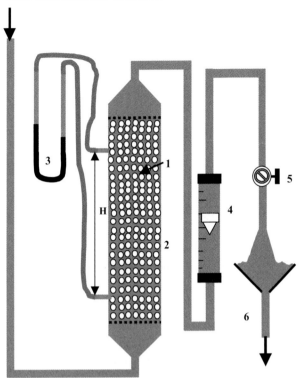

Figure 6.11 Device with a packed bed one-phase flow for the measurement of the pressure drop. 1: glass spherical particles (diameter = 1.5 mm), 2: glass column (diameter = 30 mm, H = 0.1 m), 3: differential manometer, 4: flow meter, 5: control valve, 6: water collector.

We will use these data to obtain a general relationship between the pressure drop per unit height of packed bed and the other variables. To search for an actual solution to this problem, we begin by performing a dimensional analysis, which can be realized without any experiment. We will assume that the pressure drop per unit height of packed bed, $\Delta p/H$, is a function of the equivalent packed body diameter, d_e, the fluid density, ρ, the fluid viscosity, η, and the mean packed bed fluid velocity, w.

The equivalent packed body diameter, d_e, is related to the bed holdup, ϵ, and specific packed surface, s, as well as via the relationship $d_e = 4\epsilon/\sigma$. For the packed bed from spherical bodies ($\epsilon = 0.44$ and $\sigma = 6/d_p$), the equivalent packed body diameter depends only on the sphere diameter. The mean internal packed bed fluid velocity represents the ratio between the fictive velocity and the packed bed porosity (ϵ). According to these data, we can write:

$$\frac{\Delta p}{H} = f(d_e, \rho, \eta, w)$$

the application of the pi theorem yields:

$$\frac{\Delta p}{\rho w^2}\frac{d_e}{H} = \Phi\left(\frac{wd_e\rho}{\eta}\right)$$

The simplest way to obtain $\dfrac{\Delta p}{\rho w^2}\dfrac{d_e}{H}$ for various $Re = \dfrac{wd_e\rho}{\eta} = \dfrac{4w_f\rho}{\sigma\eta}$ is to vary the fictive velocity.

Based on the data given in Table 6.5, we can calculate the values for both pi terms. The results obtained are given in Table 6.6. A plot of these pi terms can now be made as a function of the Reynolds number. The results are shown in Fig. 6.12.

Table 6.6 Packed bed pressure drop in dimensionless terms.

$\dfrac{\Delta p}{\rho w^2}\dfrac{d_e}{H}$	7	3.5	2.33	1.75	3.65	3.50	3.42	3.33	3.25	3.18
$\dfrac{wd_e\rho}{\eta}$	10	20	30	40	50	60	70	80	90	100

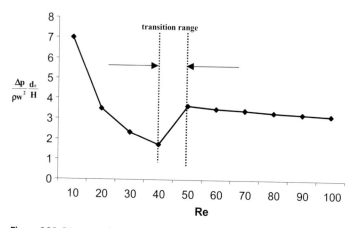

Figure 6.12 Dimensionless packed bed pressure drop versus Reynolds number.

The correlation appears to be quite good and shows that the valid range is divided into two parts. The first part corresponds to a Reynolds number lower than 40, which corresponds to the laminar flow range in the packed bed. The apparent turbulent flow range in the packed bed is obtained in the second part, corresponding to Reynolds numbers greater than 50 [6.32].

A wrong correlation may be due either to important experimental errors or to the omission of an important variable. The curve shown in Fig. 6.12 represents the general relationship between the pressure drop and the other factors for Reynolds numbers between 10 and 100. For this range of Reynolds number, as far as the provided independent variables (d_e, ρ, η, w) are the only important parameters, it is not necessary to repeat the test for other packed beds or fluids. In order to determine the power of the relationships of pi terms, the data from Table 6.5 allow the identification of the next equations:

$$\frac{\Delta p}{\rho w^2} \frac{d_e}{H} = \frac{70}{Re} \quad \text{for Re} <40$$

and:

$$\frac{\Delta p}{\rho w^2} \frac{d_e}{H} = \frac{8}{(Re)^{0.2}} \quad \text{for Re} >40$$

It is then not difficult to write:

$$\Delta p = \zeta \frac{H\sigma}{8\varepsilon^3} w_f^2 \rho \ , \ \zeta = 140/Re \quad \text{for Re} <40; \ \zeta = 16/(Re)^{0.2} \text{ for Re} >40;$$

$$Re = (4w_f\rho)/(\sigma\eta)$$

which represents the Javoronkov procedure for packed bed pressure drop on phase flow calculation [6.32]. This so-called Javoronkov procedure is based on numerous experimental results similar to the type used in this example.

6.7.3
Data Correlation for Problems with More than Two Dimensionless Groups

When the number of pi groups involved in the dimensional problems increases, it becomes more difficult to organize the experimental research, to display the results in a convenient graphical form and to determine a specific empirical equation describing the phenomenon. If we accept that a power relationship between the pi groups is validated for all experimental ranges or for clearly identified portions of a range, we can easily identify the coefficients that characterize this relationship.

In this case, we can use a special experiment planning characterized by the repetition of a classical second order planning, where the centre of the plan is changed to cover a large range of each pi group and to discover the possible breaks in the state of dependent pi groups. In the previous chapter, it was shown that the majority of the functional dependences can be reduced to a multiple linear regression. If we propose for the relationships between the pi groups equations different from powers then they can be identified by using the particularized system of equations of the multiple regression. For most of the problems involving heat and mass transfer, the dependent pi groups are represented by Nusselt and Sherwood numbers, the independent pi groups characterizing the flow by the Reynolds or Grashof numbers and the media properties by the Prandtl and Schmidt numbers.

Changes in the fundamental flow mechanism are expected if large ranges of the pi groups characterizing the flow are considered. Consequently, a transition zone in the state of pi dependent groups has to be observed. This case is illustrated in Fig. 6.13. We have to specify that the general dependences considered in this figure cannot be extended to heat and mass transfers associated with phase transformations, because this case is more complicated. In complicated systems it is often more feasible to use models to predict specific characteristics of the system rather than to try to develop general correlations. If we extend the situation shown by the figure below to more than three terms, we obtain a very complicated problem where a graphical representation and a suitable empirical equation become intractable.

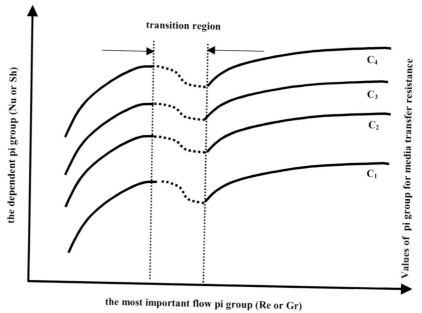

Figure 6.13 The graphical presentation of data for chemical engineering problems involving three pi dimensionless groups.

6.8
Physical Models and Similitude

To validate models based on transfer equations or stochastic models and, especially, to develop a coherent and planned experimental investigation of the studied process, the researcher has to imagine and build up a reduced scale experimental installation (laboratory device or model, LM). The goal using this reduced scale pilot plant is to obtain the experimental data necessary to validate the models.

Major chemical engineering projects involving structures, tray or packed columns, reactors, separators, heat exchangers and heaters, reservoirs and special deposits, fluid pumping as well as compressing devices, frequently involve the use of small scale studies using laboratory scale devices. According to the context, the term "laboratory model", "engineering model", "physical model", "laboratory pilot unit" or "small prototype" may be used. It is important to note that, as explained above, the term LM concerns one small pilot unit, which is different from the term "mathematical model" currently used throughout this book.

The pilot laboratory units, which are generally a different size, may involve different fluids, and often operate under different conditions (temperature, pressure, velocity, etc.). These units are frequently designed so that the parameters can be varied independently. The idea is not only to facilitate the study of the influence of the different process parameters but also to include the complexity of industrial scale units in the study.

Until now, the classical way to scale up industrial plants using laboratory scale units was very tedious and laborious: it consisted in systematically studying all the influencing parameters and operating conditions. In addition, many works were based on the use of different increasing scale. Consequently, methods concerning the characterization by a more rapid jump from laboratory small pilot to larger scale plants began to be developed [6.22, 6.33, 6.34]. These methods are therefore based on mathematical model simulations with incomplete laboratory experimental data for one actual problem [6.35]. There is, of course, an inherent danger in the use of models if the predictions realized are not correctly validated, because they can be erroneous and it may not be possible to detect errors until the industrial size plant is found not to perform as predicted. It is, therefore, imperative to have a properly designed and tested model, as well as correctly interpreted results. This is the basic question of the similitude theory: "To what extent can experimental data be relied upon when the dimensions of the experimental devices increase or decrease?"

In the following sections we will present some procedures and examples which show how LM can be designed in order to have a similar behaviour evolution for different device scales such as laboratory device (small scale) and prototype units (medium scale).

6.8.1
The Basis of the Similitude Theory

Which mathematical models or designing procedures were used to build the Egyptian pyramids or the gothic cathedrals? Since the ancient times, geometric scaling-up procedures have been used. These rules are based on similitude laws, which are still used today.

It is recognized that a phenomenon which occurs in an apparatus or a plant at different scales (various geometrical dimensions) presents the same evolution for all scales only if the conditions of the geometric similarity (geometric similitude), material similarity (material similitude), dynamic similarity (dynamic similitude) are respected and if the phenomenon shows the same initial state in all cases. The parametric description of a phenomenon occurring at laboratory and prototype scales is given in Table 6.7. In this case, we consider that the initial state of the phenomenon is identical for both scales.

Table 6.7 Determinant general parameters for the phenomenon evolution (two dimension scales).

Name of parameters	Symbolic notations for the laboratory model	Symbolic notations for the prototype
1 Characteristic geometric parameters (geometric dimensions of apparatus or plant)	$l_0, l_1, l_2, ...l_n$	$L_0, L_1, L_2, ...L_n$
2 Characteristic material parameters (specific properties of the materials used in experiments: density, viscosity, etc.)	$c_0, c_1, c_2, ...c_r$	$C_0, C_1, C_2, ...C_r$
3 Characteristic parameters for the phenomenon dynamics (flow velocities, mixing flow, heating rate, mass transfer rate, reaction rate, etc.)	$\tau_0, = \tau_1, \tau_2, ...\tau_q$	$T_0, T_1, T_2, ...T_q$

With respect to the evolution of the phenomenon considered in Table 6.7, if a_g, a_c and a_τ are the scaling factors (the coefficients that multiply the laboratory model parameters in order to obtain the value of the prototype's parameters) then these can be written as:

$$a_g = \frac{L_0}{l_0} = \frac{L_1}{l_1} = \frac{L_2}{l_2} = ... = \frac{L_n}{l_n} \tag{6.198}$$

$$a_c = \frac{C_0}{c_0} = \frac{C_1}{c_1} = \frac{C_2}{c_2} = ... = \frac{C_r}{c_r} \tag{6.199}$$

$$a_\tau = \frac{T_0}{\tau_0} = \frac{T_1}{\tau_1} = \frac{T_2}{\tau_2} = ... = \frac{T_q}{\tau_q} \tag{6.200}$$

It is obvious that $a_c = 1$ or $c_0 = C_0$, $c_1 = C_1$, $c_2 = C_2$, $c_r = C_r$ when the same materials are used for the LM and for the prototype unit.

Equations (6.198)–(6.200) can be arranged to show dimensionless or dimensional ratios, which express only the LM or the prototype. These dimensionless or dimensional relations are called similitude simplexes when they result from the same type of parameters and similitude multiplexes when they are composed of different types of parameters.

Using l_0 and L_0, c_0 and C_0 and τ_0 and T_0 as the characteristic parameters for the laboratory device and prototype geometry, used materials and phenomenon dynamics respectively we transform the preceding relationships as:

$$i_{g1} = \frac{l_1}{l_0} = \frac{L_1}{L_0} \ , \ i_{g2} = \frac{l_2}{l_0} = \frac{L_2}{L_0} \ , \ .... , \ i_{gn} = \frac{l_n}{l_0} = \frac{L_n}{L_0} \tag{6.201}$$

$$i_{c1} = \frac{c_1}{c_0} = \frac{C_1}{C_0} \ , \ i_{c2} = \frac{c_2}{c_0} = \frac{C_2}{C_0} \ , \ .... , \ i_{cr} = \frac{c_r}{c_0} = \frac{C_r}{C_0} \tag{6.202}$$

$$i_{\tau 1} = \frac{\tau_1}{\tau_0} = \frac{T_1}{T_0} \ , \ i_{\tau 2} = \frac{\tau_2}{\tau_0} = \frac{T_2}{T_0} \ , \ \dots\dots, \ i_{\tau q} = \frac{\tau_q}{\tau_0} = \frac{T_q}{T_0} \tag{6.203}$$

where $i_{g1}, i_{g2},\dots, i_{gn}$ are the geometric simplexes, $i_{c1}, i_{c2},\dots, i_{cr}$ are the dimensional material simplexes or multiplexes and $i_{\tau 1}, i_{\tau 2}, \dots i_{\tau q}$ are the dynamic dimensional multiplexes. Then we can postulate that "the phenomenon occurring in the laboratory device and in the prototype present the same evolution only when all $i_{g1}, i_{g2},\dots, i_{gn}, i_{c1}, i_{c2},\dots, i_{cr}, i_{\tau 1}, i_{\tau 2}, \dots i_{\tau q}$ are the same ($i_{g1}, i_{g2},\dots, i_{gn}, i_{c1}, i_{c2},\dots, i_{cr}, i_{\tau 1}, i_{\tau 2}, \dots i_{\tau q}$ stay unchanged when the dimensions of the model increase or decrease)"

Because all the dimensional material and dynamic multiplexes are reported only for the LM or for the prototype, we can combine these r*q dimensional multiplexes and the characteristic geometric parameters (so we have r*q+1 dimensional terms) to formulate new dimensionless independent multiplexes. It is not difficult to observe that these independent dimensionless multiplexes are the dimensionless pi groups that characterize the evolution of the phenomenon. If $\Pi_1, \Pi_2,\dots\Pi_s$ represent the dimensionless groups that characterize the evolution of the phenomenon in the laboratory device or in the prototype, we transform the similitude postulate into the next new statement:

"One phenomenon occurring in two differently scaled devices (models) presents the same evolution only if the dimensionless pi groups characterizing the phenomenon have the same values. In other words, we have similitude if the dimensionless pi groups characterizing the phenomenon stay unchanged when the dimension of the device (model) changes".

The theory of the models can be readily developed using the principles of dimensional analysis. It has been shown that any given problem can be described in terms of a set of pi terms as:

$$\Pi_1 = \Phi(\Pi_2, \Pi_3, \dots \Pi_s) \tag{6.204}$$

Once this relationship is formulated, all we need to know is the general nature of the physical phenomenon and variables. Specific values for variables (size of components, fluid proprieties, etc.) are not needed to perform the dimensional analysis. This relationship could be applied to any system, if it is governed by the same variables and laws. If Eq. (6.204) describes the behaviour of a laboratory device, a similar relationship can be written for evolution of the phenomenon in the prototype:

$$\Pi_{1p} = \Phi(\Pi_{2p}, \Pi_{3p}, \dots \Pi_{sp}) \tag{6.205}$$

where the subscript p shows that this is the case of the evolution of the phenomenon in the model prototype.

The pi terms can be developed so that Π_{1p} contains the variables that have to be predicted from the observation made on the laboratory apparatus. Therefore, if the prototype is designed and operated, in relationship to the LM, under the following conditions:

$$\Pi_{2p} = \Pi_2, \; \Pi_{3p} = \Pi_3, \; ...., \Pi_{sp} = \Pi_s \qquad (6.206)$$

then, with the presumption that the form of function Φ is the same for the LM and for the prototype, it follows that:

$$\Pi_{1p} = \Pi_1 \qquad (6.207)$$

Equation (6.207) indicates that the measured value of Π_1 for the LM will be identical to the corresponding Π_1 for the prototype as long as the other pi terms are similar.

The conditions specified by Eq. (6.206) provide the conditions required to design the model, also called similarity requirements or modeling laws. The same analysis could be carried out for the governing differential equations or the partial differential equation system that characterize the evolution of the phenomenon (the conservation and transfer equations for the momentum). In this case the basic theorem of the similitude can be stipulated as: "A phenomenon or a group of phenomena which characterizes one process evolution, presents the same time and spatial state for all different scales of the plant only if, in the case of identical dimensionless initial state and boundary conditions, the solution of the dimensionless characteristic equations shows the same values for the internal dimensionless parameters as well as for the dimensionless process exits".

As an example of this methodology, we can consider the problem of determining the heat loss of a rectification column which is placed perpendicularly to a fluid flowing at the velocity w. (see Fig. 6.14). The dimensional analysis of this problem shows that:

$$\alpha = f(d, H, \eta, \rho, \lambda, w)$$

where α represents the heat transfer coefficient, d, the column diameter, H, the column height, λ, the fluid thermal, while η and ρ show the fluid viscosity and density conductivity and w, represents the incident air velocity. Application of the pi theorem gives:

$$\frac{\alpha d}{\lambda} = \Phi\left(\frac{H}{d}, \frac{wd\rho}{\eta}\right) \qquad (6.208)$$

We are now concerned with the design of a laboratory device which can be used to predict the heat loss on a different-sized prototype.

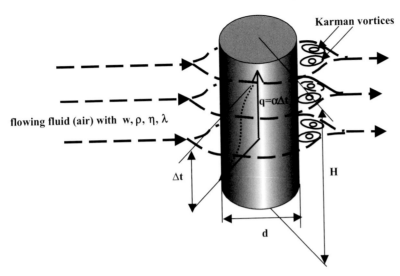

Figure 6.14 Heat loss in a rectification column (flow around the column).
Fluid characteristics: w: velocity, η: viscosity, ρ: density, λ: thermal conductivity.
Column characteristics: d: diameter, H: height. Δt: differential temperature
between the air and the column surface.

Since the relationship expressed by Eq. (6.208) applies to both prototype and
laboratory models, we can assume that for the prototype, a similar relationship
could be written:

$$\frac{a_p d_p}{\lambda_p} = \Phi\left(\frac{H_p}{d_p}, \frac{w_p d_p \rho_p}{\eta_p}\right) \tag{6.209}$$

the design conditions, or similarity requirements are therefore:

$$\frac{H}{d} = \frac{H_p}{d_p}, \quad \frac{w d \rho}{\eta} = \frac{w_p d_p \rho_p}{\eta_p} \tag{6.210}$$

the size of the laboratory device is obtained from the first requirement which indicates that:

$$d = \frac{H}{H_p} d_p \tag{6.211}$$

we can then establish the height ratio H/H_p, and the diameter of the laboratory
device d is fixed in accordance with Eq. (6.210).

The second similarity requirement indicates that the LM and the prototype
must be operated at the same Reynolds number. The required velocity for the laboratory model is obtained from the relationship:

$$w = \frac{\eta}{\eta_p} \frac{\rho_p}{\rho} \frac{d_p}{d} w_p \tag{6.212}$$

Note that this model design requires not only geometric scaling, as specified in Eq. (6.211), but also the correct scaling of the velocity in accordance with Eq. (6.212). This result is typical of most design procedures, where the scaling up is more difficult than simply scaling the geometry.

With the foregoing similarity requirement satisfied, the equation for the prediction of the column heat loss is:

$$\frac{\alpha_p d_p}{\lambda_p} = \frac{\alpha d}{\lambda}$$

or

$$\alpha_p = \alpha \frac{d}{d_p} \frac{\lambda_p}{\lambda}$$

or

$$Q_p = \alpha_p \pi d_p H_p \Delta t = \alpha \frac{d}{d_p} \frac{\lambda_p}{\lambda} \pi d_p H_p \Delta t = Q \left(\frac{H_p}{H}\right) \left(\frac{\lambda_p}{\lambda}\right) \tag{6.213}$$

where Q represents the heat loss of the laboratory size column that is operated with a temperature difference (Δt) between its surface and the incident flowing fluid. Once the heat loss is measured on the laboratory device, Q has to be multiplied by the ratios corresponding to the height of the columns and the conductivity of the flowing fluids in order to obtain the predicted value of the heat loss for a real column.

If we analyze the case of the rectification column which loses heat by natural convection, then we change the list of variables by considering the specific ascension force, $g\beta_t\Delta t$, as an important variable and by removing the fluid velocity, w. In this case, the application of the pi theorem shows that:

$$\frac{\alpha H}{\lambda} = \Phi\left(\frac{d}{H}, \frac{g\beta_t\Delta t H^3 \rho^2}{\eta^2}\right) \tag{6.214}$$

the design conditions, or similarity requirements are therefore:

$$\frac{d}{H} = \frac{d_p}{H_p} \ , \quad \frac{g\beta_t\Delta t H^3 \rho^2}{\eta^2} = \frac{g\beta_t\Delta t H_p^3 \rho_p^2}{\eta_p^2} \tag{6.215}$$

if the same fluid is used for both geometric scales, then the size of LM cannot be established because we obtain:

$$d = d_p \frac{H}{H_p} \ , \quad H = H_p$$

the same result will be obtained when we remove the diameter (d), from the list of variables.

This very simple example shows that sometimes it is not possible to use this scaling-up procedure. Fortunately, the majority of the problems presenting scaling-up impossibilities can be solved using other methodologies. As illustrated in this example, to achieve the similarity between the behaviour of both the laboratory model and the prototype, all the corresponding pi terms must be equated between these two scales. Usually, one or more pi terms involve ratios of important lengths (such as H/d in the foregoing example). Thus, when we equate the pi terms involving length ratios we are requiring "a complete geometric similarity" to exist between the laboratory device and the prototype. Geometric scaling could be extended to the finest feature of the system, such as surface roughness, or small protuberances on a structure, since the surface state also determines the flow pattern. When a deviation from the complete geometric scaling must be considered, a careful analysis has to be carried out. For example, the design of one new packed body, or of a new sort of tray, cannot be produced without a complete geometric scaling.

Other groups of characteristic pi terms (such as $Re = \dfrac{wd\rho}{\eta}$ in the foregoing example) involve mechanisms' ratio or forces' ratio as noted in Table 6.4. The equality of these pi terms requires the same mechanisms' or forces' ratios in laboratory devices and prototypes. Thus, for similar Reynolds numbers, the values defining the turbulent and laminar flow mechanisms have to be the same for both devices. If other pi groups are involved, such as the Froude, Weber, Archimede or Grashof numbers, a similar conclusion can be drawn; that is, the equality of these pi groups requires identical ratios of identical forces for the laboratory apparatus and for the prototype. In the case of similar pi terms, we can say that we have a "hydrodynamic similarity".

Other similarities used in chemical engineering concern the "field concentration similarity" and the "property transfer similarity". The field concentration of the property is characterized by the Prandtl, Schmidt or reaction Fourier numbers (Table 6.4). It has been shown here that if the geometric and hydrodynamic similarities are respected, the transfer similarities exist between the laboratory device and the prototype. These last two similarity conditions represent the conditions of "transfer kinetic similarity". In order to have complete similarity, we need to maintain the similarity between the geometry, dynamics and transfer kinetics between both units. If scaling up is found to be impossible, we have to ascertain whether all important variables are included in the dimensional analysis, and whether all the similarity requirements based on the resulting pi groups are satisfied.

6.8.2
Design Aspects: Role of CSD in Compensating for Significant Model Uncertainties

The experimental studies with one laboratory device, generally involve simplifying assumptions concerning the variables to be considered. In spite of the fact that the number of assumptions is less restrictive than required for mathematical models they introduce some uncertainty into the design of the device. It is, therefore, desirable to check the design experimentally whenever possible. Generally, the purpose of the LM is to predict the effects of certain changes proposed in a given prototype or in a larger-scale device. The LM has to be designed, constructed, and tested and then the predictions can be compared with the available data from larger-scale devices. If the agreement is satisfactory, then the changes allowing one to build a bigger model can be accepted with increased confidence. Another useful procedure is to run tests with a series of LM having different sizes, where one of the devices can be considered as the prototype and the others as models of this prototype. With the devices designed and operated on this basis, we also need to improve other conditions for the validity of the LM design: accurate predictions have to be made between any pair of devices, since each of them can always be considered as the model of another one.

It is important to note that a good agreement in the validation tests described above does not unequivocally indicate that the scaling up is correct, especially when the dimensions of the scales between the different LMs are significantly different from those required by the basic laboratory model. However, if the agreement between the various models is not good, it is impossible to use the same model design to predict the behaviour of the basic laboratory model.

Some designing cases show that the general ideas, which establish similarity conditions for models when we use simple corresponding pi terms, are not always able to satisfy all the known requirements. To illustrate such a case: if, for a relationship $\Pi_1 = \Phi(\Pi_2, \Pi_3, ..., \Pi_s)$ one or more similarity requirements are not respected, such as, for example: $\Pi_2 \neq \Pi_{2p}$, then it follows that the prediction relationship $\Pi_1 = \Pi_{1p}$ is not true. The models designed without satisfying all the requirements are called limited models or distorted models.

The classic example of a distorted model occurs in the study of liquid media which are mixed mechanically as described earlier in this chapter. The dimensional analysis shows that:

$$K_N = \Phi\left(\frac{H}{d}, \frac{h}{d}, \frac{b}{d}, Re, We\right)$$

where:

$$K_N = \frac{N}{d^5 n^3 \rho} \ , \quad Re = \frac{n d^2 \rho}{\eta} \ , \quad We = \frac{\rho n^2 d^3}{\sigma}$$

If we consider that heat transfer occurs during the mixing, the dimensional analysis shows that:

$$Nu = \Phi\left(\frac{H}{d}, \frac{h}{d}, \frac{b}{d}, K_N, Re, We\right) \tag{6.216}$$

where the Nusselt number is $Nu = (\alpha d)/\lambda$ where α is the heat transfer coefficient to the wall of the mixing unit and λ the thermal conductivity of the mixed fluid. If a_g is the geometric scaling factor the geometric similarity requires:

$$H_p = Ha_g \ , \quad h_p = ha_g \ , \quad b_p = ba_g$$

The similarity of the Euler number (the group K_N is a form of the Euler number for mixing) requires:

$$\frac{N_p}{\rho_p n_p^3} = \frac{N}{\rho n^3} a_g^5$$

or

$$N_p = N \frac{\rho_p}{\rho} \left(\frac{n_p}{n}\right)^3 a_g^5$$

The similarity of the Reynolds number requires:

$$\frac{n_p^2 \rho_p}{\eta_p} = \frac{n^2 \rho}{\eta} \frac{1}{a_g^2} \quad \text{or} \quad \frac{n_p}{n} = \left(\frac{\eta_p}{\eta} \frac{\rho}{\rho_p}\right)^{1/2} \frac{1}{a_g}$$

Whereas the similarity of the Weber number requires:

$$\frac{\rho_p n_p^2}{\sigma_p} = \frac{\rho n^3}{\sigma} \frac{1}{a_g^3} \quad \text{or} \quad \frac{n_p}{n} = \left(\frac{\rho}{\rho_p} \frac{\sigma_p}{\sigma}\right)^{1/2} \frac{1}{a_g^{3/2}}$$

Since the scale of the speed of rotation $\left(\frac{n_p}{n}\right)$ is expressed by two relationships, their combination could be written as:

$$\left(\frac{\eta_p}{\eta} \frac{\sigma}{\sigma_p}\right)^{1/2} = a_g^{1/2} \tag{6.217}$$

When we use the same fluid for the LM and the prototype, we obtain $a_g=1$ which is unacceptable. Apparently, with a different fluid, it may be possible to satisfy this design condition but it may be quite difficult, if not impossible, to find a suitable model fluid, particularly for a small scale unit. When the identity requirement of the Weber numbers is eliminated, we obtain a distorted model, which gives a good approach to intensive mixing (a large Reynolds number shows that inertia forces are dominant in the mixing process). A distorted model could also be obtained when the identity requirements of the Reynolds numbers are eliminated. Then these models are good for the description of cases in which the mechanical mixing is slow but intensive because the forces at the surface are important.

Distorted models can be used successfully, but the interpretation of the results obtained using this type of model is obviously difficult compared to the true models for which all similarity requirements are obtained. The success of using distorted models is dependent on the skill and experience of the investigator responsible for the design of the model and on the interpretation of the experimental data produced with the model. In many cases, the distorted models are associated with one or more uncertainties and the use of their data in the scaling-up design of the complex processes must be appreciated and compensated using an adequate control system design (CSD).

6.8.2.1 Impact of Uncertainties and the Necessity for a Control System Design

While designing complex systems, we can basically encounter two types of uncertainties. In the first, we know that the system will work but it is difficult to determine its scale. Typical examples of these are the uncertainties in physical properties such as heat transfer coefficients, mass transfer coefficients, tray efficiency etc. [6.36–6.39]. A judicious over design will solve the uncertainties. If we extend the observation, we can appreciate that their main impact concerns the capacity of the plant. Under-sized equipment will bottleneck the process. The important point is that a judicious over-size design is relatively cheap, while retrofitting to relieve bottlenecks is expensive in both low-cost capacity and capital cost. This is the more critical type of uncertainty, which, if not adequately addressed, can lead to a total process failure. We can illustrate this case with important examples extracted from the experience of some chemical industries [6.40] when the scaling-up of one plant ended up in a great fiasco. Even if the plant was not entirely abandoned, extensive and expensive modifications were required to operate it.

However, this problem can be easily avoided through straightforward concurrent design. One might also argue that building a large pilot plant or a small demonstration plant could minimize some of these difficulties and risk involved. For example, the Exxon case, in which an expensive demonstration plant was built to demonstrate the new Fischer-Tropsch process, is well known [6.40]. This type of expensive unit can be justified in some special cases but will not always be necessary.

Pilot plants are not often designed to provide the essential information for a scaling up. Instead, they are operated to demonstrate a single steady state that gives acceptable results. They are seldom meant to investigate the impact of the process variables, which is essential for safe scaling up and control design. There is no great difference between designing a large pilot plant or a commercial plant. In both cases we have to make certain that the design can deal with the risks of scaling up.

To avoid any misunderstanding, we would like to emphasize that our main concern here includes such critical parts of the plant as new chemical reactors, processes or some complex separations that cannot be reliably modelled from a limited set of experiments carried out in a laboratory or on a small-scale pilot plant. For the remaining and more classical plant devices we can use modern simulators that provide all sorts of mathematical models and that have had a tremendous impact on the modern designing of plants. At the same time, we must note that the scaling-up based on combining complete complex mathematical models with experimental data from small-sized laboratory units [6.41–6.47] begins to be frequently used for the needs of the design and also for better operation of existing plants.

Nevertheless, we have to specify that the scaling up of one process must be considered as a complex problem with not only controlled goals determined by the dominant process variables but also independent degrees of freedom. Here, the efficiency of the process, the process modelling and design procedure have to be identified, computed, changed if necessary and used in order to perform an efficient process control. For a process scaling-up or design, the following objectives of control have to be considered:

1. Allow the system to meet the specifications. Allow on-line change of the specifications.
2. Stabilize the system.
3. Compensate for changes in feedstock, in properties of the catalyst (if the core process is a catalytic reaction) and unknown dynamic and persistent perturbations.
4. Allow compensation for uncertainties in the design and scaling-up of the unit.

To achieve these goals, both dynamic and steady state controls are required. For most chemical plants, the control to meet specifications is the primary objective. The capacity to stabilize and reject perturbations is essential to achieve the goals mentioned above.

Before proceeding to the designing methodology itself, it will be helpful to review and define some of the principles and concepts of partial control. We are concerned with the control of a system in which the number of process variables to be controlled is higher than the number of variables which are manipulated to realize this control. If all the process variables have to be controlled according to exact set points, the process has to abandoned or the design modified in order to provide the requisite number of manipulated variables. However, it is quite often the case that many of these variables need only be controlled within prescribed limits, hence the terminology of partial control.

Dominant variables are characterized by observation, which shows that they exert a strong influence on many of the other interesting process variables. The operating temperature of the reactor is a typical example, because changes in this parameter generally modify the reactant conversion and product composition. Thus, by controlling the dominant variables, we can maintain the other process variables of concern within their prescribed limits. A variable is dominant for the stability if, by controlling it, we exert an effect on the system stability. This is particularly important if the system is an unstable open loop. It is very important to identify which variables are dominant in the laboratory. Strictly speaking, each apparatus presents its dominant variables. For example, potential dominant variable candidates for a catalytic reactor are: the temperature, pressure, space velocity, catalyst activity, and the properties of the reactants.

While the impact of dominant variables on process outputs can be identified and measured in the laboratory, this is not always true for stability. In this case, we have to rely on the availability of used models to identify possible dominant variables. It is clear that all dominant variables for stability are also dominant variables for outputs.

In conventional control (design control, evolution control), the number of degrees of freedom is considered as the number of available manipulated variables. We define the practical degrees of freedom as the number of dominant variables that can be controlled independently. The ability to have an impact on the outputs and stability of one chemical system is limited by the number of independently controllable dominant variables available in the design. The design of a control system is one part of the whole design and is completely dependent on the scaling-up process. In the situation with one system of control, the choice of the dominant variables and the independent degrees of freedom is deemed to be sufficient if they provide the management with adequate constraints and also result in a good stabilization of the system.

For the design of any complex system, including or not a chemical reactor, we do not need a complete model but rather minimal information of the model, which strongly depends on the design itself. The laboratory identification of all dominant variables is essential, together with sufficient data on their impact on the most important outputs of the process. This is essential for a safe scaling-up and to produce a preliminary model. It is also important to know how the mini-

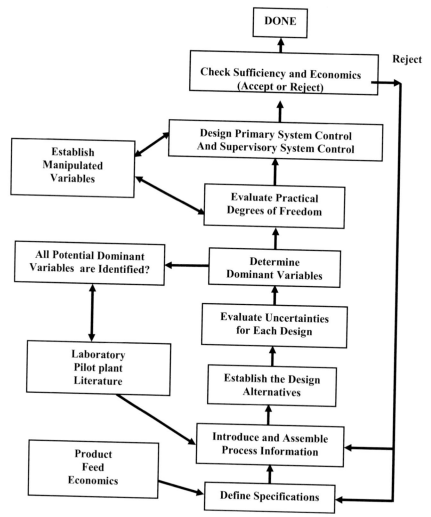

Figure 6.15 Possible structure for the design of a chemical fabrication.

mal information of the model determines the choice of the basic design and the control design.

In all cases the design key features are:
- the identification of the control system goals (specifications),
- the generation of the model information that adequately characterizes the system,
- the identification of the dominant process variables,
- the determination of the effective degrees of freedom,
- the determination of the control structure.

Figure 6.15 shows the general detailed structure of the working steps for the design of a scaling-up and control unit for a chemical fabrication. It is observable that we can generate a number of alternate process designs, either sequentially or in parallel. Then, one must also develop and evaluate the best partial structure control for each of these designs. The design which represents the best compromise between cost and controllability in the face of uncertainty could be considered as the final design.

6.9
Some Important Particularities of Chemical Engineering Laboratory Models

Generally, the chemical engineering processes include steps where interface transfer with or without a chemical reaction is dominant. In these cases the surface of transfer is one the parameters which controls the transfer efficiency. Some of the various technical solutions which have been developed to increase the surface of transfer are:
- the use of packed beds of small bodies for the differential transfer apparatus involved in the phase contacting procedure,
- the development of the highly efficient tray equipped with devices that produce small bubbles or drops with dense and uniform spatial distribution.
- the use of catalytic fixed beds with catalytic pellets with a diameter not exceeding 10–15 mm as well as fluidized beds with catalysts in powder form.

For all the examples given above, the analysis of the characteristic geometric length shows that this dimension is very small (diameter of packed body, bubble or drop diameter, catalyst particle diameter). In this section, we have shown that both the laboratory plant and the scaling-up apparatus (plant) have to be designed to work with the same characteristic geometric lengths. We can conclude that the relationship $l/l_p = 1$ is necessary in the case of the geometric scaling-up of laboratory models. In addition, since the majority of the laboratory models are designed to use real fluids (those that will be used in the extended model) in their tests, we can appreciate that we have scales of the same unitary value for all materials.

If we analyze the fundamental dynamics aspects of the fluids contacting in the laboratory device, we can easily observe that the velocities and other dynamic parameters of the specific phases, could be identical to those of the extended model. If we neglect the wall effects, which, in the case of LM could be important, we can easily conclude that the dimensionless pi terms that characterize the dynamics of the process present the same values for the laboratory plant and for the extended model. Taking these observations into consideration, we can see that LMs do not require a scaling up of the data and information obtained when we want to use them on experimental investigations of a physico-chemical process. This means that the relationships, the curves and the qualitative observations obtained with an LM could be directly applicable to larger devices.

We cannot finish without presenting some uncertainties, which show the differences between the LMs and their real homologue:

- the chemical processes are frequently studied at laboratory scale taking into account only the critical parts. This means that the elements are considered as new or unknown. Then, the studied problems may be presented when computed and experimented parts are assembled.
- The stationary time of laboratory models is quite small when compared with the corresponding time for extended models. This fact introduces important uncertainties in the capacity of the experimental data to predict the states of the system not covered by experimentation (in, for example, concentration and temperature fields).
- Frequently the experimental models are tested with boundary conditions that are not identical to those of the extended model. This is sustained by the different evolution of the temperature and concentration fields described above.

If we can ensure the control and compensation of these uncertainties, then we can appreciate the enormous importance of the experimental chemical engineering research for developing new processes or for modernizing those that are already in use.

References

6.1 E. Buckingham, *Phys. Rev.* **1914**, *4*, 345.

6.2 M. Vaschy, *Anal. Telegraph.* **1892**, *III*, 11.

6.3 G. Murphy, *Similitude in Engineering*, Ronald Press, New York, 1951.

6.4 H. K. Langhahar, *Dimensional Analysis and Theory of Models*, John Wiley, New York, 1951.

6.5 K. I. Sedov, *Similarity and Dimensional Methods in Mechanics*, Academic Press, New York, 1959.

6.6 E. A. Bratu, *Process and Installations for Chemical Industry*, Polytechnic Institute of Bucharest, Bucharest, 1959.

6.7 D. C. Ipsen, *Units, Dimensions and Dimensionless Number*, McGraw-Hill, New York, 1960.

6.8 S. J. Kline, *Similitude and Approximation Theory*, McGraw Hill, New York, 1965.

6.9 R. C. Pankhurst, *Dimensional Analysis and Scale Factors*, Chapman and Hall, London, 1964.

6.10 V. J. Koglund, *Similitude Theory and Applications*, International Textbook, Scranton, 1973.

6.11 R.B. Bird, W. E. Stewart, E. N. Lightfoot, *Transport Phenomena*, John Wiley, New York, 1960.

6.12 H. Schlichting, *Boundary Layer-Theory*, McGraw Hill, NewYork, 1979.

6.13 E. S. Taylor, *Dimensional Analysis for Engineers*, Clarendon Press, Oxford, 1974.

6.14 E de St Q., Isaacson, M de St Q. Isaacson, *Dimensional Methods in Engineering and Physics*, John Wiley, New York, 1975.

6.15 D. J. Schuring, *Scale Models in Engineering*, Pergamon Press, New York, 1977.

6.16 T. Dobre, O. Floarea, *Momentum Transfer*, Matrix-Rom, Bucharest, 1997.

6.17 J. Monson, P. Young , F. Okisihi, *Fundamentals of Fluid Mechanics*, John Wiley, New York, 1998.

6.18 O. Floarea, R. Dima, *Mass Transfer Operations and Specific Devices*, Didactic and Pedagogic Editure, Bucharest, 1984.

6.19 G. B. Tatterson, *Fluid Mixing and Gas Dispersion in Agitated Tanks*, McGraw-Hill, New York, 1991.

6.19 E. A. Bratu, *Processes and Apparatus for Chemical Industry*, Technical Book, I and II, Bucharest, 1969, 1970.

6.20 M. Zlokarnik, *Chem. Eng. Sci.* **1998**, *53* (17), 3023.

6.21 M. Zlokarnik, *Scale-up in Chemical Engineering*, JohnWiley, New York, 2000.

6.22 C. W. Robert, J. A. Melvin (Eds.), *Handbook of Chemistry and Physics* , CRC Press, Boca Raton, FL, 1982.

6.23 E. Ruckenstein, E. Dutkai, *Chem. Eng. Sci.* **1968**, *23*, 1365.

6.24 R. Z. Tudose, A. Lungu, *Genie Chim.* **1971**, *104*, 1769.

6.25 E. A. Bratu, F. Gothard, *Br. Chem. Eng.* **1971**, *16* (8), 691.

6.26 O. Floarea, G. Jinescu, *Intensive Processes for Transfer Unit Operations* , Technical Book, Bucharest, 1975.

6.27 V. P. Luikov, *Fundamentals of Heat and Mass Transfer* , Izd. Himia, Kiev, 1978.

6.28 T. Dobre, Engineering Elements of Chemical Surface Finishing, Matrix-Rom, Bucharest, 1999.

6.29 H. Hadamard, *Comp. Rend. Acad. Sci.* **1911**, *952*, 1735.

6.30 R. Rybczynski, *Bull. Cracovie*, **1911**, *A*, 40.

6.31 K. F. Pavlov, P. G. Romankov, A. N. Noskov, *Unit Operations for Technological Chemistry*, Mir ,Moscow, 1982.

6.32 W. Hoyle (Ed.)., *Pilot Plants and Scale-Up of Chemical Processes*, Royal Society of Chemistry, Cambridge, 1999

6.33 E.A. Bratu, *Unit Operations for Chemical Engineering*, Technical Book , Bucharest, 1983

6.34 E. Ruckenstein, *Adv. Chem. Eng.* **1987**, *13*, 16.

6.35 L. Francisco, C. Francesc, *Ind. Eng. Chem. Res.* **1999**, *38* (7), 2747.

6.36 F. Magelli, *Ind. Eng. Chem. Res.* **1998**, *37* (4), 1528.

6.37 M. M. Hupa, *Ind. Eng. Chem. Res.* **1997**, *36* (12), 5439.

6.38 M. A. Alaa-Eldin, *Ind. Eng. Chem. Res.* **1997**, *36* (11), 4549.

6.39 R. Shinnar, B. Dainson, I. H. Rinard, *Ind. Eng. Chem. Res.* **2000**, *39* (1), 103.

6.40 Z. Rudolf, G. J. Alan, *Ind. Eng. Chem. Res.* **2000**, *39* (7), 2392.

6.41 R. Krisna, J. M. van Baten, *Chem. Eng. J.* **2001**, *82*, 247.

6.42 K. Kazdobin, N. Shwab, S. Tsapakh, *Chem. Eng. J.* **2000**, *79*, 203.

6.43 V. Novozhilov, *Prog. Energ. Comb. Sci.* **2001**, *27* (6), 611.

6.44 J. G. Sanchez Marcano, T. T. Tsotsis, *Catalytic Membrane and Membrane Reactors*, Wiley-VCH, Weinheim, 2002.

6.45 E. B. Cummings, S. K. Grifffits, R. H. Nilson, P. H. Paul, *Anal. Chem.*, **2000**, *72* (11), 2526.

6.46 G. J. S. van der Gulik, J. G. Wijers, J. T. F. Keurentjes, *Ind. Eng. Chem. Res.* **2001**, *40* (22), 4731.

Index

Chemical Engineering. Tanase G. Dobre and José G. Sanchez Marcano
Copyright © 2007 WILEY-VCH Verlag GmbH & Co. KGaA, Weinheim
ISBN: 978-3-527-30607-7